Coquitlam Open Learning

(Grades 9 - 12)
380 MONTGOMERY STREET
COQUITLAM, BC   V3K 5G2

024

AUTHORS

Robert Alexander

Brendan Kelly

Contributing Writers

Paul Atkinson
Superintendent of Instruction
Waterloo Board of Education
Waterloo, Ontario

Maurice Barry
Vista School District
Newfoundland

Fred Crouse
Annapolis Valley Regional
  School Board
Berwick, Nova Scotia

Garry Davis
Saskatoon Public Board
  of Education
Saskatoon, Saskatchewan

George Gadanidis
Durham Board of Education
Whitby, Ontario

Liliane Gauthier
Saskatoon Board of Education
Saskatoon, Saskatchewan

Florence Glanfield
Professor of Mathematics
  Education
University of Saskatchewan
Saskatoon, Saskatchewan

Jim Nakamoto
Sir Winston Churchill
  Secondary School
Vancouver, British Columbia

Linda Rajotte
Georges P. Vanier
  Secondary School
Courtenay, British Columbia

Elizabeth Wood
National Sport School
Calgary Board of Education
Calgary, Alberta

Rick Wunderlich
Shuswap Junior
  Secondary School
Salmon Arm, British Columbia

Toronto

DEVELOPMENTAL EDITORS
Claire Burnett
Lesley Haynes
Charlotte Urbanc

EDITORS
Mei Lin Cheung
Julia Cochrane
Annette Darby

RESEARCHER
Rosina Daillie

DESIGN/PRODUCTION/PERMISSIONS
Pronk&Associates

ART DIRECTION
Pronk&Associates/Joe Lepiano

ELECTRONIC ASSEMBLY & TECHNICAL ART
Pronk&Associates

Acknowledgments appear on page 629.

Canadian Cataloguing in Publication Data

Alexander, Bob, 1941 –
    Addison-Wesley mathematics 12

Western Canadian ed.
Includes index.
ISBN 0–201-34629–X

1. Mathematics.   I.Kelly, B. (Brendan), 1943 – .   II. Davis, Garry, 1963 – .   III. Title.

QA39.2.K448 1999   510   C98-930674-7

ClarisWorks is a trademark of Claris Corporation.
Claris is a registered trademark of Claris Corporation.
Microsoft and Windows are either registered trademarks or trademarks of Microsoft Corporation.
Macintosh is a registered trademark of Apple Computer, Inc. Graphmatica is a trademark of kSoft, Inc.

ISBN 0–201–34629–X

This book contains recycled product and is acid free.
Printed and bound in Canada

4 5 6–GG–04 03 02

REVIEWERS/CONSULTANTS

Joseph F. Prokopow

Margaret Barbour Collegiate
The Pas, Manitoba

Lana Rinn

Mathematics Consultant
Transcona-Springfield School Division #12
Winnipeg, Manitoba

Darryl Smith

Mathematics Department Head
Austin O'Brien High School
Edmonton, Alberta

Don Smith

Aden Bowman Collegiate
Saskatoon, Saskatchewan

Mila Stout

Swan Valley Regional Secondary School
Swan River, Manitoba

Mike Wierzba

Mathematics Coordinator
Etobicoke Board of Education
Etobicoke, Ontario

CONTENTS Mathematics 12

CONTENTS

Welcome to Addison-Wesley Mathematics 12
Western Canadian Edition

This book is about mathematical thinking, mathematics in the real world, and using technology to enhance mathematical understanding. We hope it helps you see that mathematics can be useful, interesting, and enjoyable.

These introductory pages illustrate how your student book works.

Mathematical Modelling

Each chapter begins with a provocative problem. Once you have explored relevant concepts, you use mathematical modelling to illustrate and solve the problem.

Consider This Situation is the first step in a four-stage modelling process. This page introduces the chapter problem. You think about the problem and discuss it in general terms.

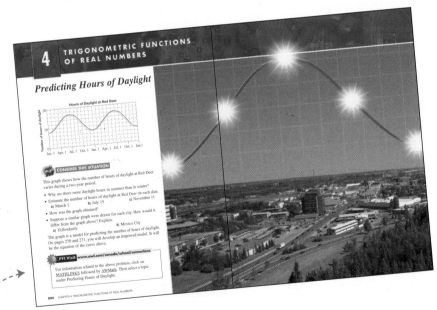

FYI Visit refers you to our web site, where you can connect to other sites with information related to the chapter problem.

You revisit the chapter problem in a Mathematical Modelling section later in the chapter.

Develop a Model suggests a graph or table, a formula or pattern, or some approximation that can represent the situation. Related exercises help you construct and investigate the model.

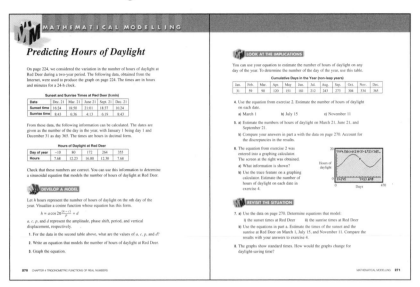

Look at the Implications encourages you to relate your findings to the original problem. What is the significance of your results?

Revisit the Situation invites you to criticize the model you developed. Does it provide a reasonable representation of the situation? Can you refine the model to obtain a closer approximation to the situation? Can you develop another problem that fits a similar model?

Short **Mathematical Modelling** boxes echo the fourth stage of the modelling process. Each box leads you back to an earlier problem to reflect on the validity of the solution, and to consider alternative ways to model the problem.

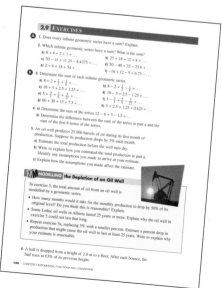

Concept Development

Here's how a typical lesson in your student book works.

By completing **Investigate**, you discover the thinking behind new concepts.

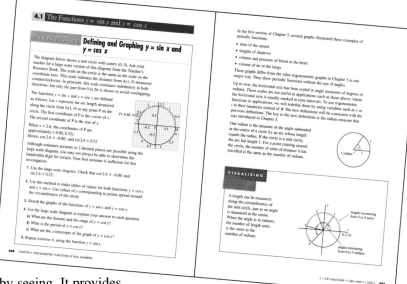

Visualizing helps you learn by seeing. It provides another way to understand and remember concepts.

Examples with full solutions provide you with a model of new methods.

Discussing the Ideas helps you clarify your understanding by talking about the preceding explanations and examples.

Communicating the Ideas helps you confirm that you understand important concepts. If you can explain it or write about it, you probably understand it!

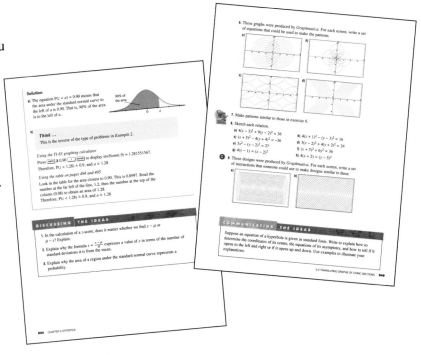

Exercises, including end-of-chapter **Review** and **Cumulative Review** exercises, help you reinforce your understanding.

There are three levels of **Exercises** in the sections in this book.

A exercises involve the simplest skills of the lesson.

B exercises usually require several steps, and they may involve applications or problem solving.

C exercises are more thought-provoking. They may call on previous knowledge or foreshadow upcoming work.

Technology is incorporated into exercises in several ways. Special logos tell you when an exercise requires the use of technology.

This logo tells you that you need a graphing calculator to complete this exercise.

This logo tells you that you need a graphing calculator, a computer with graphing software, or grid paper to complete this exercise.

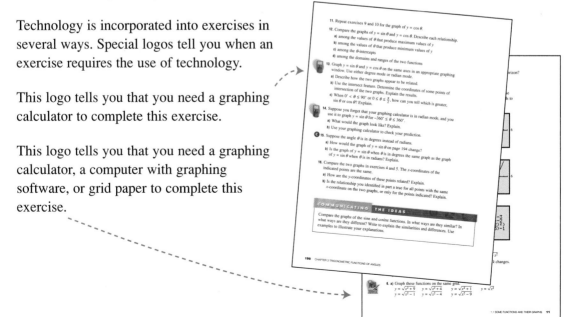

Technology

The graphing calculator and the computer are tools for learning and doing mathematics in ways that weren't possible a few years ago.

For some computer activities, you use computer applications such as spreadsheets.

This computer logo tells you that the exercise involves a spreadsheet.

Other computer activities require the use of a computer database. The CD logo tells you that the data are available from *Addison-Wesley Mathematics 12 Template and Data Kit*. These databases provide authentic data for you to analyze. The *Template and Data Kit* also provides every spreadsheet in your student book.

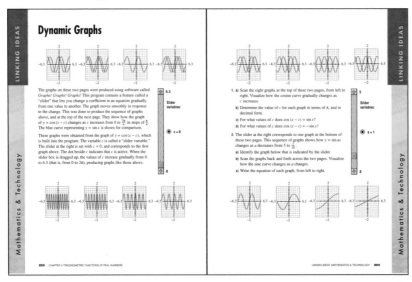

You need ClarisWorks™, Microsoft Works®, or Microsoft® Office® 95 or Office 97 to use *Addison-Wesley Mathematics 12 Template and Data Kit*.

Computer applications are also featured in **Linking Ideas: Mathematics & Technology**. Completing these activities will show you an efficient and effective use of technology.

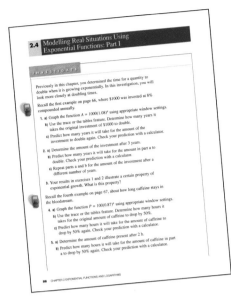

The graphing calculator is a powerful mathematical tool. It does not replace the need to develop good graphing skills, but it can enhance your mathematical understanding.

In selected Investigates and exercises, the calculator helps you explore patterns graphically, numerically, and symbolically. You develop skills to connect these different ways of looking at a situation.

Exploring with a Graphing Calculator presents a series of exercises that involve the graphing calculator and relate to a specific mathematical concept.

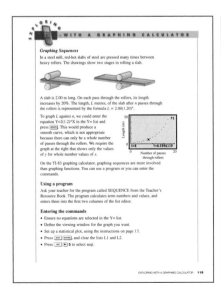

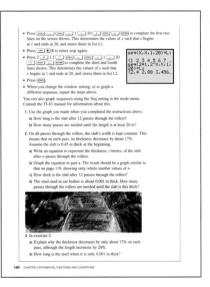

Here are some other sections you will encounter as you work through your student book.

Linking Ideas

These sections show how mathematics topics relate to other school subjects or to the world outside school.

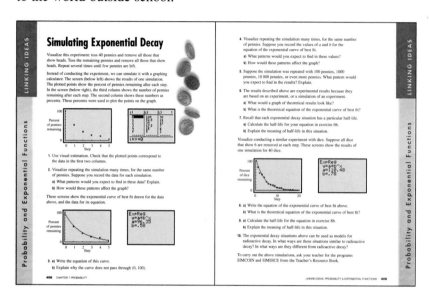

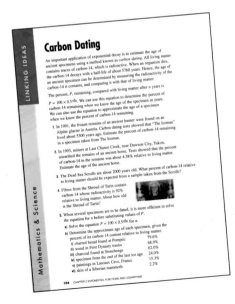

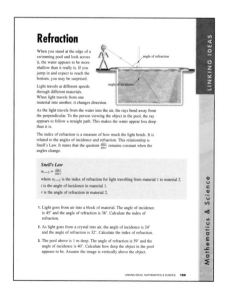

Problem Solving

Special **Problem Solving** sections introduce problems and puzzles, and their solutions. The accompanying exercises offer problems that may involve extensions to the introductory problem.

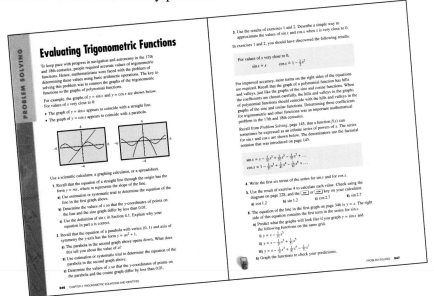

Mathematics Files

These short sections present key mathematical concepts that form part of the foundation for upcoming lessons. Selected **Mathematics Files** may present an alternative method to one already taught, or a specialized method for solving a problem.

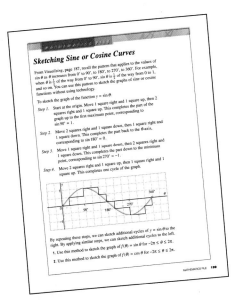

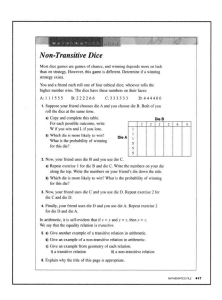

When Will Women's Pay Equal Men's?

 CONSIDER THIS SITUATION

The excerpt on this page is from a newspaper article published in 1998. The headline predicts that women's pay could equal men's in the year 2020. Is there enough information in the excerpt to confirm this prediction?

- If your answer is yes, carry out calculations to confirm the prediction.
- If your answer is no, what additional information do you need to carry out the calculation? Why do you need that information?
- Why is women's pay less than men's?
- What are some reasons why the gap between women's pay and men's pay is decreasing?

On pages 60 and 61, you will develop mathematical models to predict when women's pay will equal men's. Your models will be functions.

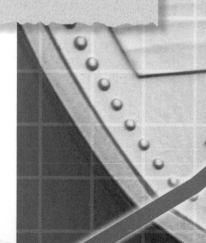

Trend sees 2020 vision of women's pay
BY JOHN KETTLE
Special to the Globe and Mail

The 1996 Canadian census showed that the pay cheques of full-year, full-time female workers rose again as a percentage of what similarly employed men were paid. Women took home an average of $30,000 in 1995, the last complete year before the census, compared with men at $42,000.

That works out to just 71 percent of men's pay. But as recently as 1980, full-year, full-time female workers were paid only 64 percent of what men made on average.

 FYI Visit www.awl.com/canada/school/connections

For information related to the above problem, click on MATHLINKS, followed by AWMath. Then select a topic under When Will Women's Pay Equal Men's?

Relating Equations and Graphs

This chapter contains several investigations in which you will explore what happens to the graph of a function when you make certain systematic changes to its equation. You can conduct these investigations using:

- a graphing calculator, such as the TI-82 or TI-83
- a computer with software such as *Zap-a-Graph* or *Graphmatica*
- grid paper

In principle, it does not matter which method you use. Graphing calculators and computers are more efficient. However, you need additional components (for example, a TI-Graph Link) and a printer to print the graphs. Using grid paper may seem tedious, but this method works well if you carefully choose the points to plot. You will also have a record of the results.

This symbol will accompany some activities and exercises in this book. When you see it, you should use a graphing calculator, a computer with graphing software, or grid paper. Whichever method you use, make sure you have a record of the results on paper. The following examples illustrate these methods using quadratic functions, which you studied previously.

Example 1

Draw a graph to investigate the effect of changing a in the equation $y = ax^2$.

Solution

Using a graphing calculator

Set the window and enter equations such as $y = \pm x^2$, $y = \pm 2x^2$, and $y = \pm 0.5x^2$. The result at the right appears on the screen, but it disappears when other functions are graphed or when the calculator is turned off.

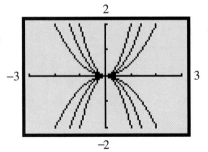

Copy the graph onto plain paper.

Example 2

Draw a graph to investigate the effect of changing q in the equation $y = x^2 + q$.

Solution

Using a computer

Set the grid range on both axes. Enter
equations such as $y = x^2$, $y = x^2 \pm 1$, and
$y = x^2 \pm 2$. A graph like this appears on the
screen, but it disappears when other functions
are graphed, the program is closed, or when
the computer is shut down.

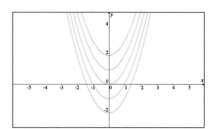

Either print the graph or copy it onto plain paper.

Example 3

Draw a graph to investigate the effect of changing p in the equation $y = (x - p)^2$.

Solution

Using grid paper

Make tables of values for equations such as $y = x^2$, $y = (x - 4)^2$, and $y = (x + 2)^2$.
Use values of x that produce the same values of y. Draw the graphs on grid paper.

$y = x^2$

x	y
−2	4
−1	1
0	0
1	1
2	4

$y = (x - 4)^2$

x	y
2	4
3	1
4	0
5	1
6	4

$y = (x + 2)^2$

x	y
−4	4
−3	1
−2	0
−1	1
0	4

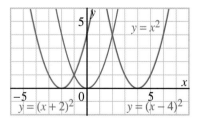

1. Write to explain the results in each of *Examples 1*, *2*, and *3*. Explain why the equation
   $y = x^2 - 2$ corresponds to motion in the negative y-direction, but the equation $y = (x - 4)^2$
   corresponds to motion in the positive x-direction.

In previous grades, you studied certain kinds of functions. The graphs of these functions have characteristic shapes. By transforming the graphs in certain ways, we can use them to solve practical problems. Here is a brief summary of some of the kinds of functions you encountered in previous grades.

Linear Functions

The graph of the simplest linear function, $y = x$, is a straight line with slope 1 through the origin. By changing the slope and the vertical intercept, we can use the graph to model situations involving fixed and variable costs.

Yearbook printing costs

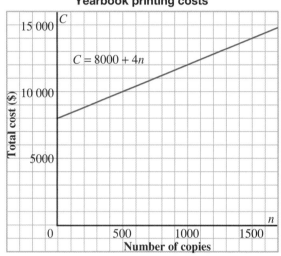

$C = 8000 + 4n$

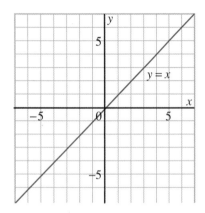

$y = x$

Quadratic Functions

The graph of the simplest quadratic function, $y = x^2$, is a parabola with vertex $O(0, 0)$ and axis of symmetry the y-axis. By transforming the graph of the parabola, we can use it to model situations involving projectiles.

Height of falling object

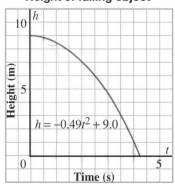

$h = -0.49t^2 + 9.0$

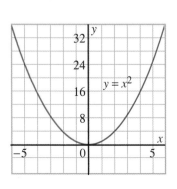

$y = x^2$

Reciprocal Functions

The graph of the simplest reciprocal function, $y = \frac{1}{x}$, is
a hyperbola whose asymptotes are the coordinate axes.
By transforming the part of the graph that is in the first
quadrant, we can use it to model certain kinds of investment
problems.

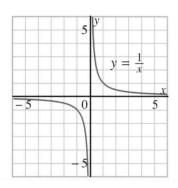

Investment required to earn $100 interest in 1 year

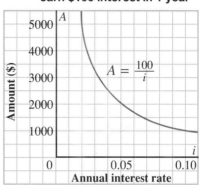

$$A = \frac{100}{i}$$

Square-Root Functions

A simple square-root function is $f(x) = \sqrt{x}$. Since there are
real values for $\sqrt{x}$ only when $x \geq 0$, the graph has an
unusual property: it starts at a fixed point $O(0, 0)$ and
extends in only one direction. The graph is part of a
parabola on one side of its axis of symmetry. By
transforming the graph, we can use it as a model to solve
problems involving the distance to the horizon.

Distance to the horizon

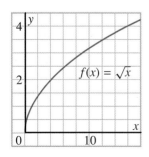

$f(x) = \sqrt{x}$

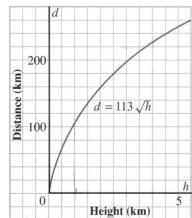

$$d = 113\sqrt{h}$$

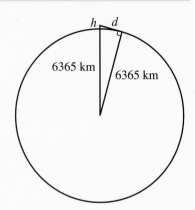

This diagram illustrates the distance, d kilometres, to the horizon as seen from a height, h kilometres, above Earth. You can express d as a function of h as follows.

Step 1. Apply the Pythagorean Theorem to show that $d = \sqrt{h^2 + 12\ 730h}$.

Step 2. Since h^2 is usually very small compared with $12\ 730h$, it will not make much difference if we delete h^2 from the equation.
Hence, $d \doteq \sqrt{12\ 730h}$, or $d \doteq 113\sqrt{h}$

This equation (and its graph on page 7) is a model for determining the distance to the horizon from a height h kilometres above the ground.

- What other assumptions were made to develop the model?
- Suppose you want h to be in metres. How would you change the equation? Explain.

In this chapter, and later chapters, we will investigate other functions that have graphs with characteristic shapes. We shall determine how the coefficients in the equations of the functions are related to the graphs. You can use the results to sketch the graphs of functions.

Example 1

a) Graph the function $y = \sqrt{25 - x^2}$.

b) Describe the graph, and account for its shape.

c) Determine the domain and the range of the function.

Solution

a) *Using a graphing calculator*

Since $25 - x^2$ cannot be negative, the possible values of x are from -5 to 5. The possible values of y are from 0 to 5. Use window settings $-9 \le x \le 9$, $-6 \le y \le 6$. Enter the equation $Y1= \sqrt{(25-X^2)}$.

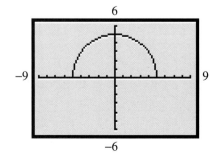

Using grid paper

Make a table of values, plot the ordered pairs on a grid, then draw a smooth curve through them.

x	y
0	5.0
±1	4.9
±2	4.6
±3	4.0
±4	3.0
±5	0

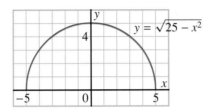

b) The graph appears to be a semicircle. To explain why it is a semicircle, let $P(x, y)$ be any point on the graph.

We know $y = \sqrt{25 - x^2}$.

Square each side.

$y^2 = 25 - x^2$, or $x^2 + y^2 = 25$

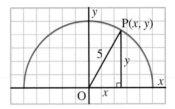

This equation expresses the distance OP to be 5 units. That is, P lies on a circle with centre O and radius 5. Since $y \geq 0$, all points on the graph must be above the x-axis. Hence, the graph is a semicircle.

c) The domain is the set of all possible values of x. This is the set of real numbers between -5 and 5, inclusive. The range is the set of all possible values of y. This is the set of real numbers between 0 and 5, inclusive.

Example 2

Graph the function $f(x) = |x|$.

Solution

Recall that $|x|$ is the absolute value of x, and is defined to equal x if $x \geq 0$, and $-x$ if $x < 0$. When $x \geq 0$, the graph consists of the line defined by $y = x$. When $x \leq 0$, the graph consists of the line defined by $y = -x$.

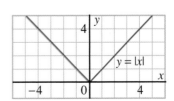

1. Refer to the graph on page 6 that shows the cost of printing a yearbook.

   a) In the equation $C = 8000 + 4n$, explain what the numbers 8000 and 4 represent.

   b) Describe how the graph would change if these numbers change.

2. Refer to the graph on page 6 that shows the height of a falling object.

   a) What was the initial height of the object?

   b) Describe how the graph would change if the initial height were different.

3. Refer to the graph on page 7 that shows the investment required to earn $100 interest in one year. Describe how the graph would change for other amounts of interest earned in one year.

4. Refer to the graph on page 7 that shows the distance to the horizon from different heights. Describe how the graph would change for planets other than Earth.

5. In *Example 1*, explain how both the domain and the range can be determined using either the graph or the equation of the function.

1.1 EXERCISES

A 1. Use the quadratic function on page 6 that shows the height of an object above the ground after it has fallen from 9.0 m.

   a) Calculate the heights above the ground after 1 s, 2 s, and 4 s.

   b) Calculate how far the object fell during each time.
   　i) the first second　**ii)** the first two seconds　**iii)** the first four seconds

   c) Use the results from part b. Suppose the time of falling is doubled. What happens to the distance the object falls?

   d) Calculate the time required for the object to hit the ground.

2. Use the reciprocal function on page 7 that shows the investment required to earn $100 interest in one year at different interest rates.

   a) Calculate the amounts required to earn $100 interest at 2%, 4%, and 8%.

   b) Suppose the interest rate is doubled. What happens to the investment required to earn $100 interest?

3. Use the square-root function on page 7 that shows showing the distance to the horizon from different heights.

   a) Calculate the distances to the horizon from heights of 1 km, 2 km, and 4 km.

b) Suppose the height is doubled. What happens to the distance to the horizon?

c) Suppose the distance to the horizon is doubled. What happens to the height?

B **4.** A graphing calculator produced these five graphs. The equations of these graphs are shown on the last screen. Write the equation that corresponds to each graph. Explain each choice.

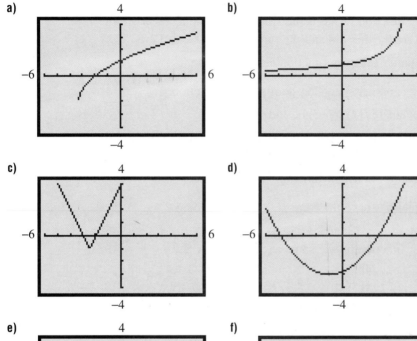

a)

b)

c)

d)

e)

f)

```
Plot1  Plot2  Plot3
\Y1=5/(6-X)
\Y2=.2(X+1)²-3
\Y3=√(3X+10)-2
\Y4=√(4-(X-2)²)
\Y5=abs(2X+5)-1
\Y6=
\Y7=
```

5. a) Graph these functions on the same grid.

$$y = \sqrt{16 - x^2} \qquad y = \sqrt{9 - x^2} \qquad y = \sqrt{4 - x^2} \qquad y = \sqrt{1 - x^2}$$

b) Write to describe how the graph of $y = \sqrt{k - x^2}$ changes as k changes.

c) What special case occurs when $k = 0$?

d) For what values of k is the function $y = \sqrt{k - x^2}$ defined?

6. a) Graph these functions on the same grid.

$$y = \sqrt{x^2 + 9} \qquad y = \sqrt{x^2 + 4} \qquad y = \sqrt{x^2 + 1} \qquad y = \sqrt{x^2}$$
$$y = \sqrt{x^2 - 1} \qquad y = \sqrt{x^2 - 4} \qquad y = \sqrt{x^2 - 9}$$

b) Write to describe how the graph of $y = \sqrt{x^2 + k}$ changes as k changes.

c) What special case occurs when $k = 0$?

d) For what values of k is the function $y = \sqrt{x^2 + k}$ defined?

7. The *identity function* is a special case of a linear function. It is defined by $i(x) = x$.

   a) Evaluate $i(3)$, $i(4.5)$, and $i(-2.7)$.

   b) Graph the identity function. What is the slope of its graph? What are the intercepts? What are the domain and the range?

   c) Suggest why the name "identity function" is appropriate.

8. A *constant function* is another special case of a linear function.

   a) Given $f(x) = 5$, evaluate $f(2), f(-3.4)$, and $f(5)$.

   b) Graph the function in part a. What is the slope of its graph? What are the intercepts? What are the domain and the range?

   c) Suggest why the name "constant function" is appropriate.

9. Functions whose equations have the form $f(x) = x^a$ are called *power functions*.

   a) Graph each power function on the same axes, for $0 \le x \le 2$.

     i) $f(x) = x^{0.5}$ **ii)** $f(x) = x^{0.75}$ **iii)** $f(x) = x^{1.0}$

     iv) $f(x) = x^{1.5}$ **v)** $f(x) = x^{2.0}$ **vi)** $f(x) = x^{2.5}$

   b) Describe how the graph of $f(x) = x^a$ changes as a varies, when $a > 0$ and $x \ge 0$.

C **10.** Graph each equation.

   a) $|y| = x$ **b)** $|y| = |x|$ **c)** $|x| + |y| = 1$

   d) $|x| - |y| = 1$ **e)** $|x + y| = 1$ **f)** $|x - y| = 1$

11. Graph each function. Describe and account for the similarities and differences in the graphs.

   a) $y = \sqrt[3]{25 - x^2}$ **b)** $y = \sqrt[3]{25 - x^3}$

12. In the modelling box on page 8, we omitted h^2 from the equation. Investigate how this affects the calculated distances for various values of h.

COMMUNICATING THE IDEAS

Write to explain what is meant by a function. Include some examples to illustrate different ways to represent a function.

WITH A GRAPHING CALCULATOR

Lines and Curves of Best Fit

Suppose an amount of money, P dollars, is invested at an interest rate, i percent, compounded annually. The accumulated amount, A dollars, after n years is given by the formula $A = P(1 + i)^n$. A common application of this formula is to determine the time required for an investment to double. Suppose we invest $1000 at 6% compounded annually. To determine the number of years for the investment to double, we must find a value of n so that $1000 \times 1.06^n = 2000$. By systematic trial, we find that $1000 \times 1.06^{12} \doteq 2012.196$. Hence, $1000 will double in about 12 years. Using the same method, we obtain the following data, rounded to the nearest year.

Annual interest rate (*i*%)	2	4	6	8	10	12
Doubling time (*n* years)	35	18	12	9	7	6

We use a graphing calculator to plot the data, and to calculate and graph the equations of functions that might be used as models for the situation. The screens and keystrokes below are from the TI-83 calculator.

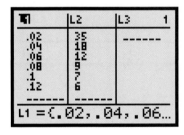
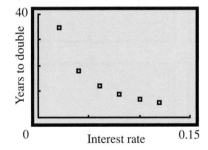

Displaying a scatterplot

- *Clear the graphing screen.*
  Make sure that no equations are selected in the Y= list.

- *Define the viewing window.*
  Choose values so the data will appear on the screen.

- *Enter the data in the list editor.*
  Press ⬚STAT⬚ **1**. Enter the interest rates in the first list and the doubling times in the second list.

- *Define a statistical plot.*
  Press ⬚STAT PLOT⬚ **1**. In the menu that appears, make sure that On is selected in the first line, the first graph type is selected in the second line, and lists L1 and L2 appear in the third and fourth lines.

- *Graph the data.*
  Press GRAPH.

Graphing lines and curves of best fit

- *Choose the desired function type.*
  Press STAT ▶ to obtain the menu at the right
  (you must scroll down to see all of it). This is the
  STAT CALC menu.

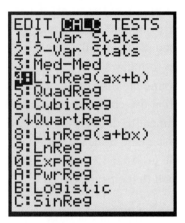

Start with the third line in the menu. There are 10
different function types from which to choose. The
results of three choices are shown below. In each
case, the first screen shows the calculated results,
and the second screen shows the graph of the
corresponding function.

4:LinReg(ax+b)   5:QuadReg   6:CubicReg

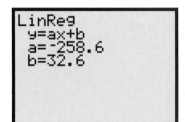   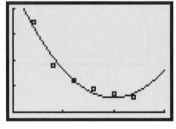   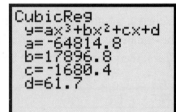

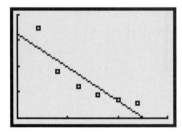   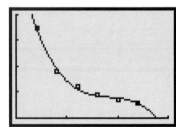

These calculations are called *regressions*. The graphs show the straight
line, the parabola, and the cubic curve that come closest to the plotted
points, as determined by the calculator. As the graphs suggest, none of
these choices seems appropriate as a model for the data.

To fit a curve to data, the first step is to select a suitable type of function.
This involves comparing patterns in the plotted points with features of
particular functions. For the graph on page 13, as i becomes larger, the
points come closer to the horizontal axis; as i becomes smaller, the points
come closer to the vertical axis. The graph of the reciprocal function
$y = \frac{1}{x}$ has this property. This suggests that the reciprocal function might
be a suitable one to model the data.

The reciprocal function does not appear in the list of function types in the STAT CALC menu. However, recall that $\frac{1}{x}$ can be written as a power, x^{-1}. This suggests that A:PwrReg should be chosen from the menu.

Press [ALPHA] **A** [ENTER] to select PwrReg.

You will get a screen, similar to those on page 14, that shows the general equation y=a*x^b, which means $y = ax^b$. The values of a and b will appear on the screen.

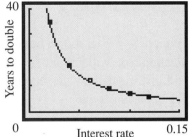

- *Graph the curve.*
  Substitute the values of a and b into the equation $y = ax^b$ and enter it in the Y= list manually. Or, use the calculator to enter the equation directly, by pressing [Y=] [CLEAR] [VARS] **5** [▶] [▶] **1**. Then press [GRAPH].

The graph shows the power function that comes closest to the plotted points. You should have found that its equation is $n \doteq 0.73x^{-0.99}$, which is very close to $n = 0.73x^{-1}$, or $n = \frac{0.73}{x}$. This equation of the reciprocal function models the number of years to double at different interest rates.

1. Use the reciprocal model. Estimate the number of years to double at each rate.

   a) 3%   **b)** 5.25%   **c)** 8.5%   **d)** 15%

2. In the screens on page 14, the parabola and the cubic graph come close to the plotted points. Explain why quadratic and cubic functions do not seem appropriate as models for the data.

3. A simple pendulum consists of a mass suspended from a fixed point. The *period* of the pendulum is the time for one complete swing. In an experiment to determine how the period of a pendulum is related to its length, these data were obtained.

Length (m)	0	0.5	1.0	1.5	2.0	2.5	3.0
Period (s)	0	1.4	2.0	2.5	2.8	3.2	3.5

   a) Graph the data.

   b) Choose an appropriate function type from the menu on page 14. Explain why your choice is appropriate.

   c) Determine the equation of the curve of best fit of the type chosen in part b.

   d) Use the equation to graph the curve.

1.2 Translating Graphs of Functions

To use graphs of functions in applied problems, we must be able to change the graphs and the equations to fit the quantities in the situation. To do this, we must be able to relate the changes in the equations to the changes in the graphs.

INVESTIGATE Transforming Function Graphs Part I

The graph of $y = x^3$ is shown below, with its table of values. In this investigation, you will graph other related equations on the same grid. You will investigate how the changes in the graphs are related to the changes in the equations.

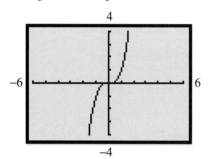

Comparing the graphs of $y = f(x)$ and $y = f(x - k)$

1. a) Graph the function $y = x^3$. Use the window settings above.

b) Predict what the graph of $y = (x - 2)^3$ would look like, compared with the graph of $y = x^3$.

c) Graph these two equations. Did the result agree with your predictions?

d) Use the trace or tables feature. Determine the x-coordinates of points on the two curves that have the same y-coordinate. How are the x-coordinates related?

2. In exercise 1b, change the constant 2 to other numbers, and repeat. Use both positive and negative constants. Are the results what you expected?

3. Would you have obtained the same results if you had started with the graph of $y = x$, or with the graph of $y = x^2$ and replaced x with $x - 2$? Explain.

4. a) What information about the graph of $y = f(x - k)$ does k provide?

    b) What happens to the graph of $y = f(x - k)$ when k is changed? Consider both positive and negative values of k.

    c) Explain why the changes in part b occur.

5. For each list, sketch the graphs of the functions on the same axes.

    a) $y = \sqrt{x - 1}$ **b)** $y = 2(x - 2) + 3$ **c)** $f(x) = \sqrt{25 - (x - 3)^2}$

        $y = \sqrt{x}$ $y = 2x + 3$ $f(x) = \sqrt{25 - x^2}$

        $y = \sqrt{x + 1}$ $y = 2(x + 2) + 3$ $f(x) = \sqrt{25 - (x + 3)^2}$

Comparing the graphs of $y = f(x)$ and $y - k = f(x)$

6. a) Graph the function $y = x^3$. Use the window settings on page 16.

    b) Predict what the graph of $y - 2 = x^3$ would look like, compared with the graph of $y = x^3$.

    c) Graph these two equations. Did the result agree with your predictions?

    d) Use the trace or tables feature to examine how the y-coordinates of points on the two graphs which have the same x-coordinate are related.

7. In exercise 6b, change the constant 2 to other numbers, and repeat. Use both positive and negative constants. Are the results what you expected?

8. Would you have obtained the same results if you had started with the graph of $y = x$, or with the graph of $y = x^2$? Explain.

9. a) What information about the graph of $y - k = f(x)$ does k provide?

    b) What happens to the graph of $y - k = f(x)$ when k is changed? Consider both positive and negative values of k.

    c) Explain why the changes in part b occur.

10. For each list, sketch the graphs of the functions on the same axes. Solve the first and third equations for y.

    a) $y - 1 = \sqrt{x}$ **b)** $y - 2 = (2x + 3)$ **c)** $y - 3 = \sqrt{25 - x^2}$

        $y = \sqrt{x}$ $y = 2x + 3$ $y = \sqrt{25 - x^2}$

        $y + 1 = \sqrt{x}$ $y + 2 = (2x + 3)$ $y + 3 = \sqrt{25 - x^2}$

In a previous grade, you found that when you change the equation $y = x^2$, corresponding changes occur to the graph. For example, if x is replaced with $x - 2$, the equation becomes $y = (x - 2)^2$, whose graph is 2 units to the right of $y = x^2$ (below left). To see why, let P and Q be corresponding points with the same y-coordinate on the two graphs. To determine the y-coordinate of P, square the x-coordinate. To determine the y-coordinate of Q, subtract 2 from the x-coordinate and square the result. To produce the same y-coordinate as for P, the x-coordinate of Q must be 2 *greater than* the x-coordinate of P. Hence, the graph of $y = (x - 2)^2$ is translated 2 units right, relative to the graph of $y = x^2$. The translated graph is called the *image* of the original graph.

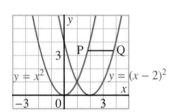
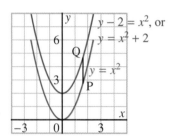

A similar analysis applies to the y-coordinates. The graph of $y - 2 = x^2$ (above right) is translated 2 units up relative to $y = x^2$. The y-coordinates of points on the graph of $y - 2 = x^2$ are 2 units *greater than* the y-coordinates of corresponding points on the graph of $y = x^2$. Instead of writing the equation of the image as $y - 2 = x^2$, we usually write it as $y = x^2 + 2$.

Similar results occur with other functions. For example, the graph of $y = \dfrac{1}{x - 2}$ is 2 units to the right of the graph of $y = \dfrac{1}{x}$. Also, the graph of $y - 2 = \dfrac{1}{x}$, or $y = \dfrac{1}{x} + 2$, is 2 units above the graph of $y = \dfrac{1}{x}$.

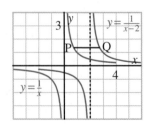
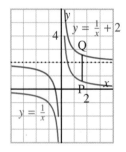

These properties apply to the graphs of all functions. You will use them many times in this book. Since they are so useful, we call them the *translation tools*.

Vertical Translation Tool

When you replace y with $y - k$ in the equation of a function $y = f(x)$, its graph is translated vertically.

If $k > 0$, the translation is up.
If $k < 0$, the translation is down.

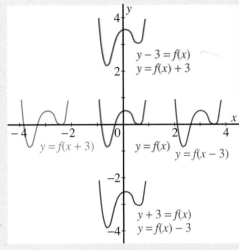

Horizontal Translation Tool

When you replace x with $x - k$ in the equation of a function $y = f(x)$, its graph is translated horizontally.

If $k > 0$, the translation is to the right.
If $k < 0$, the translation is to the left.

Example 1

a) Sketch the graph of the function $y = |x + 2| - 4$ for $-7 \le x \le 3$.

b) What are the domain and range of the function?

c) What are the x- and y-intercepts of the graph?

Solution

a) Sketch the graph of $y = |x|$. Use the horizontal translation tool. The graph of $y = |x + 2|$ is obtained by translating the graph of $y = |x|$ two units left. Use the vertical translation tool. The graph of $y = |x + 2| - 4$ is obtained by translating the graph of $y = |x + 2|$ four units down.

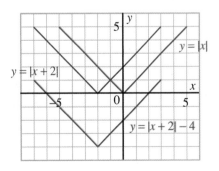

b) From the graph, the domain is the set of all real numbers. The range is the set of real numbers greater than or equal to -4.

c) From the graph, the x-intercepts are -6 and 2. The y-intercept is -2.

Example 2

a) Graph the function $y = \sqrt{9 - (x - 3)^2}$.

b) Explain the nature of the graph.

c) Identify the domain, range, and intercepts.

Solution

a) Enter the equation Y1=√(9–(X–3)²).

b) The graph is a semicircle with centre
$(3, 0)$ and radius 3, obtained by
translating the semicircle defined
by $y = \sqrt{9 - x^2}$ three units right.

c) The domain is defined by $0 \le x \le 6$.
The range is defined by $0 \le y \le 3$.
The x-intercepts are 0 and 6.
The y-intercept is 0.

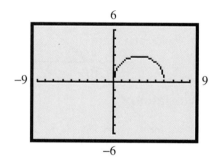

DISCUSSING THE IDEAS

1. Explain why the graph of $y = \dfrac{1}{x - 2}$ is located 2 units to the right of the graph
of $y = \dfrac{1}{x}$, and not 2 units to the left.

2. Refer to *Example 1*.

   a) Explain how to determine the domain and the range from the equation.

   b) Explain how to determine the x- and y-intercepts from the equation.

3. Repeat exercise 2, referring to *Example 2*.

1.2 EXERCISES

 1. Describe what happens to the graph of a function if you make these changes
to its equation.

   a) Replace x with $x + 3$.   b) Replace x with $x - 4$.

   c) Replace y with $y + 3$.   d) Replace y with $y - 4$.

   e) Replace x with $x - 2$ and y with $y - 2$.

   f) Replace x with $x + 5$ and y with $y + 1$.

2. Describe what happens to the equation of a function if you make these changes to its graph.

a) Translate the graph 5 units right.

b) Translate the graph 8 units left.

c) Translate the graph 2 units right and 7 units up.

d) Translate the graph 4 units left and 5 units down.

3. The graph of a function $y = f(x)$ is transformed as described below. The equation of its image has the form $y = f(x - c) + d$. Determine the value of c and d for each transformation.

a) Translate the graph 4 units down.

b) Translate the graph 4 units right.

c) Translate the graph 3 units left and 2 units up.

d) Translate the graph 8 units right and 6 units down.

4. Describe how the graph of $y = x^2$ compares to the graph of each function.

a) $y = x^2 - 3$ **b)** $y = (x - 3)^2$ **c)** $y = (x + 3)^2$

5. Describe how the graph of each function compares to the graph of $y = |x|$.

a) $y = |x| + 4$ **b)** $y = |x + 4|$ **c)** $y = |x - 4|$

6. Describe how the graph of the second function compares to the graph of the first function.

a) $y = \sqrt{x}$ $y = \sqrt{x - 5}$

b) $y = \frac{1}{2}x - 1$ $y = \frac{1}{2}(x - 3) - 1$

c) $y = \frac{1}{x}$ $y = \frac{1}{x + 1}$

7. A graphing calculator produced the three graphs below. The equations of the functions are shown on the fourth screen. Write the equation that corresponds to each graph. Explain each choice.

a)

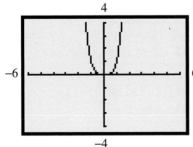

b)

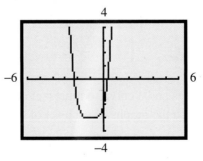

c)

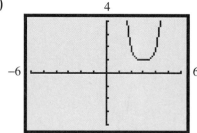

d)

```
Plot1  Plot2  Plot3
\Y1=X^4
\Y2=(X-3)^4+1
\Y3=(X+1)^4-3
\Y4=
\Y5=
\Y6=
\Y7=
```

8. Use the graphs in exercise 7 as a guide. Sketch the graph of each function.

   a) $y = x^4 - 2$
   b) $y = (x + 3)^4$
   c) $y = (x + 3)^4 + 2$
   d) $y = (x - 2)^4 + 3$

9. Sketch the graphs of these functions on the same grid.

   a) $y = \sqrt{x} - 2$
   b) $y = \sqrt{x + 3}$
   c) $y = \sqrt{x + 3} + 2$
   d) $y = \sqrt{x - 2} + 3$

10. Write to explain how you sketched the graphs in exercise 8 or 9.

B 11. For each list, sketch the graphs of the functions on the same grid.

   a) $y = 2x$ **b)** $y = x^2$ **c)** $y = \sqrt{x}$
       $y = 2(x - 3)$ $y = (x - 3)^2$ $y = \sqrt{x - 3}$
       $y = 2(x + 3)$ $y = (x + 3)^2$ $y = \sqrt{x + 3}$

12. Describe how the graph of each function is related to the first graph in the corresponding part of exercise 11.

   a) $y = 2(x - 5) + 3$ **b)** $y = (x - 5)^2 + 3$ **c)** $y = \sqrt{x - 5} + 3$

13. For each function $f(x)$ defined by the diagrams below, sketch $f(x), f(x - 2)$, and $f(x + 4)$.

   a)
   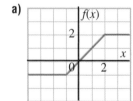

   b)
   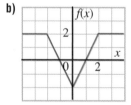

   c)
   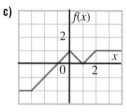

14. Draw coordinate axes and a curve to represent any function $f(x)$. On the same diagram, sketch the graphs of each set of functions. Describe the changes.

   a) $f(x - 2)$ $f(x) - 2$ $f(x - 2) - 2$
   b) $f(x + 3) - 1$ $f(x - 3) + 1$ $f(x + 5) + 3$

15. Later in this book, you will study the graph of the function $y = \log x$.
    a) Describe how the graph of $y = \log(x - 2) + 7$ is related to the graph of $y = \log x$.
    b) Write to explain why you can answer part a without knowing anything about the graph of $y = \log x$.

16. Predict what the graphs of the equations in each list would look like. Use your graphing calculator to check your predictions.

    a) $y = |x - 1|$
    $y = |x|$
    $y = |x + 1|$

    b) $y = \dfrac{1}{x - 2}$
    $y = \dfrac{1}{x}$
    $y = \dfrac{1}{x + 2}$

    c) $y = \sqrt{16 - (x - 4)^2}$
    $y = \sqrt{16 - x^2}$
    $y = \sqrt{16 - (x + 4)^2}$

    d) $y = |x + 3| + 1$
    $y = |x| + 1$
    $y = |x - 3| + 1$

    e) $y = \dfrac{1}{x - 2} + 4$
    $y = \dfrac{1}{x} + 4$
    $y = \dfrac{1}{x + 2} + 4$

    f) $y = \sqrt{4 - (x - 2)^2} + 3$
    $y = \sqrt{4 - x^2} + 3$
    $y = \sqrt{4 - (x + 2)^2} + 3$

17. At an interest rate, i percent, the principal, P dollars, that must be invested now to have \$100 in one year is $P = \dfrac{100}{1 + i}$.
    a) Explain why this equation is correct.
    b) Describe some ways in which the graph of P against i would be:
       i) similar to the graph on page 7
       ii) different from the graph on page 7

18. An important example of the horizontal translation of a graph occurred in 1787. The French chemist, Jacques Charles, investigated the effect on the volume of a gas when the temperature is increased. The pressure of the gas is constant. The results for a sample of air are shown below.

Temperature (°C)	Volume (mL)
0	152
20.0	163
40.0	174
60.0	185
80.0	197
100.0	208

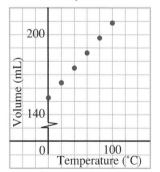

Volume of a gas as a function of temperature

The points appear to lie on a straight line. This line could be extended to the left until it intersects the temperature axis.

a) Visualize what the points on the line represent. What would the intersection of the line and the temperature axis represent?

b) Graph the data and determine the equation of the line of best fit.

c) Find the temperature corresponding to a volume of 0 mL.

d) Suppose the line in part b is translated to the right so that it passes through the origin. Determine its equation.

 MODELLING the Relation between Volume and Temperature

Exercise 18 illustrates Charles' Law in physics. This states that if the pressure is kept constant, the volume of a fixed mass of gas varies directly as the Kelvin temperature.

• Explain why it is impossible for the volume to vary directly as the Celsius temperature.

• What is the significance of translating the graph to the right in exercise 18d?

C 19. A function $f(x)$ has the following property. When its graph is translated 3 units right and 2 units up, the result coincides with the graph of $f(x)$.

a) Find an example of a function with this property.

b) How many functions like this are there?

20. In the lumber industry, the length of a log is measured in feet, its diameter in inches, and its volume in units called board feet. Various methods are used to estimate the volume that can be obtained from a log, allowing for waste and shrinkage. One rule of thumb is to subtract 4 inches from the diameter of the log, square the result, and multiply by the length of the log in feet.

a) Write an equation to express the volume, V board feet, of a log as a function of its diameter, d inches, and its length, L feet.

b) Graph the functions corresponding to $L = 12$, $L = 16$, and $L = 20$ for $26 \le d \le 36$.

COMMUNICATING THE IDEAS

Write a paragraph to explain why changing the value of k in the equation of a function $y = f(x - k)$ translates its graph right or left, and how you can tell which direction it moves. Include graphs as part of your explanation.

Transforming Function Graphs Part II

In this investigation, you will continue
the investigations of the previous
section making other changes to the
equations of functions. You will use
the linear function $y = \frac{1}{4}x + 3$,
which has the graph shown.

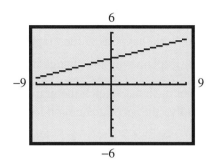

Comparing the graphs of $y = f(x)$ and $y = f(-x)$

1. a) Graph the function $y = \frac{1}{4}x + 3$. Use the window
settings above.

b) Predict what the graph of $y = \frac{1}{4}(-x) + 3$ would look like, compared
with the graph of $y = \frac{1}{4}x + 3$.

c) Graph these two equations. Did the result agree with your
predictions?

d) Use the trace or tables feature. Determine the x-coordinates of
points on the two graphs that have the same y-coordinate. How are
the x-coordinates related?

2. Would you have obtained the same results if you had started with the
graph of $y = x$, or with the graph of $y = x^2$? Explain.

3. a) What happens to the graph of $y = f(x)$ when x is replaced with $-x$?

b) Explain why the change in part a occurs.

4. For each list, sketch the graphs of the functions on the same axes.

a) $y = 2x + 3$	**b)** $y = x^3$	**c)** $y = \sqrt{x}$
$y = 2(-x) + 3$	$y = (-x)^3$	$y = \sqrt{-x}$

Comparing the graphs of $y = f(x)$ and $-y = f(x)$

5. a) Graph the function $y = \frac{1}{4}x + 3$. Use the window settings above.

b) Predict what the graph of $-y = \frac{1}{4}x + 3$ would look like compared
with the graph of $y = \frac{1}{4}x + 3$.

c) Graph these two equations. Did the result agree with your predictions?

d) Use the trace or tables feature. On the two graphs, examine the relationship between the y-coordinates of points that have the same x-coordinate.

6. Repeat exercise 2.

7. a) What happens to the graph of $y = f(x)$ when y is replaced with $-y$?

b) Explain why the change in part a occurs.

8. For each list, sketch the graphs of the functions on the same axes. Solve the second equation for y.

a) $y = 2x + 3$
$-y = 2x + 3$

b) $y = x^3$
$-y = x^3$

c) $y = \sqrt{x}$
$-y = \sqrt{x}$

Comparing the graphs of $y = f(x)$ and $x = f(y)$

9. a) Graph the function $y = \frac{1}{4}x + 3$. Use the window settings on page 25.

b) Predict what the graph of $x = \frac{1}{4}y + 3$ would look like, compared with the graph of $y = \frac{1}{4}x + 3$.

c) Graph these two equations. (To graph $x = \frac{1}{4}y + 3$ on a graphing calculator, you will have to solve this equation for y.) Did the result agree with your predictions?

d) Use the trace or tables feature. Examine how the coordinates of points on the two graphs are related. Are the results what you expected?

10. Repeat exercise 2.

11. a) What happens to the graph of $y = f(x)$ when x and y are interchanged?

b) Explain why the change in part a occurs.

12. For each list, sketch the graphs of the equations on the same axes. Solve the second equation for y. Do all the equations represent functions? Explain.

a) $y = 2x + 3$
$x = 2y + 3$

b} $y = x^3$
$x = y^3$

c) $y = \sqrt{x}$
$x = \sqrt{y}$

Consider the function $y = \sqrt{x}$, whose graph is shown on page 7. If x is replaced with $-x$, the equation becomes $y = \sqrt{-x}$. In *Investigate*, you should have found that the graph is reflected in the y-axis. To see why, observe that the values of x that satisfy the second equation are the opposites of the values that satisfy the first equation. On the two graphs, P and Q are corresponding points with the same y-coordinate. The x-coordinate of Q is the opposite of the x-coordinate of P. Hence, the graph of $y = \sqrt{-x}$ is the reflection of the graph of $y = \sqrt{x}$ in the y-axis. A similar analysis applies if y is replaced with $-y$. The graph of $-y = \sqrt{x}$, or $y = -\sqrt{x}$, is the reflection of the graph of $y = \sqrt{x}$ in the x-axis because the y-coordinate of R is the opposite of the y-coordinate of P.

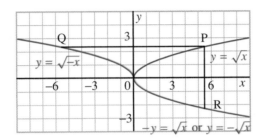

If x and y are interchanged in the equation $y = \sqrt{x}$, the equation becomes $x = \sqrt{y}$. In *Investigate*, you should have found that the graph is reflected in the line defined by $y = x$. When P and S are corresponding points on the two graphs, the x-coordinate of one is the y-coordinate of the other. The equation $x = \sqrt{y}$ represents the *inverse* of the function $y = \sqrt{x}$. When the equation of the function is $y = f(x)$, the equation of the inverse is $x = f(y)$.

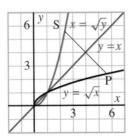

These properties apply to the graphs of all functions. We call them the *reflection tools*.

y-Axis Reflection Tool

When you replace x with $-x$ in the equation of a function $y = f(x)$, its graph is reflected in the y-axis.

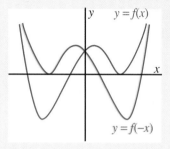

Reflection in the y-axis, or horizontal reflection

x-Axis Reflection Tool

When you replace y with $-y$ in the equation of a function $y = f(x)$, its graph is reflected in the x-axis.

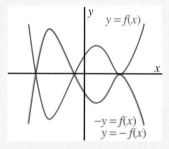

Reflection in the x-axis, or vertical reflection

y = x Reflection Tool

When you interchange x and y in the equation of a function $y = f(x)$, its graph is reflected in the line $y = x$.

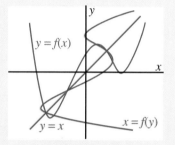

Reflection in the line $y = x$

Example 1

Each graph is the reflection of the other in the y-axis. Write the equation of the graph in red.

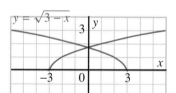

Solution

Use the y-axis reflection tool.
The equation of the graph is $y = \sqrt{3 - (-x)}$, or $y = \sqrt{3 + x}$.

Example 2

a) Graph the function $f(x) = \dfrac{1}{x^2 + 1}$.

b) Explain the appearance of the graph.

c) Use the graph of this function as a guide. Sketch the graphs of $y = f(-x)$, $y = -f(x)$, and $x = f(y)$.

d) Which of the graphs in part c represent functions? Explain.

Solution

a) *Using a graphing calculator*   **a)** *Using grid paper*

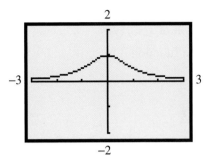

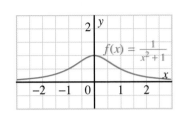

b) Since $x^2 \geq 0$, the expression $\dfrac{1}{x^2 + 1}$ is positive for all real values of x.
Since $x^2 + 1$ has a minimum value of 1 when $x = 0$, $\dfrac{1}{x^2 + 1}$ has a maximum value of 1 when $x = 0$. Hence, all points on the graph lie between the x-axis and the line $y = 1$, and there is a maximum point at $(0, 1)$.
Replacing x with $-x$ has no effect on the graph. Hence, the graph is symmetrical about the y-axis.

c) Use the reflection tools to sketch the graphs of $y = f(-x)$, $y = -f(x)$, and $x = f(y)$.

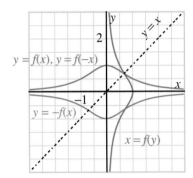

d) Recall the vertical line test, which states that if no two points on a graph can be joined by a vertical line, the graph represents a function. The graphs of $y = f(-x)$ and $y = -f(x)$ represent functions. The graph of $x = f(y)$ does not represent a function.

In *Example 2*, the points on the graph come closer and closer to the x-axis, yet never reach it. Recall that the x-axis is an asymptote of $f(x)$.

In *Example 2*, the equation $x = f(y)$ does not represent a function. In other situations, this equation can represent a function, and we can write its equation in a form solved for y, as $y = f^{-1}(x)$.

Example 3

a) Graph the function $f(x) = \frac{1}{8}x^3 + 1$.

b) Use the graph of this function as a guide. Sketch the graph of $x = f(y)$.

c) Use function notation to write the equation of the inverse function.

Solution

a) *Using a graphing calculator*   a) *Using grid paper*

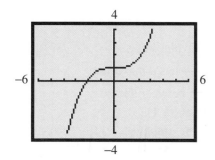

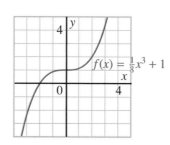

b) Use the $y = x$ reflection tool to sketch the graph of $x = f(y)$. This is the graph of $y = f^{-1}(x)$.

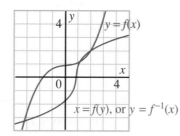

c) The equation of the given function can be written as $y = \frac{1}{8}x^3 + 1$.
The equation of the inverse is $x = \frac{1}{8}y^3 + 1$.
Solve this equation for y.

$$\frac{1}{8}y^3 = x - 1$$
$$y^3 = 8(x - 1)$$
$$y = 2\sqrt[3]{x - 1}$$

Hence, $f^{-1}(x) = 2\sqrt[3]{x - 1}$

1. In the second graph on page 27, the equation $x = \sqrt{y}$ represents the inverse of the function $y = \sqrt{x}$.

   a) Is the inverse a function? Explain.

   b) What other way is there to write the inverse?

2. Refer to the display on page 28.

   a) Explain why a reflection in the y-axis is called a horizontal reflection.

   b) Explain why a reflection in the x-axis is called a vertical reflection.

3. In *Example 2*, explain why the points on the graph never reach the x-axis, regardless of their distances from the origin.

1.3 EXERCISES

 1. Describe what happens to the graph of a function if you make each change to its equation.

   a) Replace x with $-x$.

   b) Replace y with $-y$.

   c) Replace x with $-x$ and y with $-y$.

   d) Interchange x and y.

2. Describe what happens to the equation of a function if you make each change to its graph.

   a) Reflect the graph in the x-axis.

   b) Reflect the graph in the y-axis.

   c) Reflect the graph in the line $y = x$.

   d) Reflect the graph in both axes.

3. Write to explain the difference between reflecting a graph horizontally and reflecting it vertically.

4. Describe how the graph of $y = 2x + 1$ compares to the graph of each function.

   a) $y = 2(-x) + 1$

   b) $-y = 2x + 1$

   c) $x = 2y + 1$

5. Each graph is a reflection of the other in the x-axis. Write the equation of each coloured graph.

a)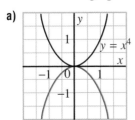
b)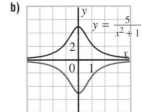
c)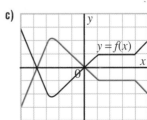

6. Each graph is a reflection of the other in the *y*-axis. Write the equation of each coloured graph.

a)

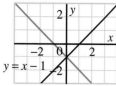

b) $y = \sqrt{4 - x}$

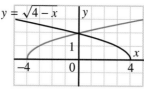

c) $y = g(x)$

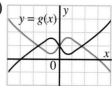

7. Each graph is a reflection of the other in the line $y = x$. Write the equation of each coloured graph.

a)

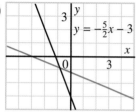

$y = -\frac{5}{2}x - 3$

b)

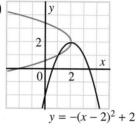

$y = -(x - 2)^2 + 2$

c) $y = \frac{1}{x + 2} - 1$

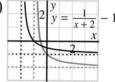

8. A graphing calculator produced these three graphs. The equations of the functions are shown on the fourth screen. Write the equation that corresponds to each graph. Explain each choice.

a)

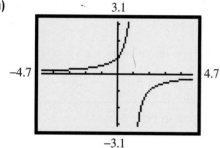

b)

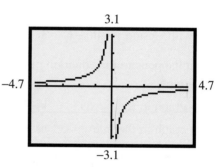

c)

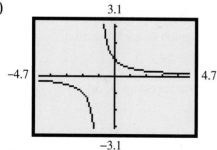

d)

```
Plot1  Plot2  Plot3
\Y1=-1/X
\Y2=1/(1+X)
\Y3=1/(1-X)
\Y4=
\Y5=
\Y6=
\Y7=
```

9. For each graph in exercise 8, write the equations of the asymptotes.

10. Use the graphs in exercise 8 as a guide. Sketch the graph of each function.

a) $y = \frac{-1}{1 + x}$

b) $y = \frac{-1}{1 - x}$

B **11.** Sketch the graphs of each set of equations on the same grid. Which equations represent functions?

a) $y = |x - 2|$ $y = |-x - 2|$ $y = -|x - 2|$ $x = |y - 2|$

b) $y = (x + 2)^3$ $y = (-x + 2)^3$ $y = -(x + 2)^3$ $x = (y + 2)^3$

12. Choose exercise 11 a or b. Write to explain how you sketched the graphs.

13. For each function $f(x)$ defined by the diagrams below, sketch the graph of $f(-x), -f(x), f^{-1}(x),$ and $f^{-1}(f(x))$.

a)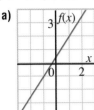

b)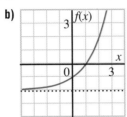

c)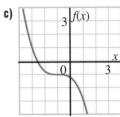

14. a) Explain why the equation of the function graphed at the right is $y = \sqrt{25 - (x - 5)^2}$.

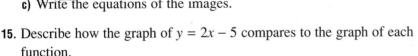

  b) Sketch the graph after it has been reflected in each line.

   i) the y-axis

   ii) the x-axis

   iii) $y = x$

  c) Write the equations of the images.

15. Describe how the graph of $y = 2x - 5$ compares to the graph of each function.

a) $y = 2(-x) - 5$ b) $-y = 2x - 5$ c) $x = 2y - 5$

16. Draw coordinate axes and a curve to represent any function $f(x)$. On the same diagram, sketch the graph of each relation. Describe the changes.

a) $y = f(-x)$ b) $y = -f(x)$ c) $y = -f(-x)$ d) $x = f(y)$

17. A function $f(x)$ with the property $f(-x) = f(x)$ is called an *even function*.

  a) Find two examples of even functions, then sketch their graphs.

  b) What special property does the graph of an even function have? Why does it have this property?

18. A function $f(x)$ with the property $f(-x) = -f(x)$ is called an *odd function*.

  a) Find two examples of odd functions, then sketch their graphs.

  b) What special property does the graph of an odd function have? Why does it have this property?

19. Recall that a polynomial function of degree n is a function whose equation can be written as $f(x) = a_n x^n + a_{n-1} x^{n-1} + a_{n-2} x^{n-2} + \cdots + a_1 x + a_0$, where n is a whole number. What conditions must be satisfied by n and the numerical coefficients in each case?

   a) A polynomial function $f(x)$ is an even function.

   b) A polynomial function $f(x)$ is an odd function.

20. Later in this book, you will study the graph of the function $y = \tan x$.

   a) Describe how the graph of $y = \tan(-x)$ is related to the graph of $y = \tan x$.

   b) Write to explain why you can answer part a without knowing anything about the graph of $y = \tan x$.

21. These graphs were produced by *Graphmatica*. Write a set of equations that could be used to make each pattern.

   a) 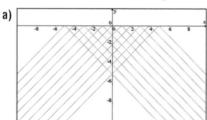   b)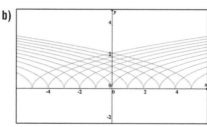

22. a) Make patterns like those in exercise 21.

   b) Use a different function to make a similar pattern of your own.

C 23. Find an example of a function that gives the same result when its graph is reflected in the y-axis as it does when its graph is reflected in the x-axis. Use the equation of the function to show that the two graphs are the same.

24. a) Use the reflection and translation tools to sketch the graph of each function.

   i) $y = 3 - \sqrt{x}$ ii) $y = \sqrt{3 - x}$

   b) Write to explain how you sketched the graphs in part a.

   c) Describe what happens to the graph of a function $y = f(x)$ if x is replaced with $k - x$.

COMMUNICATING THE IDEAS

Write a paragraph to explain why replacing x with $-x$ in the equation of a function $y = f(x)$ reflects its graph in the y-axis. Include graphs as part of your explanation.

1.4 Stretching Graphs of Functions

INVESTIGATE **Transforming Function Graphs Part III**

In this investigation, you will complete the investigations in the previous sections by making other changes to the equations of functions. You will begin with the function $y = \dfrac{1}{x^2 + 1}$, which has the graph shown.

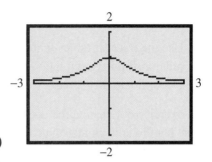

Comparing the graphs of $y = f(x)$ and $y = f(kx)$

1. **a)** Graph the function $y = \dfrac{1}{x^2 + 1}$. Use the window settings above.

   b) Predict what the graph of $y = \dfrac{1}{(2x)^2 + 1}$ would look like, compared with the graph of $y = \dfrac{1}{x^2 + 1}$.

   c) Graph these two equations. Did the result agree with your predictions?

   d) Use the trace or tables feature. Determine the x-coordinates of points on the two curves that have the same y-coordinate. How are the x-coordinates related?

2. Repeat exercise 1 using other coefficients of x. Use coefficients that are both greater than 1 and less than 1. Are the results what you expected?

3. Would you have obtained the same results if you had started with the graph of $y = x$, or with the graph of $y = x^2$ and replaced x with $2x$? Explain.

4. **a)** What information about the graph of $y = f(kx)$ does k provide?

   b) What happens to the graph of $y = f(kx)$ when k is changed?

   c) Explain why the changes in part b occur.

5. For each list, sketch the graphs of the functions on the same axes. Simplify the equations if necessary.

   a) $y = 2(2x) + 3$
   $y = 2x + 3$
   $y = 2\left(\frac{1}{2}x\right) + 3$

   b) $y = |2x|$
   $y = |x|$
   $y = |\frac{1}{2}x|$

   c) $y = \sqrt{25 - (2x)^2}$
   $y = \sqrt{25 - x^2}$
   $y = \sqrt{25 - \left(\frac{1}{2}x\right)^2}$

Comparing the graphs of $y = f(x)$ and $ky = f(x)$

6. a) Graph the function $y = \dfrac{1}{x^2 + 1}$. Use the window settings on page 35.

   b) Predict what the graph of $2y = \dfrac{1}{x^2 + 1}$ would look like, compared with the graph of $y = \dfrac{1}{x^2 + 1}$.

   c) Graph these two equations. Did the result agree with your predictions?

   d) Use the trace or tables feature. On the two graphs, examine the relationship between the y-coordinates of points that have the same x-coordinate.

7. In exercise 6b, change the coefficient 2 to other numbers, and repeat. Use coefficients greater than 1, coefficients between 0 and 1, and negative coefficients. Are the results what you expected?

8. Would you have obtained the same results if you had started with the graph of $y = x$, or with the graph of $y = x^2$? Explain.

9. a) What information about the graph of $ky = f(x)$ does k provide?

   b) What happens to the graph of $ky = f(x)$ when k is changed?

   c) Explain why the changes in part b occur.

10. For each list, sketch the graphs of the functions on the same axes.

a) $2y = 2x + 3$	**b)** $2y = \lvert x \rvert$	**c)** $2y = \sqrt{25 - x^2}$
$y = 2x + 3$	$y = \lvert x \rvert$	$y = \sqrt{25 - x^2}$
$\frac{1}{2}y = 2x + 3$	$\frac{1}{2}y = \lvert x \rvert$	$\frac{1}{2}y = \sqrt{25 - x^2}$

Consider the graph of $y = x^2$ once again. If x is replaced with $2x$, the equation becomes $y = (2x)^2$, or $y = 4x^2$. Although the graph of this function is a vertical expansion of the graph of $y = x^2$, it is also a horizontal compression of the graph. On the two graphs, let P and Q be corresponding points with the same y-coordinate. To determine the y-coordinate of P, we square the x-coordinate. To determine the y-coordinate of Q, we double the x-coordinate, then square the result. To produce the same y-coordinate as for P, the x-coordinate of Q must be *one-half* the x-coordinate of P. Hence, the graph of $y = (2x)^2$ is compressed horizontally relative to the graph of $y = x^2$.

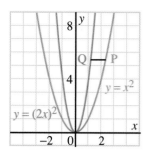 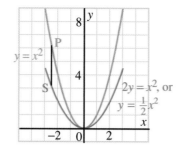

A similar analysis applies to the y-coordinates. The graph of $2y = x^2$ is compressed vertically relative to the graph of $y = x^2$ because the y-coordinate of a point on the graph of $2y = x^2$ is *one-half* the y-coordinate of the corresponding point on the graph of $y = x^2$. Instead of writing the equation of the image as $2y = x^2$, we usually write it as $y = \frac{1}{2}x^2$.

The same reasoning applies to any function, as illustrated in the following graph. Replacing x with $2x$ compresses the graph horizontally; replacing x with $\frac{1}{2}x$ expands it horizontally. Replacing x with $-2x$ compresses it horizontally and reflects it in the y-axis. Replacing x with $-\frac{1}{2}x$ expands it horizontally and reflects it in the y-axis.

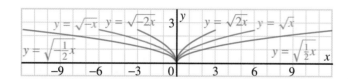

Similarly, replacing y with $2y$ compresses the graph vertically; replacing y with $\frac{1}{2}y$ expands it vertically. Replacing y with $-2y$ compresses it vertically and reflects it in the x-axis. Replacing y with $-\frac{1}{2}y$ expands it vertically and reflects it in the x-axis.

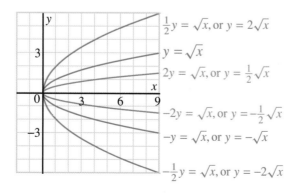

These properties apply to the graphs of all functions. They are the *stretching tools.*

Vertical Stretching Tool

When you replace y with ky in the equation of a function $y = f(x)$, its graph is expanded or compressed vertically.

If $0 < k < 1$, there is an expansion.
If $k > 1$, there is a compression.

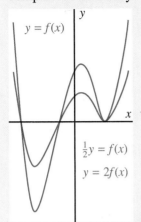

$\frac{1}{2}y = f(x)$

$y = 2f(x)$

$y = f(x)$

$2y = f(x)$

$y = \frac{1}{2}f(x)$

$y = f(x)$

$-2y = f(x)$

$y = -\frac{1}{2}f(x)$

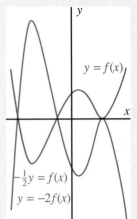

$-\frac{1}{2}y = f(x)$

$y = -2f(x)$

If $k < 0$, there is a reflection in the x-axis as well as the expansion or compression.

Horizontal Stretching Tool

When you replace x with kx in the equation of a function $y = f(x)$, its graph is expanded or compressed horizontally.

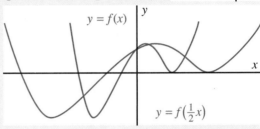

$y = f(x)$

$y = f\left(\frac{1}{2}x\right)$

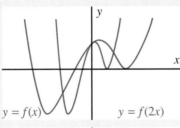

$y = f(x)$

$y = f(2x)$

If $0 < k < 1$, there is an expansion.
If $k > 1$, there is a compression.

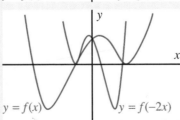

$y = f(x)$

$y = f(-2x)$

If $k < 0$, there is a reflection in the y-axis as well as the expansion or compression.

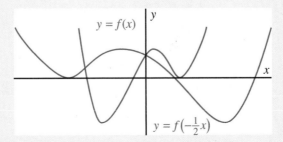

$y = f(x)$

$y = f\left(-\frac{1}{2}x\right)$

Example 1

Sketch the graph of each function.

a) $y = \sqrt{25 - (2x)^2}$

b) $y = \sqrt{25 - \left(\frac{1}{2}x\right)^2}$

Solution

> **Think...**
>
> In each case, start with the graph of $y = \sqrt{25 - x^2}$, which is a semicircle with radius 5, centred at the origin, and above the x-axis.

a) In $y = \sqrt{25 - (2x)^2}$, the coefficient of x is greater than 1. According to the horizontal stretching tool, the graph of $y = \sqrt{25 - x^2}$ is compressed horizontally by a factor of $\frac{1}{2}$.

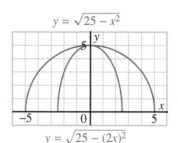

b) In $y = \sqrt{25 - \left(\frac{1}{2}x\right)^2}$, the coefficient of x is less than 1. The graph of

$y = \sqrt{25 - x^2}$ is expanded horizontally by a factor of 2.

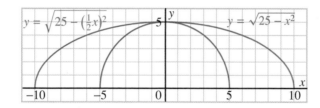

Example 2

a) Sketch the function $f(x) = x(x - 4)$.

b) Write the equations of the functions $f(2x)$ and $-f\left(-\frac{1}{2}x\right)$.

c) On the same axes as in part a, sketch the graphs of the functions $y = f(2x)$ and $y = -f\left(-\frac{1}{2}x\right)$.

Solution

a)

> ### Think...
>
> If $y = x(x - 4)$, then $y = 0$ when $x = 0$ or when $x = 4$. Hence, $f(x)$ is a quadratic function with zeros 0 and 4.

The graph of $f(x)$ is a parabola opening up, passing through $(0, 0)$ and $(4, 0)$, with axis of symmetry halfway between them.

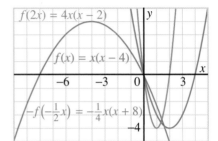

b) Since $f(x) = x(x - 4)$,

$f(2x) = 2x(2x - 4)$, or $4x(x - 2)$, and

$-f(-\frac{1}{2}x) = -(-\frac{1}{2}x)(-\frac{1}{2}x - 4)$, or $-\frac{1}{4}x(x + 8)$.

c) Use the stretching tools. To sketch the graph of $f(2x)$, compress the graph of $f(x)$ horizontally by a factor of $\frac{1}{2}$. To sketch the graph of $-f(-\frac{1}{2}x)$, expand the graph of $f(x)$ horizontally by a factor of 2, then reflect it in the y-axis and in the x-axis.

In *Example 2*, the compressions and expansions are relative to the y-axis, and not relative to the axis of symmetry of the parabola.

DISCUSSING THE IDEAS

1. a) Explain why replacing x with $2x$ compresses a graph horizontally.

b) Explain why replacing x with $\frac{1}{2}x$ expands a graph horizontally.

2. a) Explain why replacing y with $2y$ compresses a graph vertically.

b) Explain why replacing y with $\frac{1}{2}y$ expands a graph vertically.

3. Refer to *Example 2*.

a) Why are the compressions and expansions relative to the y-axis and not relative to the axis of symmetry of the parabola?

b) In part b, after substituting to determine $f(2x)$ and $-f(-\frac{1}{2}x)$, the expressions were simplified further. Explain how the simplified forms of the expressions confirm that the sketches are correct.

4. Compare the stretching tools with the translation tools. In what ways are they similar? In what ways are they different?

1.4 EXERCISES

A 1. Describe what happens to the graph of a function if you make each change to its equation.

a) Replace x with $3x$.

b) Replace y with $3y$.

c) Replace x with $\frac{1}{2}x$.

d) Replace y with $\frac{1}{2}y$.

e) Replace x with $-2x$.

f) Replace y with $-2y$.

g) Replace x with $2x$ and y with $2y$.

h) Replace x with $-3x$ and y with $4y$.

2. Describe what happens to the equation of a function if you make each change to its graph.

a) Expand horizontally by a factor of 4.

b) Compress horizontally by a factor of $\frac{1}{2}$ and reflect in the y-axis.

c) Expand vertically by a factor of 5 and reflect in the x-axis.

d) Compress vertically by a factor of $\frac{2}{3}$.

3. The graph of a function $y = f(x)$ is transformed as described below. The equation of its image has the form $y = af(bx)$. Determine the values of a and b for each transformation.

a) Compress by a factor of $\frac{1}{3}$ horizontally and a factor of $\frac{1}{2}$ vertically.

b) Expand by a factor of 2 horizontally, by a factor of 3 vertically, and reflect in the x-axis.

c) Compress by a factor of $\frac{1}{2}$ horizontally, expand by a factor of 4 vertically, and reflect in the y-axis.

d) Expand by a factor of 4 horizontally, compress by a factor of $\frac{1}{2}$ vertically, and reflect in both axes.

4. In exercise 3, does it matter in which order the transformations are applied? Explain.

5. a) Graph the function $y = \sqrt{x}$.

b) Determine the equation of the image of $y = \sqrt{x}$ in each case.
i) Its graph is compressed horizontally by a factor of $\frac{1}{4}$.
ii) Its graph is expanded vertically by a factor of 2.

c) Compare the results in part b. Explain how it is possible that the same image graph can be obtained by the two different transformations.

d) Give an example of a graph whose images would not be the same under the two transformations in part b.

6. Describe how the graph of each function compares to the graph of $y = |x|$.

a) $y = |2x|$ b) $y = |\frac{1}{2}x|$ c) $2y = |x|$ d) $\frac{1}{2}y = |x|$

7. Describe how the graph of each function compares to the graph of $y = \frac{1}{x}$.

a) $y = \frac{1}{2x}$ b) $y = \frac{1}{\frac{1}{2}x}$ c) $2y = \frac{1}{x}$ d) $\frac{1}{2}y = \frac{1}{x}$

8. Describe how the graph of the second function compares to the graph of the first function.

a) $y = \sqrt{x}$ $y = \sqrt{\frac{2}{3}x}$

b) $y = 6x - 3$ $y = 6\left(-\frac{3}{2}x\right) - 3$

c) $y = x^2 - 4x$ $y = \left(\frac{1}{2}x\right)^2 - 4\left(\frac{1}{2}x\right)$

9. The period, T seconds, of a pendulum is given approximately by the formula $T = 2\sqrt{L}$, where L metres is its length.

a) Sketch a graph of the function for reasonable values of L.

b) What are the domain and the range of the function in part a?

MODELLING the Period of a Pendulum

From physics, the formula relating the period, T seconds, and length, L metres, of a pendulum is $T = 2\pi\sqrt{\frac{L}{g}}$, where $g = 9.8$ m/s^2. This formula can be verified experimentally, and proved using physics principles.

- Check that this formula simplifies to $T \doteq 2\sqrt{L}$.
- Calculate the percent error in the period if $2\sqrt{L}$ is used instead of $2\pi\sqrt{\frac{L}{g}}$.
- Why do you think the number π appears in the physics formula?

10. A graphing calculator produced these three graphs. The equations of the functions are shown on the fourth screen. Write the equation that corresponds to each graph. Explain each choice.

a)
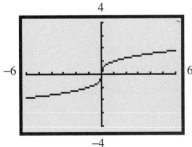

b)
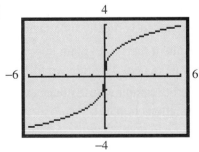

c)

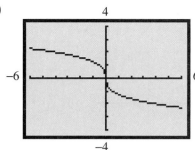

d)

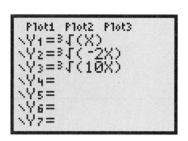

11. Use the graphs in exercise 10 as a guide. Sketch the graph of each function.

a) $y = \sqrt[3]{-x}$ **b)** $y = \sqrt[3]{-\frac{1}{2}x}$ **c)** $y = \sqrt[3]{\frac{1}{10}x}$

12. Sketch the graphs of these functions on the same grid.

a) $y = \sqrt{-x}$ **b)** $y = \sqrt{-\frac{1}{2}x}$ **c)** $y = \sqrt{\frac{1}{10}x}$

13. Write to explain how you sketched the graphs in exercise 11 or 12.

B 14. For each list, sketch the graphs of the functions on the same grid.

a) $y = |x + 1|$ **b)** $y = x^2 + 1$ **c)** $y = x^3$

$y = |2x + 1|$ $y = (2x)^2 + 1$ $y = (2x)^3$

$y = |\frac{1}{2}x + 1|$ $y = \left(\frac{1}{2}x\right)^2 + 1$ $y = \left(\frac{1}{2}x\right)^3$

15. For each list, describe how the graphs of the functions are related to the graphs in the corresponding part of exercise 14.

a) $y = |-x + 1|$ **b)** $y = (-x)^2 + 1$ **c)** $y = (-x)^3$

$y = |-2x + 1|$ $y = (-2x)^2 + 1$ $y = (-2x)^3$

$y = |-\frac{1}{2}x + 1|$ $y = \left(-\frac{1}{2}x\right)^2 + 1$ $y = \left(-\frac{1}{2}x\right)^3$

16. The graph of the function $y = \sqrt{9 - (x - 3)^2}$ is shown.

a) Sketch the graph after each transformation.

  i) Expand horizontally by a factor of 2.

  ii) Compress horizontally by a factor of $\frac{1}{3}$.

  iii) Expand vertically by a factor of 2.

  iv) Compress vertically by a factor of $\frac{1}{3}$.

b) Write the equations of the images.

17. Later in this book, you will study the graph of the function $y = \sin x$.

a) Describe how the graph of $y = \sin 3x$ is related to the graph of $y = \sin x$.

b) Write to explain why you can answer part a without knowing anything about the graph of $y = \sin x$.

18. For each function $f(x)$ defined by the diagrams below, sketch $f(2x)$, $f(0.5x)$, $f(-2x)$, and $f(-0.5x)$.

a)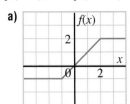

b)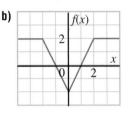

c)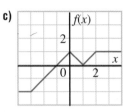

19. Draw coordinate axes and a curve to represent any function $f(x)$. On the same diagram, sketch the graph of each function. Describe the changes.

a) $y = f(2x)$ $y = 2f(x)$ $y = f(-2x)$ $y = -2f(x)$

b) $y = f\left(\frac{2}{3}x\right)$ $y = \frac{2}{3}f(x)$ $y = f\left(-\frac{2}{3}x\right)$ $y = -\frac{2}{3}f(x)$

20. For each list, predict what the graphs of the equations would look like. Use a graphing calculator to check your predictions.

a) $y = 2(2x) + 1$

$y = 2x + 1$

$y = 2\left(\frac{1}{2}x\right) + 1$

$y = 2\left(-\frac{1}{2}x\right) + 1$

$y = 2(-x) + 1$

$y = 2(-2x) + 1$

c) $y = 2(2x)^2 + 1$

$y = 2x^2 + 1$

$y = 2\left(\frac{1}{2}x\right)^2 + 1$

b) $y = \sqrt{2 - 2x}$

$y = \sqrt{2 - x}$

$y = \sqrt{2 - \left(\frac{1}{2}x\right)}$

$y = \sqrt{2 - \left(-\frac{1}{2}x\right)}$

$y = \sqrt{2 - (-x)}$

$y = \sqrt{2 - (-2x)}$

d) $y = \sqrt{2 - (2x)^2}$

$y = \sqrt{2 - x^2}$

$y = \sqrt{2 - \left(\frac{1}{2}x\right)^2}$

COMMUNICATING THE IDEAS

Write a paragraph to explain why changing the value of k in the equation of a function $y = f(kx)$ expands or compresses the graph. In your paragraph, explain how you can tell whether the value of k represents an expansion or a compression. Include graphs as part of your explanation.

Combining Transformations

In preceding sections, you learned that when x is replaced with $x - k$ or y is replaced with $y - k$ in the equation of a function $y = f(x)$, its graph is translated. When x is replaced with kx or y is replaced with ky, the graph is expanded or compressed. In this investigation, you will determine what happens when translations and expansions or compressions are applied in succession.

Horizontal Transformations

1. Sketch the graph of the function $y = x^2$.

2. Visualize replacing x with $x - 4$, then replacing x with $2x$.

   a) Write the equation of the new function.

   b) Sketch the graph of the new function.

   c) Describe what happened to the graph of the original function.

3. Visualize replacing x with $2x$, then replacing x with $x - 4$. Repeat exercise 2a, b, and c.

4. Compare the results of exercises 2 and 3.

   a) What general conclusion can you make about what happens when a horizontal translation and a horizontal expansion or compression are applied to the graph of a function?

   b) Why can you make this conclusion using only one example?

Vertical Transformations

5. Repeat exercise 1.

6. Visualize replacing y with $y - 3$, then replacing y with $\frac{1}{2}y$. Repeat exercise 2a, b, and c.

7. Visualize replacing y with $\frac{1}{2}y$, then replacing y with $y - 3$. Repeat exercise 2a, b, and c.

8. Compare the results of exercises 6 and 7.

   a) What general conclusion can you make about what happens when a vertical translation and a vertical expansion or compression are applied to the graph of a function?

   b) Why can you make this conclusion using only one example?

Combining Horizontal and Vertical Transformations

In the preceding exercises, you used these transformations:

HT	horizontal translation 4 units right
HC	horizontal compression by a factor of $\frac{1}{2}$
VT	vertical translation 3 units up
VE	vertical expansion by a factor of 2

When some of these transformations are applied to a graph, the order does not matter. For example, the two translations can be applied in either order. When other transformations are applied, the order does matter. For example, you discovered in the preceding exercises that horizontal translations and compressions give different results depending on the order.

9. Suppose the horizontal compression and the vertical expansion are applied to the graph of $y = x^2$. Does the result depend on the order in which these two transformations are applied? Explain, using both equations and graphs.

10. Suppose all four transformations are applied to the graph of $y = x^2$ in the order indicated. Write the equation of the new function and sketch its graph.

   a) HC, HT, VE, VT

   b) HC, VE, HT, VT

   c) HT, HC, VT, VE

   d) HT, VT, HC, VE

11. Suppose all four transformations listed above are applied to the graph of $y = x^2$.

   a) In how many different orders can four transformations be applied?

   b) How many different results are possible after four transformations have been applied? Explain.

12. In exercises 9, 10, and 11, you used the function $y = x^2$. Would your answers to these exercises be the same if you had used any function $y = f(x)$? Explain.

In the first two parts of *Investigate*, you should have found that when translations are combined with expansions or compressions, the result depends on the order in which the transformations are applied. Unless otherwise stated, we will always apply the expansions or compressions before applying the translations.

Vertical Expansion or Compression Followed by a Vertical Translation

Suppose the graph of the function $y = x^2$ is expanded vertically by a factor of 2. To find the new equation, replace y with $\frac{1}{2}y$. The equation becomes $\frac{1}{2}y = x^2$, or $y = 2x^2$.

Suppose the resulting graph is translated 3 units up. To find the new equation, replace y with $y - 3$. The equation becomes $y - 3 = 2x^2$, or $y = 2x^2 + 3$.

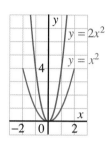

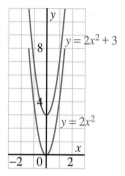

Horizontal Expansion or Compression Followed by a Horizontal Translation

Suppose the graph of the function $y = x^2$ is compressed horizontally by a factor of $\frac{1}{2}$. To determine the new equation, replace x with $2x$. The equation becomes $y = (2x)^2$.

Suppose the resulting graph is translated 8 units right. To determine the new equation, replace x with $x - 8$. The equation becomes $y = (2(x - 8))^2$.

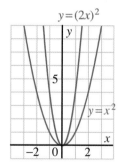

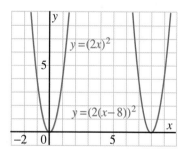

Example 1

Sketch the graph of each function.　**a)** $y = \sqrt{3(x + 6)}$　**b)** $y = \sqrt{3x + 6}$

Solution

a)

> **Think...**
>
> Start with $y = \sqrt{x}$. Replace x with $3x$ to obtain $y = \sqrt{3x}$, then replace x with $x + 6$ to obtain $y = \sqrt{3(x + 6)}$.

Step 1. Sketch the graph of $y = \sqrt{x}$. Compress this graph by a factor of $\frac{1}{3}$ horizontally. The result is the graph of $y = \sqrt{3x}$.

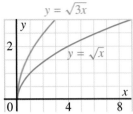

Step 2. Translate the graph of $y = \sqrt{3x}$ 6 units left. The result is the graph of $y = \sqrt{3(x + 6)}$.

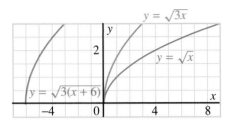

b)

> ***Think…***
>
> Rewrite the equation in a form similar to the equation in part a.

$y = \sqrt{3x + 6}$ becomes $y = \sqrt{3(x + 2)}$

Step 1. Sketch the graph of $y = \sqrt{x}$. Compress this graph by a factor of $\frac{1}{3}$ horizontally. The result is the graph of $y = \sqrt{3x}$.

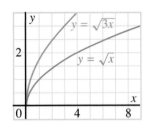

Step 2. Translate the graph of $y = \sqrt{3x}$ two units left. The result is the graph of $y = \sqrt{3(x + 2)}$. This is also the graph of $y = \sqrt{3x + 6}$.

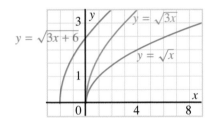

Horizontal Tools — Compress (or Expand), Then Translate

Suppose you replace x with $2x$ in the equation of a function $y = f(x)$ to obtain $y = f(2x)$, then replace x with $x - 8$ to obtain $y = f(2(x - 8))$. Its graph is compressed by a factor of $\frac{1}{2}$, then translated 8 units right.

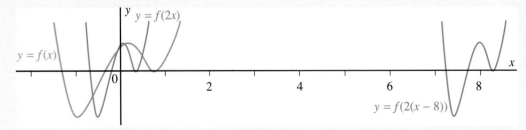

Example 2

This graphing calculator screen shows the graphs of $y = |x|$ and $y = 3|2(x - 5)| + 1$.

a) Use the equations. Describe how the graph of the second function is related to the graph of the first function.

b) Sketch graphs to verify your answer in part a.

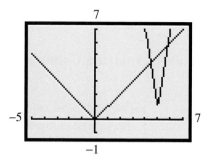

Solution

a) Determine the horizontal transformations. Start with $y = |x|$, replace x with $2x$ to obtain $y = |2x|$, then replace x with $x - 5$ to obtain $y = |2(x - 5)|$. There is a horizontal compression by a factor of $\frac{1}{2}$, then a translation 5 units right.

Determine the vertical transformations. The expression $|2(x - 5)|$ is multiplied by 3, then 1 is added. There is a vertical expansion by a factor of 3, then a translation 1 unit up.

b) *Either:*

Compress the graph of $y = |x|$ horizontally by a factor of $\frac{1}{2}$, and translate it 5 units right. Then expand it vertically by a factor of 3, and translate it 1 unit up.

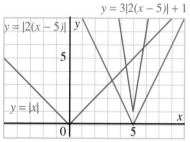

Or:

Compress the graph of $y = |x|$ horizontally by a factor of $\frac{1}{2}$, and expand it vertically by a factor of 3. Then translate it 5 units right and 1 unit up.

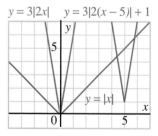

When we combine vertical and horizontal transformations, we usually apply the expansions and compressions first, then apply the translations. These are the *combined tools*.

Combined Tools — Applying Compressions and Expansions before Translations

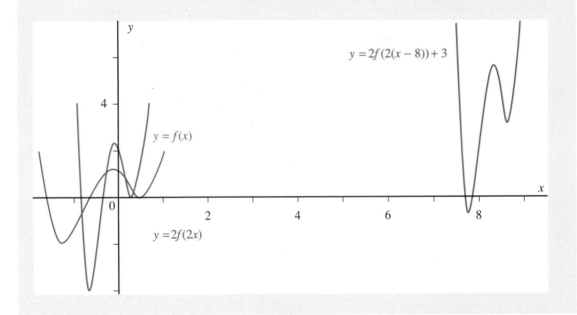

Vertical expansion by a factor of 2 and horizontal compression by a factor of $\frac{1}{2}$, …

$$y = 2f(2(x - 8)) + 3$$

… then translate 8 units right and 3 units up.

$y = 2f(2(x - 8)) + 3$

$y = f(x)$

$y = 2f(2x)$

DISCUSSING THE IDEAS

1. Refer to the graph of $y = (2x)^2$ on page 47 (bottom left). This graph looks like a vertical expansion of the graph of $y = x^2$. Explain how its equation could be written to show this.

2. Describe some other ways to complete *Example 2*.

1.5 EXERCISES

A 1. Describe what happens to the graph of a function when you make each change to its equation.

    a) Replace x with $3x$, then replace x with $x - 2$.

    b) Replace x with $x - 2$, then replace x with $3x$.

    c) Replace x with $-x$, then replace x with $x + 4$.

    d) Replace x with $x + 4$, then replace x with $-x$.

2. Describe what happens to the graph of a function when you make each change to its equation.

    a) Replace x with $2x$ and y with $2y$.

    b) Replace x with $2x$, then y with $2y$, then x with $x - 6$.

    c) Replace x with $\frac{1}{2}x$, y with $-3y$, then y with $y - 4$.

    d) Replace x with $\frac{1}{3}x$ and y with $\frac{1}{2}y$, then x with $x + 1$ and y with $y + 2$.

3. Describe what happens to the equation of a function when you make each change to its graph.

    a) Compress horizontally by a factor of $\frac{1}{4}$ and vertically by a factor of $\frac{1}{2}$.

    b) Compress horizontally by a factor of $\frac{1}{4}$, then translate 3 units right.

    c) Expand horizontally by a factor of 2, then translate 3 units left.

    d) Compress horizontally by a factor of $\frac{1}{4}$, then translate 5 units right.

B 4. For each list, sketch the graphs on the same grid.

    a) $y = x$ **b)** $y = x^2$ **c)** $y = \sqrt{x}$

         $y = 2x$ $y = (2x)^2$ $y = \sqrt{2x}$

         $y = 2(x + 6)$ $y = (2(x + 6))^2$ $y = \sqrt{2(x + 6)}$

5. Choose one list from exercise 4. Write to explain how you sketched the graphs.

6. This calculator screen shows the graphs of $y = \sqrt{x}$ and $y = \sqrt{4x + 8}$.

    a) Use the equations. Explain how the second graph is related to the first.

    b) Sketch graphs to verify your answer in part a.

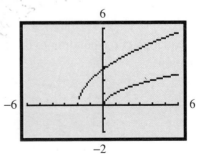

7. **a)** Graph the functions $y = x^2$ and $y = \left(\frac{1}{2}(x - 4)\right)^2$.

    b) Use the equations. Explain how the second graph is related to the first.

    c) Sketch graphs to verify your answer in part b.

8. Repeat exercise 7, using each pair of functions.

a) $y = \frac{1}{x}$, $y = \frac{1}{2(x-3)}$ **b)** $y = \sqrt{x}$, $y = \sqrt{2x-5}$ **c)** $y = |x|$, $y = |\frac{2}{3}x - 6|$

The graph of the function $f(x) = \sqrt{4 - (x-2)^2}$
is shown. Use this function in exercises 9 and 10.

9. a) Sketch the image of $f(x)$ after:

   i) a horizontal compression by a factor of $\frac{1}{2}$

   ii) a vertical expansion by a factor of 3

   iii) both transformations in parts i and ii have been applied

 b) Write the equations of the three images in part a.

10. a) Sketch the image of $f(x)$ after:

   i) a horizontal expansion by a factor of 2

   ii) a horizontal translation 3 units left

   iii) the expansion in part i, then the translation in part ii

   iv) the translation in part ii, then the expansion in part i

 b) Write the equations of the four images in part a.

11. The calculator screen shows the graphs
of these two functions:

$y = \frac{1}{x^2 + 1}$ and $y = \frac{3}{(2(x-3))^2 + 1} + 2$

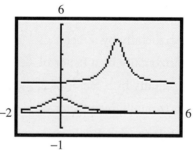

 a) Use the equations. Explain how the
second graph is related to the first.

 b) Sketch graphs to verify your answer
in part a.

C **12.** When we apply horizontal expansions or compressions and translations to
the graph of a function $y = f(x)$, we write equations such as $y = f(2x - 5)$
or $y = f(2(x - 5))$, depending on which transformation is applied first.
However, when we apply vertical expansions or compressions and
translations, we usually write equations such as $y = 3f(x) + 4$.

 a) In $y = 3f(x) + 4$, in which order are the two transformations applied?

 b) Write the equation corresponding to the other order of applying the
transformations.

COMMUNICATING THE IDEAS

Suppose you are given the equation of a function that involves horizontal translations and
expansions or compressions. Write to explain how you can tell whether the translation is
applied first or whether the expansion or compression is applied first. Use an example to
illustrate your explanation.

What Is a Line of Best Fit?

Refer to the first graph on page 14, that shows the line of best fit calculated for the data on page 13. You may have wondered how the equation of a line of best fit is defined mathematically. To understand the definition, follow these steps.

Step 1. Visualize a line with equation $y = mx + b$. This line typically does not pass through any of the data points.

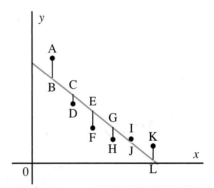

Step 2. For each data point, calculate the difference between its y-coordinate and the y-coordinate of the point on the line that is vertically below or above. On this diagram, these differences are represented by segments AB, CD, EF, GH, IJ, and KL.

Step 3. Square the differences in Step 2 and add the results.

The line of best fit is defined as the one for which the sum in Step 3 is a minimum. This means that the sum $AB^2 + CD^2 + EF^2 + GH^2 + IJ^2 + KL^2$ is to be as small as possible. There is only one line whose equation satisfies this condition. Mathematicians have determined formulas for the slope, m, and the y-intercept, b, of this line. These formulas have been programmed into the calculator's processor chip. Similar formulas exist for other functions, such as those listed in the STAT CALC menu on page 14. These formulas were used to calculate the values of the coefficients in the three screens below the menu.

1. Use the data on page 13. List the coordinates of the points on the scatterplot: A, D, F, H, I, and K.

2. Use the equation in the LinReg screen on page 14. Calculate the sum $AB^2 + CD^2 + EF^2 + GH^2 + IJ^2 + KL^2$.

3. In the LinReg equation, change a and/or b to other numbers. Calculate the sum $AB^2 + CD^2 + EF^2 + GH^2 + IJ^2 + KL^2$ for your new equation. Verify that it is greater than the result in exercise 1.

4. The method of determining the equation of the line of best fit is called the *method of least squares*. Suggest why this is an appropriate name for the method.

Recall that $|x| = x$ if $x \geq 0$, and $|x| = -x$ if $x < 0$. In *Example 2* on page 9, we used this definition to graph the function $f(x) = |x|$. If $y = f(x)$ is any function, we can use the definition of absolute value to sketch the graph of the related function, $y = |f(x)|$.

Example 1

a) Graph the function $y = |x^2 - 4|$.

b) Describe the graph, and account for its shape.

c) Determine the domain and range of the function.

Solution

a) Enter the equation Y1=abs(X²−4).
(To enter abs(, press [MATH] [▶] 1.)

b) The graph is formed from the parabola $y = x^2 - 4$ by reflecting in the *x*-axis the part that is below the *x*-axis.

c) The domain is the set of all real numbers. The range is the set of all real numbers greater than or equal to 0.

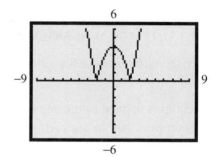

To sketch the graph of $y = |f(x)|$, we reflect in the *x*-axis the part of the graph of $y = f(x)$ that is below the *x*-axis. This is the *absolute value tool*.

Absolute Value Tool

To graph $y = |f(x)|$, first graph $y = f(x)$. Then take any parts of the graph that are below the *x*-axis and reflect them in the *x*-axis.

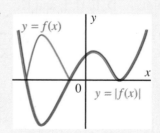

Example 2

Sketch the graph of $y = |2x - 3|$.

Solution

Sketch the graph of $y = 2x - 3$, which is a line with slope 2 and y-intercept -3. Reflect the part of the line that is below the x-axis in the x-axis.

$y = |2x - 3|$

To sketch the reciprocal of a function, we use the properties of reciprocals.

Properties of Reciprocal Functions

$f(x)$	$\dfrac{1}{f(x)}$
is less than -1	is between -1 and 0
-1	-1
is between -1 and 0	is less than -1
0	a vertical asymptote may exist
is between 0 and 1	is greater than 1
1	1
is greater than 1	is between 0 and 1

Example 3

a) Given $f(x) = 2x - 3$, sketch the graphs of $y = f(x)$ and of $y = \dfrac{1}{f(x)}$.

b) Identify the asymptotes of the graph of $y = \dfrac{1}{f(x)}$.

Solution

a) The graph of $y = f(x)$ is a straight line with slope 2 and y-intercept -3. For each point on the line, visualize a corresponding point such that the y-coordinates of the two points are reciprocals. Points above the line $y = 1$ are transformed between this line and the x-axis. The farther the original point is above the line $y = 1$, the closer the transformed point is to the x-axis. Points between the x-axis and the line $y = 1$ are transformed above the line $y = 1$. The closer the point is to the x-axis, the farther the transformed point is above the line $y = 1$. The point on the x-axis cannot be transformed. Points between the x-axis and the line $y = -1$ are transformed below the line $y = -1$. Points below the line $y = -1$ are transformed between this line and the x-axis.

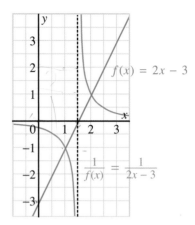

Draw a smooth curve through the plotted points. There are two distinct branches, separated by the value of x for which $2x - 3 = 0$, that is, $x = \frac{3}{2}$.

b) The x-axis is a horizontal asymptote.
The line defined by $x = \frac{3}{2}$ is a vertical asymptote.

The method of sketching the graph of the reciprocal of a function used in *Example 3* is the *reciprocal tool*.

Reciprocal Tool

To graph $y = \frac{1}{f(x)}$, first graph $y = f(x)$. For all points above the line $y = 1$, estimate the positions of the corresponding points on the graph of $y = \frac{1}{f(x)}$.

These are between the x-axis and the line $y = 1$; points farther from the line $y = 1$ correspond to points closer to the x-axis.

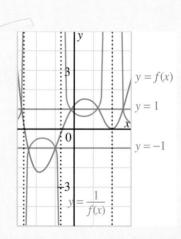

Do the same for points between the x-axis and the line $y = 1$. The corresponding points are above the line $y = 1$; points closer to the x-axis correspond to points farther from the line $y = 1$.

Repeat for points below the x-axis.

1. In *Example 1* and *Example 2*, what other ways are there to sketch the graphs?

2. Is the range of every absolute value function the set of real numbers greater than or equal to 0? Explain.

3. In the solution of *Example 3*, explain why the two branches in the graph of $y = \dfrac{1}{2x - 3}$ are not joined.

4. Refer to the graph of $y = \dfrac{1}{f(x)}$ at the bottom of page 56.

   a) Identify the vertical asymptotes, and explain why each occurs.

   b) Do there appear to be any horizontal asymptotes? Explain.

5. Consider any function $y = f(x)$. Explain your answer to each question.

   a) Is the domain of $y = |f(x)|$ always the same as the domain of $y = f(x)$?

   b) Is the domain of $y = \dfrac{1}{f(x)}$ always the same as the domain of $y = f(x)$?

   c) Could the range of $y = |f(x)|$ be the same as the range of $y = f(x)$?

   d) Could the range of $y = \dfrac{1}{f(x)}$ be the same as the range of $y = f(x)$?

1.6 EXERCISES

A 1. Copy the graph of each function $y = f(x)$ on grid paper. Sketch the graph of $y = |f(x)|$ on the same grid.

a)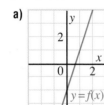

b)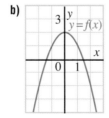

c)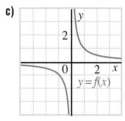

2. Copy the graph of each function $y = f(x)$ on grid paper. Sketch the graph of $y = \dfrac{1}{f(x)}$ on the same grid and identify the asymptotes.

a)

b)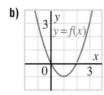

c)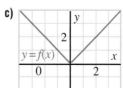

3. A graphing calculator produced these five graphs. The equations of the functions are shown on the last screen. Write the equation that corresponds to each graph. Explain each choice.

a)

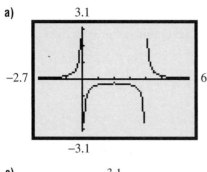

b)

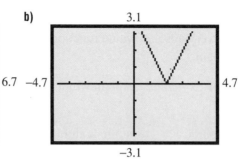

c)

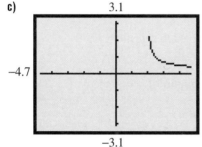

d)

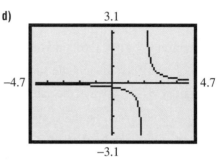

e)

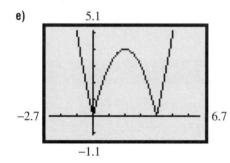

f)

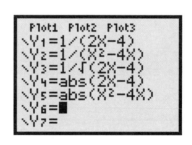

B 4. Graph each pair of functions on the same grid.

a) $y = x - 1$ and $y = |x - 1|$ b) $y = x^2 - 1$ and $y = |x^2 - 1|$ c) $y = x^3 - 1$ and $y = |x^3 - 1|$

5. Choose one pair of functions from exercise 4. Write to explain how you graphed the functions.

6. Graph each pair of functions on the same grid. Identify the asymptotes of the reciprocal functions.

a) $y = x - 1$ and $y = \frac{1}{x - 1}$ b) $y = x^2 - 1$ and $y = \frac{1}{x^2 - 1}$ c) $y = x^3 - 1$ and $y = \frac{1}{x^3 - 1}$

7. Choose one pair of functions from exercise 6. Write to explain how you graphed the functions.

8. Sketch the graphs of each set of functions on the same grid.

a) $y = 2x + 5$ $y = |2x + 5|$ $y = \frac{1}{2x + 5}$

b) $y = x^2 + 5x$ $y = |x^2 + 5x|$ $y = \dfrac{1}{x^2 + 5x}$

c) $y = \sqrt{4 - x^2}$ $y = \dfrac{1}{\sqrt{4 - x^2}}$

9. Graph each function. Compare your graph with your sketches in exercise 8c. Explain the similarities and differences.

a) $y = \sqrt{|4 - x^2|}$ 　　　　　　　**b)** $y = \dfrac{1}{\sqrt{|4 - x^2|}}$

10. For each function $f(x)$ defined by the diagrams below, sketch $|f(x)|$ and $\dfrac{1}{f(x)}$.

a) 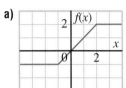 　**b)** 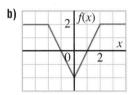 　**c)**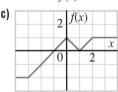

11. Graph each pair of functions. Explain the similarities and differences.

a) $y = |3x + 6|$ 　　　　　$y = 3|x| + 6$

b) $y = \sqrt{x}$ 　　　　　　$y = \sqrt{|x|}$

C **12.** Graph each equation and determine if it represents a function. Explain.

a) $|y| = x$ 　　**b)** $|y| = x + 1$ 　　**c)** $|y| = |x|$ 　　**d)** $|y| = |x| + 1$

13. What happens to the graph of a function when you make each replacement in its equation? Support each answer with an example.

a) Replace x with $|x|$. 　　　　　**b)** Replace y with $|y|$.

c) Replace x with $\dfrac{1}{x}$. 　　　　　**d)** Replace y with $\dfrac{1}{y}$.

14. Suppose you are given the graph of a function $y = f(x)$.

a) Explain the meaning of $y = \dfrac{1}{|f(x)|}$ and $y = |\dfrac{1}{f(x)}|$.

b) Would the graphs of the two functions in part a be the same or different? Explain both algebraically and graphically.

c) Explain the meaning of $y = \dfrac{1}{\sqrt{f(x)}}$ and of $y = \sqrt{\dfrac{1}{f(x)}}$.

d) Would the graphs of the two functions in part c be the same or different? Explain both algebraically and graphically.

COMMUNICATING THE IDEAS

Suppose you are given the graph of a function $y = f(x)$. Write to explain how the graphs of $y = |f(x)|$ and $y = \dfrac{1}{f(x)}$ are related to the graph of $y = f(x)$.

When Will Women's Pay Equal Men's?

On page 2, we considered the problem of predicting when women's pay will equal men's pay. The newspaper article on that page suggested that this could occur around the year 2020. Here is more information from the article.

As the chart shows, the improvement in the comparison between men and women has been slow but steady. If you run trend lines through the data of the past four censuses, you get a better idea of how long it will take before women's pay is equal to men's.

The flatter trend line on the chart is simply a straight line. In this case, in the usage of statisticians, it "explains" 97 percent of the variations in the data. In other words, it doesn't fit the actual data exactly, but it's a good approximation.

The other trend line, based on a more complicated equation (a second-order polynomial), explains 99.6 percent of the variations — a pretty solid fit.

Neither trend line carries any sort of guarantee. The slow, straight-line trend indicates that pay equality is more than 60 years away. The shorter, jazzier line says, based on current trends, equality will be reached by 2020, roughly one generation away.

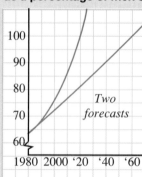

**Women's pay
as a percentage of men's**

*Two
forecasts*

 DEVELOP A MODEL

The graph illustrates linear and quadratic models based on expressing women's pay as a percent of men's. From the last four censuses, these percents are given.

Women's pay as a percent of men's

Year	1980	1985	1990	1995
Percent	64.1	64.5	68.1	71.4

In this model, you will graph the data and determine the equations of the line and the parabola of best fit.

1. **a)** Clear the graphing screen and define an appropriate viewing window.

   b) Enter the above data in the first two columns of the list editor.

   c) Define a statistical plot, then graph the data.

2. **a)** Press STAT ▶ to obtain the STAT CALC menu.

 b) Press **4** VARS ▶ ENTER **1** ENTER. The calculator will determine the equation of the line of best fit and copy it to the first line in the Y= list.

 c) Press STAT ▶ **5** VARS ▶ ENTER **2** ENTER. The calculator will determine the equation of the parabola of best fit and copy it to line 2 in the Y= list.

 d) Use the trace or intersect functions to predict when women's pay will be 100% of men's pay. Do your results agree with those in the article?

3. Repeat, using some of the other function types in the STAT CALC menu.

 **LOOK AT THE IMPLICATIONS**

The models in the article are based on expressing women's pay as a percent of men's pay. However, we could use other models that express men's pay as a percent of women's pay.

4. **a)** Visualize a similar graph that expresses men's pay as a percent of women's pay. Describe how it would differ from the graph above.

 b) Modify the data on page 60 so that the second line of the table shows men's pay as a percent of women's pay.

 c) Repeat exercises 2 and 3 using the modified data.

5. Compare the results with those for the previous models. Explain any discrepancies.

 **REVISIT THE SITUATION**

To avoid the discrepancies in exercise 5, we can use models based on incomes.

6. **a)** Use these data to model women's and men's incomes.

 b) Use the models to predict when pay equality will occur.

7. Does every model predict that equality will occur? Explain.

Annual Income for Full-Time Workers

Year	Females	Males
1980	18 000	28 100
1985	20 700	32 100
1990	24 500	36 000
1995	30 000	42 000

Visualizing Foreign Exchange Rates

The financial section of a newspaper often contains graphs that show daily variations in the relative values of different currencies. Use only these two graphs to answer the questions below.

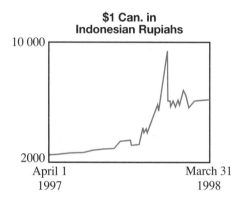

$1 Can. in Indonesian Rupiahs

10 000

2000

April 1   March 31
1997   1998

$1 U.S. in Canadian Dollars

1.47

1.36

April 1   March 31
1997   1998

1. What information does each graph show?

2. In the fall and winter of 1997–98, the economy collapsed in Asia.

  a) Explain how this is shown in the first graph.

  b) Both graphs show a series of line segments extending from the bottom to the top during the same time period. Explain why the second graph does not show a similar collapse of the Canadian economy.

3. a) Describe how the two graphs could be drawn to make it more obvious that the economy collapsed in Asia but not in Canada.

  b) Would the graphs you described in part a be a more accurate representation of the information than the graphs above? Explain.

4. Visualize the graphs described below. Sketch each graph, showing its essential features.

  a) a graph showing one Indonesian rupiah in Canadian dollars

  b) a graph showing one Canadian dollar in U.S. dollars

5. Visualize a graph showing the Indonesian rupiah against the U.S. dollar. Describe how it would be similar to and different from the first graph above.

1. The length, L metres, of a steel girder is $L = 8.5 + 1.2 \times 10^{-5}(T - 20)$, where T is the temperature in degrees Celsius.

   a) Calculate the lengths of the girder when the temperature are 10°C, 20°C, and 30°C.

   b) Visualize a graph of L against T. Describe the graph.

   c) Explain what the numbers 8.5 and 1.2×10^{-5} in the equation represent.

2. These graphs were produced by *Graphmatica*. Write a set of equations that could be used to make patterns like these.

   a) 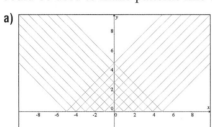   b)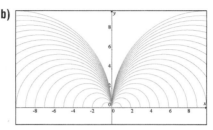

3. What happens to the graph of a function if you make each replacement in its equation? Support each answer with an example.

   a) Replace x with $-x$ and y with $-y$.

   b) Replace x with $-y$ and y with $-x$.

4. In a study of wind speeds near a large city, it was found that wind speed increases with height above the ground. The wind speed is represented by v metres per second at height h metres. The equations expressing v as a function of h are given.

   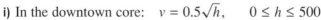

      i) In the downtown core: $v = 0.5\sqrt{h}$, $0 \le h \le 500$
     ii) In the suburbs: $v = 0.6\sqrt{h}$, $0 \le h \le 350$
   iii) In nearby rural areas: $v = 0.7\sqrt{h}$, $0 \le h \le 200$

   a) Sketch the graphs of these functions on the same grid.

   b) What are the domain and the range of each function?

5. For each list, sketch the graphs on the same grid.

   a) $y = x^2$

       $y = 3(2x)^2$

       $y = 3(2(x - 4))^2 + 1$

   b) $y = |x|$

       $y = 3|2x|$

       $y = 3|2(x - 4)| + 1$

   c) $y = \dfrac{1}{x}$

       $y = \dfrac{3}{2x}$

       $y = \dfrac{3}{2(x - 4)} + 1$

6. Later in this book, you will study the graph of the function $y = \cos x$. Describe how the graph of $y = 4\cos(2(x - 3)) + 1$ is related to the graph of $y = \cos x.$

2 EXPONENTIAL FUNCTIONS AND LOGARITHMS

Modelling the Growth of Computer Technology

CONSIDER THIS SITUATION

The astounding growth of computer technology in the last 30 years has been described in different ways. The quotations on pages 64 and 65 appear in a consulting firm's handbook to help employees adjust to change.

- Discuss the quotations.

- Major changes in society that affect the average person in significant ways are called "revolutions." In what ways has the Information Revolution affected the average person?

- Since 1800, what other "revolutions" similar to the Information Revolution have occurred? In what ways did they affect the average person?

- Could the second quotation be true?

On pages 102 and 103, you will develop mathematical models to describe the growth of computer technology during the latter part of the 20th century.

1. Development of the integrated circuit (invented in the late 1950s) has permitted an ever-increasing amount of information to be processed or stored on a single microchip. This is what has driven the Information Revolution.

FYI Visit www.awl.com/canada/school/connections

For information related to the above problem, click on <u>MATHLINKS</u>, followed by <u>AWMath</u>. Then select a topic under Modelling the Growth of Computer Technology.

2. If we had similar progress in automotive technology, today you could buy a Lexus for about $2. It would travel at the speed of sound, and go about 1000 km on a thimble of gas.

3. Today's average consumers wear more computing power on their wrists than existed in the entire world before 1961.

2.1 Introduction to Exponential Functions

Exponents were originally defined as a notation for repeated multiplication. Repeated multiplication occurs frequently in applications involving growth and decay.

Compound Interest

Suppose you make a long-term investment of $1000 at a fixed interest rate of 8% compounded annually. The amount, A dollars, of your investment after n years is represented by the equation $A = 1000(1.08)^n$. In this equation, n is a natural number because it indicates the number of years.

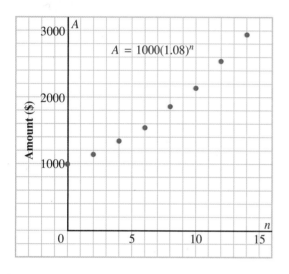

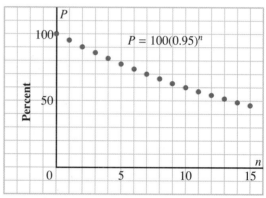

Light Transmission through Glass

Several layers of glass are stacked together. Each layer reduces the light passing through it by 5%. The percent, P, of light passing through n layers is represented by the equation $P = 100(0.95)^n$. Again, n is a natural number because it indicates the number of layers of glass.

Growth of British Columbia's Population

The population, P million, of British Columbia can be modelled by the equation $P = 2.76(1.022)^n$, where n is the number of years since 1981.

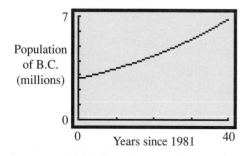

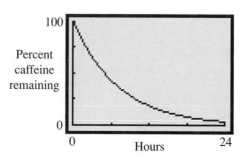

How Long Caffeine Stays in Your Bloodstream

Coffee, tea, cola, and chocolate contain caffeine. When you consume caffeine, the percent, P, left in your body can be modelled as a function of the elapsed time, n hours, by the equation $P = 100(0.87)^n$.

Comparing the Equations

The situations on page 66 are examples of exponential growth and exponential decay. Consider the similarities in the equations of the functions.

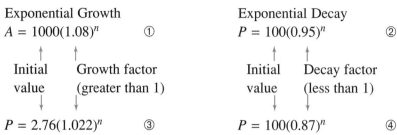

Exponential Growth
$A = 1000(1.08)^n$ ①

| Initial value | Growth factor (greater than 1) |

Exponential Decay
$P = 100(0.95)^n$ ②

| Initial value | Decay factor (less than 1) |

$P = 2.76(1.022)^n$ ③ $P = 100(0.87)^n$ ④

In each equation, the variable in the expression on the right side appears in an exponent. Functions, whose defining equations have this property, are exponential functions.

> An *exponential function* has an equation that can be written in the form $y = Ab^x$, where A and b are constants, and $b > 0$.

In some applications of exponential functions, there is an underlying principle from which the equation is derived. Equations ① and ② above were obtained in this way. In other applications, the equation is found using empirical data. Equations ③ and ④ were determined by calculating the equation of the exponential curve of best fit, see pages 72, 73. The screens below show the plotted points and the calculated results used to determine equation ③. The second screen shows that the equation of the exponential curve of best fit is $y \doteq 2.76(1.022)^x$.

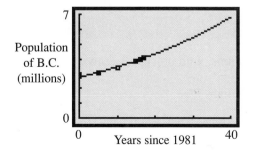

Population of B.C. (millions)

Years since 1981

```
ExpReg
y=a*b^x
a=2.757864618
b=1.022317678
```

Equations determined in this way will be provided in many examples and exercises in this chapter. One of these equations occurs in the *Example*.

Example

The first artificial satellites were put in orbit in the late 1950s. The cumulative mass, c tonnes, of all the satellites in orbit can be modelled as a function of n, the number of years since 1960, by the equation $c = 120 \times 1.12^n$.

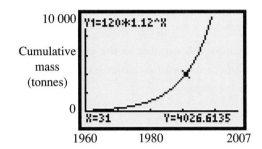

a) Graph the equation.

b) Estimate the cumulative mass in 1969, the year when a person first walked on the moon.

c) According to the model, when did the cumulative mass become 4000 t?

Solution

a) Use appropriate window settings:
$0 \leq x \leq 47; -1000 \leq y \leq 10\,000$

b) Substitute 9 for n in the equation.
$c = 120 \times 1.12^9$
$\doteq 333$
In 1969, the cumulative mass was about 333 t.

c) Substitute 4000 for c in the equation.
$4000 = 120 \times 1.12^n$
Trace to the point where the y-coordinate is closest to 4000. This occurs when the x-coordinate is 31. According to the model, the mass became 4000 t in 1960 + 31, or 1991.

In the *Example* part c, we solved the equation $1.20 \times 1.12^n = 4000$ by graphing and tracing. This is an example of an *exponential equation*. In the next two sections, we will develop a more efficient method to solve exponential equations.

DISCUSSING THE IDEAS

1. Explain why the first two graphs on page 66 show points, with no curves drawn.

2. Determine the domain and range of each function on page 66.

3. a) Explain the underlying principles from which equations ① and ② on page 67 were obtained.

 b) Do you think there would be an underlying principle for obtaining equation ③ or equation ④? Explain.

2.1 EXERCISES

If you have a graphing calculator, use it to draw the graphs in these exercises.

A 1. Use the graph on page 66 that shows the growth of $1000 at 8%.

   a) Estimate the value of the investment after 5 years.

   b) Estimate how many years it takes for the investment to grow to $2500.

   c) Describe how both the graph and the equation change in each case.
   i) The original investment is greater than, or less than, $1000.

   ii) The interest rate is greater than, or less than, 8%.

2. Use the equation and graph on page 66 that shows the growth of British Columbia's population.

   a) Use the equation to estimate the annual rate of growth, as a percent.

   b) Describe how both the graph and the equation change in each case.
   i) The growth rate increases.   ii) The growth rate decreases.

3. **a)** Display the graph on page 66 that shows the growth of British Columbia's population.

   b) Trace to estimate the population in 2011.

   c) Trace to estimate when the population might reach 6 million.

MODELLING British Columbia's Population Growth

On page 66 and in exercises 2 and 3, the exponential function and the graph are models of the growth of the population of British Columbia.

* What assumption is made about the population growth?
* Give some reasons why the actual population of British Columbia might be different from the numbers predicted by the model.
* How do you think the annual growth rate of 2.2% was determined?

4. Use the graph on page 66 that shows the percent of light transmitted through panes of glass.

   a) Estimate how many panes are needed before only 50% of the light passes through.

   b) The graph was drawn for clear glass. For frosted glass, each layer reduces the light passing through it by 10%. Describe how both the graph and the equation change for this glass.

5. Use the equation and graph on page 66 that show how long caffeine stays in your bloodstream. Describe how both the graph and the equation may change for pregnant women, who require a much longer time to metabolize caffeine than other adults.

6. a) Display the graph on page 66 that shows how long caffeine stays in your bloodstream.

   b) Trace to determine how long it takes until only 20% of the caffeine remains.

B **7.** Suppose you invest $1000 at 8% compounded semi-annually. The value, A dollars, of your investment after n years is represented by the equation $A = 1000(1.04)^{2n}$.

   a) Graph the equation for $0 \leq n \leq 10$.

   b) Estimate the value of the investment after 5 years.

   c) Estimate how many years it takes the investment to grow to $2500.

   d) How does the graph in part a compare with the graph on page 66? Explain.

8. Compare exercises 7b, c and 1a, b. Write to explain the similarities and differences in each case.

   a) between your answers to parts a and b in the two exercises

   b) between the equations of the two functions

9. The number of Canadian cellular phone subscribers, s, has grown according to the formula $s = 130\ 000 \times 1.45^n$, where n is the number of years since 1987.

   a) Graph the equation for $0 \leq n \leq 15$.

   b) Use the equation to predict the number of phone subscribers in 2002. Identify any assumptions used.

   c) Estimate when the number of phone subscribers reached 10 million.

   d) According to the equation, what is the average annual rate of increase in the number of phone subscribers since 1987?

10. The growth of the Internet is measured by the number of computers offering information, called *hosts*. In 1994, there were about 3.5 million hosts. Since then, the number has been growing at about 84% annually.

   a) Write an equation that expresses the number of hosts, h million, as a function of n, the number of years since 1994.

   b) Graph h as a function of n.

   c) At this rate of growth, estimate the year when the number of hosts reached 100 million.

11. For every metre below the surface of water, the light intensity is reduced by 2.5%.

    a) Write an equation to express the percent, P, of light remaining as a function of the depth, d metres, below the surface.

    b) Graph P as a function of d.

    c) Describe how the graph is similar to, and different from, the graph on page 66 that shows the percent of light transmitted through panes of glass.

    d) Determine the light intensity at a depth of 10 m.

    e) At what depth is the light intensity reduced to 50% of the intensity at the surface?

C 12. In 1998, a newspaper article about how people are using the Internet to buy and sell goods and services contained these graphs.

In 1997 the market totalled $13.1-billion (U.S.) ...

Rest of world, 1.4%
Asia, 1.1%
Japan, 3.1%
Canada, 5.4%

Western Europe, 7.7%

United States, 81.3%

by 2002 the market will grow to $435.1-billion.

Rest of world, 9.4%
Asia, 7.9%

Japan, 6%

Canada, 2.2%

Western Europe, 12.8%
United States, 61.8%

Annual growth is forecast at 102 per cent from 1997 to 2002

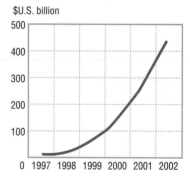

$U.S. billion

500
400
300
200
100

0 1997 1998 1999 2000 2001 2002

    a) Use the information above the circle graphs to check the statement above the broken-line graph.

    b) Calculate the predicted growth rate for Canada from 1997 to 2002.

    c) Which region has the greatest predicted growth rate from 1997 to 2002? Calculate the predicted growth rate for this region.

COMMUNICATING THE IDEAS

Your friend missed today's mathematics lesson and telephones you to ask about it. How would you explain to him, over the telephone, the meaning of an exponential function?

Exponential Curves of Best Fit

This table shows how Canada's national debt has been growing since 1970. Since 1995, the Canadian government has taken steps to slow the growth of the national debt.

Year	1970	1975	1980	1985	1990	1995
National debt ($ billions)	36	55	111	251	407	596

We can use a graphing calculator to plot the data, and to calculate and graph the equation of an exponential curve of best fit to model the data. The screens and keystrokes are from the TI-83 calculator.

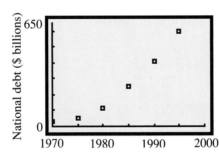

Follow these steps.
- Clear the Y= list.
- Define an appropriate viewing window.
- Press [STAT] **1**, and enter the data to be plotted in the first two columns.
- Press [STAT PLOT], and set up the first plot.
- Press [GRAPH] to graph the data.
- Press [STAT] [▶]. Scroll down and select "ExpReg."
- Press [Y=] [VARS] **5** [▶] [▶] **1**. The calculator determines the equation of the exponential function of best fit and copies it into the Y= list beside Y1.
- Press [GRAPH] to see the plotted points and the graph of the exponential function of best fit.

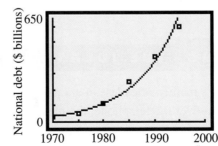

1. a) Use the data above to produce the screens above.

   b) Write an equation that models the national debt as an exponential function of the number of years since 1970.

c) According to the model, at what annual rate has the national debt been increasing since 1970?

d) What debt does the model predict for 1995? Suggest some reasons why the national debt in 1995 was considerably less than the prediction.

2. Cars kept in heated garages tend to rust more quickly than those kept outside, because most chemical reactions take place more rapidly at high temperatures than at low temperatures. This table gives the times for a certain chemical reaction to take place at different temperatures.

Temperature, $T°C$	−30	−10	0	12	31
Time, t seconds	71	18	8.4	3.8	1.0

a) Graph the data. Determine the equation of the exponential curve of best fit, and graph the curve with the data.

b) According to the model, how long does the reaction require at 20°C?

c) Suppose the temperature increases by 10°C. What happens to the time for the reaction to take place?

3. The 1999 *Canadian Global Almanac* contains these data.

a) Graph the data. Determine the equation of each exponential curve of best fit, and graph each curve with the data.

b) According to the equations, predict the population of each province in 2011.

c) Predict when the population of each province might become 6 million.

Population (millions)

Year	Alta.	B.C.
1981	2.294	2.824
1986	2.431	3.004
1991	2.593	3.373
1996	2.781	3.882
1997	2.836	3.964
1998	2.913	4.014

 **MODELLING** **Population Growth**

The equations you found in exercise 3 are models for the populations of Alberta and British Columbia, based on data from 1981 to 1998.

- Do the models give accurate estimates of the populations during 1981 to 1998? Explain.
- What factors might contribute to the breakdown of the models in the future?
- Explain how you can tell which province was growing more rapidly.
- According to the models, what was the average annual rate of growth for the population of each province?

2.2 Defining a Logarithm

INVESTIGATE The [LOG] Key on a Calculator

1. Find out what the [LOG] key on your calculator does. Try a wide variety of numbers such as these. Record the results for later reference.

   a) Numbers selected at random, for example:
   3, 65, 239, 4772

   b) Powers of 10, for example:
   10, 100, 1000, 10 000, …   1, 0.1, 0.01, 0.001, …

   c) Multiples of 10, 100, 1000, for example:
   20, 200, 2000, …   30, 300, 3000, …

   d) Numbers with the same significant digits, for example:
   2.43, 24.3, 243, 2430, …   0.243, 0.0243, 0.002 43, …

   e) Square roots of powers of 10, for example:
   $\sqrt{10}, \sqrt{100}, \sqrt{1000}, …$ $\sqrt{1}, \sqrt{0.1}, \sqrt{0.01}, …$

   f) Zero and negative numbers, for example:
   0, –2, –3, –10

2. For any positive input number x, the [LOG] key calculates an output number represented by $\log x$.

   a) Explain how $\log x$ is related to x.

   b) What single word best describes what a logarithm is?

In Section 2.1, you encountered problems such as these.

Example, page 68
What year did the cumulative mass of satellites in orbit reach 4000 t?

In the solution, the answer was obtained by solving this equation.
$$4000 = 120 \times 1.12^n \quad ①$$

Exercise 4a, page 69
How many panes are needed before only 50% of the light passes through?

In your solution, you would have solved this equation.
$$50 = 100 \times 0.95^n \quad ②$$

Equations ① and ② can be solved to any accuracy using graphing technology. However, exponential equations occur so frequently in applications that a way has been devised to solve them directly. This method involves logarithms. In this section and the next, we define a logarithm and develop some of its properties. You will then be able to use logarithms to solve equations ① and ②, and other similar exponential equations.

In *Investigate*, you determined logarithms of 10, 100, and 1000, and obtained these results, which are also on the calculator screen.

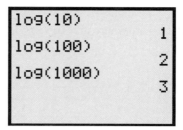

$$\log 10 = 1$$
$$\log 100 = 2$$
$$\log 1000 = 3$$

These examples suggest that log x is defined as the exponent that 10 has when x is written as a power of 10. The expression "log x" is read "the logarithm of x."

Example 1

Use a calculator to determine each logarithm, if possible. Explain the meaning of each result, and check it without using the [LOG] key.

a) log 250   **b)** log 0.64   **c)** log (–5)

Solution

a) Press: [LOG] 250 [ )] [ENTER] to display 2.397940009.
log 250 $\doteq$ 2.397 94
This means that $10^{2.397\ 94} \doteq 250$.
To check, press 10 [^] 2.39794 [ENTER] to display 249.999995.

b) log 0.64 $\doteq$ –0.193 82
This means that $10^{-0.193\ 82} \doteq 0.64$.
To check, press 10 [^] –0.19382 [ENTER] to display .6400000383.

c) Pressing [LOG] –5 [ )] [ENTER] results in an error message.
Hence, log (–5) is not defined as a real number. The reason is that there is no power of 10 that equals –5.

In *Example 1* and throughout the text, the keystrokes are for the TI-83 graphing calculator. If you have a different calculator, consult the manual to determine similar keystrokes.

The restriction to base 10 is not necessary, and logarithms can be defined with any positive base. Some examples of base 2 logarithms are shown at the right.

$$\log_2 2 = 1$$
$$\log_2 4 = 2$$
$$\log_2 8 = 3$$

These examples show that $\log_2 x$ is defined as the exponent that 2 has when x is written as a power of 2. Similarly, $\log_a x$ is defined as the exponent that a has when x is written as a power of a ($a > 0, a \neq 1$).

$\log_a x = y$ means $a^y = x$, where $x > 0,\ a > 0,\ a \neq 1$

Example 2

Determine each logarithm.

a) $\log_5 25$ **b)** $\log_7 \sqrt{7}$ **c)** $\log_a a$

Solution

a) $\log_5 25$

> ***Think ...***
> What exponent should 5 have to make 25?

Since $25 = 5^2$, then $\log_5 25 = 2$

b) $\log_7 \sqrt{7}$

> ***Think ...***
> What exponent should 7 have to make $\sqrt{7}$?

Since $\sqrt{7} = 7^{\frac{1}{2}}$, then $\log_7 \sqrt{7} = \frac{1}{2}$

c) $\log_a a$

> ***Think ...***
> What exponent should a have to make a?

Since $a = a^1$, then $\log_a a = 1$

VISUALIZING   **A Logarithm Is an Exponent**

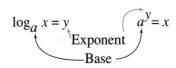

$\log_a x = y$ $a^y = x$

Exponent

Base

Example 3

a) Write each expression in exponential form.

   i) $\log_2 16 = 4$ ii) $\log_2 0.5 = -1$ iii) $\log_x y = z$

b) Write each expression in logarithmic form.

   i) $3^4 = 81$ ii) $3^{-2} = \frac{1}{9}$ iii) $a^b = c$

Solution

a) i) $\log_2 16 = 4$ ii) $\log_2 0.5 = -1$ iii) $\log_x y = z$
   $2^4 = 16$ $2^{-1} = 0.5$ $x^z = y$

b) i) $3^4 = 81$ ii) $3^{-2} = \frac{1}{9}$ iii) $a^b = c$

   $\log_3 81 = 4$ $\log_3\left(\frac{1}{9}\right) = -2$ $\log_a c = b$

DISCUSSING THE IDEAS

1. Why can negative numbers not have logarithms?

2. How can you remember how to convert from exponential form to logarithmic form, and vice versa?

2.2 EXERCISES

A 1. Use your calculator to determine each logarithm to 5 decimal places. Explain the meaning of each result, and check it without using the ⬛ LOG ⬛ key.

   a) $\log 23.6$ b) $\log 180$ c) $\log 5100$

   d) $\log 0.45$ e) $\log 0.072$ f) $\log 0.0029$

2. For exercise 1, write to explain why the first three logarithms are positive and the last three are negative.

3. Determine the logarithms in each list.

   a) $\log 1$ b) $\log_3 1$ c) $\log_a 1$
      $\log 10$ $\log_3 3$ $\log_a a$
      $\log 100$ $\log_3 9$ $\log_a a^2$
      $\log 1000$ $\log_3 27$ $\log_a a^3$
      $\vdots$ $\vdots$ $\vdots$
      $\log 10^n$ $\log_3 3^n$ $\log_a a^n$

B 4. Evaluate each logarithm.

a) $\log_2 16$

b) $\log_2 4$

c) $\log_3 27$

d) $\log_5 25$

e) $\log_5\left(\frac{1}{5}\right)$

f) $\log_7 7$

g) $\log_3 1$

h) $\log_4 64$

5. Choose two expressions in exercise 4. Write to explain how you determined the logarithms.

6. Write in exponential form.

a) $\log_2 8 = 3$

b) $\log_2 32 = 5$

c) $\log_6 36 = 2$

d) $\log_5 625 = 4$

e) $\log_{16} 4 = \frac{1}{2}$

f) $\log_2\left(\frac{1}{4}\right) = -2$

7. Write in logarithmic form.

a) $7^2 = 49$

b) $3^5 = 243$

c) $2^0 = 1$

d) $5^{-1} = 0.2$

e) $2^{-3} = \frac{1}{8}$

f) $25^{0.5} = 5$

8. Refer to exercise 3. The numbers following the logarithm symbols are powers of 10, 3, and a with positive exponents.

a) Write similar lists of logarithms involving powers of 10, 3, and a with negative exponents.

b) Determine the logarithms in the lists in part a.

9. Suppose you record the first digits of many numbers randomly selected from the front pages of newspapers, stock-market prices, census data, or street addresses. Most people think that the first digits 1, 2, 3, ..., 9 would occur about the same number of times. However, this is not true. In the 1990s, it was proved that in such a set of data, the probability that the first non-zero digit of a number is n is:

P(first digit is n) $= \log\left(1 + \frac{1}{n}\right)$, $n = 1, 2, 3, ..., 9$

For example, the probability that the first digit is 1 is $\log\left(1 + \frac{1}{1}\right)$, or $\log 2$.

This relationship is known as Benford's Law. It has important applications to the design of computers and the detection of fraudulent accounting.

a) Calculate the probability that the first digit is each number.

i) 2

ii) 3

iii) 4

iv) 5

v) 6

vi) 7

vii) 8

viii) 9

b) Graph P(first significant digit is n) against n.

COMMUNICATING THE IDEAS

Write to explain what is meant by a logarithm. Use some examples to illustrate your explanation.

2.3 The Laws of Logarithms

Since a logarithm is an exponent, we can write the laws of exponents in logarithmic form. Recall that logarithms were introduced to develop a method to solve exponential equations. Since exponential equations contain powers, we will begin with powers.

Suppose we evaluate the base-10 logarithms of some powers of 2, to 5 decimal places. Notice the patterns in the results at the right. All the logarithms of the powers of 2 are multiples of the logarithm of 2. For example, $\log 8 = 3 \times 0.301\ 03$, or $\log 2^3 = 3 \log 2$

$\log 2 \doteq 0.301\ 03$
$\log 4 \doteq 0.602\ 06$
$\log 8 \doteq 0.903\ 09$
$\log 16 \doteq 1.204\ 12$
$\log 32 \doteq 1.505\ 15$
$\log 64 \doteq 1.806\ 18$

We can use the definition of a logarithm to explain this result.

Start with: $\log 2 = k$

This means: $2 = 10^k$

Cube each side: $2^3 = (10^k)^3$

$2^3 = 10^{3k}$

This means: $\log 2^3 = 3k$

Hence, $\log 2^3 = 3 \log 2$

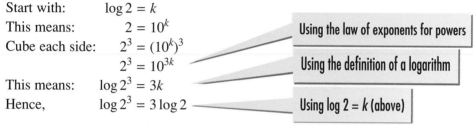

Using the law of exponents for powers

Using the definition of a logarithm

Using $\log 2 = k$ (above)

This is an example of the law of logarithms for powers. This law applies to any power and any base for which logarithms are defined. The law of logarithms for powers has the form $\log_a x^n = n \log_a x$. To prove the law, we repeat the above steps, replacing 2 with x, 3 with n, and using base-a logarithms.

Start with: $\log_a x = s$

This means: $x = a^s$

Raise each side to the nth power:

$x^n = (a^s)^n$

$x^n = a^{ns}$

This means: $\log_a x^n = ns$

Hence, $\log_a x^n = n \log_a x$

Law of Logarithms for Powers

If x and n are real numbers, and $x > 0$, then

$\log_a x^n = n \log_a x \qquad a > 0,\ a \neq 1$

The law of logarithms for powers provides a way to remove a variable from the position of an exponent. Hence, we can use it to solve exponential equations.

Example 1

Solve the equation $3^x = 20$ to 5 decimal places, and check.

Solution

$3^x = 20$
Take the base-10 logarithm of each side.

$\log 3^x = \log 20$

$x \log 3 = \log 20$ ——— Using the law of logarithms for powers

$x = \dfrac{\log 20}{\log 3}$ Press: [LOG] 20 [ )] [÷] [LOG] 3 [ )] [ENTER]

$\doteq 2.726\ 833\ 028$

To 5 decimal places, $x = 2.726\ 83$

To check, determine $3^{2.726\ 83}$. The result is 19.999 933 47.

To solve an exponential equation, take the logarithm of each side.

We can also use the law of logarithms for powers to determine the logarithm of any positive number to any positive base.

Example 2

Determine $\log_5 9$ to 5 decimal places, and check.

Solution

> **Think...**
> To find $\log_5 9$ means to find the exponent that 5 has when 9 is expressed as a power of 5.

Let $9 = 5^x$
Take the logarithm of each side.

$\log 9 = \log 5^x$

$\log 9 = x \log 5$ ——— Using the law of logarithms for powers

$x = \dfrac{\log 9}{\log 5}$ Press: [LOG] 9 [ )] [÷] [LOG] 5 [ )] [ENTER]

$\doteq 1.365\ 212\ 389$

To 5 decimal places, $\log_5 9 = 1.365\ 21$

To check, determine $5^{1.365\ 21}$. The result is 8.999 965 396.

At the beginning of Section 2.2, we stated that we would develop a method to solve exponential equations such as $120 \times 1.12^n = 4000$ and $100 \times 0.95^n = 50$. The expressions on the left sides of these equations are products. Hence, when we take the logarithm of each side of these equations we have to determine the logarithm of a product.

Suppose we evaluate the base-10 logarithms of 2, 20, 200, 2000, and 20 000. Notice the patterns in the results. Each logarithm is the sum of a natural number and the logarithm of 2. For example, $\log 2000 = 3 + 0.301\ 03$, or $\log(1000 \times 2) = \log 1000 + \log 2$

$$\log 2 \doteq 0.301\ 03$$
$$\log 20 \doteq 1.301\ 03$$
$$\log 200 \doteq 2.301\ 03$$
$$\log 2000 \doteq 3.301\ 03$$
$$\log 20\ 000 \doteq 4.301\ 03$$

We can use the definition of a logarithm to explain this result.

Start with: $\qquad \log 2 = k$

This means: $\qquad 2 = 10^k$

Multiply each side by 1000, or 10^3.

$$1000 \times 2 = 10^3 \times 10^k$$
$$1000 \times 2 = 10^{3+k}$$

> Using the law of exponents for multiplication

This means:
$$\log(1000 \times 2) = 3 + k$$

> Using the definition of a logarithm

Hence, $\log(1000 \times 2) = 3 + \log 2$

$$\log(1000 \times 2) = \log 1000 + \log 2$$

> Using $\log 2 = k$ (above)

This is an example of the law of logarithms for multiplication. This law applies to any product and any base for which logarithms are defined. The law of logarithms for multiplication has the form $\log_a xy = \log_a x + \log_a y$.

To prove the law, we repeat the above steps.

Start with: $\qquad \log_a x = s \qquad \log_a y = t$

This means: $\qquad x = a^s \qquad y = a^t$

Multiply the left and right sides of these two equations.

$$xy = a^s \times a^t$$
$$xy = a^{s+t}$$

This means: $\quad \log_a xy = s + t$

Hence, $\qquad \log_a xy = \log_a x + \log_a y$

Law of Logarithms for Multiplication

If x and y are positive real numbers, then

$$\log_a xy = \log_a x + \log_a y \qquad a > 0,\ a \neq 1$$

To solve an exponential equation such as equation ① on page 74, we take the logarithm of each side and use the law of logarithms for multiplication.

Example 3

Solve the equation $120 \times 1.12^n = 4000$ to 1 decimal place.

Solution

$$120 \times 1.12^n = 4000$$

Take the logarithm of each side.

$$\log(120 \times 1.12^n) = \log 4000$$
$$\log 120 + \log 1.12^n = \log 4000 \quad \text{— Using the law of logarithms for multiplication}$$
$$\log 120 + n\log 1.12 = \log 4000$$
$$n\log 1.12 = \log 4000 - \log 120 \quad \text{— Using the law of logarithms for powers}$$
$$n = \frac{\log 4000 - \log 120}{\log 1.12}$$

Press: $\boxed{\text{LOG}}$ 4000 $\boxed{)}$ $\boxed{-}$ $\boxed{\text{LOG}}$ 120 $\boxed{)}$ $\boxed{\text{ENTER}}$ $\boxed{\div}$ $\boxed{\text{LOG}}$ 1.12 $\boxed{)}$ $\boxed{\text{ENTER}}$

$$n \doteq 30.941\ 485\ 71$$

To 1 decimal place, $n = 30.9$

In *Examples 1* and *3*, logarithms were used as a tool to solve exponential equations. You will often use this tool for the rest of this chapter.

There is a law of logarithms for division that can be proved in the same way as the law of logarithms for multiplication (see exercise 15).

Law of Logarithms for Division

If x and y are positive real numbers, then

$$\log_a\left(\frac{x}{y}\right) = \log_a x - \log_a y \qquad a > 0,\ a \neq 1$$

Example 4

Simplify.

a) $\log 25 + \log 4$

b) $\log 64 - \log 640$

c) $\log_3 6 + \log_3 1.5$

d) $\log_5 50 - \log_5 0.4$

Solution

a) $\log 25 + \log 4 = \log(25 \times 4)$
$= \log 100$
$= 2$

b) $\log 64 - \log 640 = \log\left(\frac{64}{640}\right)$
$= \log 0.1$
$= -1$

c) $\log_3 6 + \log_3 1.5 = \log_3(6 \times 1.5)$
$= \log_3 9$
$= 2$

d) $\log_5 50 - \log_5 0.4 = \log_5\left(\frac{50}{0.4}\right)$
$= \log_5 125$
$= 3$

DISCUSSING THE IDEAS

1. Why is taking the logarithm of each side the first step to solve an exponential equation?

2. Compare the solution of *Example 3* with the solution of the same equation in the *Example*, page 68. Are the results the same? Explain.

3. Solve *Example 3* by first writing $1.12^n = \frac{4000}{120}$, then taking the logarithm of each side. Explain why the result is the same as in the solution of *Example 3*.

2.3 EXERCISES

A 1. Determine the logarithms in each list, to 3 decimal places. Explain the pattern in the results.

   a) log 5, log 50, log 500, log 5000, log 50 000

   b) log 5, log 25, log 125, log 625, log 3125

2. Evaluate the expressions in each list. Explain the results.

   a) log 1 + log 36
   log 2 + log 18
   log 3 + log 12
   log 4 + log 9
   log 6 + log 6

   b) log 2 – log 1
   log 4 – log 2
   log 8 – log 4
   log 16 – log 8
   log 32 – log 16

3. Create other patterns similar to those in exercise 2.

4. Evaluate the expressions in each list and compare the results. Explain.

   a) log 2, log 4, log 8, log 16, log 32

   b) log 2, 2 log 2, 3 log 2, 4 log 2, 5 log 2

5. Create other patterns similar to those in exercise 4.

6. Simplify.
   a) $\log_6 9 + \log_6 4$ **b)** $\log_5 15 - \log_5 3$ **c)** $\log_4 2 + \log_4 32$
   d) $\log_2 48 - \log_2 6$ **e)** $\log_3 54 - \log_3 2$ **f)** $\log_3 9 + \log_3 9$

B   **7.** Solve each equation to 5 decimal places, and check.
   a) $2^x = 11$ **b)** $3^x = 17$ **c)** $6^x = 5$

8. a) Solve the equations in each list.
   i) $3^x = 2$, $3^x = 4$, $3^x = 8$, $3^x = 16$
   ii) $3^x = 2$, $9^x = 2$, $27^x = 2$, $81^x = 2$

   b) How are the roots of the equations in each list in part a related? Explain.

9. Solve.
   a) $5^{x-1} = 9$ **b)** $2^{x+3} = 6$ **c)** $7^{-x} = 3$
   d) $3^{1-x} = 5$ **e)** $\left(\frac{1}{8}\right)^x = 25$ **f)** $5^{3x} = 41$

10. Express:
   a) 7 as a power of 3   **b)** 5 as a power of 2   **c)** 29 as a power of 2
   d) 77 as a power of 8   **e)** 3 as a power of 0.5   **f)** 0.45 as a power of 6

11. Determine each logarithm to 5 decimal places, and check.
   a) $\log_3 5$ **b)** $\log_2 12$ **c)** $\log_6 55$
   d) $\log_2 3$ **e)** $\log_2 20$ **f)** $\log_2 5$

12. Write the expressions in each list in terms of $\log x$.
   a) $\log x$, $\log x^2$, $\log x^3$, $\log x^4$, $\log x^5$
   b) $\log x$, $\log 10x$, $\log 100x$, $\log 1000x$, $\log 10\,000x$

13. Alex invests \$50 000 at an interest rate of 7% compounded monthly. Laura invests \$40 000 at 9.5% compounded annually. After how many years will the two investments be equal in value?

14. a) Determine each quotient to 5 decimal places.
   i) $\dfrac{\log_2 20}{\log_2 3}$ **ii)** $\dfrac{\log_6 20}{\log_6 3}$ **iii)** $\dfrac{\log_8 20}{\log_8 3}$

   b) Refer to *Example 1*, where the solution contains the line $x = \dfrac{\log 20}{\log 3}$. Compare the quotients in part a with this value of x. What do you notice?

   c) In part b, you should have found that the quotients are the same, regardless of the base used for the logarithms. Explain why this is true.

   d) The $\boxed{\text{LN}}$ key on your calculator determines logarithms of numbers to a base different from 10. Use this key to solve the equation in *Example 1*.

15. Prove the law of logarithms for division:

$$\log_a\left(\frac{x}{y}\right) = \log_a x - \log_a y \quad (a > 0, a \neq 1)$$

16. Prove the law of logarithms for roots: $\log_a \sqrt[n]{x} = \frac{1}{n} \log_a x \quad (a > 0, a \neq 1)$

17. Simplify.

a) $\log_4 48 + \log_4\left(\frac{2}{3}\right) + \log_4 8$

b) $\log_8 24 + \log_8 4 - \log_8 3$

c) $\log_9 36 + \log_9 18 - \log_9 24$

d) $\log_4 20 - \log_4 5 + \log_4 8$

e) $\log_3 \sqrt{45} - \log_3 \sqrt{5}$

f) $\log_2 \sqrt{5} - \log_2 \sqrt{40}$

18. Write in terms of $\log x$ and $\log y$.

a) $\log 1000xy$

b) $\log x^2 y$

c) $\log \frac{x}{y^2}$

19. x and y are two positive numbers. How are $\log x$ and $\log y$ related in each case?

a) $y = 10x$

b) $y = 100x$

c) $y = 1000x$

d) $y = \frac{1}{x}$

e) $y = x^2$

f) $y = \frac{1}{x^2}$

C **20.** In 1938, the physicist Sir Arthur Eddington estimated the number of particles in the universe to be 33×2^{259}. He called this number the *cosmical number*. This number is so large that you may get an error message when you try to evaluate it with a calculator. You can use logarithms to find out something about the number.

a) Determine the logarithm of the cosmical number.

b) How many digits are there in the number? Explain.

c) What are the first few digits of the number? Explain.

21. This statement was in a television news report.

In one day, one bacterium can produce 76 billion trillion descendants.

a) Calculate to confirm or refute this statement.

b) What assumptions are you making in part a? Are these assumptions reasonable? Explain.

COMMUNICATING THE IDEAS

Write to explain a method to solve an exponential equation and why it works. Include a couple of examples in your explanation.

2.4 Modelling Real Situations Using Exponential Functions: Part I

Previously in this chapter, you determined the time for a quantity to double when it is growing exponentially. In this investigation, you will look more closely at doubling times.

Recall the first example on page 66, where $1000 was invested at 8% compounded annually.

1. a) Graph the function $A = 1000(1.08)^n$ using appropriate window settings.

   b) Use the trace or the tables feature. Determine how many years it takes the original investment of $1000 to double.

   c) Predict how many years it will take for the amount of the investment to double again. Check your prediction with a calculator.

2. a) Determine the amount of the investment after 3 years.

   b) Predict how many years it will take for the amount in part a to double. Check your prediction with a calculator.

   c) Repeat parts a and b for the amount of the investment after a different number of years.

3. Your results in exercises 1 and 2 illustrate a certain property of exponential growth. What is this property?

Recall the fourth example on page 67, about how long caffeine stays in the bloodstream.

4. a) Graph the function $P = 100(0.87)^n$ using appropriate window settings.

   b) Use the trace or the tables feature. Determine how many hours it takes for the original amount of caffeine to drop by 50%.

   c) Predict how many hours it will take for the amount of caffeine to drop by 50% again. Check your prediction with a calculator.

5. a) Determine the amount of caffeine present after 2 h.

   b) Predict how many hours it will take for the amount of caffeine in part a to drop by 50% again. Check your prediction with a calculator.

c) Repeat parts a and b for the amount of caffeine present after a different number of hours.

6. Your results in exercises 4 and 5 illustrate a certain property of exponential decay. What is this property?

In Sections 2.1 and 2.3, you learned to solve exponential equations. You can use this skill to solve problems in exponential growth and decay situations.

Example 1

The population, P million, of Alberta can be modelled by the equation $P = 2.28(1.014)^n$, where n is the number of years since 1981. Assume that this pattern continues. Determine when the population of Alberta might become 4 million.

Solution

Substitute 4 for P in $P = 2.28(1.014)^n$.
$2.28(1.014)^n = 4$

Solution 1
Take the logarithm of each side.

$\log 2.28 + n \log 1.014 = \log 4$
$n \log 1.014 = \log 4 - \log 2.28$
$n = \dfrac{\log 4 - \log 2.28}{\log 1.014}$
$\doteq 40.432$

Solution 2
$1.014^n = \dfrac{4}{2.28}$
Take the logarithm of each side.
$n \log 1.014 = \log\left(\dfrac{4}{2.28}\right)$
$n = \dfrac{\log\left(\dfrac{4}{2.28}\right)}{\log 1.014}$
$\doteq 40.432$

Alberta's population might become 4 million in 40 years after 1981; that is, in the year 2021.

Calculations similar to those in *Example 1* occur in situations involving doubling times.

Exponential Growth and Doubling Time

When $1000 is invested at 8% per annum, the amount A dollars after n years is given by $A = 1000(1.08)^n$. In *Investigate*, you should have found it takes about 9 years for the investment to double. To see why, substitute $A = 2000$.

$2000 = 1000(1.08)^n$

$1.08^n = 2$

Take the logarithm of each side.

$n \log 1.08 = \log 2$

$$n = \frac{\log 2}{\log 1.08}$$

$$n \doteq 9.006\ 468$$

Hence, it takes approximately 9 years for the investment to double. This doubling time applies at all times during the growth of the investment, not just for the original investment.

VISUALIZING   **Doubling Time for Exponential Growth**

Visualize any point P on the graph of $A = 1000(1.08)^n$. Q is the point on the graph 9 units to the right of P. The second coordinate of Q is double the second coordinate of P.

Visualize P and Q moving along the graph, always 9 units apart horizontally. The second coordinate of Q will always be double the second coordinate of P.

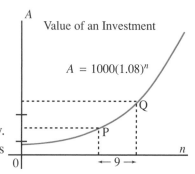

Value of an Investment

$A = 1000(1.08)^n$

A similar property applies to any function $y = Ab^x$ that models exponential growth ($b > 1$). When something is growing exponentially, it will become twice as large within a fixed length of time. The doubling time depends only on the base b, and not on the value of A.

Example 2

In 1995, Canada's population was 29.6 million, and was growing at about 1.24% per year. Estimate the doubling time for Canada's population growth.

Solution

Let P represent the population in millions.
Let n represent the number of years since 1995. Then,

$$P = 29.6 \times 1.0124^n$$

To determine the doubling time, substitute 2×29.6 for P in the equation.

$$2 \times 29.6 = 29.6 \times 1.0124^n$$
$$1.0124^n = 2$$

Take the logarithm of each side.

$$n \log 1.0124 = \log 2$$
$$n = \frac{\log 2}{\log 1.0124}$$
$$n \doteq 56.245$$

The doubling time is about 56 years.

Stating the doubling time is just another way to describe the growth rate of something that is growing exponentially. For example, in *Example 2*, instead of saying that the population is increasing at 1.24% per year, we could say that the doubling time is 56 years. We can also use the doubling time to write the equation of the model in a different form.

The population after:

56 years is	29.6×2^1	Each exponent is the number
112 years is	29.6×2^2	of years divided by 56.
168 years is	29.6×2^3	
$\vdots$	$\vdots$	
n years is	$29.6 \times 2^{\frac{n}{56}}$	

Hence, this is another equation that models the population after n years.

$$P = 29.6 \times 2^{\frac{n}{56}}$$

Time elapsed

Doubling time

Initial population

The equation above is equivalent to the equation $P = 29.6 \times 1.0124^n$ in *Example 2*. The only difference is that the equation above is expressed with base 2 instead of base 1.0124.

 MODELLING Canada's Population Growth

The information in *Example 2* is based on Statistics Canada data for 1971–1995.

- What assumption about the population growth is being made?
- From 1991 to 1995, the average annual growth rate was 1.30%. How would this affect the doubling time? Suggest why the growth rate was higher in the 1990s.

Exponential Decay and Half-Life

When you consume caffeine, the percent, P, left in your body after n hours is represented by $P = 100(0.87)^n$. In *Investigate*, you should have found that it takes about 5 h for the amount of caffeine to drop by 50%. This applies at all times, not just for the original amount of caffeine consumed. We say that the *half-life* for caffeine in the bloodstream is about 5 h.

VISUALIZING   **Half-Life for Exponential Decay**

Visualize any point P on the graph of $P = 100(0.87)^n$. Q is the point on the graph 5 units to the right of P. The second coordinate of Q is one-half the second coordinate of P.

Visualize P and Q moving along the graph, always 5 units apart horizontally. The second coordinate of Q will always be one-half the second coordinate of P.

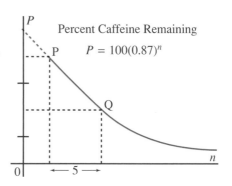

A similar property applies to any function $y = Ab^x$ that models exponential decay ($0 < b < 1$). When something decays exponentially, it will become one-half as large within a fixed length of time. The half-life depends only on the base b, and not on the value of A.

Example 3

Consider the equation $P = 100(0.87)^n$ that models the percent of caffeine in your body n hours after consumption. Write this equation as an exponential function with $\frac{1}{2}$ as the base instead of 0.87.

Solution

> *Think...*
>
> To change the base of the equation from 0.87 to 0.5, write 0.87 as a power of 0.5.

Let $0.87 = 0.5^x$.

Take the logarithm of each side.

$$\log 0.87 = x \log 0.5$$
$$x = \frac{\log 0.87}{\log 0.5}$$
$$\doteq 0.200\ 91$$

Hence, $0.87 \doteq 0.5^{0.2}$

Replace 0.87 in the equation $P = 100(0.87)^n$ with $0.5^{0.2}$.

$P \doteq 100(0.5^{0.2})^n$, or

$P \doteq 100(0.5)^{0.2n}$

In *Example 3*, recall that the half-life is 5 years. The coefficient $(0.2,$ or $\frac{1}{5})$ in the exponent in the equation is the reciprocal of the half-life (5). Hence, we can write the equation in this form.

$$P = 100 \times 0.5^{\frac{n}{5}}$$

Time elapsed

Half-life

Percent remaining

Initial value

Radioactive isotopes of certain elements decay with a characteristic half-life. You can use an equation similar to that above to calculate the percent remaining after a certain time, or to calculate the time for the radioactivity to reduce to a certain level.

Example 4

In April 1986, there was a major nuclear accident at the Chernobyl power plant in Ukraine, part of the former Soviet Union. The atmosphere was contaminated with quantities of radioactive iodine-131, which has a half-life of 8.1 days. How long did it take for the level of radiation to reduce to 1% of the level immediately after the accident?

Solution

Let P represent the percent of the original radiation that was present d days after the accident. Use the above equation.

$$P = 100 \times 0.5^{\frac{d}{8.1}}$$

Substitute 1 for P.
$$1 = 100 \times 0.5^{\frac{d}{8.1}}$$
$$0.5^{\frac{d}{8.1}} = 0.01$$

Take the logarithm of each side.
$$\frac{d}{8.1} \times \log 0.5 = -2$$
$$d = \frac{-2 \times 8.1}{\log 0.5}$$
$$\doteq 53.815$$

It took about 54 days for the level of radiation to reduce to 1% of the level immediately after the accident.

DISCUSSING THE IDEAS

1. Compare the two solutions to *Example 1*. Explain the similarities and the differences.

2. Refer to the doubling-time calculation on page 88. Explain why the doubling time of 9 years applies at all times during the growth of the investment.

3. Refer to the solution to *Example 2*. Explain why we did not multiply the expression 2×29.6 in the first step of the solution.

4. a) Explain why the doubling time for exponential growth is constant.
   b) Explain why the half-life for exponential decay is constant.

2.4 EXERCISES

 1. Suppose you invest $200 at 6% compounded annually. How many years would it take for your investment to grow to each amount?

   a) $300   b) $400   c) $600

2. According to Statistics Canada data for 1971–1995, the infant mortality rate in Canada has been declining dramatically. Assume this trend continues. An equation modelling this situation is $D = 6 \times 0.96^n$, where D represents the number of deaths per 1000 children under age 1, and n represents the number of years since 1995.

   a) Calculate when the infant mortality rate might become each rate.
      i) 3 per 1000   ii) 2 per 1000   iii) 1 per 1000
   b) Give some reasons why the infant mortality rate is declining.

B 3. In 1995, the population of Calgary was 828 500, and was increasing at the rate of 2.2% per year.

   a) Write an equation to represent the population of Calgary, P, as a function of the number of years, y, since 1995.

   b) Calculate how many years it would take for the population to double.

   c) Calculate when the population could become 1 million.

   d) Write the equation in part a as an exponential function with base 2.

4. According to the manual, the battery in a cellular phone loses 2% of its charge each day. Assume the battery is 100% charged.

   a) Write an equation to represent the percent charge, P, as a function of the number of days, d, since the battery was charged.

   b) Determine the number of days until the battery is only 50% charged.

   c) Write the equation in part a as an exponential function with base 0.5.

5. Calculate the number of years for an investment of $1000 to double at an interest rate of 7.2% for each compounding period.

   a) annually   b) semi-annually   c) monthly   d) daily

6. Calculate the number of years for an investment of $1000 to double at each interest rate.

   a) 7.2% compounded annually   b) 7.07% compounded semi-annually

   c) 6.97% compounded monthly   d) 6.95% compounded daily

7. Write to explain the similarities and differences between exercises 5 and 6.

8. For every metre below the surface of water, the intensity of three colours of light is reduced as shown.

Colour	Percent reduction (per metre)
Red	35%
Green	5%
Blue	2.5%

   a) For each colour, write an equation to express the percent, P, of surface light as a function of the depth, d metres.

   b) For each colour, determine the depth at which about one-half the light has disappeared.

   c) Write each equation in part a as an exponential function with base 2.

   d) For all practical purposes, the light has disappeared when the intensity is only 1% of that at the surface. At what depth would this occur for each colour?

9. Polonium-210 is a radioactive element with a half-life of 20 weeks. From a sample of 25 g, how much would remain after each time?

   a) 30 weeks   b) 14 weeks   c) 1 year   d) 511 days

10. The half-life of sodium-24 is 14.9 h. Suppose a hospital buys a 40-mg sample of sodium-24.

   a) How much of the sample will remain after 48 h?

   b) How long will it be until only 1 mg remains?

11. The table shows some substances that are present in radioactive waste from the nuclear power industry.

Substance	Half-Life
iodine-131	8 days
strontium-90	28 days
nickel-59	76 000 years
iodine-129	16 000 000 years

   Choose one substance.

   a) Draw a graph to show the percent, P, remaining during the first 5 half-lives.

   b) What percent of the substance remains after the first 5 half-lives?

   c) The "hazardous life" of a radioactive substance is from 10 to 20 half-lives. What percent of the original radiation will be present at the end of the hazardous life?

12. In *Example 2* and the discussion following it, two exponential equations were given to represent the growth of the Canadian population.

$$P = 29.6 \times 1.0124^n \qquad P = 29.6 \times 2^{\frac{n}{56}}$$

   a) Express 1.0124 as a power of 2. Explain how the result shows that the first equation can be converted to the second equation.

   b) Express 2 as a power of 1.0124. Explain how the result shows that the second equation can be converted to the first equation.

   c) Explain why the two equations are equivalent.

C 13. The total amount of arable land in the world is about 3.2×10^9 ha. About 0.4 ha of land is required to grow food for one person.

   a) Assume a 1998 population of 6 billion and a constant growth rate of 1.5%. Determine when the demand for arable land exceeds the supply.

   b) Compare the effect of each scenario on the result of part a.
   i) reducing the growth rate by one-half to 0.75%
   ii) doubling the productivity of the land
   iii) reducing the growth rate by one-half and doubling the productivity

COMMUNICATING THE IDEAS

Write to explain how to calculate the doubling time in an exponential growth situation and the half-life in an exponential decay situation. Use examples to illustrate your explanation.

2.5 Modelling Real Situations Using Exponential Functions: Part II

In Section 2.4, exponential growth and decay situations were expressed in terms of doubling time and half-life. Stating these times is just another way to describe the growth rate or the decay rate. However, these times do not have to be times to double or times to divide by 2. For example, we occasionally see statements similar to this, in magazines and newspapers.

In favourable breeding conditions, the population of a swarm of desert locusts can multiply tenfold every 20 days.

We can modify the equation on page 89 to write an equation to model the population, P, of the swarm after n days.

$$P = P_0 \times 10^{\frac{n}{20}}$$

Time elapsed
Time to multiply P_0 by 10
Initial population

The symbol, P_0, is used for the initial population, since this is not given. Even though the populations are not known, we can still use the model to compare the populations at two different times.

Example 1

A swarm of desert locusts can multiply tenfold every 20 days.

a) How many times as great is the swarm after 12 days than after 3 days?

b) Calculate how long it takes for the population of the swarm to double.

Solution

a) Use the equation above, where P_0 represents the population after 3 days. Since 12 days is 9 days later, the population, P, at that time is:

$$P = P_0 \times 10^{\frac{9}{20}} \qquad \text{Press: } 10 \boxed{\wedge} \boxed{(} 9 \boxed{\div} 20 \boxed{)} \boxed{\text{ENTER}}$$
$$\doteq 2.818 P_0$$

After 12 days, the swarm is about 2.8 times as great as after 3 days.

b) Suppose the population doubles in n days. Use the equation $P = P_0 \times 10^{\frac{n}{20}}$.

$2P_0 = P_0 \times 10^{\frac{n}{20}}$

$10^{\frac{n}{20}} = 2$

Take the base-10 logarithm of each side.

$\frac{n}{20} \times \log 10 = \log 2$

$\frac{n}{20} = \log 2$

$n = 20 \log 2$

$\doteq 6.02$

The population doubles approximately every 6 days.

Calculations similar to those in *Example 1* are used in many applications of exponential functions. A common example is the use of exponential functions in the Richter scale for comparing the intensities of earthquakes. The intensity of an earthquake is measured by the amount of ground motion as recorded on a seismograph. When we use the Richter scale, we do not need to know the intensities. The scale compares the intensities of two earthquakes using this principle.

Each increase of 1 unit in magnitude on the Richter scale represents a 10-fold increase in intensity as measured on a seismometer.

In 1976, the Italy earthquake had a magnitude 6.5. The Guatemala earthquake that year had a magnitude 7.5, which is 1 unit greater. This means that the Guatemala earthquake was 10 times as intense as the Italy earthquake. Similarly, the 1964 Alaska earthquake was 10×10, or 100 times as intense as the Italy earthquake, and $10 \times 10 \times 10$, or 1000 times as intense as the 1983 Columbia earthquake. But how do we compare the intensities of earthquakes such as the 1964 Alaska earthquake and the 1989 San Francisco earthquake, whose magnitudes do not differ by a whole number?

The Richter Scale

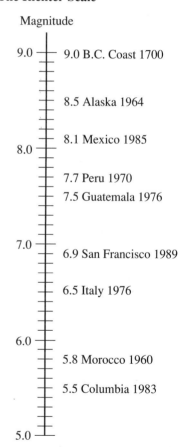

We can modify the equation on page 95 to represent this situation.
- The intensity of an earthquake, I, corresponds to the population of the swarm, P.
- The increase in Richter magnitude, R, corresponds to the elapsed time, n.
- The increase in Richter magnitude of 1 unit for every tenfold increase in intensity corresponds to the time to multiply by 10.

We obtain the equation below. It represents the intensity, I, of an earthquake that has a Richter magnitude R units greater than that of an earthquake with intensity I_0.

$$I = I_0 \times 10^R \quad \longleftarrow \text{Increase in Richter magnitude}$$

$\uparrow$

Given intensity

Example 2

a) How many times as intense as the 1989 San Francisco earthquake was the 1964 Alaska earthquake?

b) Calculate the magnitude of an earthquake that is twice as intense as the 1989 San Francisco earthquake.

Solution

a) The magnitudes of the earthquakes were 6.9 and 8.5. The magnitude of the Alaska earthquake was $8.5 - 6.9$, or 1.6 units greater than that of the San Francisco earthquake. Use the equation $I = I_0 \times 10^R$, where I_0 represents the intensity of the San Francisco earthquake. The intensity, I, of the Alaska earthquake was:

$I = I_0 \times 10^{1.6}$
$\doteq 39.811 I_0$

The Alaska earthquake was about 40 times as intense as the San Francisco earthquake.

b) Use the equation $I = I_0 \times 10^R$, where I_0 represents the intensity of the San Francisco earthquake.

$2I_0 = I_0 \times 10^R$
$10^R = 2$
$R = \log 2$
$\doteq 0.301$

An earthquake that is twice as intense as the 1989 San Francisco earthquake has a magnitude 0.3 units greater than the magnitude of the San Francisco earthquake. An earthquake that is twice as intense has a magnitude 7.2.

The Richter scale is an example of a *logarithmic scale*. In a logarithmic scale, each increase of 1 unit on the scale corresponds to multiplication by a constant. On the Richter scale, this constant is 10. Other examples of logarithmic scales are in the exercises.

1. Refer to *Visualizing*, page 88. Describe how the diagram could be modified in each case.

   a) to represent a population that grows tenfold every 20 days

   b) to represent the Richter scale

2. Explain why the term "logarithmic scale" is appropriate for a scale in which each increase of 1 unit corresponds to multiplication by a constant.

2.5 EXERCISES

B Growth of Populations

1. The population of a swarm of locusts can multiply tenfold in 3 weeks. Suppose there are 2000 locusts now. How many will there be in each time?

   a) 3 weeks   b) 6 weeks   c) 9 weeks

2. The population of a swarm of insects can multiply fivefold in 4 weeks. Let P_0 represent the population at time $t = 0$.

   a) Write an expression to represent the population after each number of weeks.

   i) 4   ii) 5   iii) 6   iv) 7   v) 8

   b) How many times as great is the population after 6 weeks as it is after 4 weeks?

3. The population of a nest of ants can multiply threefold in 5 weeks. After 8 weeks, how many times as great is the population as it is after 5 weeks?

4. The population of a colony of bacteria can double in 25 min. After one hour, how many times as great is the population as it is after 25 min?

Earthquake Magnitudes

5. How many times as intense is an earthquake with a magnitude 8 than one with each magnitude?

   a) 7   b) 6   c) 5   d) 4   e) 3   f) 2

6. On July 26, 1986, an earthquake with magnitude 5.5 hit California. On the next day, a second earthquake with magnitude 6.2 hit the same region. How many times as intense as the first earthquake was the second earthquake?

7. Most of Canada's earthquakes occur along the west coast. Three British Columbia earthquakes are listed.

Some British Columbia Earthquakes		
Date	Place	Magnitude
June 23, 1946	Near Campbell River	7.3
Aug. 22, 1949	Queen Charlotte Islands	8.1
June 24, 1997	Southwestern B.C.	4.6

a) How many times as intense was:

   i) the 1946 earthquake as the 1997 earthquake?

   ii) the 1949 earthquake as the 1997 earthquake?

   iii) the 1949 earthquake as the 1946 earthquake?

b) Why do earthquakes occur along Canada's west coast?

8. Choose one part of exercise 7a. Write to explain how you compared the intensities of the two earthquakes.

9. On page 96, find examples of two earthquakes in each case.

a) One is 2 times as intense as the other.

b) One is 4 times as intense as the other.

c) One is 8 times as intense as the other.

d) One is 16 times as intense as the other.

10. For each decrease of 1 unit in magnitude, earthquakes are about 6 or 7 times as frequent. In a given year, how should the number of earthquakes with magnitudes between 4.0 and 4.9 compare with the number of earthquakes with magnitudes between each pair of numbers?

a) 5.0 and 5.9   b) 6.0 and 6.9   c) 7.0 and 7.9

11. For each increase of 1 unit in magnitude of an earthquake, there is a thirty-one-fold increase in the amount of energy released. Use the scale on page 96.

a) How many times as much energy was released by:

   i) the 1976 Guatemala earthquake as by the 1976 Italy earthquake?

   ii) the 1964 Alaska earthquake as by the 1976 Italy earthquake?

b) How many times as much energy was released by the 1700 earthquake off the British Columbia coast as by the 1983 Colombia earthquake?

12. The 1976 Italy earthquake released approximately 10^{14} J (joules) of energy. How much energy was released by each earthquake?

a) Guatemala 1976   b) Alaska 1964   c) B.C. coast 1700

Loudness of Sounds

13. The loudness of a sound is measured in *decibels* (dB). Every increase of 10 dB represents a tenfold increase in loudness. For example, the increase from the hum of a refrigerator to an air conditioner is 20 dB. This is 2 increases of 10 dB, so the increase in loudness is 10^2, or 100. Hence, an air conditioner is 100 times as loud as a refrigerator. How many times as loud as:

    a) a conversation is a chain saw?

    b) a whisper is a chain saw?

    c) the rustle of a leaf is a rock group?

14. A research was conducted to test the noise levels in school gymnasiums. Loudness levels ranged from 75 dB to 115 dB. How many times as loud as a noise at 75 dB is a noise at 115 dB?

15. Let L_0 represent the loudness of a sound at 0 dB, the threshold of hearing.

    a) Write an equation to express the loudness, L, of a sound with decibel level, d.

    b) How many times as loud as a heavy truck is an air conditioner?

    c) How many times as loud as a refrigerator hum is average street traffic?

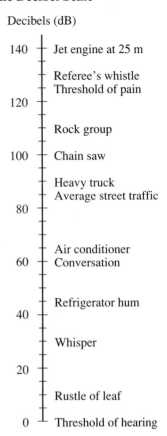

The Decibel Scale

Decibels (dB)

- 140 — Jet engine at 25 m
- Referee's whistle
- Threshold of pain
- 120
- Rock group
- 100 — Chain saw
- Heavy truck
- Average street traffic
- 80
- Air conditioner
- 60 — Conversation
- Refrigerator hum
- 40
- Whisper
- 20
- Rustle of leaf
- 0 — Threshold of hearing

16. The loudness level of a heavy snore is 69 dB.

    a) How many times as loud as a conversation is a heavy snore?

    b) How many times as loud as a whisper is a heavy snore?

17. When a person listens to very loud sounds for prolonged periods of time, the hearing can be impaired. An 8-h exposure to a 90-dB sound is considered acceptable. For every 5-dB increase in loudness, the acceptable exposure time is reduced by one-half.

    a) What is the acceptable exposure time for each noise?
       i) a rock group playing at 110 dB   ii) a referee's whistle at 130 dB

    b) Write an equation to express the acceptable exposure time as a function of the sound level.

    c) Graph your equation in part b.

Acidity and Alkalinity of Solutions

18. In chemistry, the pH scale measures the acidity or alkalinity of a solution. Most pH values range from 0 to 14, with 7 being neutral. Above 7, each 1-unit increase in pH represents a tenfold increase in alkalinity. Below 7, each 1-unit decrease in pH represents a tenfold increase in acidity.

pH Scale

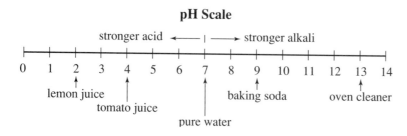

a) How many times as acidic as tomatoes is lemon juice?

b) How many times as acidic as pure water is lemon juice?

c) How many times as alkaline as baking soda is oven cleaner?

d) How many times as alkaline as pure water is oven cleaner?

19. In spring, the pH value of a stream dropped from 6.5 to 5.5 during a 3-week period in April.

a) How many times as acidic did the stream become?

b) Why would this happen in April?

20. The pH values of milk range from 6.4 to 7.6. Pure water has a pH value of 7.0.

a) How many times as acidic is milk with a pH value of 6.4 than pure water?

b) How many times as alkaline is milk with a pH value of 7.6 than pure water?

21. Water in a swimming pool should have a pH value of 7.5, with values between 7.2 and 7.8 being acceptable.

a) How many times as alkaline is a pool with a pH value of 7.8 than one with a pH value of 7.2?

b) The human eye has a pH value of 7.5. Determine the pH value of something that is twice as alkaline as the human eye.

c) Write to explain how the alkaline level of the human eye relates to the range of acceptable alkaline levels in a swimming pool.

COMMUNICATING THE IDEAS

Write to explain what is meant by a logarithmic scale. Include some examples of logarithmic scales in your explanation. Choose one scale and explain how it is used.

Modelling the Growth of Computer Technology

On page 64, you discussed some quotations that describe the growth of computer technology. Other quotations can be used to develop mathematical models to describe this growth. You can use these models to help you decide whether the second quotation on page 65 is reasonable.

Use the information in this quotation for the exercises below.

> Between 1960 and 1970, the number of components on a chip doubled each year from 1 in 1960 to 1000 in 1970. Since 1970, the number of components on a chip has doubled every year and a half, reaching ... 1 000 000 000 in 1992.

1. Consider the years from 1960 to 1970.
   a) Verify the first sentence in the quotation.
   b) Write an equation to represent the number of components, C, on a chip in terms of the number of years, n, since 1960, where $n \leq 10$.

2. Consider the years from 1970 to 1992.
   a) Show that the last part of the second sentence in the quotation is not correct.
   b) The first part of the second sentence in the quotation is generally accepted as true in the computer industry.
      i) Write an equation to represent the number of components, C, on a chip in terms of the number of years, n, since 1970, where $n \leq 22$.
      ii) Your equation is a mathematical model to describe the growth of computer technology. Correct the second sentence in the quotation by changing 1992 to the year when there would be 1 000 000 000 components on a chip, as predicted by the model.

3. In 1998, the Merced chip contained 14 000 000 components. Is this number consistent with your model in exercise 2? Explain.

 LOOK AT THE IMPLICATIONS

The quotations below also appear in the handbook. Assume these quotations still apply.

Microchips are doubling in performance power every 18 months.	Computer power is now 8000 times cheaper than it was 30 years ago.

4. The first statement above is generally accepted as true in the computer industry.

    a) In 1972, a Motorola chip could execute 60 000 instructions per second. Write an equation to represent the number of instructions, I, a chip can execute in terms of the number of years, n, since 1972.

    b) Is this statement consistent with your model? Explain.
    In 1990, a Motorola chip could execute 200 million instructions per second.

5. Use the information in the second statement above.

    a) Calculate how the current cost of computer power compares with its cost:
      i) 20 years ago   **ii)** 10 years ago

    b) Predict how the current cost of computer power might compare with its cost:
      i) 10 years from now   **ii)** 20 years from now

 **REVISIT THE SITUATION**

Recall the second quotation on page 65. A Lexus was not available 30 years ago, but the Cadillac is a comparable car.

6. In 1971, a Cadillac cost about $7000. It could travel at 150 km/h and used about 1 L of gas for 8 km. Use this information to determine whether the second quotation on page 65 is reasonable. You will have to make some decisions about the quantities to compare and how to compare them. Identify any assumptions you make.

7. Give some reasons why there has not been the same progress in automotive technology in the latter part of the 20th century as there has been in computer technology.

2.6 Analyzing the Graphs of Exponential Functions

In earlier sections of this chapter, you graphed examples of exponential functions. The simplest exponential function has an equation of the form $y = a^x$, where a is a constant. For example, this is the graph of the exponential function $f(x) = 2^x$.

x	y
−3	$\frac{1}{8}$
−2	$\frac{1}{4}$
−1	$\frac{1}{2}$
0	1
1	2
2	4
3	8

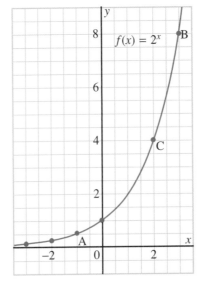

The graph illustrates these properties of the function $f(x) = 2^x$.

Vertical intercept

Let $x = 0$.
Then, $f(0) = 2^0$
$\qquad\quad = 1$
The vertical intercept is 1.

Horizontal intercept

Let $f(x) = 0$.
Then, $2^x = 0$
This equation has no real solution since $2^x > 0$ for all real values of x. Hence, there is no horizontal intercept.

Domain and range

Since we can define 2^x for all real values of x, the domain is the set of real numbers. Since there is a value of 2^x for each positive real value of x, the range is the set of positive real numbers.

Asymptote

When x decreases through negative values, the points on the graph come closer and closer to the x-axis, but never reach it. The x-axis is an asymptote.

Law of Exponents

Select any two points on the curve, such as A$(-1, \frac{1}{2})$ and B$(3, 8)$.

Add their *x*-coordinates.

$-1 + 3 = 2$

Multiply their *y*-coordinates.

$\frac{1}{2} \times 8 = 4$

The results are the coordinates of another point C$(2, 4)$ on the graph. Adding the *x*-coordinates and multiplying the *y*-coordinates of two points on the graph produces the coordinates of another point on the graph.

Example 1

Sketch the graphs of these functions on the same grid.

a) $f(x) = 2^x$ **b)** $g(x) = 1.5^x$ **c)** $h(x) = 0.5^x$

Solution

All three graphs pass through the point $(0, 1)$.

a) The graph of $f(x) = 2^x$ is the same as that on page 104.

b) If $x > 0$, then $1.5^x < 2^x$. In the first quadrant, the graph of $g(x) = 1.5^x$ lies below the graph of $f(x) = 2^x$. If $x < 0$, then $1.5^x > 2^x$. In the second quadrant, the graph of $g(x) = 1.5^x$ lies above the graph of $f(x) = 2^x$.

c) If $x > 0$, 0.5^x is less than 1. Also, as x increases, 0.5^x becomes closer and closer to 0. If $x < 0$, 0.5^x becomes larger and larger, without limit.

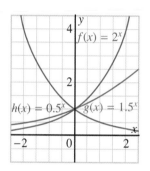

The graphs in *Example 1* illustrate properties of the graph of the exponential function $f(x) = b^x$.

Properties of the Graph of the Exponential Function $f(x) = b^x$ *(b > 0)*

If $b > 1$, the graph goes up to the right.

If $0 < b < 1$, the graph goes down to the right.

Vertical intercept: 1
Horizontal intercept: none

Domain: all real numbers
Range: all positive real numbers

Asymptote: *x*-axis

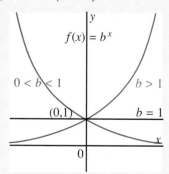

Using the Tools from the Functions Toolkit

In *Example 1c*, 0.5^x can be written as $(2^{-1})^x$, or 2^{-x}. Hence, the equation of the function is $h(x) = 2^{-x}$. This means that we could use the *y*-axis reflection tool from Chapter 1. When we use this tool, the graph of $y = 2^{-x}$ is the reflection of the graph of $y = 2^x$ in the *y*-axis.

Earlier sections of this chapter contained many examples of functions whose equations have the form $y = Ab^{kx}$. We can use the stretching tools from Chapter 1 to visualize how these graphs are related to the graphs of functions whose equations have the form $y = b^x$. For example, on page 66, we graphed the function $P = 2.76(1.022)^n$ to model the growth of British Columbia's population since 1981. Visualize how the graph of $P = 2.76(1.022)^n$ coincides with a vertical expansion of the graph of $y = 1.022^x$ by a factor of 2.76.

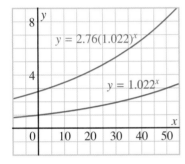
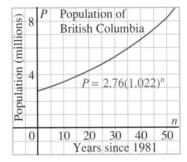

Similarly, on page 66, we graphed the function $P = 100(0.87)^n$ to model the percent of caffeine in your bloodstream after consumption. Although the scales are different in the two graphs, we can still visualize how the graph of $P = 100(0.87)^n$ coincides with a vertical expansion of the graph of $y = 0.87^x$ by a factor of 100.

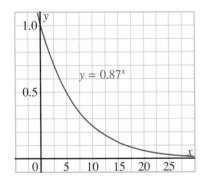
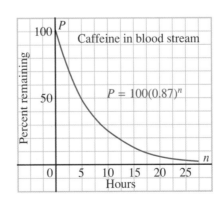

Example 2

a) Graph the functions $y = 3^x$ and $y = 2(3^x - 1)$ on the same axes.

b) Identify the domain, range, asymptotes, and intercepts of each graph.

Solution

a) Graph $y = 3^x$, translate it 1 unit down, then expand it vertically by a factor of 2.

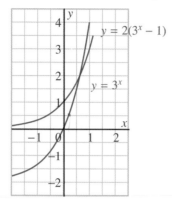

b) For each graph, the domain is the set of real numbers.
For $y = 3^x$, the range is the set of positive real numbers, and the x-axis is an asymptote.
For $y = 2(3^x - 1)$, which can be written $y = 2(3^x) - 2$, the range is the set of real numbers greater than -2, and the line $y = -2$ is an asymptote.
The graph of $y = 3^x$ does not have an x-intercept. Its y-intercept is 1.
The graph of $y = 2(3^x - 1)$ has x-intercept 0 and y-intercept 0.

Example 3

a) Graph the functions $y = 10^x$, $y = 10^{\frac{x}{3}}$, and $y = 2^x$ on the same screen.

b) What do you notice about the graphs of $y = 10^{\frac{x}{3}}$ and $y = 2^x$?

c) Explain why the graph of $y = 10^{\frac{x}{3}}$ appears to be close to the graph of $y = 2^x$.

Solution

a) Enter the equations Y1=10^X, Y2=10^(X/3), and Y3=2^X. On the TI-83, use the graph style icons in the Y= list to distinguish the three graphs as shown below. The thick curve represents $y = 10^x$. The thin curve represents $y = 10^{\frac{x}{3}}$. The dotted curve represents $y = 2^x$.

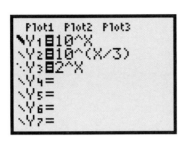

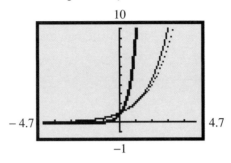

b) The graphs of $y = 10^{\frac{x}{3}}$ and $y = 2^x$ appear to be approximately the same, especially for small values of x.

c) The graph of $y = 10^{\frac{x}{3}}$ is a horizontal expansion of the graph of $y = 10^x$ by a factor of 3. It is expanded by almost the right amount to make it coincide with the graph of $y = 2^x$.

Example 3 suggests the graph of $y = 10^x$ can be expanded horizontally until its image coincides with the graph of $y = 2^x$. The expansion factor is approximately 3. In exercise 10, you will determine the expansion factor accurately. The graphs of exponential functions with different bases are always related this way.

VISUALIZING   **Horizontal Stretches of Exponential Graphs**

The graph of any exponential function, $y = a^x$, can be made to coincide with the graph of an exponential function with a different base, $y = b^x$. Use an appropriate horizontal expansion or compression.

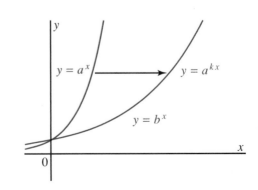

1. In *Example 1b*, for $x > 0$, the graph of $g(x) = 1.5^x$ lies below that of $f(x) = 2^x$. How can we tell how far below to draw it?

2. In *Example 2*, the coefficient 2 caused a vertical expansion. In *Example 3*, the coefficient $\frac{1}{3}$ in the exponent caused a horizontal expansion. Explain why both coefficients caused expansions, although one coefficient is greater than 1 and the other coefficient is less than 1.

3. In *Example 3*, what other way can we obtain the graph of $y = 2(3^x - 1)$ from the graph of $y = 3^x$?

4. In *Example 3*, we found that when the graph of $y = 10^x$ expands horizontally by a factor of 3, the image graph is close to the graph of $y = 2^x$. Would the expansion factor that makes the graph of $y = 10^x$ coincide with the graph of $y = 2^x$ be greater than 3 or less than 3? Explain.

5. For the graphs in *Visualizing*, is k greater than 1 or less than 1? Explain.

2.6 EXERCISES

 1. Identify which graph best represents each function.

a) $f(x) = 3^x$

b) $g(x) = 10^x$

c) $h(x) = \left(\frac{3}{4}\right)^x$

d) $k(x) = \left(\frac{1}{4}\right)^x$

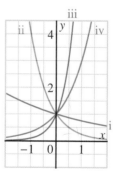

2. a) Graph $y = 10^x$ and $y = 2^x$ on the same axes.

b) Use your graph in part a to sketch the graph of $y = 5^x$.

c) Use a graphing calculator to confirm your sketch in part b.

3. a) Use the results of exercise 2. Sketch the graphs of $y = \left(\frac{1}{2}\right)^x$, $y = \left(\frac{1}{5}\right)^x$, and $y = \left(\frac{1}{10}\right)^x$ on the same axes.

b) Use a graphing calculator to confirm your sketch in part b.

4. a) Sketch a sequence of graphs to illustrate how the graph of $y = a^x$ changes as a increases through all positive values.

b) Visualize how the properties on page 105 apply to each graph in part a.

B **5. a)** Suppose $a > 0$. Visualize what the graph of $y = a^x$ looks like as a becomes larger and larger, and as a becomes smaller and smaller.

b) Predict what the graphs of the equations in each list would look like.

   i) $y = 2^x$, $y = 3^x$, $y = 4^x$, $\cdots$, $y = 100^x$

   ii) $y = \left(\frac{1}{2}\right)^x$, $y = \left(\frac{1}{3}\right)^x$, $y = \left(\frac{1}{4}\right)^x$, $\cdots$, $y = \left(\frac{1}{100}\right)^x$

c) Use a graphing calculator to check your predictions in part b.

6. For each list, predict what the graphs of the equations would look like. Check your predictions.

a) $y = 1 \times 2^x$
$y = 2 \times 2^x$
$y = 4 \times 2^x$
$y = 8 \times 2^x$
$\vdots$
$y = 2^{10} \times 2^x$

b) $y = 1 \times 2^x$
$y = \frac{1}{2} \times 2^x$
$y = \frac{1}{4} \times 2^x$
$y = \frac{1}{8} \times 2^x$
$\vdots$
$y = \frac{1}{2^{10}} \times 2^x$

7. For each list, predict what the graphs of the equations would look like. Check your predictions.

a) $y = 2^x$, $y = 2^{x+1}$, $y = 2^{x+2}$, $y = 2^{x+3}$

b) $y = 2^x$, $y = 2^{x-1}$, $y = 2^{x-2}$, $y = 2^{x-3}$

8. Compare the results of exercises 6 and 7. Write to describe and explain any similarities you observe.

9. The temperature of hot chocolate is recorded as it cools down. Analysis of the data shows that the temperature, T degrees Celsius, of the hot chocolate can be expressed as a function of time, t minutes, by the equation $T = 79 \times 0.85^t + 20$.

a) Graph the equation using appropriate values of t.

b) Determine the temperature of the hot chocolate after 5 min.

c) Hot chocolate is safe to drink at 40°C. How long did it take the hot chocolate to cool down to 40°C?

d) Use the tools from the functions toolkit in Chapter 1. Explain how the graph of $T = 79 \times 0.85^t + 20$ is related to the graph of $y = 0.85^x$.

Hot and cold objects cool down or warm up to the temperature of their surroundings. Newton's Law of Cooling in physics states that the difference between the temperature of a hot substance and the temperature of the surroundings is an exponential function of the time it has been cooling.

- Explain how the equation $T = 79 \times 0.85^t + 20$ in exercise 9 illustrates Newton's Law of Cooling.

- What does the number 20 in the equation represent? What does the number 79 represent? What does the number 0.85 represent?

- A similar principle applies when a cold substance warms up to the temperature of its surroundings. What changes would be needed in the equation above to represent a substance warming up instead of cooling down? Explain.

10. In *Example 3*, when k is approximately 0.333, the graph of $y = 10^{kx}$ comes close to the graph of $y = 2^x$.

   a) Express 2 as a power of 10.

   b) What is the value of k such that the graph of $y = 10^{kx}$ coincides with the graph of $y = 2^x$? Explain.

   c) What is the expansion factor corresponding to your answer to part a?

11. Refer to exercise 2.

   a) Express as a power of 5:   i) 10   ii) 2

   b) Write the equations of the graphs in exercise 2 using base 5.

12. Recall the equation $P = 29.6 \times 2^{\frac{n}{56}}$ from page 89, which models the growth of Canada's population, in terms of the doubling time, 56 years. Use the stretching tools from Chapter 1. Explain how the graph of this function is related to the graph of $y = 2^x$.

13. Recall the equation $P = 100 \times 0.5^{\frac{n}{5}}$ from page 91, which models the percent of caffeine in your body n hours after consumption. Use the stretching tools from Chapter 1. Explain how the graph of this function is related to the graph of $y = 0.5^x$.

14. a) Graph the function $f(x) = 2^x$ for $-3 \le x \le 3$.

   b) On the same grid as in part a, sketch the graph of each function.
      i) $y = f(x) - 1$ ii) $y = f(x - 1)$ iii) $y = f(x + 1)$
      iv) $y = f(0.5x)$ v) $y = f(2x)$ vi) $y = f(-x)$

15. Consider the function $y = 2^{3(x-2)}$.

    a) Graph the function.

    b) Identify the domain and the range of the function.

    c) Write the equations of any asymptotes.

    d) Determine the intercepts of the function.

16. Repeat exercise 15 for each function.

    a) $y = 2^{4(x-3)}$ **b)** $y = 3^{2x-5}$ **c)** $f(x) = 2^{3(x+2)}$ **d)** $f(x) = 10^{3x+2}$

17. Recall the function on page 66 that models British Columbia's population, P million, by the equation $P = 2.76(1.022)^n$, where n is the number of years since 1981.

    a) Change the equation so that n represents the number of years since 1991.

    b) Change the equation so that n represents the year.

    c) What do the vertical intercepts of the graphs of the equations in parts a and b represent?

C **18.** Let (x_1, y_1) and (x_2, y_2) be any two points on the graph of $y = 2^x$. Prove that each point with the given coordinates is also on the graph.

    a) $(x_1 + x_2, y_1 y_2)$ **b)** $\left(x_1 - x_2, \dfrac{y_1}{y_2}\right)$ **c)** $(2x_1, y_1{}^2)$ **d)** $\left(-x_1, \dfrac{1}{y_1}\right)$

19. When police arrived at the scene of a murder at 10 P.M., they found that the victim's temperature was 27°C. One hour later, it was 25°C. Determine the probable time of death. Assume the room temperature was constant at 20°C and normal body temperature is 37°C.

20. a) Believing that hot water freezes faster than cold water, a person puts a tray of water at 95°C into the freezer, where the temperature is a constant −10°C. It takes 90 min for the water to reach 0°C. Determine how long it would take a tray of water at 20°C to reach 0°C in the same freezer.

    b) Does hot water freeze faster than cold water? Explain.

21. Suppose cream is added to a cup of coffee at the time the coffee is poured. Alternatively, suppose the coffee is allowed to cool first before the cream is added. Determine if it makes any difference to the temperature of the coffee after 10 min. Make any necessary assumptions to complete this problem.

COMMUNICATING THE IDEAS

Write to explain why increasing the value of a in the equation $y = a \times 2^{bx}$ expands the graph vertically, but increasing the value of b compresses the graph horizontally, where both $a > 0$ and $b > 0$. Include some graphs in your explanation.

2.7 Geometric Sequences and Exponential Functions

Recall the first two functions on page 66 concerning the investment of $1000 at 8% compounded annually and the percent of light passing through layers of glass. The graph of each function comprises discrete points because the variable n in each function represents a whole number. Consider the values of the first function, $A = 1000(1.08)^n$, for the first few values of n.

Table of values for $A = 1000(1.08)^n$

n	0	1	2	3	4	5
A	1000.00	1080.00	1166.40	1259.71	1360.49	1469.33

$\times 1.08 \quad \times 1.08 \quad \times 1.08 \quad \times 1.08 \quad \times 1.08$

The sequence of numbers in the second row is a geometric sequence. Each term after the first term is formed by multiplying by the same number, 1.08 (apart from rounding). Recall that in a geometric sequence, the ratio formed by dividing any term by the preceding term is a constant, called the common ratio.

These are geometric sequences:

$$2, 10, 50, 250, \ldots \qquad 12, 6, 3, 1.5, 0.75, \ldots$$

$$3, -12, 48, -192, \ldots \qquad a, ar, ar^2, ar^3, \ldots$$

When you know the first term and the common ratio of a geometric sequence, you can determine any other term.

Example 1

In the geometric sequence 6, 12, 24, …, determine each term. **a)** t_{10} **b)** t_n

Solution

The first term is 6 and the common ratio is $\frac{12}{6}$, or 2.

a) t_{10} can be found by starting with the first term and multiplying by the common ratio 9 times.

$$t_{10} = 6 \times 2^9$$
$$= 3072$$

b) An expression for t_n can be found by multiplying the first term by $(n - 1)$ common ratios.

$$t_n = 6 \times 2^{n-1}$$

In the general geometric sequence, the first term is represented by a and the common ratio by r. The first few terms and the general term are:
$a, ar, ar^2, ar^3, \ldots, ar^{n-1}, \ldots$

The general term of a geometric sequence is given by $t_n = ar^{n-1}$, where a is the first term, r is the common ratio, and n is the term number.

Example 2

Consider the geometric sequence 3, 6, 12, 24, ….

a) Determine the 14th term.

b) Which term is 384?

Solution

Use the formula for the general term: $t_n = ar^{n-1}$

a) Substitute $a = 3$, $r = 2$, and $n = 14$.
$$t_{14} = 3 \times 2^{13}$$
$$= 24\,576$$
The 14th term is 24 576.

b) Substitute $a = 3$, $r = 2$, and $t_n = 384$.
$$384 = 3 \times 2^{n-1}$$
Divide each side by 3.
$$128 = 2^{n-1}$$
Since $128 = 2^7$, then $n - 1 = 7$
Hence, $n = 8$
384 is the 8th term of the sequence.

Recall the table of values for the function $A = 1000(1.08)^n$ on page 113. The values of A form a geometric sequence with first term 1000 and common ratio 1.08. The general term is $t_n = 1000(1.08)^{n-1}$. Compare this equation with the equation for A on page 113. The equations are the same except for the exponent. The reason for this discrepancy is that the n in the equation for A counts the number of years, while the n in the equation for t_n counts the number of terms of the sequence.

Table of values for $t_n = 1000(1.08)^{n-1}$

n	1	2	3	4	5	6
t_n	1000.00	1080.00	1166.40	1259.71	1360.49	1469.33

We can use the tools from the functions toolkit in Chapter 1 to explain how the graphs of A and t_n are related. Visualize drawing the graphs of A and t_n on the same grid. Although different variables are used in the equations, the graph of $t_n = 1000(1.08)^{n-1}$ is the image of the graph of $A = 1000(1.08)^n$ when it is translated 1 unit right.

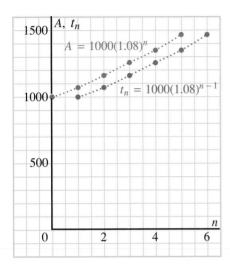

1. Suppose you know some terms of a geometric sequence. Explain how you can determine the common ratio.

2. In *Example 2b*, we used $128 = 2^7$ to solve the equation $128 = 2^{n-1}$. How can you solve this equation if you do not realize that $2^7 = 128$?

2.7 EXERCISES

A 1. Is each sequence geometric? If it is, what is the common ratio?

a) 1, 2, 4, 8, 16, ...

b) 2, 4, 6, 10, 16, ...

c) 4, –2, 1, $-\frac{1}{2}$, $\frac{1}{4}$, ...

d) 0.6, 0.06, 0.006, ...

e) –3, 2, 7, 12, 17, ...

f) 1, $-\frac{1}{3}$, $\frac{1}{9}$, $-\frac{1}{27}$, ...

2. State the common ratio, then list the next 3 terms of each geometric sequence.

a) 1, 3, 9, 27 ...

b) 5, –15, 45, –135, ...

c) 3, 6, 12, 24, ...

d) 6, 2, $\frac{2}{3}$, $\frac{2}{9}$, ...

e) 36, 9, $\frac{9}{4}$, $\frac{9}{16}$, ...

f) $\frac{1}{2}$, –2, 8, –32, ...

3. Bacteria grow by division, and it is possible for this to occur every 20 min. One bacterium splits into two, 20 min later these split to become four, 20 min later these become eight, and so on. Suppose one bacterium is put into a culture in a dish at 9 A.M. and the culture is covered by bacteria at 6 P.M.

a) When was the culture half covered?

b) When would the bacteria in the culture realize that they were running out of space?

c) Suppose three more cultures became available at 6 P.M. When would all four cultures be covered by bacteria?

4. You have 2 parents 1 generation ago, 4 grandparents 2 generations ago, 8 great-grandparents 3 generations ago, and so on.

a) Determine how many ancestors you had each number of generations ago.

   i) 5   ii) 10   iii) 20   iv) 40

b) Assuming that one generation lasts about 25 years, then 40 generations ago was about 1000 years ago. Explain why your answer to part a iv is much greater than the world population of about 300 million at that time.

5. Compare the equation of the exponential function $y = Ab^x$ with the formula for the nth term of a geometric sequence, $t_n = ar^{n-1}$. Why does -1 occur in the exponent of the second equation but not in the first?

B 6. Is each sequence geometric? Explain.

a) Camera shutter speeds (seconds):

   $4, 2, 1, \frac{1}{2}, \frac{1}{4}, \frac{1}{8}, \frac{1}{15}, \frac{1}{30}, \frac{1}{60}, \frac{1}{100}, \frac{1}{125}, \frac{1}{250}, \frac{1}{500}, \frac{1}{1000}, \frac{1}{2000}$

b) Frequencies of a piano's A notes in hertz (cycles per second):

   $27.5, 55, 110, 220, 440, 880, 1760, 3520$

7. For each geometric sequence, write a formula for t_n, then use it to determine each indicated term.

a) $2, 4, 8, 16, \ldots; t_{10}$ b) $5, 10, 20, 40, \ldots; t_{13}$ c) $-3, 15, -75, 375, \ldots; t_8$

d) $12, 6, 3, \frac{3}{2}, \ldots; t_{12}$ e) $6, -2, \frac{2}{3}, -\frac{2}{9}, \ldots; t_9$ f) $3, 18, 108, 648, \ldots; t_7$

8. Because insects have exoskeletons, they grow in stages, at the time of each moult. In one study, a biologist measured the head width, w millimetres, of the caterpillar of a swallowtail butterfly at each of 5 steps in its growth. This gave five data points, shown on the screen. The equation of the exponential curve of best fit is $w \doteq 0.42(1.52)^x$.

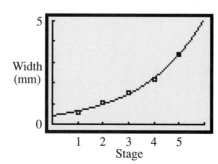

a) Use the equation to estimate the width of the head of the caterpillar at each stage.

b) Show that the data in part a form a geometric sequence.

c) What is the common ratio of the sequence? What does it represent?

9. The function on page 66 shows the percent of light passing through layers of glass is modelled by the equation $P = 100(0.95)^n$.

a) Write the first term and the common ratio of the geometric sequence corresponding to this situation.

b) Write the general term of the geometric sequence.

c) Explain why the exponent in the equation in part a is different from the exponent in the equation of the function.

10. A ball is dropped from a height of 2.0 m. After each bounce, it rises to 63% of its previous height.

a) Write the first term and the common ratio of the geometric sequence corresponding to this situation.

b) Write the general term of the geometric sequence.

c) What height does the ball reach after 5 bounces?

d) After how many bounces does the ball reach a height of only 20 cm?

e) Graph bounce height against the number of bounces.

MODELLING Bounce Heights

In exercise 10, the height of a bouncing ball after each bounce is modelled by a geometric sequence.

- According to this model, would the ball stop bouncing? Explain.
- Explain why the ball eventually stops bouncing.

11. These are the aperture markings on a camera lens.

1.4   2   2.8   4   5.6   8   11   16   22

They form a geometric sequence, but the numbers have been rounded for convenience. Determine the common ratio as accurately as possible. Explain.

12. In △ABZ, the length of AB is 1 unit, and ∠B = 90°. From B, perpendiculars BC, CD, DE, … are drawn alternately to AZ and to BZ as shown. Let x and y represent the lengths of BC and AC, respectively.

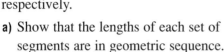

a) Show that the lengths of each set of segments are in geometric sequence.
  i) the segments AB, BC, CD, DE, …
  ii) the vertical segments AB, CD, EF, …
  iii) the inclined segments BC, DE, FG, …

b) Find other segments in this triangle whose lengths are in geometric sequence.

13. About 100 years ago, mathematicians devised some strange figures, which have been revived by modern computer graphics. The *Sierpinski triangle* is one example. The first four steps in the construction of the Sierpinski triangle are shown. Visualize these steps continuing forever. After each step, the figure comes closer and closer to the Sierpinski triangle.

a) What percent of the original triangle remains after each step?

b) How many steps are needed until only 1% of the original triangle remains?

C **14.** Determine the geometric sequence with each property.

a) The sum of the first 2 terms is 3. The sum of the next 2 terms is $\frac{4}{3}$.

b) The sum of the 3rd and 4th terms is 36. The sum of the 4th and 5th terms is 108.

15. Determine if it is possible to find 3 numbers that form both an arithmetic sequence and a geometric sequence. Explain.

COMMUNICATING THE IDEAS

Write to explain the similarities and differences between a geometric sequence and an exponential function.

Graphing Sequences

In a steel mill, red-hot slabs of steel are pressed many times between heavy rollers. The drawings show two stages in rolling a slab.

A slab is 2.00 m long. On each pass through the rollers, its length increases by 20%. The length, L metres, of the slab after n passes through the rollers is represented by the formula $L = 2.00(1.20)^n$.

To graph L against n, we could enter the equation Y=2(1.2)^X in the Y= list and press GRAPH. This would produce a smooth curve, which is not appropriate because there can only be a whole number of passes through the rollers. We require the graph at the right that shows only the values of y for whole number values of x.

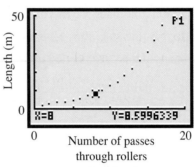

On the TI-83 graphing calculator, graphing sequences are more involved than graphing functions. You can use a program or you can enter the commands.

Using a program

Ask your teacher for the program called SEQUENCE from the Teacher's Resource Book. The program calculates term numbers and values, and enters them into the first two columns of the list editor.

Entering the commands

- Ensure no equations are selected in the Y= list.
- Define the viewing window for the graph you want.
- Set up a statistical plot, using the instructions on page 13.
- Press STAT ENTER, and clear the lists L1 and L2.
- Press LIST ▶ 5 to select seq(.

- Press [X,T,θ,n] [,] [X,T,θ,n] [,] 1 [,] 20 [ )] [STO▶] [L₁] [ENTER] to complete the first two lines on the screen shown. This determines the values of *x* such that *x* begins at 1 and ends at 20, and stores them in list L1.

- Press [LIST] [▶] 5 to select seq(again.

- Press 2 [×] 1.2 [^] [X,T,θ,n] [,] [X,T,θ,n] [,] 1 [,] 20 [ )] [STO▶] [L₂] [ENTER] to complete the third and fourth lines shown. This determines the values of *x* such that *x* begins at 1 and ends at 20, and stores them in list L2.

```
seq(X,X,1,20)→L₁
{1 2 3 4 5 6 7 …
seq(2*1.2^X,X,1,
20)→L₂
{2.4 2.88 3.456…
```

- Press [GRAPH].

- When you change the window setting, or to graph a different sequence, repeat the steps above.

You can also graph sequences using the Seq setting in the mode menu. Consult the TI-83 manual for information about this.

1. Use the graph you made when you completed the instructions above.

    a) How long is the slab after 12 passes through the rollers?

    b) How many passes are needed until the length is at least 20 m?

2. On all passes through the rollers, the slab's width is kept constant. This means that on each pass, its thickness decreases by about 17%. Assume the slab is 0.45 m thick at the beginning.

    a) Write an equation to represent the thickness, *t* metres, of the slab after *n* passes through the rollers.

    b) Graph the equation in part a. The result should be a graph showing only whole number values of *n*.

    c) How thick is the slab after 12 passes through the rollers?

    d) The steel used in car bodies is about 0.001 m thick. How many passes through the rollers are needed until the slab is this thick?

3. In exercise 2:

    a) Explain why the thickness decreases by only about 17% on each pass, although the length increases by 20%.

    b) How long is the steel when it is only 0.001 m thick?

2.8 Geometric Series and Exponential Functions

Some families succeed in tracing their roots back ten generations or more. When you go back ten generations, how many ancestors could there be in your family tree?

Every person has 2 parents, 4 grandparents, 8 great-grandparents, and so on. The number of ancestors through ten generations is:

$$2 + 4 + 8 + 16 + 32 + 64 + 128 + 256 + 512 + 1024$$

This expression is an example of a *geometric series* because it indicates the terms of a geometric sequence are to be added. Instead of adding the ten numbers, the sum can be found as shown:

Let S represent the sum: $S = 2 + 4 + 8 + \ldots + 512 + 1024$ ①

Multiply by the common ratio 2: $2S = 4 + 8 + \ldots + 512 + 1024 + 2048$ ②

Subtract ① from ②: $S = -2 + 2048$

$$S = 2046$$

The sum of the first ten terms of the series is 2046. Going back through ten generations, each person has 2046 ancestors.

This method can be used to calculate the sum of any number of terms of a geometric series.

Example 1

Determine the sum of the first 9 terms of the geometric series $2 + 6 + 18 + 54 + \ldots$.

Solution

> **Think …**
>
> To use the method, we need to know the 9th term.

Determine t_9 using $t_n = ar^{n-1}$.

In this series, $a = 2, r = 3$, and $n = 9$

$$t_9 = 2 \times 3^8$$
$$= 13\ 122$$

Let S represent the sum: $\qquad S = 2 + 6 + 18 + \ldots + 13\ 122$ ①

Multiply by the common ratio 3: $\qquad \underline{3S = 6 + 18 + \ldots + 13\ 122 + 39\ 366}$ ②

Subtract ① from ②: $\qquad 2S = -2 + 39\ 366$

$$= 39\ 364$$

$$S = 19\ 682$$

The sum of the first 9 terms of the series is 19 682.

The method of *Example 1* can be used to find a formula for the sum of the first n terms of the general geometric series.

Let $\qquad\qquad\qquad\qquad S_n = a + ar + ar^2 + \ldots + ar^{n-1}$ ①

Multiply by r: $\qquad\qquad\quad\ rS_n = ar + ar^2 + \ldots + ar^{n-1} + ar^n$ ②

Subtract ① from ②: $\qquad rS_n - S_n = -a + ar^n$

$$S_n(r - 1) = a(r^n - 1)$$

$$S_n = \frac{a(r^n - 1)}{r - 1} \qquad r \neq 1$$

For the general geometric series $a + ar + ar^2 + \ldots + ar^{n-1}$,

the sum of the first n terms is $S_n = \dfrac{a(r^n - 1)}{r - 1}, \qquad r \neq 1$

Example 2

Determine the sum of the first 10 terms of each geometric series.

a) $4 + 12 + 36 + 108 + \ldots$ $\qquad\qquad$ **b)** $6 + 3 + 1.5 + 0.75 + \ldots$

Solution

Use the formula $S_n = \dfrac{a(r^n - 1)}{r - 1}$.

a) Substitute $a = 4$, $r = 3$, $n = 10$. $\qquad$ **b)** Substitute $a = 6$, $r = 0.5$, $n = 10$.

$$S_{10} = \frac{4(3^{10} - 1)}{3 - 1} \qquad\qquad\qquad\qquad S_{10} = \frac{6(0.5^{10} - 1)}{0.5 - 1}$$

$$= 2(3^{10} - 1) \qquad\qquad\qquad\qquad\qquad = -12(0.5^{10} - 1)$$

$$= 118\ 096 \qquad\qquad\qquad\qquad\qquad\quad = 11.988\ 281\ 25$$

Examine the line above the answer in the solution to *Example 2a*. When you replace 10 with n, the formula for the sum of the first n terms of this sequence becomes the function $S = 2(3^n - 1)$. We can use the functions toolkit from Chapter 1 to visualize how the graph of this function is related to the graph of $y = 3^x$. We translate the graph of $y = 3^x$ down 1 unit, then expand it vertically by a factor of 2. See *Example 2* on page 107. When we draw a graph of $S = 2(3^n - 1)$, it would coincide with the graph of $y = 2(3^x - 1)$ in that example. The scales would be different and it would show only points with whole number values of x.

Similarly, for the sequence in *Example 2b*, the formula for the sum of the first n terms is $S = -12(0.5^n - 1)$. Visualize how the graph of this function coincides with the graph of $y = -12(0.5^x - 1)$. We translate the graph of $y = 0.5^x$ down 1 unit, reflect it in the x-axis, then expand it vertically by a factor of 12.

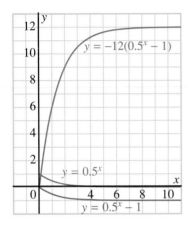
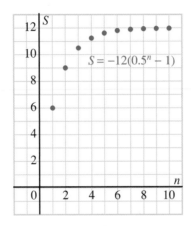

The values of S come closer and closer to 12. If we were to consider more terms of this series, the sum would come closer still to 12. Visualize the series in *Example 2b* continuing forever. We say that the *sum to infinity* is 12, and we write:

$$S_\infty = 6 + 3 + 1.5 + 0.75 + \ldots$$
$$= 12$$

This means that if we take enough terms, we can make the sum as close to 12 as we want. However, the sum never becomes 12. You will learn more about series similar to this in Section 2.9.

1. a) Why can the formula $S_n = \dfrac{a(r^n - 1)}{r - 1}$ not be used to find the sum of a geometric series with $r = 1$?

   b) When $r = 1$, what is the formula for the sum? Explain.

2. At the bottom of page 123, we stated that the values of S come closer and closer to 12. How can we tell this:

   a) from the graphs?   b) from the equations?

3. What other ways can we use the functions toolkit from Chapter 1 to visualize how the graph of $S = -12(0.5^n - 1)$ is related to the graph of $y = 0.5^x$?

4. Explain how the solution to *Example 2b* would change for this series.
   $6 - 3 + 1.5 - 0.75 + \ldots$

2.8 EXERCISES

A 1. Use the method of *Example 1* to determine the sum of each geometric series.

   a) $1 + 2 + 4 + 8 + 16 + 32$ b) $3 + 9 + 27 + 81 + 243 + 729$

   c) $2 + 8 + 32 + 128 + 512$ d) $40 + 20 + 10 + 5 + 2.5$

2. Choose one series from exercise 1. Write to explain how you determined the sum.

3. Use the formula for S_n to determine the sum of the first 5 terms of each geometric series.

   a) $2 + 10 + 50 + \ldots$ b) $4 + 12 + 36 + \ldots$

   c) $3 + 6 + 12 + \ldots$ d) $24 + 12 + 6 + \ldots$

   e) $5 + 15 + 45 + \ldots$ f) $80 - 40 + 20 - \ldots$

B 4. Consider the geometric series $4 + 12 + 36 + 108 + \ldots$.

   a) Determine the 10th term.   b) Determine the sum of the first 10 terms.

5. Repeat exercise 4 for the series $4 - 12 + 36 - 108 + \ldots$.

6. Determine the sum of the first 10 terms of each geometric series.

   a) $5 + 10 + 20 + 40 + \ldots$ b) $5 - 10 + 20 - 40 + \ldots$

   c) $1 + \dfrac{1}{3} + \dfrac{1}{9} + \dfrac{1}{27} + \ldots$ d) $1 - \dfrac{1}{3} + \dfrac{1}{9} - \dfrac{1}{27} + \ldots$

   e) $5 + \dfrac{5}{2} + \dfrac{5}{4} + \dfrac{5}{8} + \ldots$ f) $5 - \dfrac{5}{2} + \dfrac{5}{4} - \dfrac{5}{8} + \ldots$

7. A doctor prescribes 200 mg of medication on the first day of treatment. The dosage is halved on each successive day. The medication lasts for seven days. To the nearest milligram, what is the total amount of medication administered?

8. Sixty-four players enter a tennis tournament. When a player loses a match, the player drops out; the winners go on to the next round.

   a) Find as many different methods as you can to determine the total number of matches to be played until the champion is declared.

   b) Choose one method in part a. Write to explain how this method solves the problem.

9. A contest winner is given a choice of two prizes:
   Prize 1 We will give you $1 today, $2 a year from now, $4 two years from now, and so on, for 20 years. Each year you will receive twice as much as the year before.
   Prize 2 We will give you $1 000 000 today.

   a) Calculate the total value of the money the winner receives if she chooses Prize 1.

   b) Which prize would you choose? Explain.

10. Here are 3 levels in a school trip telephoning tree.

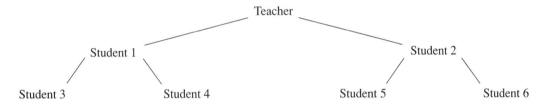

   a) At what level are 64 students contacted?

   b) How many students are contacted at the 8th level?

   c) By the 8th level, how many students in total have been contacted?

   d) By the nth level, how many students in total have been contacted?

   e) Suppose there are 300 students in total. By what level will all students have been contacted?

11. Suppose your class has a class set of programmable graphing calculators, such as the TI-83. Suppose one calculator has a program stored in memory, and this program is to be downloaded to the rest of the calculators. Write a set of instructions for a group of students to download the program to the rest of the calculators. State clearly how many students are needed to download, and include an estimate of the total time it would take to download the programs. Make any reasonable assumptions.

12. Calculators such as the TI-92 and computer programs such as *Theorist* and *Maple* can perform algebraic computation. This screen shows the results of some polynomial multiplications performed by the TI-92.

```
F1▼     F2▼    F3▼   F4▼    F5      F6
 ┌─ Algebra Calc Other PrgmIO Clear  a-z...
■ expand((r − 1)·(r + 1))                    r² − 1
■ expand((r − 1)·(r² + r + 1))               r³ − 1
■ expand((r − 1)·(r³ + r² + r + 1))          r⁴ − 1
■ expand((r − 1)·(r⁴ + r³ + r² + r + 1))
                                             r⁵ − 1
MAIN          RAD EXACT          FUNC 4/30
```

a) Predict the expansion of $(r − 1)(r^5 + r^4 + r^3 + r^2 + r + 1)$.

b) Explain why all the products have the form $r^n − 1$.

c) Use the information in the screen. Write an expression for the sum of each geometric series.

   i) $1 + r + r^2$ ii) $1 + r + r^2 + r^3$ iii) $1 + r + r^2 + r^3 + r^4$

d) Use the patterns in part c. Predict the formula for the sum of the series $1 + r + r^2 + \ldots + r^{n-1}$.

e) Explain how the above results could be used to predict the formula for the sum of the series $a + ar + ar^2 + \ldots + ar^{n-1}$.

C 13. Let $S_n = 1 + \frac{1}{2} + \frac{1}{4} + \frac{1}{8} + \ldots + \frac{1}{2^{n-1}}$.

a) Prove algebraically that S_n is less than 2 for all values of n.

b) Explain why S_n becomes closer and closer to 2 as n becomes larger and larger.

c) Draw a diagram to illustrate why $S_n < 2$.

14. Let $S_n = 1 + \frac{1}{3} + \frac{1}{9} + \frac{1}{27} + \ldots + \frac{1}{3^{n-1}}$.

a) Prove algebraically that S_n is less than 1.5 for all values of n.

b) Explain why S_n becomes closer and closer to 1.5 as n becomes larger and larger.

c) Draw a diagram to illustrate why $S_n < 1.5$.

COMMUNICATING THE IDEAS

Write to explain how geometric series and exponential functions are related. Use examples to illustrate your explanation.

2.9 Infinite Geometric Series

Consider the two geometric series below. The first term in each series is 3 and the common ratios are 4 and $\frac{1}{4}$, respectively.

$$3 + 12 + 48 + 192 + \ldots \quad \textcircled{1} \qquad 3 + \frac{3}{4} + \frac{3}{16} + \frac{3}{64} + \ldots \quad \textcircled{2}$$

Suppose we calculate the sum of n terms of each series for a few values of n.

For series ①:

$S_1 = 3$

$S_2 = 3 + 12 = 15$

$S_3 = 3 + 12 + 48 = 63$

$S_4 = 3 + 12 + 48 + 192 = 255$

$S_5 = 3 + 12 + 48 + 192 + 768 = 1023$

For series ②:

$S_1 = 3$

$S_2 = 3 + \frac{3}{4} = 3.75$

$S_3 = 3 + \frac{3}{4} + \frac{3}{16} = 3.9375$

$S_4 = 3 + \frac{3}{4} + \frac{3}{16} + \frac{3}{64} = 3.984\,375$

$S_5 = 3 + \frac{3}{4} + \frac{3}{16} + \frac{3}{64} + \frac{3}{256} = 3.996\,093\,75$

The values of S_n for series ① become larger and larger quickly, while the values of S_n for series ② come closer and closer to 4. When we consider more terms of this series, the sum would come closer still to 4. To see why, consider the expression for S_n for this series.

$$S_n = \frac{a(r^n - 1)}{r - 1}$$

$$= \frac{3\left(\left(\frac{1}{4}\right)^n - 1\right)}{\frac{1}{4} - 1} \times \frac{-1}{-1}$$

$$= \frac{3\left(1 - \left(\frac{1}{4}\right)^n\right)}{1 - \frac{1}{4}}$$

$$= 4\left(1 - \left(\frac{1}{4}\right)^n\right)$$

As n gets larger, $\left(\frac{1}{4}\right)^n$ gets closer and closer to 0, but never equals 0. Hence, S_n gets closer and closer to 4. We say that the *sum to infinity* of the series is 4. This means that if we take enough terms, we can make the sum as close to 4 as we please. However, the sum never becomes 4.

Example 1

Consider the infinite geometric series $4 - \frac{4}{5} + \frac{4}{25} - \ldots$.

a) Explain why the series has a sum to infinity.

b) Determine the sum to infinity.

Solution

a) For this series, $a = 4$ and $r = -\frac{1}{5}$

$$S_n = \frac{a(r^n - 1)}{r - 1}$$

$$= \frac{4\left(\left(-\frac{1}{5}\right)^n - 1\right)}{-\frac{1}{5} - 1} \times \frac{-1}{-1}$$

$$= \frac{4\left(1 - \left(-\frac{1}{5}\right)^n\right)}{\frac{6}{5}}$$

$$= \frac{10}{3}\left(1 - \left(-\frac{1}{5}\right)^n\right)$$

The series has a sum to infinity because the expression $\left(-\frac{1}{5}\right)^n$ gets closer to 0 as n gets larger.

b) The sum to infinity is $\frac{10}{3}$.

We apply similar reasoning to the general infinite geometric series with first term a and common ratio r.

$$a + ar + ar^2 + ar^3 + \ldots + ar^{n-1} + \ldots$$

For this series,

$$S_n = \frac{a(r^n - 1)}{r - 1} \quad (r \neq 1)$$

$$= \frac{a(r^n - 1)}{r - 1} \times \frac{-1}{-1}$$

$$= \frac{a(1 - r^n)}{1 - r}$$

$$= \frac{a}{1 - r}(1 - r^n)$$

If $|r| < 1$, then r^n becomes smaller and smaller as n increases. We can make r^n as small as we wish by making n large enough. Hence, if $|r| < 1$, the sum of the infinite geometric series is:

$$S = \frac{a}{1 - r}(1 - 0)$$

$$= \frac{a}{1 - r}$$

The sum of the infinite geometric series $a + ar + ar^2 + \ldots + ar^{n-1} + \ldots$

is $S = \frac{a}{1-r}$, provided that $|r| < 1$.

Example 2

Which infinite geometric series has a sum? What is the sum?

a) $4 - 6 + 9 - 13.5 + \ldots$ 　　　　　 b) $6 + 2 + \frac{2}{3} + \frac{2}{9} + \ldots$

Solution

a) $4 - 6 + 9 - 13.5 + \ldots$
For this series, $r = -1.5$; since $|r| > 1$, the series has no sum.

b) $6 + 2 + \frac{2}{3} + \frac{2}{9} + \ldots$
For this series, $r = \frac{1}{3}$; since $|r| < 1$, the series has a sum. Use the formula.

$$S = \frac{a}{1-r}$$
$$= \frac{6}{1 - \frac{1}{3}}$$
$$= \frac{6}{\frac{2}{3}}$$
$$= 9$$

The sum of the series is 9.

In *Example 2b*, the terms of the series become smaller and smaller, and the series has a sum to infinity. Every geometric series whose terms become smaller and smaller has a sum to infinity. However, there are other kinds of series whose terms become smaller and smaller that do not have sums to infinity.

DISCUSSING THE IDEAS

1. On page 128, and in the solution to *Example 1*, did we have to multiply the expression for S_n by $\frac{-1}{-1}$? Explain.

2. In the solution to *Example 2b*, can we be certain that the sum is exactly 9? Explain.

3. The condition that the infinite geometric series $a + ar + ar^2 + \ldots + ar^{n-1} + \ldots$ has a sum is $|r| < 1$. Explain why this condition depends only on the value of r and not on the value of a.

2.9 EXERCISES

A 1. Does every infinite geometric series have a sum? Explain.

2. Which infinite geometric series have a sum? What is the sum?
   a) $8 + 4 + 2 + 1 + \ldots$
   b) $27 + 18 + 12 + 8 + \ldots$
   c) $20 - 15 + 11.25 - 8.4375 + \ldots$
   d) $50 - 40 + 32 - 25.6 + \ldots$
   e) $2 + 6 + 18 + 54 + \ldots$
   f) $-16 + 12 - 9 + 6.75 - \ldots$

B 3. Determine the sum of each infinite geometric series.
   a) $8 + 2 + \frac{1}{2} + \frac{1}{8} + \ldots$
   b) $8 - 2 + \frac{1}{2} - \frac{1}{8} + \ldots$
   c) $10 + 5 + 2.5 + 1.25 + \ldots$
   d) $10 - 5 + 2.5 - 1.25 + \ldots$
   e) $5 + \frac{5}{3} + \frac{5}{9} + \frac{5}{27} + \ldots$
   f) $5 - \frac{5}{3} + \frac{5}{9} - \frac{5}{27} + \ldots$
   g) $60 + 30 + 15 + 7.5 + \ldots$
   h) $5 + 2.5 + 1.25 + 0.625 + \ldots$

4. a) Determine the sum of the series $12 - 6 + 3 - 1.5 + \ldots$

   b) Determine the difference between the sum of the series in part a and the sum of the first 8 terms of the series.

5. An oil well produces 25 000 barrels of oil during its first month of production. Suppose its production drops by 5% each month.

   a) Estimate the total production before the well runs dry.

   b) Write to explain how you estimated the total production in part a. Identify any assumptions you made to arrive at your estimate.

   c) Explain how the assumptions you made affect the estimate.

MODELLING the Depletion of an Oil Well

In exercise 5, the total amount of oil from an oil well is modelled by a geometric series.

• How many months would it take for the monthly production to drop by 50% of its original level? Do you think this is reasonable? Explain.

• Some Leduc oil wells in Alberta lasted 25 years or more. Explain why the oil well in exercise 5 could not last that long.

• Repeat exercise 5a, replacing 5% with a smaller percent. Estimate a percent drop in production that might cause the oil well to last at least 25 years. Write to explain why your estimate is reasonable.

6. A ball is dropped from a height of 2.0 m to a floor. After each bounce, the ball rises to 63% of its previous height.

a) What is the total vertical distance the ball has travelled after 5 bounces?

b) Estimate the total vertical distance the ball travels before it comes to rest.

c) Write to explain how you estimated the distance in part b. Identify any assumptions you made to arrive at your estimate.

d) Explain how your assumptions affect the estimate.

7. Refer to exercise 13, page 118. Prove that the sum of the areas of the triangles removed is equal to the area of the original triangle.

C 8. When mathematicians devised the *snowflake curve* about 100 years ago, they were surprised to learn that it has an unusual property. You can discover this property. The first four steps in the construction of the snowflake curve are shown, starting with an equilateral triangle with sides 24 mm long. Visualize these steps continuing forever. After each step, the figure comes closer and closer to the snowflake curve.

a) Determine, if possible, the perimeter of:
   i) each figure below
   ii) the snowflake curve that results from continuing the steps forever

b) Determine, if possible, the area of:
   i) each figure below
   ii) the snowflake curve that results from continuing the steps forever

c) What unusual property does the snowflake have? Explain.

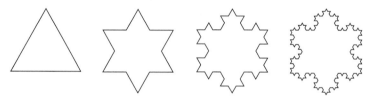

9. Refer to exercise 12, page 118. Assume the construction of the perpendiculars continues indefinitely.

a) Estimate the total length of each set of segments.
   i) the segments AB, BC, CD, DE, …
   ii) the vertical segments AB, CD, EF, …
   iii) the inclined segments BC, DE, FG, …

b) Estimate the total lengths of the segments you identified in exercise 12b.

COMMUNICATING THE IDEAS

Write to explain what it means when we say that an infinite geometric series has a sum even though it is not possible to add the terms. Use some examples to illustrate your explanation.

Sigma Notation

There is a special notation that is used to represent a series. For example, the geometric series $3 + 6 + 12 + 24 + 48 + 96$ has 6 terms, with first term 3 and common ratio 2. The general term is $t_n = 3(2)^{n-1}$.

Each term in the series can be expressed in this form.

$$t_1 = 3(2)^{1-1} \quad t_2 = 3(2)^{2-1} \quad t_3 = 3(2)^{3-1}$$
$$t_4 = 3(2)^{4-1} \quad t_5 = 3(2)^{5-1} \quad t_6 = 3(2)^{6-1}$$

The series is the sum of all these terms, and is represented as shown.

The sum of ... $\longrightarrow \displaystyle\sum_{k=1}^{6} 3(2)^{k-1} \longleftarrow$ ... all numbers of the form $3(2)^{k-1}$ ...

... for integral values of k from 1 to 6.

The symbol Σ is the capital Greek letter sigma, which corresponds to S, the first letter of the word "sum." When Σ is used as shown above, it is called *sigma notation*. In sigma notation, k is frequently used as the variable under the Σ sign and in the expression following it. Any letter can be used, as long as it is not used elsewhere.

1. Write the series corresponding to each expression.

   a) $\displaystyle\sum_{k=1}^{6} (k + 4)$ b) $\displaystyle\sum_{k=1}^{5} 2^k$ c) $\displaystyle\sum_{k=1}^{5} (-2)^k$ d) $\displaystyle\sum_{s=1}^{6} 5(2)^{s-1}$

2. Write each series using sigma notation.

   a) $1 + \frac{1}{2} + \frac{1}{4} + \frac{1}{8} + \frac{1}{16} + \frac{1}{32}$ b) $2 - 6 + 18 - 54 + 162 - 486 + 1458$

3. Determine the sum of each series.

   a) $\displaystyle\sum_{k=1}^{6} 3(2)^{k-1}$ b) $\displaystyle\sum_{k=1}^{6} 3(-2)^{k-1}$ c) $\displaystyle\sum_{k=1}^{6} 2(3)^{k-1}$ d) $\displaystyle\sum_{k=1}^{6} 2(-3)^{k-1}$

4. Determine each sum.

   a) $\displaystyle\sum_{k=1}^{\infty} \left(\frac{1}{10^k} \right)$ b) $\displaystyle\sum_{k=1}^{\infty} \left(\frac{3}{10^k} \right)$ c) $\displaystyle\sum_{k=1}^{\infty} \left(\frac{6}{10^k} \right)$ d) $\displaystyle\sum_{k=1}^{\infty} \left(\frac{9}{10^k} \right)$

5. Determine the sum of the series $\displaystyle\sum_{i=1}^{n} (-1)^i$ in each case.

   a) n is even.   b) n is odd.

6. $\prod$, the capital Greek letter P, is the first letter of the word "product." Create some examples to show what pi notation could mean.

2.10 Logarithmic Functions

Many examples of exponential equations were given in earlier sections of this chapter. We can use logarithms to solve any of these equations for the variable in the exponent. Solving an equation is one of the steps to determine the equation of the inverse of a function. For example, consider the function $f(x) = 2^x$ on page 104. To find the equation of the inverse, let $y = 2^x$ and follow these steps.

Step 1.   Solve for *x*. $x = \log_2 y$

Step 2.   Interchange *x* and *y*. $y = \log_2 x$

The equation of the inverse of $y = 2^x$ is $y = \log_2 x$.

We can graph the inverse of any function by reflecting its graph in the line $y = x$. This is equivalent to interchanging the coordinates of the points on the graph. For example, the graph below shows the function $f(x) = 2^x$ and its inverse. Since the graph of the inverse satisfies the vertical line test, the inverse is a function, and we can write its equation as $f^{-1}(x) = \log_2 x$. A table of values for this function is shown.

x	y
$\frac{1}{8}$	−3
$\frac{1}{4}$	−2
$\frac{1}{2}$	−1
1	0
2	1
4	2
8	3

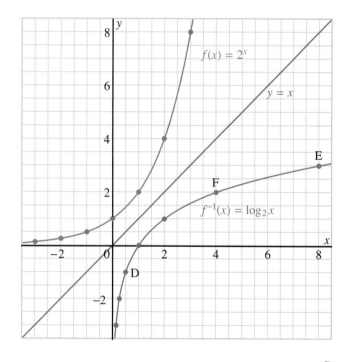

The graph illustrates these properties of the function $f^{-1}(x) = \log_2 x$.

Vertical intercept

Let $x = 0$.

$f^{-1}(0) = \log_2 0$

Let $\log_2 0 = y$, or $2^y = 0$.

There is no value of y such that $2^y = 0$.

Hence, $\log_2 0$ is undefined.

There is no vertical intercept.

Horizontal intercept

Let $f^{-1}(x) = 0$.

Then $\log_2 x = 0$, or $2^0 = x$

Hence, $x = 1$.

The horizontal intercept is 1.

Domain and range

Since we can define $\log_2 x$ for all positive values of x, the domain is the set of positive real numbers.

The range is the set of real numbers.

Asymptote

When y decreases through negative values, the points on the graph come closer and closer to the y-axis, but never reach it. The y-axis is an asymptote.

Law of Logarithms

Select any two points on the curve, such as $D(\frac{1}{2}, -1)$ and $E(8, 3)$.

Multiply their x-coordinates.

$\frac{1}{2} \times 8 = 4$

Add their y-coordinates.

$-1 + 3 = 2$

The results are the coordinates of another point $F(4, 2)$ on the graph.

Multiplying the x-coordinates and adding the y-coordinates of two points on the graph produce the coordinates of another point on the graph.

Example 1

a) Sketch a graph to show the exponential function $f(x) = \left(\frac{1}{3}\right)^x$ and its inverse on the same grid.

b) Write the equation of the inverse function.

Solution

a) When x is very large and positive, $f(x)$ is very small and positive.

When x is negative and has a large absolute value, $f(x)$ is very large.

Also, $f(0) = 1$.

Reflect the graph of $y = \left(\frac{1}{3}\right)^x$ in the line $y = x$.

The image is the graph of the inverse.

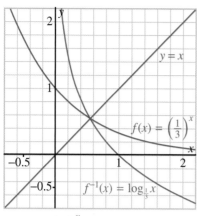

b) Let $y = \left(\frac{1}{3}\right)^x$. Solve for x to obtain $x = \log_{\frac{1}{3}} y$. Interchange x and y to

obtain $y = \log_{\frac{1}{3}} x$. The equation of the inverse function is $f^{-1}(x) = \log_{\frac{1}{3}} x$.

The graphs in the above examples illustrate properties of the graph of the logarithmic function $f(x) = \log_b x$.

> ### *Properties of the Graph of the Logarithmic Function*
> ### $f(x) = \log_b x \ (b > 0, \ b \neq 1)$
>
> When $b > 1$, the graph goes up to the right.
> When $0 < b < 1$, the graph goes down to the right.
>
> Vertical intercept: none
> Horizontal intercept: 1
>
> Domain: all positive real numbers
> Range: all real numbers
>
> Asymptote: y-axis
>
>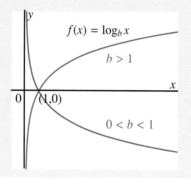

Logarithmic functions can arise in practical situations. In some problems, an approximate equation can be obtained from a curve of best fit using empirical data.

Example 2

To determine the number of plant species in a lawn, the numbers of species in square patches with different areas were recorded. The results showed that the number of plant species, s, contained in a patch of lawn with area a square metres was modelled by the equation $s \doteq 10.7 \log a + 8.1$.

a) Graph the equation.

b) Estimate the number of plant species in a patch with area 100 m^2.

c) Estimate the area of lawn that contains 20 plant species.

Solution

a) Graph the equation using appropriate window settings.

b) Substitute 100 for a in the equation.

$s = 10.7 \log 100 + 8.1$

$\quad = 10.7 \times 2 + 8.1$

$\quad = 29.5$

There are approximately 30 plant species in a patch with area 100 m^2.

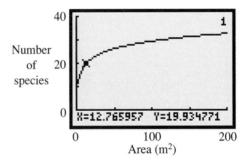

c) Trace to the point where the y-coordinate is closest to 20. According to the model, a patch with area approximately 13 m^2 contains 20 species.

We can use the stretching tools from Chapter 1 to visualize how the graph in *Example 2* is related to the graph of $y = \log x$. Visualize how the graph of $s = 10.7 \log a + 8.1$ coincides with a vertical expansion of the graph of $y = \log x$ by a factor of 10.7, followed by a vertical translation of 8.1 units.

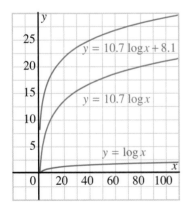

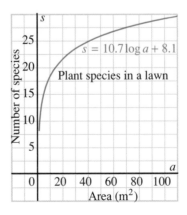

A logarithmic function can be obtained from any exponential function given earlier in this chapter.

Example 3

On page 66, the exponential function $P = 2.76(1.022)^n$ models the growth of British Columbia's population since 1981.

a) Solve the equation for n to express n as a function of P.

b) Graph the function in part a.

Solution

a) $P = 2.76(1.022)^n$

Take the base-10 logarithm of each side.

$$\log P = \log 2.76 + n \log 1.022$$

$$n \log 1.022 = \log P - \log 2.76$$

$$= \log\left(\frac{P}{2.76}\right)$$

$$n = \frac{1}{\log 1.022} \times \log\left(\frac{P}{2.76}\right)$$

$$n \doteq 106 \log\left(\frac{P}{2.76}\right)$$

b) Enter the equation
$Y1=106\log(X/2.76)$.

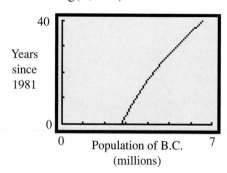

Years since 1981

Population of B.C. (millions)

Compare the results of *Example 3* with the function $P = 2.76(1.022)^n$ and its graph on page 66. The function $n \doteq 106 \log\left(\frac{P}{2.76}\right)$ presents the same information in a different form.

DISCUSSING THE IDEAS

1. Compare the properties of the exponential and logarithmic functions on pages 105 and 135. Explain why $b \neq 1$ in the properties of the logarithmic function, but not in the properties of the exponential function.

2. Recall exercise 17, page 112.

   a) Explain how the equation that expresses n as a function of P in *Example 3* could be used to complete this exercise.

   b) Explain why the equation in *Example 3* is useful.

3. Use the tools from the functions toolkit in Chapter 1. Explain how the graph in *Example 3* is related to the graph of $y = \log x$.

A 1. Write the equation of the inverse of each exponential function.

a) $y = 10^x$ b) $y = 3^x$ c) $f(x) = 7^x$

d) $f(x) = 0.4^x$ e) $g(x) = \left(\frac{3}{2}\right)^x$ f) $h(x) = 15^x$

2. Write the equation of the inverse of each logarithmic function.

a) $y = \log x$ b) $y = \log_2 x$ c) $f(x) = \log_6 x$

3. a) Sketch a sequence of graphs to illustrate how the graph of $y = \log_a x$ changes as a increases through all positive values, $a \ne 1$.

b) Visualize how the properties on page 135 apply to each graph in part a.

B 4. a) Graph $y = 10^x$ and $y = 2^x$ on the same axes.

b) Use your graph in part a. Sketch the graphs of $y = \log_{10} x$ and $y = \log_2 x$ on another set of axes.

c) Sketch the graph of $y = \log_5 x$ on the axes in part b.

5. For each list, predict what the graphs of the equations would look like. Check your predictions.

a) $y = \log x$
   $y = 2\log x$
   $y = \log 2x$

b) $y = \log x$
   $y = \frac{1}{2}\log x$
   $y = \log \frac{1}{2}x$

c) $y = \log x$
   $y = \log(x - 2)$
   $y = \log(x + 2)$

6. For each list, sketch the graphs on the same grid.

a) $y = \log x$
   $y = 2\log x$
   $y = 2\log(x + 3)$

b) $y = \log x$
   $y = \log 2x$
   $y = \log 2(x + 3)$

c) $y = \log x$
   $y = \log(x + 3)$
   $y = \log(2x + 3)$

7. Two researchers measured people's average walking speed in communities that ranged in size from about 350 people to over 2 million. They found that the larger the community, the faster people walk, and the differences in speed can be quite large. From the data, the walking speed, v metres per second, in a community with population, p, is modelled by the function $v = 0.26\log p + 0.028$.

a) Graph v against p for appropriate values of p.

b) What walking speed does the model predict for a community with each population?

i) 350 ii) 50 000 iii) 2 million

c) Use the model to predict the size of a community in which the walking speed exceeds 1 m/s.

MODELLING the Pace of Life

Most people would agree that the pace of life in large cities is faster than in villages. The logarithmic function in exercise 7 can be considered a model of the pace of life.

- Do you think average walking speed is a reasonable way to measure the pace of life?
- In what other ways could you measure the pace of life?

8. Psychologists perform experiments to determine how quickly people forget. In one experiment, lists of words were read to people to determine how many words these people could remember at a later time. The result was scored as a percent, called the retention score. The retention score, S, was expressed as a function of time, t minutes, by the equation $S = -18 \log t + 84$.

   a) Graph S against t for appropriate values of t.

   b) What was the retention score after each time?
      i) 30 min   ii) 1 h   iii) 24 h   iv) 1 week

   c) After how many minutes was the score 50%?

9. Choose either the equation in exercise 7 or the equation in exercise 8. Write to explain how the graph of the equation is related to the graph of $y = \log x$.

10. On page 66, the amount, A dollars, of an investment of $1000 growing at 8% compounded annually is expressed as a function of the number of years, n, by the equation $A = 1000(1.08)^n$.

    a) Solve this equation for n, thus expressing n as a logarithmic function of A.

    b) Calculate the value of n for each value of A and explain the result.
       i) $1250   ii) $350

    c) Graph the function in part a for $0 < A \leq 1250$. Compare your graph with the graph on page 66.

    d) State the domain and the range of the function.

11. The population of Vancouver in 1996 was 1 832 000. According to data in the *1999 Canadian Global Almanac*, the average annual growth rate during the previous 40 years was 2.5%.

    a) Write an equation to express the population, P, as a function of n, the number of years since 1996.

    b) Solve this equation for n.

    c) Determine the value of n for each value of P.
       i) 2 000 000   ii) 3 000 000

    d) Graph the functions in parts a and b. How are these functions related?

12. a) Graph the function $f(x) = \log x$ for $0 < x \leq 10$.

b) On the grid in part a, sketch the graph of each function.

    i) $y = f(x) - 1$ **ii)** $y = f(x - 1)$ **iii)** $y = f(x + 1)$

    iv) $y = f(0.5x)$ **v)** $y = f(2x)$ **vi)** $y = f(-x)$

13. Consider the function $y = 4 \log 3(x - 2)$.

    a) Graph the function.

    b) Identify the domain and the range of the function.

    c) Write the equations of any asymptotes.

    d) Determine the intercepts of the function.

14. Repeat exercise 13 for each function.

    a) $y = 5 \log 4(x - 3)$ **b)** $y = -3 \log (2x - 5)$

    c) $f(x) = 2 \log 3(x - 2)$ **d)** $f(x) = 8 \log (3x + 2)$

15. Solve each equation for x, thus expressing x as a logarithmic function of y.

    a) $y = 5 \times 2^x$ **b)** $y = 1.3 \times 10^x$ **c)** $y = 8.2 \times 1.03^x$

    d) $y = 6.4\left(\frac{1}{2}\right)^x$ **e)** $y = 3.5(2.7)^x$ **f)** $y = 2.75\left(\frac{2}{3}\right)^x$

16. Given the exponential function $f(x) = 3^x$, graph $y = f(x)$ and $y = f^{-1}(x)$ on the same grid.

17. Graph each function and its inverse on the same grid.

    a) $f(x) = 2^x$ **b)** $g(x) = \left(\frac{2}{3}\right)^x$ **c)** $y = \log_3 x$

Ⓒ 18. Let (x_1, y_1) and (x_2, y_2) be any two points on the graph of $y = \log_2 x$.

    a) Write the coordinates of points corresponding to those in exercise 18, page 112, that should be on the graph of $y = \log_2 x$.

    b) Choose one point in part a. Prove that this point lies on the graph of $y = \log_2 x$.

19. In *Example 1*, the graphs of $y = \left(\frac{1}{3}\right)^x$, $y = \log_{\frac{1}{3}} x$, and $y = x$ are shown.

Determine the coordinates of their point of intersection.

COMMUNICATING THE IDEAS

Your friend missed today's mathematics lesson and phones to ask you about it. How would you explain, over the telephone, what logarithmic functions are and why they are useful?

Expressing Exponential Functions Using the Same Base

In this chapter, you encountered many examples of exponential functions that arise in real situations. Recall two important features of these functions.

- The equations usually had the form $y = Ab^x$, and the base b was different in each situation.

- The graph of an exponential function with one base can be made to coincide with the graph of an exponential function with a different base using an appropriate horizontal expansion or compression (see *Visualizing*, page 108).

Mathematicians find it is often useful to express the equations of different exponential functions using the same base. Any base can be used, but there is an advantage to using one particular number for the base. This number is represented by the letter e. It is so useful for a base that your calculator has $\boxed{e^x}$ and $\boxed{\text{LN}}$ keys to evaluate powers of e and logarithms to base e. These logarithms are *natural logarithms*.

1. a) Use the $\boxed{e^x}$ key to determine e^1. The result is the value of e.

   b) Determine each value.

   i) e^2 ii) e^3 iii) e^0 iv) e^{-1} v) e^{-2}

2. Use the $\boxed{\text{LN}}$ key. Determine each value and explain its meaning.

   a) $\ln 1$ b) $\ln 2$ c) $\ln e$ d) $\ln 10$ e) $\ln 0.5$

3. Solve each equation. Explain the results.

   a) $e^x = 1$ b) $e^x = 2$ c) $e^x = e$ d) $e^x = 10$ e) $e^x = 0.5$

4. a) Graph the functions $y = e^x$ and $y = \ln x$. Trace to confirm the results of exercises 1, 2, and 3.

   b) How is the graph of $y = \ln x$ related to the graph of $y = e^x$? Explain.

By horizontally expanding or compressing the graph of $y = e^x$, we can make it coincide with the graph of any other exponential function. For example, we can expand it to coincide with the graph of $y = 2^x$. To find the equation of the image, express 2 as a power of e. That is, let $2 = e^k$. The solution of this equation is $\ln 2$, or about 0.693. Hence, the graph of $y \doteq e^{0.693x}$ coincides with the graph of $y = 2^x$. Similarly, the graph of $y \doteq e^{2.303x}$ coincides with the graph of $y = 10^x$.

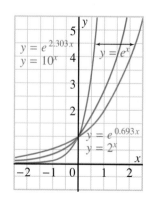

From page 66, recall the equation $P = 2.76(1.022)^n$ that models the growth of the population of British Columbia since 1981. We can express this equation in terms of base e instead of base 1.022. To do this, let $1.022 = e^k$, where k is to be determined.

5. a) Solve the equation $1.022 = e^k$ for k. Use the result to write the equation $P = 2.76(1.022)^n$ in the form $P = 2.76e^{kn}$, where k is a constant.

   b) Use both $P = 2.76(1.022)^n$ and the equation you determined in part a. Substitute 30 for n to predict British Columbia's population in 2011. Are the results the same?

   c) Graph the equation $P = 2.76(1.022)^n$ and the equation you obtained in part a on the same axes. Are the two graphs the same?

In exercise 5a, you should have obtained the equation $P \doteq 2.76e^{0.022n}$. The stretching tools from the functions toolkit in Chapter 1 show how the graph of this function is related to the graph of $y = e^x$. Visualize expanding the graph of $y = e^x$ horizontally until it becomes $y = e^{0.022x}$, then expanding it vertically.

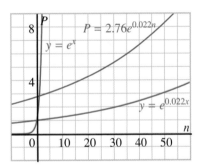
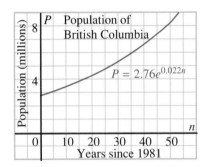

6. Compare the graphs above with the graphs on page 106. In what ways are they similar? In what ways are they different?

In exercise 5a, you obtained this equation.

$$P = 2.76\, e^{0.022n}$$

Initial value   Growth rate

The constant in the exponent is 0.022, which is the growth rate, 2.2%. This is the advantage of using base e. When an exponential function is expressed with base e, the constant in the exponent is the growth rate. Hence, e is the natural base to use in problems involving exponential growth and decay. This is the reason why base-e logarithms are called natural logarithms.

There is another advantage to using base e. The value of k in exercise 5a was not exactly 0.022. To 6 decimal places, it was 0.021 761. This slight discrepancy is caused by the way e is defined in higher mathematics. The definition assumes the population grows continuously, and all the "new" people are not added at once at the end of each year. In this case, the growth rate is *instantaneous*. In the example, the instantaneous rate of growth is 0.021 761, while the annual rate is 0.022. In some applications, the difference may not be significant. Since a rigorous development of instantaneous rates of growth requires calculus, we will ignore its effect.

7. Liberia has one of the world's fastest growing populations. In 1995, its population was 2.5 million and it is increasing at 3.3% annually.

   a) Express the population of Liberia as a base-e exponential function of the number of years, n, since 1995.

   b) At this rate, in how many years will the population be 10 million?

   c) What factors could contribute to the breakdown of this model?

8. Recall the equation $P = 100(0.87)^n$ from page 67 that models the percent of caffeine in your bloodstream n hours after consumption.

   a) Write this equation using base e.

   b) Graph the equation in part a. Compare the graph with that on page 66.

   c) Use the stretching tools. Explain how the graph of the equation in part a is related to the graph of $y = e^x$.

9. The altitude of an aircraft can be determined by measuring the air pressure. In the stratosphere (between 12 000 m and 30 000 m), the pressure, P kilopascals, is expressed as a function of the altitude, h metres, by the equation $P = 130e^{-0.000\ 155h}$.

   a) What is the pressure at an altitude of 20 000 m?

   b) What is the altitude when the pressure is 8.5 kPa?

   c) Derive an equation to express the altitude as a function of the pressure.

10. A rule of thumb to estimate the time for an investment to double is to divide 70 by the interest rate. For example, an investment at 8% will double in approximately $\frac{70}{8}$, or 9, years. Use an exponential function with base e to explain why the rule of thumb works.

Evaluating Exponential Functions

When you press the keys [10^x], [y^x], or [e^x], your calculator uses methods developed 300 years ago by mathematicians who required the values of exponential functions to solve applied problems. The problem of computing the values of exponential functions was an important mathematical problem in the 17th and 18th centuries. The key to solving this problem was to make a connection between the graphs of exponential functions and the graphs of polynomial functions.

For example, the graph of $y = e^x$ is shown in each screen below. For values of x very close to 0, the graph of $y = e^x$ appears to coincide:

- with a straight line

- with a parabola

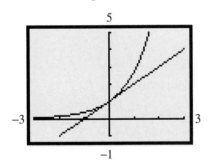

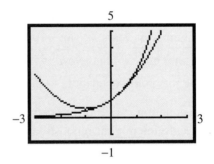

Use a scientific calculator, a graphing calculator, or a spreadsheet.

1. Recall that the equation of a straight line can be written in the form $y = mx + b$, where m is the slope and b the y-intercept.

   a) Use estimation or systematic trial. Determine the equation of the line in the first graph above.

   b) Determine the values of x so that the y-coordinates of points on the line and the graph of $y = e^x$ differ by less than 0.01.

2. Recall that the equation of a parabola can be written in the form $y = ax^2 + bx + c$.

   a) Use estimation or systematic trial. Determine the equation of the parabola in the second graph.

   b) Determine the values of x so that the y-coordinates of points on the parabola and the graph of $y = e^x$ differ by less than 0.01.

3. Use the results of exercises 1 and 2. Describe a way to approximate the value of e^x when x is very close to 0.

In exercises 1 and 2, you should have discovered this result.

For values of x very close to 0, $e^x \doteq 1 + x$

A better approximation is $e^x \doteq 1 + x + \frac{1}{2}x^2$

For improved accuracy, more terms are required on the right side of the equation for e^x. Mathematicians determined these terms in the 17th century. They proved that, under certain conditions, a function $f(x)$ could be expressed using an infinite series similar to the infinite geometric series you studied earlier in this chapter. The series for e^x is given below. The denominators in the coefficients use a special notation called *factorial notation*. The factorial sign ! that follows a natural number means the product of all natural numbers up to and including the number. For example, $4! = 4 \times 3 \times 2 \times 1$, or 24

$$e^x = 1 + x + \frac{1}{2!}x^2 + \frac{1}{3!}x^3 + \frac{1}{4!}x^4 + \frac{1}{5!}x^5 + \frac{1}{6!}x^6 + \frac{1}{7!}x^7 + \ldots$$

4. Simplify each factorial.

a) $2!$ b) $3!$ c) $5!$ d) $6!$ e) $7!$

5. Write the first eight terms of the series for e^x without using the factorial sign.

6. a) Predict what the graphs will look like if you graph $y = e^x$ and these functions on the same grid.

i) $y = 1 + x + \frac{1}{2!}x^2 + \frac{1}{3!}x^3 + \frac{1}{4!}x^4$

ii) $y = 1 + x + \frac{1}{2!}x^2 + \frac{1}{3!}x^3 + \frac{1}{4!}x^4 + \frac{1}{5!}x^5$

iii) $y = 1 + x + \frac{1}{2!}x^2 + \frac{1}{3!}x^3 + \frac{1}{4!}x^4 + \frac{1}{5!}x^5 + \frac{1}{6!}x^6$

b) Graph the functions to check your predictions.

7. Use the series for e^x to determine each value to 3 decimal places. Check using the $\boxed{e^x}$ key.

a) e^0 b) e^1 c) e^{-1} d) e^2 e) e^{-2}

8. By definition, $0! = 1$ and $1! = 1$

a) Write the series for e^x using sigma notation.

b) Suggest why both $0!$ and $1!$ are defined to be equal to 1.

9. Modify the series for e^x to obtain a series for 10^x. Use the $\boxed{10^x}$ key to check some values.

INVESTIGATE Solving Exponential Equations

In this investigation, you will solve the exponential equation $2^{x+1} = 8^{2x}$ in different ways.

1. Solve the equation by taking the base-10 logarithm of each side.

2. Solve the equation by taking the base-2 logarithm of each side.

3. Solve the equation by writing 8 as 2^3 so that both sides of the equation can be written as powers of 2.

4. Solve the equation by graphing $y = 2^{x+1}$ and $y = 8^{2x}$ on the same screen, and finding the coordinates of the point of intersection.

5. Solve the equation by graphing $y = 2^{x+1} - 8^{2x}$ and finding the x-intercept.

6. Investigate how the above five solutions to the equation would change if the equation had been $2^{x+1} = 7^{2x}$.

In this chapter, we have solved exponential equations using technology. Although the roots were determined accurately, they are only approximate roots. In this section, we will develop strategies to solve exponential equations without technology and to determine the roots exactly.

Example 1

Solve for x. $3^{x+1} = 27^{2x}$

Solution

> **Think ...**
> The bases are powers of 3. Write both sides as powers with the same base.

$$3^{x+1} = 27^{2x}$$
$$3^{x+1} = (3^3)^{2x}$$
$$3^{x+1} = 3^{6x}$$

Since the bases are the same, the exponents are equal.

$$x + 1 = 6x$$
$$x = \frac{1}{5}$$

In *Example 1*, we could have solved the equation by taking the logarithm of each side. Compare the two solutions below with the solution above.

Solution 2 (using base-3 logarithms)

$$3^{x+1} = 27^{2x}$$

Take the base-3 logarithm of each side.

$$(x + 1) \log_3 3 = 2x \log_3 27 \qquad ①$$
$$(x + 1)(1) = 2x(3)$$
$$x + 1 = 6x$$
$$x = \frac{1}{5}$$

Solution 3 (using base-10 logarithms)

$$3^{x+1} = 27^{2x}$$

Take the base-10 logarithm of each side.

$$(x + 1) \log 3 = 2x \log 27 \qquad ②$$
$$(x + 1) \log 3 = 2x \log 3^3$$
$$(x + 1) \log 3 = 2x(3 \log 3)$$
$$(x + 1) \log 3 = 6x \log 3$$

Divide each side by $\log 3$.

$$x + 1 = 6x$$
$$x = \frac{1}{5}$$

These solutions are essentially the same, and are also essentially the same as the solution to *Example 1*. Since a logarithm is an exponent, Solutions 2 and 3 are different ways to work with exponents in the original equation.

To see why, consider equation ① in Solution 2. Both sides contain the base-3 logarithm of the corresponding sides in the given equation; that is, the exponents the two sides have when written as powers of 3. Hence, Solution 2 is just another way to solve the equation by writing both sides as powers of 3.

Consider equation ② in Solution 3. Both sides contain the base-10 logarithm of the corresponding sides of the given equation; that is, the exponents the two sides have when written as powers of 10. Hence, Solution 3 is a way to solve the equation by writing both sides as powers of 10.

This analysis applies to any equation. Solving an exponential equation by taking the base-10 logarithm of each side is a convenient way to write both sides of the equation as powers with the same base, 10.

Example 2

Solve in exact form. $2(3)^x = 6^{x+1}$

Solution

> **Think ...**
>
> The bases are not powers of each other. Use base-10 logarithms. This amounts to writing both sides as powers of 10.

$$2(3)^x = 6^{x+1}$$

Take the base-10 logarithm of each side.

$\log 2 + x\log 3 = (x + 1)\log 6$ ——————— Laws of logarithms for multiplication and powers

$\log 2 + x\log 3 = x\log 6 + \log 6$

$x(\log 3 - \log 6) = \log 6 - \log 2$

$x\log \frac{1}{2} = \log 3$ ——————— Law of logarithms for division

$x = \dfrac{\log 3}{\log 0.5}$

In *Example 2*, the expression $\dfrac{\log 3}{\log 0.5}$ represents the exact value that satisfies the equation. It is the root of the equation. We can determine an approximate solution by evaluating the logarithms.

$x = \dfrac{\log 3}{\log 0.5}$ 　　　Press: [LOG] 3 [ )] [÷] [LOG] 0.5 [ )] [ENTER]

$\doteq -1.584\ 962\ 5$

To 3 decimal places, the solution is $x \doteq -1.585$.

DISCUSSING　THE IDEAS

1. Compare the equation in *Example 1* with the equation in *Investigate*. Then compare the 3 solutions to *Example 1* with your solution to exercises 1 to 3 in *Investigate*. Explain the similarities and the differences in the equations and the solutions.

2. In *Example 1* and the follow-up discussion, three solutions were given to the equation $3^{x+1} = 27^{2x}$. Investigate how these solutions would change for the equation $3^{x+1} = 7^{2x}$.

3. Investigate how the solution to *Example 2* would change for each equation.

   a) $2(3)^x = 9^{x+1}$ 　　　　　　　　b) $2(3)^x = 7^{x+1}$

2.11 EXERCISES

A 1. Solve.

a) $2^{x+1} = 4$ b) $2^{x-1} = 8$ c) $3^{x-5} = 9$ d) $5^{x+3} = 25$

e) $10^{x+1} = 1000$ f) $4^{x+2} = 16$ g) $2^{2x+1} = 8$ h) $3^{2-x} = 9$

i) $5^{3x-2} = 25$ j) $9^{x+1} = 9$ k) $9^{1-2x} = 81$ l) $4^{x+2} = 32$

B 2. Consider the equation $5^{3x+2} = 25^{2x}$. Solve the equation in each way.

a) by expressing each side as a power of 5

b) by taking the base-5 logarithm of each side

c) by taking the base-10 logarithm of each side

3. Consider the equation $8^{x+1} = 64^{x-1}$. Solve the equation in each way.

a) by expressing each side as a power of 2

b) by taking the base-2 logarithm of each side

c) by taking the base-10 logarithm of each side

4. Write to explain some other ways to solve the equation in exercise 3.

5. Solve.

a) $9^{x+1} = 3$ b) $9^{x+1} = 27$ c) $9^{x+1} = 81$

d) $3^{3x+1} = 3^{2x}$ e) $3^{3x+1} = 9^{2x}$ f) $3^{3x+1} = 27^{2x}$

6. Solve.

a) $3 \times 2^x = 12$ b) $5 \times 2^x = 40$ c) $10 \times 3^x = 270$

d) $3(6)^x = 108$ e) $4(7)^x = 4$ f) $2(4)^x = 1$

7. Solve and check.

a) $2(18)^x = 6^{x+1}$ b) $2(12)^x = 6^{x+1}$ c) $2(3)^x = 5^{x-1}$

8. Solve in exact form.

a) $3^x = 4^x$ b) $2(3)^x = 4^x$ c) $3^x = 2(4)^x$

d) $3(2)^x = 8^{x-1}$ e) $3(2)^x = 12^{x-1}$ f) $3(2)^x = 9^{x-1}$

9. Solve for x in terms of a, b, and c.

a) $ab^x = c^{x+1}$ b) $ab^x = c^{x+2}$ c) $ab^x = c^{x+3}$

COMMUNICATING THE IDEAS

Explain why solving an exponential equation by taking the logarithm of each side is equivalent to solving the equation by expressing both sides as powers with the same base.

2.12 Logarithmic Equations and Identities

To solve equations involving logarithms, we use the definition of a logarithm and the laws of logarithms.

Example 1

Solve each equation, and check.

a) $\log_5 (x - 3) + \log_5 x = \log_5 10$ **b)** $\log_8 (x - 2) + \log_8 (x - 4) = 1$

Solution

a) $\log_5 (x - 3) + \log_5 x = \log_5 10$

$\log_5 x(x - 3) = \log_5 10$ [Law of logarithms for multiplication]

Each side is the logarithm to base 5 of a certain quantity.
Hence, the quantities themselves are equal.

$$x(x - 3) = 10$$
$$x^2 - 3x - 10 = 0$$
$$(x - 5)(x + 2) = 0$$
$$x = 5 \text{ or } x = -2$$

Check.

When $x = 5$:   L.S. $= \log_5 2 + \log_5 5$ R.S. $= \log_5 10$

                            $= \log_5 10$

Hence, 5 is a root.
When $x = -2$, the first expression on the left side of the original equation is not defined. Hence, -2 is an extraneous root.
The only root of the equation is 5.

b) $\log_8 (x - 2) + \log_8 (x - 4) = 1$

$\log_8 (x - 2)(x - 4) = 1$ [Law of logarithms for multiplication]

Use the definition of a logarithm.

$(x - 2)(x - 4) = 8^1$ [Definition of a logarithm]
$$x^2 - 6x + 8 = 8$$
$$x(x - 6) = 0$$
$$x = 0 \text{ or } x = 6$$

Check.

When $x = 0$, the expressions on the left side of the given equation are not defined. Hence, 0 is an extraneous root.

When $x = 6$:   L.S. $= \log_8 4 + \log_8 2$ R.S. $= 1$

                            $= \log_8 8$

                            $= 1$

Hence, 6 is a root. The only root of the equation is 6.

Some equations are satisfied for all values of the variable for which both sides of the equation are defined. These equations are *identities*. You have encountered many examples of identities in your earlier mathematical work, but they were not called identities. Here are a few examples. Verify that these equations are satisfied for all real values of x except for the restrictions noted.

Algebraic identities $\qquad (x + 1)(x - 1) = x^2 - 1$
$\qquad\qquad\qquad\qquad\qquad 2(x + 3) = 2x + 6$
Square-root identities $\qquad \sqrt{4x} = 2\sqrt{x}, x \geq 0$
Exponential identities $\qquad x^2 \times x^3 = x^5$
Logarithmic identities $\qquad \log x^2 = 2\log x, x > 0$

Example 2

Consider the identity $\log_a \left(\frac{1}{x}\right) = -\log_a x$.

a) Verify the identity numerically when $a = 10$ and $x = 4$.

b) Determine the values of x for which each side of the identity is defined.

c) Verify the identity graphically, when $a = 10$.

d) Prove the identity for any base a and any positive value of x.

Solution

a) When $a = 10$ and $x = 4$, the identity is $\log\left(\frac{1}{4}\right) = -\log 4$.

$\log\left(\frac{1}{4}\right) = \log 0.25 \qquad\qquad -\log 4 \doteq -0.602\ 06$
$\qquad\qquad \doteq -0.602\ 06$

Since the left side equals the right side, the identity is verified.

b) The expression on the left side, $\log_a\left(\frac{1}{x}\right)$, is defined only when $\frac{1}{x} > 0$; that is, for all positive real numbers. The expression on the right side, $-\log_a x$, is also defined for all positive real numbers.

c) Graph the functions $y = \log\left(\frac{1}{x}\right)$ and $y = -\log x$. The graphs appear identical.

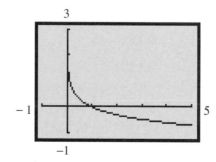

d) To prove that $\log_a\left(\frac{1}{x}\right) = -\log_a x$, prove the left side is equal to the right side.

Left side $= \log_a\left(\frac{1}{x}\right)$

$= \log_a x^{-1}$ ⟵ **Definition of x^{-1}**

$= (-1) \times \log_a x$ ⟵ **Law of logarithms for powers**

$= -\log_a x$

$=$ Right side

Since the left side simplifies to the right side, the identity is proved.

DISCUSSING THE IDEAS

1. Refer to *Example 1*.

   a) In part a, did the base of the logarithms have to be 5? Explain.

   b) From where did the extraneous roots in each equation come?

   c) In the check for part a, why is the first expression on the left side of the equation not defined when $x = -2$?

2. Explain your answer to each question.

   a) When you verify a logarithmic identity numerically, does it matter what numbers you substitute for the base a or for x?

   b) Does verifying an identity numerically prove the identity?

   c) Does verifying an identity graphically prove the identity?

3. In what other ways could you prove the identity in *Example 2*?

2.12 EXERCISES

A 1. Solve each equation.

   a) $\log x = 2$ **b)** $\log x = 5$ **c)** $\log x = -3$

   d) $\log_2 x = 3$ **e)** $\log_3 x = 3$ **f)** $\log_4 x = 3$

2. Solve each equation.

   a) $\log_2 x + \log_2 (x - 1) = \log_2 2$ **b)** $\log_3 x + \log_3 (x - 1) = \log_3 6$

   c) $\log_4 x + \log_4 (x - 1) = \log_4 12$ **d)** $\log_5 x + \log_5 (x - 1) = \log_5 20$

B **3.** Solve.

a) $\log_2 (x - 2) + \log_2 x = \log_2 3$

b) $\log_3 (x - 3) + \log_3 x = \log_3 4$

c) $\log_4 (x - 4) + \log_4 x = \log_4 5$

d) $\log_5 (x - 5) + \log_5 x = \log_5 6$

4. Solve and check.

a) $\log (x + 2) + \log (x - 1) = 1$

b) $\log (3x + 2) + \log (x - 1) = 2$

c) $2 \log (x - 1) = 2 + \log 100$

5. Solve and check.

a) $2 \log x = \log 32 + \log 2$

b) $2 \log x = \log 3 + \log 27$

c) $\log_4 (x + 2) + \log_4 (x - 1) = 1$

d) $\log_2 (x - 5) + \log_2 (x - 2) = 2$

e) $\log_2 x + \log_2 (x + 2) = 3$

f) $\log_6 (x - 1) + \log_6 (x + 4) = 2$

6. Solve and check.

a) $2 \log m + 3 \log m = 10$

b) $\log_3 x^2 - \log_3 2x = 2$

c) $\log_3 s + \log_3 (s - 2) = 1$

d) $\log (x - 2) + \log (x + 1) = 1$

e) $\log_7 (x + 4) + \log_7 (x - 2) = 1$

f) $\log_2 (2m + 4) - \log_2 (m - 1) = 3$

7. Consider the identity $\log_a \left(\dfrac{1}{x^2} \right) = -2 \log_a x$.

a) Verify the identity numerically when $a = 10$ and $x = 5$.

b) Determine the values of x for which each side of the identity is defined.

c) Verify the identity graphically.

d) Prove the identity algebraically.

8. Prove each identity, and state the value(s) of x for which the identity is true.

a) $\log (x - 1) + \log (x - 2) = \log (x^2 - 3x + 2)$

b) $\log x + \log (x + 3) = \log (x^2 + 3x)$

c) $\log (x - 5) + \log (x + 5) = \log (x^2 - 25)$

9. Prove each identity.

a) $(\log_a b)(\log_b c) = \log_a c$

b) $(\log_a b)(\log_b c)(\log_c d) = \log_a d$

C **10. a)** Prove each identity.

i) $\dfrac{1}{\log_3 10} + \dfrac{1}{\log_4 10} = \dfrac{1}{\log_{12} 10}$

ii) $\dfrac{1}{\log_3 x} + \dfrac{1}{\log_4 x} = \dfrac{1}{\log_{12} x}$

b) Use the results of part a as a guide. State a general result and prove it.

COMMUNICATING THE IDEAS

Write to explain what is meant by an identity, with some examples. Include a description of the difference between verifying an identity and proving the identity.

Carbon Dating

An important application of exponential decay is to estimate the age of ancient specimens using a method known as *carbon dating*. All living matter contains traces of carbon-14, which is radioactive. When an organism dies, the carbon-14 decays with a half-life of about 5760 years. Hence, the age of an ancient specimen can be determined by measuring the radioactivity of the carbon-14 it contains, and comparing it with that of living matter.

The percent, P, remaining, compared with living matter after n years is

$P = 100 \times 0.5^{\frac{n}{5760}}$. We can use this equation to determine the percent of carbon-14 remaining when we know the age of the specimen in years. We can also use the equation to approximate the age of a specimen when we know the percent of carbon-14 remaining.

1. In 1991, the frozen remains of an ancient hunter were found on an Alpine glacier in Austria. Carbon dating tests showed that "The Iceman" lived about 5300 years ago. Estimate the percent of carbon-14 remaining in a specimen taken from The Iceman.

2. In 1993, miners at Last Chance Creek, near Dawson City, Yukon, unearthed the remains of an ancient horse. Tests showed that the percent of carbon-14 in the remains was about 4.38% relative to living matter. Estimate the age of the ancient horse.

3. The Dead Sea Scrolls are about 2000 years old. What percent of carbon-14 relative to living matter should be expected from a sample taken from the Scrolls?

4. Fibres from the Shroud of Turin contain carbon-14 whose radioactivity is 92% relative to living matter. About how old is the Shroud of Turin?

5. When several specimens are to be dated, it is more efficient to solve the equation for n before substituting values of P.

   a) Solve the equation $P = 100 \times 0.5^{\frac{n}{5760}}$ for n.

   b) Determine the approximate age of each specimen, given the percent of its carbon-14 content relative to living matter.
i) charred bread found at Pompeii	79.6%
ii) wood in First Dynasty tombs	68.9%
iii) charcoal found at Stonehenge	62.0%
iv) specimen from the end of the last ice age	24.0%
v) paintings in Lascaux Cave, France	15.3%
vi) skin of a Siberian mammoth	2.2%

1. Let A_0 represent the acidity of a chemical with a pH value of 0.

   a) Write an equation to express the acidity, A, of a chemical with a pH value of p.

   b) The pH values of cola drinks range from 2.2 to 3.0. How many times as acidic is a drink with a pH value of 2.2 than one with a pH value of 3.0?

Exponential functions have defining equations of the form $y = Ab^x$, where $b > 0$. You should be able to calculate any of the four quantities in this equation when the other three are given.

2. What amount of money would grow to $1000 in 5 years if it is invested at 6% compounded annually?

3. In 1995, Canada's population was 29.6 million. Since 1971, the average annual growth rate was 1.24%. Estimate Canada's population in 1971.

4. An investor bought $1000 worth of Microsoft stock in 1985. It was worth $2 million in 1998. What is the equivalent annual rate of interest for an investment that grows from $1000 to $2 million in 13 years?

5. Suppose you invest $2500 at 7.25% compounded annually. How many years would it take for your investment to double?

6. A colony of bees increases by 25% every three months. How many bees should a beekeeper start with if she wishes to have 10 000 bees in 18 months?

7. Two historical purchases of land in North America are given. In each case, if the money had been invested at 6% compounded annually, what would its value be today?

   a) In 1867, the United States purchased Alaska from Russia for $7 200 000.

   b) In 1626, Manhattan Island was sold for $24.

8. In 1995, the world population was 5.8 billion. According to United Nations' forecasts, the population will grow to 8.0 billion by 2025. Suppose the prediction were true. What is the average annual growth rate?

9. Suppose the doubling time for the growth of a certain population is d years.

   a) Determine an expression for the time for the population to triple.

   b) Determine an expression for the time for the population to quadruple.

10. Solve each equation for the indicated variable.

   a) $A = P(1 + i)^n$ i) P ii) n iii) i

   b) $y = Ab^x$ i) A ii) x iii) b

Predicting Tide Levels

Botanical Beach Provincial Park is 3 km south of Port Renfrew on Vancouver Island. The park is noted for its striking geological features and abundant shore life. When the tide is out, visitors can see a wide variety of live sea creatures in tide pools. A low tide of 1.2 m or less is best for viewing.

 CONSIDER THIS SITUATION

Suppose you are planning to visit Botanical Beach on Sunday. Use the information in the table below.

	Times of Low and High Tides			
Sun	5:09 A.M.	11:19 A.M.	5:40 P.M.	11:42 P.M.
	1.19 m	2.86 m	1.11 m	2.71 m

• When is the best time to go to Botanical Beach?

• About how many hours and minutes is it from a low tide to the next high tide, or from a high tide to the next low tide?

• Predict the tide level at each time.

   a) 8:15 A.M.   **b)** 2:30 P.M.   **c)** 8:40 P.M.   **d)** 10:10 A.M.

On pages 200 to 204, you will develop a mathematical model for predicting the tide level at any time. Your model will involve graphs of trigonometric functions.

 FYI Visit **www.awl.com/canada/school/connections**

> For information related to the above problem, click on <u>MATHLINKS</u>, followed by <u>AWMath</u>. Then select a topic under Predicting Tide Levels.

Port
Renfrew

Saanich
Inlet

Jordan
River

Sooke

Victoria

0   20   40km

3.1 Introduction to Periodic Functions

In this chapter, we will describe many applications of mathematics involving quantities that change in a regular way. Applications concerning the sun and human physiology are shown in this section.

The time of sunset

In summer, the sun sets later than it does in winter. This graph shows how the time of sunset at Saskatoon varies during a two-year period. The times are from a 24-h clock, in hours and decimals of hours. For example, on June 21 the sun sets at 21.5 h. This means 21 h and 0.5×60 min, or 21 h 30 min.

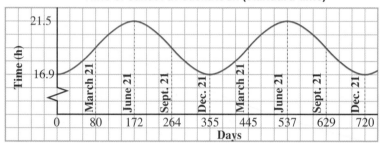

Time of sunset at Saskatoon (standard time)

Sunspots

Sunspots are dark spots that appear from time to time on the surface of the sun. The periodic variation in the number of sunspots has been recorded for hundreds of years. This graph shows how the number of sunspots varied from 1944 to 1998.

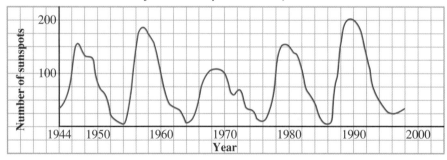

Monthly mean sunspot numbers, 1944–1998

Lengths of shadows

This graph shows how the length of the shadow of a 100-m building varies during a three-day period.

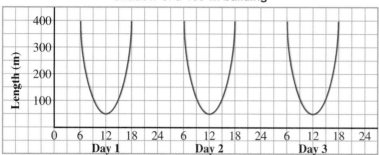

Blood pressure and volume

There are two significant phases to a heartbeat. During the systolic phase, the heart contracts and pumps blood into the arteries. This phase is marked by a sudden increase in the blood pressure and a decrease in the volume of blood in the heart. The second phase is the diastolic phase, when the heart relaxes. The blood pressure decreases and the volume of blood increases as more blood is pumped into the heart by the veins.

These graphs show how the blood pressure and volume of blood in the left ventricle of the heart vary during five consecutive heartbeats.

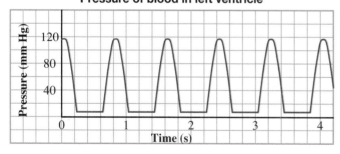

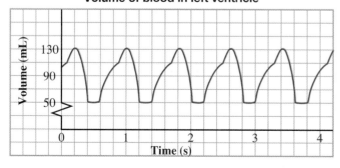

Volume of air in the lungs

The volume of air in your lungs is a periodic function of time. This graph shows how the volume of air in the lungs varies during normal breathing.

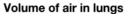

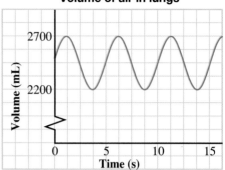

Periodic Functions

The graphs in this section suggest the meaning of a *periodic function*. The graph of a periodic function repeats in a regular way. The length of the shortest part that repeats, measured along the horizontal axis, is called the *period* of the function.

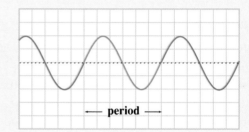

DISCUSSING THE IDEAS

1. Refer to the time of the sunset graph on page 158.

   a) What causes the variation in the time of sunset?

   b) Why are the times later in June than they are in December?

   c) Suppose a graph was drawn to show the time of sunrise at Saskatoon. How would this graph differ from the one on page 158?

2. Refer to the graph on page 158 that shows the number of sunspots. Does this graph illustrate a periodic function? Explain.

3. Refer to the graph on page 159 that shows the length of the shadow of a building.

   a) Why are the parts of the graph not connected?

   b) Suppose a graph was drawn for a shorter building. How would this graph differ from the one on page 159?

4. Refer to the graphs on pages 158 to 160. Estimate the period of each graph that represents a periodic function. Explain each estimate.

A **1.** Use the graph on page 158 that shows the time of sunset at Saskatoon.

    a) Estimate the time of the sunset on each date.
        i) February 2   **ii)** July 25   **iii)** October 30

    b) Estimate the dates when the sun sets at each time.
        i) 8 P.M.   **ii)** 7 P.M.   **iii)** 6 P.M.   **iv)** 5 P.M.

2. Use the data in this table. Suppose graphs that show sunset times were drawn for Fairbanks in Alaska and Mexico City.

	Approximate time of sunset on			
	March 21	June 21	September 21	December 21
Fairbanks	19.2 h	23.8 h	19.2 h	14.6 h
Mexico City	18.7 h	19.4 h	18.7 h	18.0 h

    a) In what ways would the graphs for these cities differ from the first graph on page 158?

    b) In what ways would they be similar?

3. The graph of monthly sunspot numbers on page 158 shows that sunspot activity increases and decreases at fairly regular intervals. Estimate the average number of years between the times when there is a maximum number of sunspots.

4. Use the graph on page 159 that shows the length of the shadow of a building.

    a) How long is the shadow at 8 A.M. and at 2 P.M.?

    b) For about how many hours during the day is the shadow longer than 100 m?

    c) In some cities, such as Singapore, the sun is directly overhead at noon. What changes would you make to the graph if it was drawn for a building in Singapore?

5. During intense physical activity, the heart beats faster to satisfy the blood's demand for more oxygen. Suppose graphs were drawn to show the variation of blood pressure and volume in this situation.

    a) How would the graphs differ from those on page 159?

    b) In what ways would the graphs be similar?

6. The graph on page 160 shows the volume of air in the lungs. How long does a person take to inhale and exhale once?

7. When the average person takes a deep breath, about 5000 mL of air is inhaled. But only about 4000 mL of this air can be exhaled. Suppose that such a breath takes twice as much time as a normal breath. A graph, similar to the one on page 160, is drawn for deep breathing. Write to explain the ways in which it would be different.

8. These graphs were made by *Graphmatica*. If a function is periodic, estimate its period. If a function is not periodic, write to explain why.

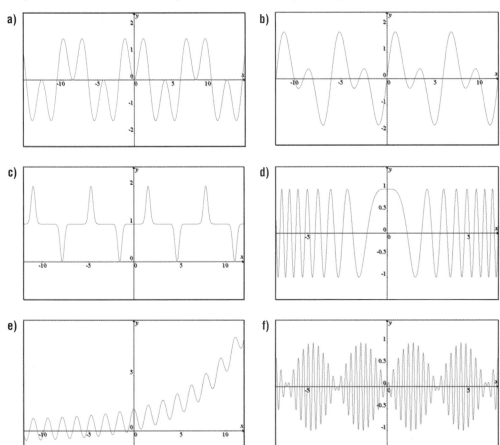

a)   b)

c)   d)

e)   f)

Write to explain the meaning of a periodic function and the period of a periodic function.

3.2 Radian Measure

The simplest example of periodic motion is motion in a circle. Recall that a degree is a unit of angle measure defined by using 360° for a complete rotation. On a calculator, you will notice another unit of angle measure called a radian. In many situations, radians are more useful than degrees. After you learn what a radian is, and how it is related to degrees, we will explain why it is a useful unit for angle measure.

INVESTIGATE **What Is a Radian?**

In this investigation, you will use a calculator to determine how many degrees there are in one radian. Instructions for the TI-83 and TI-34 calculators are given. Other calculators have similar features, but they may be accessed differently. Consult your manual if necessary.

1. To find out what one radian is, use a calculator to enter one radian, then convert it to degrees. Follow these steps.

   On the TI-83 calculator

   a) Make sure that your calculator is in degree mode. Use the mode menu to do this.

   b) Press 1 ANGLE 3 to display 1^r.

   c) Press ENTER. The number of degrees will be displayed.

   On the TI-34 calculator

   a) Press DRG until RAD is displayed. The calculator is in radian mode.

   b) Enter the number 1.

   c) Press 2nd DRG 2nd DRG. RAD in the display will change to DEG. One radian will be changed to degrees.

2. How many degrees are there in one radian?

3. You should have found that one radian is slightly less than 60°. This may seem unusual, until you realize that 3 radians are slightly less than 180°. That is, 180° is a little more than 3 radians.

   a) What number is a little more than 3 and has something to do with circles?

   b) Use your calculator to convert 180° to radians. Does the result confirm your answer to part a?

4. Your calculator may have other modes for angles. If so, find out how each mode is related to degrees.

A radian is a unit, different from a degree, for measuring angles. In *Investigate*, you may have found that one radian is approximately 57.295 78°. Instructions for converting radians to degrees were programmed into a calculator's processor chip using the following definition, which was established before calculators were invented.

One *radian* is the measure of an angle subtended at the centre of a circle by an arc equal in length to the radius of the circle.

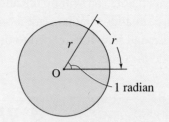

1 radian

Triangle ABC is equilateral, with rigid sides AB and BC. There is a hinge at B, and side AC is flexible.

Push point A down slightly so that AC bends outward and forms part of a circle, centred at B.

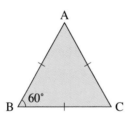

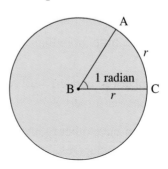

This angle is one radian.

According to the definition, one radian is slightly less than 60°. Hence, 3 radians are slightly less than 180°. That is, 180° is slightly more than 3 radians. Recall that, in the formulas for the circumference and area of a circle, $\pi \doteq 3.1416$. It turns out that 180° is π radians and we can prove this as follows.

Angle ABC is the same fraction of a complete rotation as arc AC is of the circumference.

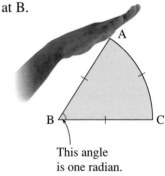

$$\frac{\angle ABC}{360°} = \frac{\text{arc AC}}{\text{circumference}}$$

$$\frac{1 \text{ radian}}{360°} = \frac{r}{2\pi r}$$

Multiply each side by 360°.

$$1 \text{ radian} = \frac{180°}{\pi}$$

Multiply each side by π.

$$\pi \text{ radians} = 180°$$

$$\pi \text{ radians} = 180°$$
$$2\pi \text{ radians} = 360°$$

π radians are 180° and this fact can be used to convert from one angle measure to the other.

Example 1

Express each angle to 2 decimal places.

a) 4 radians in degrees   **b)** 138° in radians

Solution

a) π radians = 180°

$1 \text{ radian} = \dfrac{180°}{\pi}$

$4 \text{ radians} = 4\left(\dfrac{180°}{\pi}\right)$

$\doteq 229.18°$

b) $180° = \pi$ radians

$1° = \dfrac{\pi}{180} \text{ radians}$

$138° = 138\left(\dfrac{\pi}{180}\right) \text{ radians}$

$\doteq 2.41 \text{ radians}$

We can use radian measure to derive a simple formula relating the length of an arc of a circle to the radius and the angle subtended by the arc at the centre.

Let a represent the length of an arc that subtends an angle θ radians at the centre of a circle with radius r. The ratio of the arc length to the circumference is the same as the ratio of the angle at the centre to a complete rotation. That is,

$$\frac{\text{arc length}}{\text{circumference}} = \frac{\text{angle at centre}}{2\pi}$$

$$\frac{a}{2\pi r} = \frac{\theta}{2\pi}$$

$$a = r\theta$$

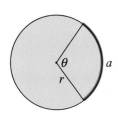

The arc length a subtended by an angle θ radians in a circle with radius r is given by the formula:

$a = r\theta$

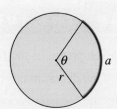

If degree measure had been used in the development of the arc length formula, we would have obtained the formula $a = \frac{\pi}{180} r\theta$ for arc length. When radian measure is used, the formula is simpler. There are other formulas involving angles in mathematics that become simpler when radians are used instead of degrees. One of these occurs in exercise 15, and many others occur in calculus. Hence, one reason for introducing radian measure is that it simplifies formulas involving angles. Another reason will be given in Chapter 4, where we will see that radians are more fundamental than degrees.

Example 2

A circle has radius 6.5 cm. Calculate the length of an arc of this circle subtended by each angle. Express each length to the nearest tenth of a centimetre.

a) 2.3 radians

b) 75°

Solution

Use the formula.

a) $a = r\theta$
$a = (6.5)(2.3)$
$= 14.95$
The arc length is 15.0 cm.

b) To use the formula $a = r\theta$, the angle must be in radians.
$180° = \pi$ radians
$1° = \frac{\pi}{180}$ radians
$75° = 75\left(\frac{\pi}{180}\right)$ radians
$\doteq 1.309$ radians
Substitute in the formula.
$a = r\theta$
$\doteq (6.5)(1.309)$
$\doteq 8.5085$
The arc length is 8.5 cm.

DISCUSSING THE IDEAS

1. Refer to the development of the formula $a = r\theta$ on page 165.

   a) Explain why the denominator on the right side of the first equation is 2π.

   b) Explain why the formula becomes $a = \frac{\pi}{180} r\theta$ when degree measure is used.

2. Explain why radian measure is useful.

A 1. Convert to radians. Express each angle in terms of π.

   a) 30°   **b)** 60°   **c)** 90°   **d)** 120°   **e)** 150°   **f)** 180°

   g) 210°   **h)** 240°   **i)** 270°   **j)** 300°   **k)** 330°   **l)** 360°

2. Convert to radians. Express each angle in terms of π.

   a) 45°   **b)** 135°   **c)** 225°   **d)** 315°

3. Choose one angle from exercise 1 or 2. Write to explain how you converted the angle to radians.

4. Convert to degrees.

   a) $\frac{\pi}{6}$ radians   **b)** $\frac{2\pi}{6}$ radians   **c)** $\frac{3\pi}{6}$ radians   **d)** $\frac{4\pi}{6}$ radians

   e) $\frac{5\pi}{6}$ radians   **f)** $\frac{6\pi}{6}$ radians   **g)** $\frac{\pi}{4}$ radians   **h)** $\frac{2\pi}{4}$ radians

   i) $\frac{3\pi}{4}$ radians   **j)** $\frac{4\pi}{4}$ radians   **k)** $\frac{5\pi}{4}$ radians   **l)** $\frac{6\pi}{4}$ radians

5. Write how many radians there are in each rotation.

   a) one full turn   **b)** one half turn   **c)** one quarter turn

6. Convert to radians. Express each angle to 2 decimal places.

   a) 100°   **b)** 225°   **c)** 57.3°   **d)** 125°   **e)** 75°   **f)** $\frac{60°}{\pi}$

   g) 65°   **h)** 24.5°   **i)** 150°   **j)** 30°   **k)** $\frac{180°}{\pi}$ **l)** 90°

7. Convert to degrees. Express each angle to 1 decimal place.

   a) 2 radians   **b)** 5 radians   **c)** 3.2 radians   **d)** 1.8 radians

   e) 0.7 radians   **f)** 1.4 radians   **g)** 6.7 radians   **h)** 2π radians

8. Convert to degrees.

   a) $\frac{\pi}{2}$ radians   **b)** $\frac{11\pi}{6}$ radians   **c)** $\frac{2\pi}{3}$ radians   **d)** $\frac{7\pi}{6}$ radians

   e) $\frac{5\pi}{3}$ radians   **f)** $\frac{3\pi}{2}$ radians   **g)** $\frac{7\pi}{4}$ radians   **h)** 2π radians

9. Choose one angle from exercise 8. Write to explain how you converted the angle to degrees.

B 10. Determine the length of the arc that subtends each angle at the centre of a circle with radius 5 cm. Express each length to 1 decimal place.

   a) 2.0 radians   **b)** 3.0 radians   **c)** 1.8 radians

   d) 6.1 radians   **e)** 4.2 radians   **f)** 0.6 radians

11. Determine the length of the arc of a circle with radius 12 cm that subtends each angle at the centre. Express each length to 1 decimal place, where necessary.

   a) 135°   **b)** 75°   **c)** 105°   **d)** 165°

   e) 240°   **f)** 180°   **g)** 310°   **h)** 200°

12. Choose one angle from exercise 10 or 11. Write to explain how you determined the arc length.

13. **a)** Write an expression for each distance from A to B.
   i) along the segment AB
   ii) along the circular arc from A to B

   b) How many times as long as the straight-line distance is the distance along the circular arc from A to B?

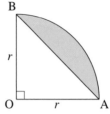

14. Determine the angle at the centre of a circle, radius 6.0 cm, for each arc length. Express each answer in degrees and radians.

   a) 3.0 cm   **b)** 7.0 cm   **c)** 12.5 cm   **d)** 16.4 cm

C 15. **a)** Derive a formula for the area, A, of a sector of a circle with radius, r, formed by an angle θ radians.

   b) Derive a similar formula when the measure of the angle is in degrees.

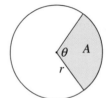

   c) Explain how the results of parts a and b justify the use of radian measure instead of degree measure.

16. When an object is moving in a circle, its *angular velocity* is the angle per unit time through which it rotates about the centre.

   a) A car tire has diameter 64 cm. Determine its angular velocity, in radians per second, when the car is travelling at 100 km/h.

   b) Write an expression for the angular velocity, in radians per second, for a car tire with diameter d centimetres when the car is travelling at x kilometres per hour.

COMMUNICATING THE IDEAS

Write to explain how one radian is defined, and how radians and degrees are related.

Refraction

When you stand at the edge of a swimming pool and look across it, the water appears to be more shallow than it really is. If you jump in and expect to reach the bottom, you may be surprised.

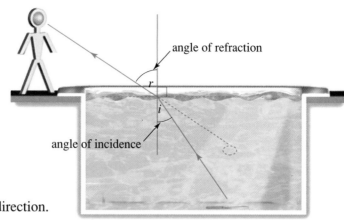

angle of refraction

angle of incidence

Light travels at different speeds through different materials. When light travels from one material into another, it changes direction.

As the light travels from the water into the air, the rays bend away from the perpendicular. To the person viewing the object in the pool, the ray appears to follow a straight path. This makes the water appear less deep than it is.

The index of refraction is a measure of how much the light bends. It is related to the angles of incidence and refraction. This relationship is Snell's Law. It states that the quotient $\frac{\sin i}{\sin r}$ remains constant when the angles change.

Snell's Law

$$n_{1 \to 2} = \frac{\sin i}{\sin r}$$

where $n_{1 \to 2}$ is the index of refraction for light travelling from material 1 to material 2.

i is the angle of incidence in material 1.

r is the angle of refraction in material 2.

1. Light goes from air into a block of material. The angle of incidence is 45° and the angle of refraction is 38°. Calculate the index of refraction.

2. As light goes from a crystal into air, the angle of incidence is 24° and the angle of refraction is 32°. Calculate the index of refraction.

3. The pool above is 1 m deep. The angle of refraction is 59° and the angle of incidence is 40°. Calculate how deep the object in the pool appears to be. Assume the image is vertically above the object.

Mathematics & Science

Motion in a circle is an example of periodic motion. To study motion in a circle, we need to define the standard position of an angle.

Let P(x, y) represent a point that moves around a circle with radius r and centre O(0, 0). P starts at the point A(r, 0) on the x-axis. For any position of P, an angle θ is defined, which represents the amount of rotation about the origin. We say that the angle θ is in *standard position*, where OA is the *initial arm* and OP the *terminal arm*. The measure of the angle may be in degrees or in radians.

If $\theta > 0$, the rotation is counterclockwise.

If $\theta < 0$, the rotation is clockwise.

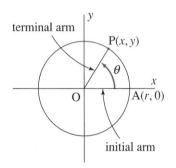
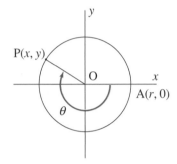

When P moves around the circle, the motion is repeated after P has rotated 360°, or 2π radians. When any angle θ is given, we can always determine other angles for which the position of P is the same. All these angles are in standard position.

Given an angle of 60°, or $\frac{\pi}{3}$ radians, …

… add 360°, or 2π radians
$60° + 360° = 420°$
$\frac{\pi}{3}$ radians $+ 2\pi$ radians
$= \frac{7\pi}{3}$ radians

… add 360°, or 2π radians again
$420° + 360° = 780°$
$\frac{7\pi}{3}$ radians $+ 2\pi$ radians
$= \frac{13\pi}{3}$ radians

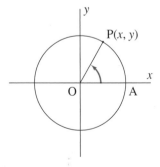
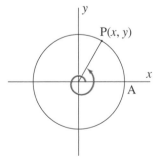
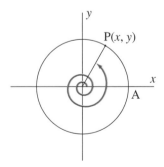

60°, or $\frac{\pi}{3}$ radians

420°, or $\frac{7\pi}{3}$ radians

780°, or $\frac{13\pi}{3}$ radians

The angles on page 170 are in standard position, and have the same terminal arm. For this reason, they are called coterminal angles. When θ is any angle in standard position, other angles that are coterminal with θ can be found by adding or subtracting multiples of 360° if θ is in degrees, or multiples of 2π radians if θ is in radians.

Coterminal Angles

- Two or more angles in standard position are *coterminal angles* when the position of P is the same for each angle.
- When θ represents any angle, then any angle coterminal with θ is represented by these expressions, where n is any integer.

  $\theta + n(360°)$, if θ is in degrees

  $\theta + n(2\pi \text{ radians})$, if θ is in radians

Example 1

For each angle:

a) $\theta = 150°$ b) $\theta = -\frac{\pi}{6}$ radians

i) Draw the angle θ in standard position.

ii) Determine two other angles coterminal with θ, and illustrate each angle with a diagram.

iii) Write an expression to represent any angle coterminal with θ.

Solution

a) i) 150° ii) $150° + 360° = 510°$ $150° - 360° = -210°$

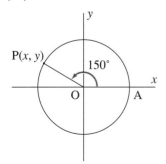 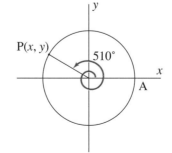 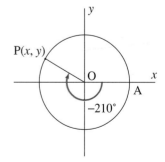

iii) Any angle coterminal with 150° is represented by the expression $150° + n(360°)$, where n is an integer.

b) i) $-\frac{\pi}{6}$ radians

ii) $-\frac{\pi}{6}$ radians $+ 2\pi$ radians $= \frac{11\pi}{6}$ radians

$-\frac{\pi}{6}$ radians $- 2\pi$ radians $= -\frac{13\pi}{6}$ radians

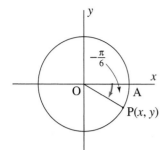 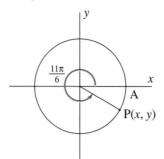 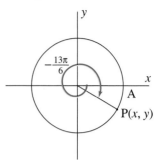

iii) Any angle coterminal with $-\frac{\pi}{6}$ radians is represented by the expression $\left(-\frac{\pi}{6} + 2\pi n\right)$ radians, where n is an integer.

Example 2

Suppose P has rotated $-830°$ about O(0, 0) from A.

a) How many complete rotations have been made?

b) In which quadrant is P located?

c) Draw a diagram to show the position of P.

Solution

a) P has made a clockwise rotation of $830°$. Since a complete rotation is $360°$, divide 830 by 360 to obtain $830 \div 360 \doteq 2.3056$.

Since the result is between 2 and 3, P has made 2 complete rotations around the circle, and part of a third rotation.

b) Two complete rotations amount to $2(360°)$, or $720°$. The additional rotation beyond $720°$ is $830° - 720°$, or $110°$.

Since $90° < 110° < 180°$, P is in the third quadrant.

c)

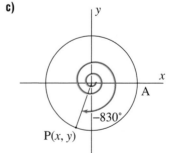

Example 3

Suppose P has rotated 3π radians about O(0, 0) from A.

a) How many complete rotations have been made?

b) In which quadrant is P located?

c) Draw a diagram to show the position of P.

Solution

a) Since a complete rotation is 2π radians, divide 3π by 2π to obtain 1.5.

   P has made one complete rotation and half of a second rotation.

b) P is on the *x*-axis, between the second and third quadrants.

c)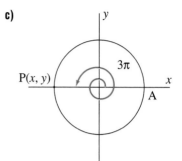

DISCUSSING   THE IDEAS

1. Suggest a reason why the positive direction for angles in standard position is defined to be counterclockwise and not clockwise.

2. The measure of any angle coterminal with an angle θ is represented by $\theta + n(360°)$ when θ is in degrees, and by $\theta + n(2\pi \text{ radians})$ when θ is in radians, where n is any integer. Explain why.

3.3 EXERCISES

(A) 1. For each angle in standard position, determine θ in degrees and in radians.

a)
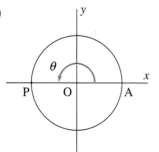

b)
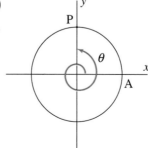

c)

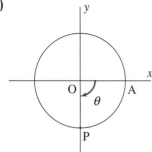

d)

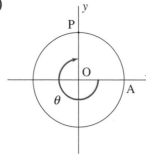

2. Sketch each angle in standard position.

a) $\theta = 50°$ **b)** $\theta = -120°$ **c)** $\theta = -165°$ **d)** $\theta = 240°$

e) $\theta = \frac{\pi}{2}$ radians **f)** $\theta = -\frac{\pi}{4}$ radians **g)** $\theta = \frac{2\pi}{3}$ radians **h)** $\theta = -\frac{3\pi}{2}$ radians

3. In each part of exercise 2, determine two angles that are coterminal with θ.

4. For each angle θ in standard position, determine two other angles that are coterminal with θ.

a)

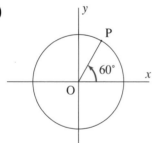

b)

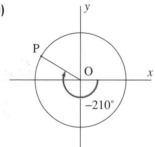

c)

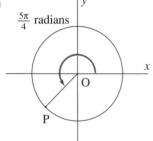

d)

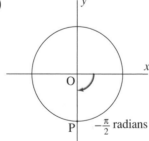

5. Determine two angles that are coterminal with each angle θ.

a) $\theta = \pi$ radians **b)** $\theta = \frac{\pi}{2}$ radians

c) $\theta = -\frac{\pi}{3}$ radians **d)** $\theta = -2\pi$ radians

6. Choose one angle from exercise 5. Write to explain how you determined two coterminal angles.

B **7.** P is a point on the terminal arm of an angle θ in standard position. Suppose P has rotated 420°.

   a) How many complete rotations have been made?

   b) In which quadrant is P located?

   c) Draw a diagram to show the position of P.

8. Repeat exercise 7 for each angle of rotation for P.

   a) 480°   **b)** −660°   **c)** 870°   **d)** −1000°

9. Sketch each angle in standard position.

   a) $\theta = 400°$ **b)** $\theta = 750°$ **c)** $\theta = -270°$ **d)** $\theta = -60°$

10. Repeat exercise 9 for each angle of rotation for P.

   a) $-\pi$ radians   **b)** $\frac{3\pi}{2}$ radians   **c)** 2π radians   **d)** $-\frac{5\pi}{2}$ radians

11. Sketch each angle in standard position.

   a) $\theta = \frac{9\pi}{2}$ radians   **b)** $\theta = \frac{10\pi}{3}$ radians

   c) $\theta = -\frac{5\pi}{4}$ radians   **d)** $\theta = -7\pi$ radians

12. Choose one angle from exercise 9, 10, or 11. Write to explain how you sketched the angle.

13. Write an expression to represent any angle coterminal with each angle θ.

   a) $\theta = -45°$ **b)** $\theta = 150°$ **c)** $\theta = 240°$ **d)** $\theta = -30°$

   e) $\theta = \pi$ radians   **f)** $\theta = -\frac{\pi}{4}$ radians   **g)** $\theta = \frac{5\pi}{2}$ radians   **h)** $\theta = -1$ radian

C **14.** Let θ represent any angle in radians. Let α represent the angle that is coterminal with θ, where $0 \leq \alpha < 2\pi$.

   a) Draw a graph to represent α as a function of θ.

   b) What are the domain, the range, and the period of the function?

COMMUNICATING THE IDEAS

P is a point on the terminal arm of an angle θ in standard position. Write to explain how you could determine the quadrant in which P is located, if you know the value of θ in:

   a) degrees   **b)** radians

3.4 The Sine and Cosine Functions of an Angle in Standard Position

In previous grades, we defined the trigonometric ratios of an angle in a right triangle. If $\angle A$ is an acute angle in a right triangle, then

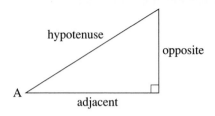

$$\sin A = \frac{\text{length of side opposite } \angle A}{\text{length of hypotenuse}}$$

$$\cos A = \frac{\text{length of side adjacent to } \angle A}{\text{length of hypotenuse}}$$

These definitions are only useful for acute angles. To study periodic motion, we extend the definitions to angles in standard position.

A point moving around a circle is an example of periodic motion because the point returns to its previous positions on each successive rotation. Visualize a point P rotating around a circle with radius 1 unit. This circle is called a *unit circle*. Suppose this circle is centred at the origin on a coordinate grid, and P is on the terminal arm of an angle θ in standard position.

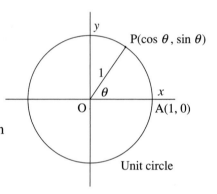

For any position of P on the unit circle, we define the first coordinate of P as the *cosine* of θ, and the second coordinate of P as the *sine* of θ. We write the coordinates of P as $(\cos \theta, \sin \theta)$. As P rotates around the circle, the values of $\cos \theta$ and $\sin \theta$ change periodically. You can use a calculator to determine these values for any given angle. Some typical values are illustrated.

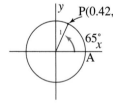 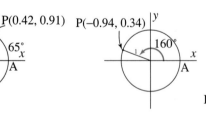 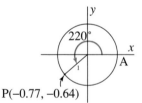 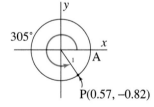

$\cos 65° \doteq 0.42$
$\sin 65° \doteq 0.91$

$\cos 160° \doteq -0.94$
$\sin 160° \doteq 0.34$

$\cos 220° \doteq -0.77$
$\sin 220° \doteq -0.64$

$\cos 305° \doteq 0.57$
$\sin 305° \doteq -0.82$

When P is in the first quadrant, ∠PON is an acute angle. According to the definitions on page 176,

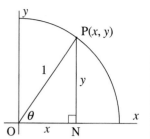

$$\sin \angle PON = \frac{\text{length of side opposite } \angle PON}{\text{length of hypotenuse}}$$

$$= \frac{y}{1}, \text{ or } y$$

$$\cos \angle PON = \frac{\text{length of side adjacent to } \angle PON}{\text{length of hypotenuse}}$$

$$= \frac{x}{1}, \text{ or } x$$

The cosine and sine of ∠PON are the coordinates of P. Hence, the right-triangle definitions you learned in previous grades are special cases of the new definitions.

Special cases of these definitions occur when the terminal arm coincides with an axis. In these cases, the angles are multiples of 90°, or $\frac{\pi}{2}$ radians.

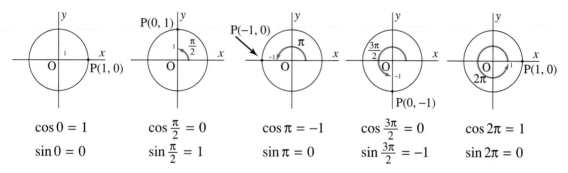

$$\cos 0 = 1$$
$$\sin 0 = 0$$

$$\cos \frac{\pi}{2} = 0$$
$$\sin \frac{\pi}{2} = 1$$

$$\cos \pi = -1$$
$$\sin \pi = 0$$

$$\cos \frac{3\pi}{2} = 0$$
$$\sin \frac{3\pi}{2} = -1$$

$$\cos 2\pi = 1$$
$$\sin 2\pi = 0$$

P(x, y) represents any point on the unit circle, forming an angle θ in standard position.

• The *cosine* of θ is the x-coordinate of P. • The *sine* of θ is the y-coordinate of P.

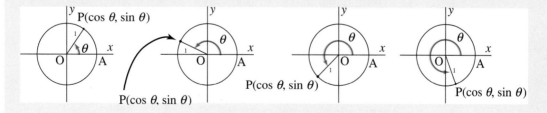

In the diagrams on page 177, the unit "radians" has been omitted. By convention, if the unit of angle measure is not present, it is understood to be radians.

Example

Determine each value to 4 decimal places.

a) cos 125° and sin 125°   b) cos 4.3 and sin 4.3

Solution

a) Ensure the calculator is in degree mode.

   cos 125° ≐ −0.5736   sin 125° ≐ 0.8192

b) Since there is no unit for the angle, the angle is in radians.
   Ensure the calculator is in radian mode.

   cos 4.3 = −0.4008   sin 4.3 = −0.9162

As P rotates around the circle, past 360° or 2π, the same values of x and y are encountered as before. Hence, in the *Example*, there are infinitely many other angles that have the same cosine as 125°, or the same sine as 4.3 radians.

These are also equal to cos 125°.   These are also equal to sin 4.3.
cos (125° + 360°), or cos 485°   sin (4.3 + 2π)
cos (125° − 360°), or cos (−235°)   sin (4.3 − 2π)

 **VISUALIZING**

As P rotates around the unit circle starting at the positive x-axis, visualize how:

- the y-coordinate of P changes from: 0 to 1 to 0 to −1 to 0, and so on

- the x-coordinate of P changes from: 1 to 0 to −1 to 0 to 1, and so on

These patterns repeat every 360°, or 2π radians.

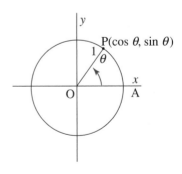

---

1. **a)** In part a of the *Example*, why is cos 125° negative and sin 125° positive?

   b) In part b of the *Example*, why are cos 4.3 and sin 4.3 both negative?

2. The sine and cosine functions are positive in some quadrants and negative in others. For each function, explain how you can remember the quadrants in which it is positive and the quadrants in which it is negative.

3. **a)** What is the maximum possible value of sin θ? What is the minimum possible value? Explain.

   b) What is the maximum possible value of cos θ? What is the minimum possible value? Explain.

4. Explain why different angles can have the same sine or cosine.

5. Explain two different ways you could check whether your calculator is in degree mode or radian mode.

3.4 EXERCISES

A 1. In each diagram, determine sin θ and cos θ.

a)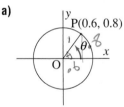

b)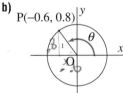

c)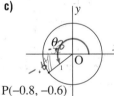

d)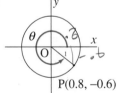

2. Use a scientific calculator in degree mode. Determine each value to 4 decimal places.

   a) sin 130°   **b)** cos 130°   **c)** sin 190°   **d)** cos 200°

   e) cos 285°   **f)** sin 325°   **g)** cos 347°   **h)** cos 534°

3. Use a scientific calculator in radian mode. Determine each value to 4 decimal places.

   a) sin 0.4   **b)** cos 0.6   **c)** sin 1.25   **d)** sin 1.8

   e) cos 3.0   **f)** cos 4.15   **g)** sin 5.39   **h)** cos 8.75

B **4.** Suppose point P starts at A(1, 0) and makes two complete rotations counterclockwise around the unit circle. Describe how the values of cos θ and sin θ change as θ increases from 0° to 720°.

5. a) Determine cos 55° to 4 decimal places.

   b) Determine three other angles that have the same cosine as 55°, and verify with a calculator.

6. a) Determine sin 125° to 4 decimal places.

   b) Determine three other angles that have the same sine as 125°, and verify with a calculator.

7. a) Determine cos 220° to 4 decimal places.

   b) Determine three other angles that have the same cosine as 220°, and verify with a calculator.

8. a) Determine sin 1.25 to 4 decimal places.

   b) Determine three other angles that have the same sine as 1.25 radians, and verify with a calculator.

9. Without using a calculator, determine if each quantity is positive or negative. Justify your answer.

   a) sin 145°   **b)** cos 160°   **c)** cos 210°   **d)** sin 255°   **e)** cos 335°

10. Choose one angle from exercise 9. Write to explain how you determined whether its sine or cosine is positive or negative.

11. These graphing calculator screens show squares whose vertices lie on a unit circle with centre O(0, 0). The lines at the bottom refer to the vertex marked with the cursor. They show the corresponding angle in standard position, in degrees, and the coordinates to the nearest thousandth. For each square:

   i) Write the coordinates of each vertex.

   ii) Write the angles in standard position, in degrees, corresponding to each vertex.

   iii) Write the cosine and the sine of each angle in part ii.

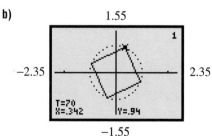

a)

b)

12. Choose one screen from exercise 11. Write to explain how you determined the coordinates of the other three vertices.

13. Use the diagram in exercise 11a. Determine each pair of values.
  a) $\cos 150°$ and $\sin 150°$ **b)** $\cos 330°$ and $\sin 330°$
  c) $\cos 60°$ and $\sin 60°$ **d)** $\cos 240°$ and $\sin 240°$

14. Use the diagram in exercise 11b. Determine each pair of values.
  a) $\cos 110°$ and $\sin 110°$ **b)** $\cos 290°$ and $\sin 290°$
  c) $\cos 20°$ and $\sin 20°$ **d)** $\cos 200°$ and $\sin 200°$

15. Choose one pair of values from exercise 13 or 14. Write to explain how you determined the cosine and sine of the angle.

16. The angle θ is in the first quadrant, and $\cos \theta = \dfrac{2}{\sqrt{5}}$.
  a) Draw a diagram to show the angle in standard position and a point P on its terminal arm.
  b) Determine possible coordinates for P.

17. Repeat exercise 16 for θ in the second quadrant, and $\cos \theta = -\dfrac{2}{\sqrt{5}}$.

18. Repeat exercise 16 for θ in the second quadrant, and $\sin \theta = \dfrac{3}{5}$.

(C) **19.** You can use a calculator to determine the sine, the cosine, or the tangent of any angle in standard position.
  a) Determine the largest angle your calculator will accept: in degrees; in radians.
  b) Are these two angles equal?

20. Triangle CBO has vertices C(−3, 0), B(−3, −7), and O(0, 0). The unit circle, centred at O(0, 0), intersects OB at P.
  a) Determine the coordinates of P, then use the coordinates to determine $\sin \theta$ and $\cos \theta$.
  b) Compare the values of $\sin \theta$ and $\cos \theta$ with those you would obtain by using the lengths of the sides of $\triangle$CBO. What do you notice?

COMMUNICATING THE IDEAS

The sine and cosine ratios were originally defined for acute angles in a right triangle. Write to explain how these ratios are defined for any angle in standard position. Include an explanation of how the extended definitions are consistent with the original definitions for acute angles. Include diagrams with your explanations.

Coordinates of Points on the Unit Circle

You can use a graphing calculator to graph points on the unit circle with centre O(0, 0). To obtain the screen at the right, follow the steps below. These steps are for the TI-82 or TI-83 calculator. Other graphing calculators should have similar capabilities.

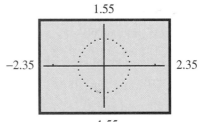

- *Set parametric graphing mode.* Press [MODE]. In the second line, select the number of decimal places desired. Select "Degree" in the third line. Select "Par" in the fourth line. (This sets the calculator to a form of graphing called parametric graphing, which is needed to draw the circle.) Select "Dot" in the fifth line.

- *Enter the equations to graph the circle.* Press [Y=]. Parametric graphing requires two equations, one for the *x*-coordinate and one for the *y*-coordinate. Enter these equations: $X_{1T} = \cos(T)$ and $Y_{1T} = \sin(T)$. To obtain T, press either [X,T,θ,n] or [ALPHA] 4. T is a variable representing the measure of the angle in standard position.

- *Set the graphing window.* Press [WINDOW]. The first three lines indicate the first and last values of T to be used, and the increment. Enter Tmin = 0, Tmax = 360, and Tstep = 10. This instructs the calculator to use angle measures from 0° to 360°, increasing in steps of 10°. Use the values of Xmin, Xmax, Ymin, and Ymax shown on the screen above.

- *Graph the circle.* Press [GRAPH]. The result should be similar to that above.

1. a) Use the trace feature to trace the plotted points. Find out how the operation of the arrow keys corresponds to the values of T indicated in the window menu.

  b) Try using other values of Tstep, Tmin, and Tmax.

2. When the trace feature is active, enter the measure of any angle in degrees. The cursor jumps to the new location and displays the corresponding coordinates. Do you think this would be a useful way to determine the cosine and sine of any angle? Explain.

3. a) Predict what the graph will look like if you use Tmin = 30, Tmax = 390, and Tstep = 90. Use a calculator to check your prediction.

  b) Without changing the window settings, predict what will happen if you use the trace feature. Use a calculator to check your prediction.

  c) Without changing the window settings, predict what will happen if you change "Dot" to "Connected" in the mode menu. Use a calculator to check your prediction.

4. Use a calculator to display these regular polygons.

a)

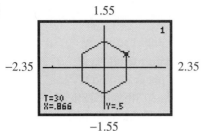

b)

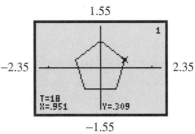

5. Use the diagram in exercise 4b. Determine each pair of values.

a) $\cos 162°$, $\sin 162°$ **b)** $\cos 198°$, $\sin 198°$ **c)** $\cos 342°$, $\sin 342°$

d) $\cos 72°$, $\sin 72°$ **e)** $\cos 108°$, $\sin 108°$ **f)** $\cos 288°$, $\sin 288°$

6. To produce the screens on page 180, the circle was graphed first and stored as a picture. Then the polygon was graphed and the picture recalled.

a) Explain why the calculator cannot graph the circle and the polygon together.

b) **i)** Produce a screen, similar to those on page 180, to show the square in different positions.

ii) Produce a screen, similar to those in exercise 4, to show a dotted circle passing through the vertices of the regular polygon.

7. In parametric graphing mode, the calculator used the equations $x = \cos t$ and $y = \sin t$ to plot the points. Explain why the plotted points lie on a unit circle with centre $O(0, 0)$.

8. Use a calculator to display these star polygons.

a)

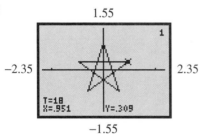

b)

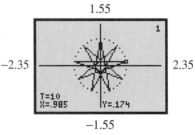

9. Use the diagram in exercise 8b. Determine each pair of values.

a) $\cos 170°$, $\sin 170°$ **b)** $\cos 190°$, $\sin 190°$ **c)** $\cos 350°$, $\sin 350°$

d) $\cos 80°$, $\sin 80°$ **e)** $\cos 100°$, $\sin 100°$ **f)** $\cos 280°$, $\sin 280°$

10. Display other star polygons like those in exercise 8.

3.5 Sine and Cosine Functions of Special Angles

When you use a calculator to evaluate trigonometric functions, the results are approximate to 10 decimal places. These approximations are accurate enough for most purposes. To determine these values, the calculator is programmed to use one of several formulas established by mathematicians in the 17th and 18th centuries (see page 347). For some special angles, the values can be determined from geometric relationships.

Angle $\frac{\pi}{4}$ and its multiples

This diagram shows the angle 45°, or $\frac{\pi}{4}$ in standard position.

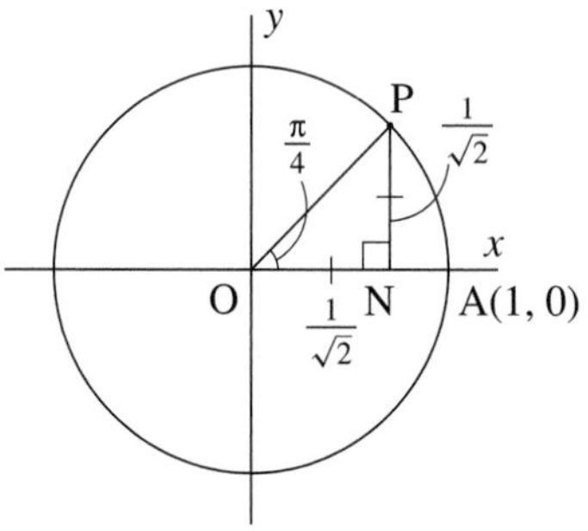

Since P is on a unit circle and PN is perpendicular to OA, then △PON is a right isosceles triangle with OP = 1.

Then ON = NP

Use the Pythagorean Theorem in △OPN.

$$ON^2 + NP^2 = OP^2$$
$$2ON^2 = 1$$
$$ON^2 = \frac{1}{2}$$
$$ON = \frac{1}{\sqrt{2}}$$

So, NP = ON = $\frac{1}{\sqrt{2}}$, or $\frac{\sqrt{2}}{2}$

The coordinates of P are $\left(\frac{1}{\sqrt{2}}, \frac{1}{\sqrt{2}}\right)$, or $\left(\frac{\sqrt{2}}{2}, \frac{\sqrt{2}}{2}\right)$.

Hence, $\cos\frac{\pi}{4} = \frac{1}{\sqrt{2}}$, or $\frac{\sqrt{2}}{2}$, and $\sin\frac{\pi}{4} = \frac{1}{\sqrt{2}}$, or $\frac{\sqrt{2}}{2}$

$$\cos\frac{\pi}{4} = \frac{1}{\sqrt{2}}, \text{ or } \frac{\sqrt{2}}{2} \qquad \sin\frac{\pi}{4} = \frac{1}{\sqrt{2}}, \text{ or } \frac{\sqrt{2}}{2}$$

We can determine the sine and cosine of any multiple of $\frac{\pi}{4}$ by drawing the right isosceles triangle in different positions. This triangle is called a 45-45-90 triangle.

Example 1

Determine $\cos \frac{3\pi}{4}$ and $\sin \frac{3\pi}{4}$.

Solution

Draw a diagram.

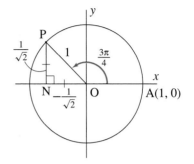

Since $\angle POA = \frac{3\pi}{4}$, then $\angle PON = \frac{\pi}{4}$

$\triangle PON$ is a 45-45-90 triangle with sides $\frac{1}{\sqrt{2}}, \frac{1}{\sqrt{2}}$, and 1.

Since P is in the second quadrant, $\cos \frac{3\pi}{4}$ is negative

and $\sin \frac{3\pi}{4}$ is positive.

Hence, $\cos \frac{3\pi}{4} = -\frac{1}{\sqrt{2}}$, or $-\frac{\sqrt{2}}{2}$, and $\sin \frac{3\pi}{4} = \frac{1}{\sqrt{2}}$, or $\frac{\sqrt{2}}{2}$

In *Example 1*, $\angle PON$ is called the reference angle. A *reference angle* is the acute angle between the terminal arm and the *x*-axis.

Angle $\frac{\pi}{6}$ and its multiples

This diagram shows the angle 30°, or $\frac{\pi}{6}$ in standard position.

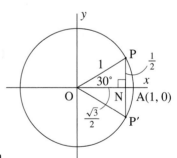

Since P is on a unit circle and PN is perpendicular to OA, then $\triangle PON$ is a 30-60-90 triangle with OP = 1.

By reflecting $\triangle PON$ in the *x*-axis, an equilateral $\triangle POP'$ is formed, in which N is the midpoint of PP'.

Hence, NP = $\frac{1}{2}$

Use the Pythagorean Theorem in $\triangle PON$ to determine the length of ON.

$$ON^2 + NP^2 = OP^2$$
$$ON^2 + \frac{1}{4} = 1$$
$$ON^2 = \frac{3}{4}$$
$$ON = \frac{\sqrt{3}}{2}$$

The coordinates of P are $\left(\frac{\sqrt{3}}{2}, \frac{1}{2} \right)$.

Hence, $\cos \frac{\pi}{6} = \frac{\sqrt{3}}{2}$ and $\sin \frac{\pi}{6} = \frac{1}{2}$

To determine the sine and cosine of 60°, or $\frac{\pi}{3}$, reflect the 30-60-90 triangle in the line $y = x$.

The coordinates of P are $\left(\frac{1}{2}, \frac{\sqrt{3}}{2}\right)$.

Hence, $\cos \frac{\pi}{3} = \frac{1}{2}$ and $\sin \frac{\pi}{3} = \frac{\sqrt{3}}{2}$

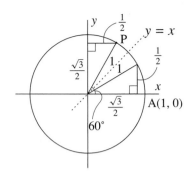

$$\cos 30° = \cos \frac{\pi}{6} = \frac{\sqrt{3}}{2} \qquad \sin 30° = \sin \frac{\pi}{6} = \frac{1}{2}$$

$$\cos 60° = \cos \frac{\pi}{3} = \frac{1}{2} \qquad \sin 60° = \sin \frac{\pi}{3} = \frac{\sqrt{3}}{2}$$

Visualize θ increasing from 0° to 90°. Since $\sin 0° = 0$ and $\sin 90° = 1$, $\sin \theta$ increases from 0 to 1. To remember that $\sin 30° = \frac{1}{2}$, observe that when θ is $\frac{1}{3}$ of the way from 0° to 90°, $\sin \theta$ is $\frac{1}{2}$ of the way from 0 to 1.

By drawing the 30-60-90 triangle in different positions, we can determine the cosine and sine of any multiple of $\frac{\pi}{6}$ and $\frac{\pi}{3}$.

Example 2

Determine $\cos \frac{4\pi}{3}$ and $\sin \frac{4\pi}{3}$.

Solution

Draw a diagram.

The reference angle is $\angle PON = \frac{4\pi}{3} - \pi$, or $\frac{\pi}{3}$.

$\triangle PON$ is a 30-60-90 triangle with sides $\frac{1}{2}$, 1, and $\frac{\sqrt{3}}{2}$.

Since P is in the third quadrant, both $\cos \frac{4\pi}{3}$ and $\sin \frac{4\pi}{3}$ are negative. Hence, $\cos \frac{4\pi}{3} = -\frac{1}{2}$ and $\sin \frac{4\pi}{3} = -\frac{\sqrt{3}}{2}$

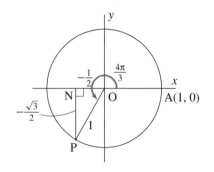

The trigonometric function values of $\frac{\pi}{4}$, or 45° and $\frac{\pi}{6}$, or 30° and their multiples are either 0, $\pm\frac{1}{2}$, ± 1, or simple expressions involving radicals. These are not the only angles whose trigonometric function values can be expressed in this way. Other examples occur in the exercises and in Chapter 5.

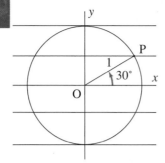

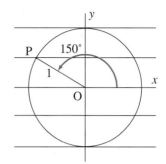

$\sin 30° = \frac{1}{2}$

θ is $\frac{1}{3}$ of the way from 0° to 90°.

$\sin \theta$ is $\frac{1}{2}$ of the way from 0 to 1.

$\sin 150° = \frac{1}{2}$

θ is $\frac{2}{3}$ of the way from 90° to 180°.

$\sin \theta$ is $\frac{1}{2}$ of the way from 1 to 0.

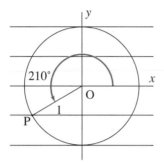

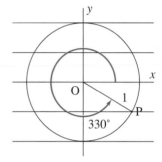

$\sin 210° = -\frac{1}{2}$

θ is $\frac{1}{3}$ of the way from 180° to 270°.

$\sin \theta$ is $\frac{1}{2}$ of the way from 0 to −1.

$\sin 330° = -\frac{1}{2}$

θ is $\frac{2}{3}$ of the way from 270° to 360°.

$\sin \theta$ is $\frac{1}{2}$ of the way from −1 to 0.

DISCUSSING THE IDEAS

1. On page 184, we showed that $\cos \frac{\pi}{4}$ and $\sin \frac{\pi}{4}$ are both equal to $\frac{\sqrt{2}}{2}$.

a) Explain why these two values should be equal.

b) Are there any other angles that have the same cosine and sine? Explain.

2. How could you check the results of *Example 1* or *Example 2*?

3. In *Visualizing*, explain how the results would change if radians were used instead of degrees.

3.5 EXERCISES

 1. What is the advantage of using geometric relationships instead of a calculator to determine the sines and cosines of special angles?

2. a) Draw a diagram to show each angle in standard position.

   i) $\frac{5\pi}{4}$ **ii)** $-\frac{7\pi}{6}$ **iii)** $\frac{7\pi}{4}$ **iv)** $\frac{4\pi}{3}$

   v) $\frac{5\pi}{6}$ **vi)** $-\frac{2\pi}{3}$ **vii)** $-\frac{3\pi}{4}$ **viii)** $\frac{5\pi}{3}$

   b) State the reference angle for each angle in part a.

3. State each exact value.

   a) $\sin \frac{\pi}{4}$ **b)** $\sin \frac{3\pi}{4}$ **c)** $\sin \left(-\frac{5\pi}{4}\right)$ **d)** $\sin \frac{7\pi}{4}$

   e) $\sin \frac{\pi}{6}$ **f)** $\sin \left(-\frac{5\pi}{6}\right)$ **g)** $\sin \frac{7\pi}{6}$ **h)** $\sin \frac{11\pi}{6}$

   i) $\sin \frac{\pi}{3}$ **j)** $\sin \frac{2\pi}{3}$ **k)** $\sin \left(-\frac{4\pi}{3}\right)$ **l)** $\sin \frac{5\pi}{3}$

4. Repeat exercise 3, replacing each sine with a cosine.

5. You know that $\sin 45° = \frac{\sqrt{2}}{2}$ and $\sin 60° = \frac{\sqrt{3}}{2}$. Look at the numbers in these expressions. The denominators are the same, and the numbers under the radical signs are 2 and 3.

   a) Find out if this is part of a pattern that includes the values of $\sin 0°$, $\sin 30°$, and $\sin 90°$.

   b) Find out if the cosine values have a similar pattern.

B **6.** Determine the exact value of each expression.

   a) $\sin \frac{7\pi}{3}$ **b)** $\cos \frac{7\pi}{3}$ **c)** $\sin \frac{9\pi}{4}$ **d)** $\cos \frac{9\pi}{4}$

   e) $\sin \frac{8\pi}{3}$ **f)** $\cos \frac{8\pi}{3}$ **g)** $\sin \frac{11\pi}{4}$ **h)** $\cos \frac{11\pi}{4}$

7. a) Triangle ABC is a 45-45-90 triangle. Verify that
   $$\sin 2A + \sin 2B + \sin 2C = 4 \sin A \sin B \sin C$$

   b) Determine if the equation in part a holds for each triangle.

   i) a 30-60-90 triangle   **ii)** an equilateral triangle

8. The graphing calculator screens on the next page show regular polygons whose vertices lie on a unit circle with centre O(0, 0). For each regular polygon:

   i) Write the coordinates of each vertex.

   ii) Write the angles in standard position, in degrees, corresponding to each vertex.

   iii) Write the cosine and the sine of each angle in part ii.

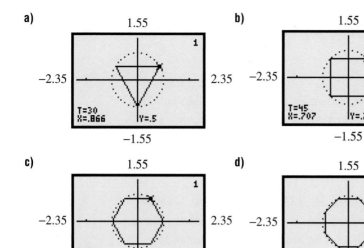

a)
1.55
1
−2.35 | 2.35
T=30
X=.866   Y=.5
−1.55

b)
1.55
1
−2.35 | 2.35
T=45
X=.707   Y=.707
−1.55

c)
1.55
1
−2.35 | 2.35
T=60
X=.5   Y=.866
−1.55

d)
1.55
1
−2.35 | 2.35
T=22.5
X=.924   Y=.383
−1.55

9. Write to explain how you determined the coordinates of the vertices of the octagon in exercise 8d.

10. a) For each polygon in exercise 8a, b, and c:

　　i) Write each angle in radians.

　　ii) Write the exact value of the cosine and sine of each angle.

　b) For the octagon in exercise 8d:

　　i) Write each angle in radians.

　　ii) Explain why you cannot write the exact value of the cosine and sine of each angle.

11. Use a calculator to verify each result.

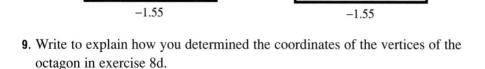

a) $\cos 15° = \dfrac{\sqrt{6} + \sqrt{2}}{4}$　　　　　**b)** $\cos 22.5° = \dfrac{\sqrt{2 + \sqrt{2}}}{2}$

c) $\cos 36° = \dfrac{\sqrt{5} + 1}{4}$　　　　　**d)** $\cos 72° = \dfrac{\sqrt{5} - 1}{4}$

C **12.** θ is an acute angle defined by a point $P(x, y)$ on a unit circle in the first quadrant. Use x and y to define the sine and cosine of each angle.

　a) $\pi + \theta$　　　　　　　　**b)** $2\pi - \theta$

COMMUNICATING THE IDEAS

Write to explain how to determine the sines and cosines of special angles, and how to remember them.

Graphing Sines and Cosines

You can use a graphing calculator to display a sequence of diagrams like the one on page 177, to show θ increasing in uniform increments. The calculator will make a table of values of θ, cos θ, and sin θ, then plot the points on a graph. The results show how the values of cos θ and sin θ vary as θ increases.

1. Ask your teacher for the program called CIRCLE from the Teacher's Resource Book. Enter the program into your calculator. When you run the program, the calculator asks for an angle increment: enter 15. When you press [ENTER], the calculator draws a circle. After the circle has been drawn, press [ENTER] several times in succession. A diagram like the one (below left) will gradually appear.

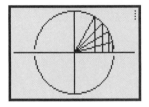
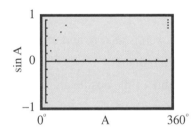

When the diagram is complete, press [ENTER] again. A menu will appear. Press 1 to select the sine graph. Press [ENTER] a few times. A graph like the one (above right) will gradually appear. This graph shows the sines plotted against the angles. When the calculator has completed the graph, press [STAT] [ENTER] to see the table of values. Then press [QUIT] to get back to the usual screens.

2. Visualize what the screens above would look like if you use a smaller increment such as 10° or 5°. Run the program to check.

3. Predict what the cosine graph would look like. Use the program to check your prediction.

4. There are two angles between 0° and 360° for which the sine and cosine values are equal. Which angles are they? Explain, using both the unit circle diagram and the graphs of $y = \sin \theta$ and $y = \cos \theta$.

Note: The CIRCLE program sets the angle mode to degrees. If you want radians, you will have to reset the mode after you have finished with the program.

Graphing $y = \sin \theta$ and $y = \cos \theta$

You can use a graphing calculator to graph the functions $y = \sin \theta$ and $y = \cos \theta$. You must be careful when you set the window. The window settings depend on whether you are using degrees or radians, and on the results desired.

Graphing $y = \sin \theta$ and $y = \cos \theta$, where θ is in degrees

In exercises 1 to 5, the calculator must be in degree mode.

1. To graph $y = \sin \theta$ for $-360° \le \theta \le 360°$, use these numbers in the window menu, and enter the equation Y1 = sin(X). Use the trace key to trace along the curve. What are the advantages and disadvantages of these window settings?

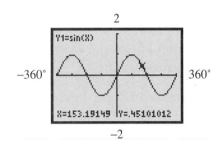

2. When tracing on the TI-83, you can make the cursor jump to the point on the curve for any value of x that is in the window.

   a) Make sure the trace function is still active. Enter 150 to display "X=150," then press ENTER. Is the value of y what you expected?

   b) Predict what other values of x should produce the same value of y as in part a. Enter these values to check your predictions.

   c) Repeat part b for values of x that produce the opposite value of y.

   d) Repeat parts a to c. Start with 120°, then with 135°.

3. a) Predict the θ-intercepts of the curve. Use the zero feature in the calculate menu to check your predictions.

   b) Predict the coordinates of the maximum and minimum points. Use the maximum and minimum features in the calculate menu to check.

On the TI-83, the display is 95 pixels wide. Allowing one pixel for the y-axis, there are 47 pixels on each side. Take advantage of this to obtain more convenient values of x when tracing.

4. a) Change the window settings to graph the function for $-470° \le \theta \le 470°$. Trace to the point where $x = 150$, and observe the value of y.

   b) Predict other values of x that should produce the same value of y. Trace to those points to check your predictions.

c) Predict values of x that should produce the opposite value of y. Check by tracing.

d) Repeat parts a to c. Start with $120°$, then with $135°$.

5. Repeat exercises 1 to 4, using $y = \cos \theta$.

Graphing $y = \sin \theta$ and $y = \cos \theta$, where θ is in radians

In exercises 6 to 10, the calculator must be in radian mode.

6. To graph $y = \sin \theta$ for $-2\pi \le \theta \le 2\pi$, use these numbers in the window menu and enter the equation Y1 = sin (X). Use the trace key to trace along the curve. What are the advantages and disadvantages of these window settings?

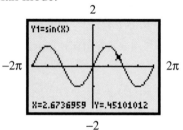

7. a) Make sure the trace function is still active. Press 5 $\boxed{\pi}$ $\boxed{\div}$ 6 to display "X=5π/6," then press $\boxed{\text{ENTER}}$. The cursor will jump to the point on the curve where $x = \dfrac{5\pi}{6}$. The coordinates of this point will be displayed in decimal form.

b) Predict what other values of x should produce the same value of y as in part a. Enter these values to check your predictions.

c) Repeat part b for values of x that produce the opposite value of y.

d) Repeat parts a to c. Start with $\dfrac{2\pi}{3}$, then with $\dfrac{3\pi}{4}$.

8. Repeat exercise 3.

To obtain more convenient values of x when tracing, use the fact that there are 47 pixels on each side of the y-axis. Since we used $-470° \le \theta \le 470°$ when θ was in degrees, we will use the corresponding interval for θ in radians.

9. a) Convert $470°$ to radians.

b) Change the window settings to accommodate the result of part a. Trace to the point where $x = \dfrac{5\pi}{6}$, and observe the value of y.

c) Predict other values of x that should produce the same value of y. Trace to those points to check your predictions.

d) Predict values of x that should produce the opposite value of y. Check by tracing.

e) Repeat parts b to d. Start by tracing to the point where $x = \dfrac{2\pi}{3}$.

10. Repeat exercises 6 to 9, using $y = \cos \theta$.

To graph the functions $y = \sin \theta$ and $y = \cos \theta$, recall their definitions. When P(x, y) is any point on the unit circle forming an angle θ in standard position, the coordinates of P are $(\cos \theta, \sin \theta)$.

Visualize P rotating counterclockwise around the circle starting at A$(1, 0)$. As θ increases, the coordinates of P change periodically. Hence, there is a periodic change in the values of $\cos \theta$ and $\sin \theta$.

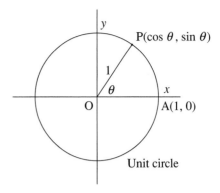

Unit circle

Graphing the function y = sin θ

Suppose θ starts at 0 and increases to π. Recall the patterns from page 187. Sin θ changes as follows.

$\theta = 0$	$\theta = \dfrac{\pi}{6}$	$\theta = \dfrac{\pi}{2}$	$\theta = \dfrac{5\pi}{6}$	$\theta = \pi$

$\sin \theta = 0$	$\sin \theta = \dfrac{1}{2}$	$\sin \theta = 1$	$\sin \theta = \dfrac{1}{2}$	$\sin \theta = 0$

We use these results to sketch the graph for $0 \le \theta \le \pi$.

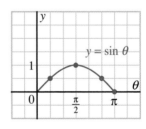

$y = \sin \theta$

Suppose θ continues from π to 2π. Then $\sin \theta$ changes as follows.

$\theta = \pi$	$\theta = \dfrac{7\pi}{6}$	$\theta = \dfrac{3\pi}{2}$	$\theta = \dfrac{11\pi}{6}$	$\theta = 2\pi$

 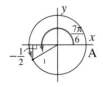

$\sin \theta = 0$	$\sin \theta = -\dfrac{1}{2}$	$\sin \theta = -1$	$\sin \theta = -\dfrac{1}{2}$	$\sin \theta = 0$

We use these results to sketch the
graph for $0 \leq \theta \leq 2\pi$.

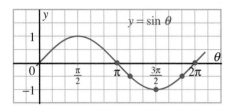

As θ continues beyond 2π, P rotates around the circle again, and the same
values of $\sin \theta$ are encountered. Hence, the graph can be continued to the right.
Similarly, the graph can be continued to the left, corresponding to a rotation in
the clockwise direction. The patterns in the graph repeat every 2π in each
direction. When θ is in radians, the period of this function is 2π.

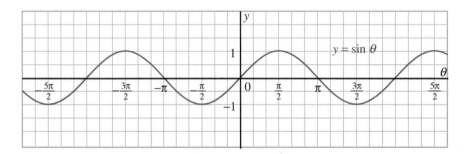

A *cycle* of a periodic function is a part of its graph from any point to the first
point where the graph starts repeating.
The *period* of a periodic function of θ is the difference in the values of θ for the
points at the ends of a cycle.

Two cycles of the function $y = \sin \theta$ are shown below. The period can be
calculated using the values of θ corresponding to the endpoints of the cycles.

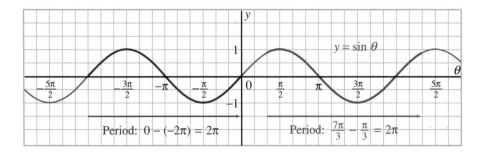

Graphing the function *y* = cos *θ*

We can graph the function $y = \cos \theta$ using the same method we used to graph the function $y = \sin \theta$. This is the result.

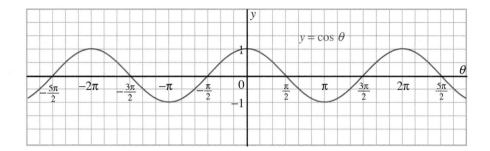

The function $y = \cos \theta$ has a period of 2π. Its graph is congruent to the graph of $y = \sin \theta$, but it is shifted horizontally so that it intersects the *y*-axis at $(0, 1)$ instead of $(0, 0)$.

Since these curves look like repeating waves, we should be able to use sine and cosine functions in applications involving quantities that rise and fall periodically. To do this, we must be able to work with them when their maximum and minimum values are different from 1 and −1, and their periods are different from 2π. This involves taking the basic graphs in this section, and expanding or compressing them in vertical or horizontal directions, as well as changing their positions relative to the axes. When these changes are made to the graphs, corresponding changes occur in the equations. In Chapter 4, we will investigate how the changes in the equations are related to the changes in the graphs.

DISCUSSING THE IDEAS

1. a) Explain why the graph of $y = \sin \theta$ is shaped like a wave.

   b) Explain why the graph of $y = \cos \theta$ has the same shape as the graph of $y = \sin \theta$.

2. Describe and account for any symmetry in the graph of each function.

   a) $y = \sin \theta$ b) $y = \cos \theta$

3. It was stated above that the graph of $y = \cos \theta$ is congruent to the graph of $y = \sin \theta$. Explain what this means.

3.6 EXERCISES

A 1. In these diagrams, graphs of $y = \sin \theta$ were started using different scales. Copy each graph onto grid paper, then extend it for the number of cycles indicated.

a) 2 cycles

b) 2 cycles

c) 1 cycle

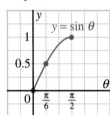

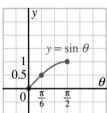

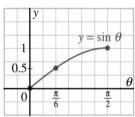

2. **Graphing the function $y = \cos \theta$**

Let P be a point on the terminal arm of an angle θ in standard position on a unit circle.

a) Suppose θ starts at 0 and increases to π. Use diagrams like those on page 193 corresponding to $\theta = 0, \frac{\pi}{3}, \frac{\pi}{2}, \frac{2\pi}{3}$, and π to determine values of $\cos \theta$. Use the results to sketch the graph of $y = \cos \theta$ for $0 \leq \theta \leq \pi$.

b) Suppose θ continues from π to 2π. Determine values of $\cos \theta$ for $\theta = \frac{4\pi}{3}$, $\frac{3\pi}{2}, \frac{5\pi}{3}$, and 2π. Use the results to continue the graph from π to 2π.

c) Continue the graph of $y = \cos \theta$ for values of θ greater than 2π and less than 0.

3. In these diagrams, graphs of $y = \cos \theta$ were started using different scales. Copy each graph onto grid paper, then extend it for the number of cycles indicated.

a) 2 cycles

b) 2 cycles

c) 4 cycles

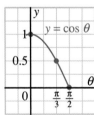

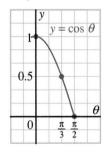

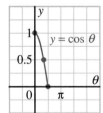

4. The graphing calculator screen (below left) shows the coordinates of one point on the graph of $y = \sin x$, where x is in degrees. Use only the information on the screen. Write the x-coordinates of four other points on the graph that have the same y-coordinate as this one.

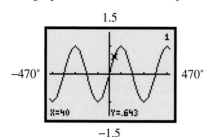

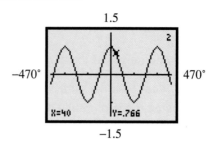

5. The graphing calculator screen (above right) shows the coordinates of one point on the graph of $y = \cos x$, where x is in degrees. Use only the information on the screen. Write the x-coordinates of four other points on the graph that have the same y-coordinate as this one.

6. Choose exercise 4 or 5. Write to explain how you determined the x-coordinates.

7. a) Graph $y = \sin x$ using a graphing window defined by $-470° \le x \le 470°$ and $-1.5 \le y \le 1.5$. Trace to the point where $x = 30°$. Observe the corresponding value of y.

b) Predict some other values of x that produce the same value of y. Trace to these points to verify your predictions.

8. a) Graph $y = \cos x$ using the same graphing window as in exercise 7. Trace to the point where $x = 30°$. Observe the corresponding value of y.

b) Predict some other values of x that produce the same value of y. Trace to these points to verify your predictions.

c) Compare your answers in part b with your answers to exercise 7b. Write to explain why some x-coordinates are the same, and why some are different.

9. Without making a table of values, draw a graph of $y = \sin \theta$ for $-2\pi \le \theta \le 2\pi$.

10. For the graph of $y = \sin \theta$:

a) What is the maximum value of y? For what values of θ does this occur?

b) What is the minimum value of y? For what values of θ does this occur?

c) What is the domain of the function?

d) What is the range of the function?

e) What are the y-intercepts?

f) What are the θ-intercepts?

11. Repeat exercises 9 and 10 for the graph of $y = \cos \theta$.

12. Compare the graphs of $y = \sin \theta$ and $y = \cos \theta$. Describe each relationship.

    a) among the values of θ that produce maximum values of y

    b) among the values of θ that produce minimum values of y

    c) among the θ-intercepts

    d) among the domains and ranges of the two functions

13. Graph $y = \sin \theta$ and $y = \cos \theta$ on the same axes in an appropriate graphing window. Use either degree mode or radian mode.

    a) Describe how the two graphs appear to be related.

    b) Use the intersect feature. Determine the coordinates of some points of intersection of the two graphs. Explain the results.

    c) When $0° < \theta \le 90°$ or $0 \le \theta \le \frac{\pi}{2}$, how can you tell which is greater, $\sin \theta$ or $\cos \theta$? Explain.

14. Suppose you forget that your graphing calculator is in radian mode, and you use it to graph $y = \sin \theta$ for $-360° \le \theta \le 360°$.

    a) What would the graph look like? Explain.

    b) Use your graphing calculator to check your prediction.

C **15.** Suppose the angle θ is in degrees instead of radians.

    a) How would the graph of $y = \sin \theta$ on page 194 change?

    b) Is the graph of $y = \sin \theta$ when θ is in degrees the same graph as the graph of $y = \sin \theta$ when θ is in radians? Explain.

16. Compare the two graphs in exercises 4 and 5. The x-coordinates of the indicated points are the same.

    a) How are the y-coordinates of these points related? Explain.

    b) Is the relationship you identified in part a true for all points with the same x-coordinate on the two graphs, or only for the points indicated? Explain.

COMMUNICATING THE IDEAS

Compare the graphs of the sine and cosine functions. In what ways are they similar? In what ways are they different? Write to explain the similarities and differences. Use examples to illustrate your explanations.

Sketching Sine or Cosine Curves

From *Visualizing*, page 187, recall the pattern that applies to the values of $\sin \theta$ as θ increases from 0° to 90°, to 180°, to 270°, to 360°. For example, when θ is $\frac{1}{3}$ of the way from 0° to 90°, $\sin \theta$ is $\frac{1}{2}$ of the way from 0 to 1, and so on. You can use this pattern to sketch the graphs of sine or cosine functions without using technology.

To sketch the graph of the function $y = \sin \theta$:

Step 1. Start at the origin. Move 1 square right and 1 square up, then 2 squares right and 1 square up. This completes the part of the graph up to the first maximum point, corresponding to $\sin 90° = 1$.

Step 2. Move 2 squares right and 1 square down, then 1 square right and 1 square down. This completes the part back to the θ-axis, corresponding to $\sin 180° = 0$.

Step 3. Move 1 square right and 1 square down, then 2 squares right and 1 square down. This completes the part down to the minimum point, corresponding to $\sin 270° = -1$.

Step 4. Move 2 squares right and 1 square up, then 1 square right and 1 square up. This completes one cycle of the graph.

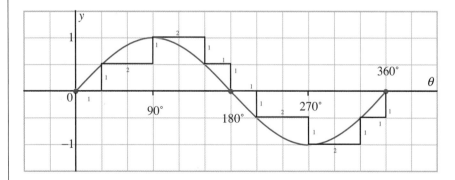

By repeating these steps, we can sketch additional cycles of $y = \sin \theta$ to the right. By applying similar steps, we can sketch additional cycles to the left.

1. Use this method to sketch the graph of $f(\theta) = \sin \theta$ for $-2\pi \le \theta \le 2\pi$.

2. Use this method to sketch the graph of $f(\theta) = \cos \theta$ for $-2\pi \le \theta \le 2\pi$.

Predicting Tide Levels

On page 156, we considered the problem of predicting tide levels at times between low and high tides. Since a sine curve is periodic with minimum and maximum points, just like the tide levels, it is appropriate to use it as a model to predict the tide level at any time. To develop the model, we will use the data from page 156, which is repeated below with times given on a 24-h clock in decimals of hours. For example, the first low tide occurs at 5:09 A.M., which is $5 + \frac{9}{60}$ h, or 5.15 h. Check that the other times have been converted correctly.

	Times of Low and High Tides			
Sun	5.15 h	11.32 h	17.67 h	23.70 h
	1.19 m	2.86 m	1.11 m	2.71 m

 **DEVELOP A MODEL**

Points corresponding to the times and levels in the table are plotted on the graph below. Visualize these points representing the maximum and minimum points on a sine curve that is stretched horizontally and vertically by the correct amounts to occupy the position shown. Hence, we will use a sine curve in our model. We do this by scaling the sine curve to fit the data represented by the graph.

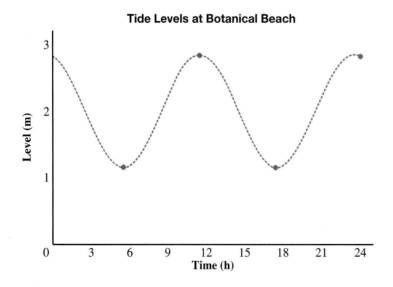

Tide Levels at Botanical Beach

The following properties of a sine curve must be present.

- The two minimum levels must be the same.
- The two maximum levels must be the same.
- All the times between consecutive minimum and maximum levels must be the same.

To ensure these propertie are present, we make some minor adjustments to the data. Changing the numbers slightly should not affect the predicted results much.

- For the minimum level, use the mean of the two minima, or 1.15 m.
- For the maximum level, use the mean of the two maxima, or 2.79 m.
- Adjust the times so they are in arithmetic sequence, as follows.

  Time from first minimum to second maximum: 23.70 h − 5.15 h = 18.55 h

  Mean time between minima and maxima: $\frac{18.55\,h}{3} \doteq 6.18$ h

  Start at 5.15 h and add 6.18 h successively to obtain these times:
  5.15 h, 11.33 h, 17.51 h, 23.69 h

The adjusted data are shown in this table. Compare these numbers with the numbers in the previous table.

	Adjusted Times of Low and High Tides			
Sun	5.15 h	11.33 h	17.51 h	23.69 h
	1.15 m	2.79 m	1.15 m	2.79 m

1. Complete this exercise to make a large graph of the adjusted data similar to the one on the next page.

   a) Use the adjusted data in the table to plot the points.

   b) Determine the time and the level halfway between the first minimum and the next maximum. Plot this point on the graph, and label it A.

   c) Determine the time and the level halfway between the second minimum and the next maximum. Plot this point on the graph, and label it B.

   d) Draw the line AB. This line is called the *central axis*.

   e) Sketch a sine curve passing through the plotted points.

   f) Recall that the period is the length of one complete cycle. Determine the period in hours, and mark it on your graph.

   g) The *amplitude* of the function is the distance from the central axis to the maximum or minimum level. Determine the amplitude in metres, and mark it on your graph.

Tide Levels at Botanical Beach

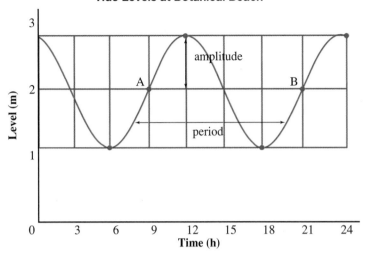

![MM] **LOOK AT THE IMPLICATIONS**

To determine the coordinates of any point D on the tide level graph, we compare it with the graph of the sine function. Visualize point A as the origin. Point C is vertically below D, at time 1:20 P.M. To calculate the tide level at 1:20 P.M., or 13.33 h, follow the steps on the next page.

Tide Graph

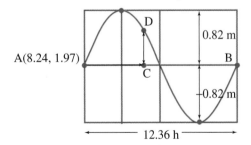

Sine Graph

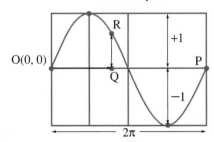

Step 1. Visualize corresponding points on the two graphs.

On the tide graph, 13.33 h corresponds to point C, with tide level at D. The corresponding points on the sine graph are Q and R.

Step 2. Express the position of C as a fraction of the period on the tide graph.

Distance from A to C on tide graph: 13.33 h − 8.24 h = 5.09 h

Hence, point C is $\frac{5.09}{12.36} \doteq 0.4118$ of the way from A to B.

Step 3. Determine the corresponding positions of Q and R on the sine graph.

Point Q is 0.4118 of the way from O to P.
Hence, the distance from O to Q is $2\pi \times 0.4118 \doteq 2.5874$.
Since $\sin 2.5874 \doteq 0.5263$, R is 0.5263 of the way from Q up to the maximum level.

Step 4. Determine the corresponding position of D on the tide graph.

On the tide graph, D is 0.5263 of the way from C up to the maximum level.
Hence, the distance from C to D is 0.5263×0.82 m $\doteq 0.43$ m.

Step 5. Use the tide level at A to determine the tide level at D.

Since the level at A is 1.97 m, the level at D is 1.97 m + 0.43 m, or 2.40 m.

2. Use this method to determine the tide level at each time. Mark the times on your graph.

   a) 8:30 A.M.   **b)** 9:30 A.M.   **c)** 2:15 P.M.   **d)** 7:20 P.M.

 REVISIT THE SITUATION

When you completed exercise 2, you probably found that you were repeating the same sequence of steps, but working with new numbers each time. If we examine the example above we can formulate a general rule to produce a shortcut for the process. In that example, the following information was given:

- the time and height corresponding to point A: 8.24 h and 1.97 m
- the period and amplitude of the tide graph: 12.36 h and 0.82 m
- the time corresponding to point C: 13.33 h

The problem was to calculate the tide level corresponding to point D. We found that it was 2.40 m. Let us examine how the given numbers 8.24, 1.97, 0.82, 12.36, and 13.33 were combined to arrive at this height.

Step

In Step 2, we subtracted 8.24 from 13.33 and divided by 12.36.

In Step 3, we multiplied by 2π and took the sine of the result.

In Step 4, we multiplied by 0.82.

In Step 5, we added 1.97.

Result

$\dfrac{13.33 - 8.24}{12.36}$

$\sin 2\pi \left(\dfrac{13.33 - 8.24}{12.36} \right)$

$0.82 \sin 2\pi \left(\dfrac{13.33 - 8.24}{12.36} \right)$

$0.82 \sin 2\pi \left(\dfrac{13.33 - 8.24}{12.36} \right) + 1.97$

Observe how the final expression is built up from the given information. The general pattern in the equation of the function is shown below.

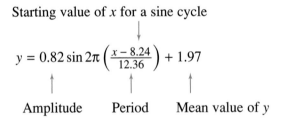

Starting value of x for a sine cycle

$$y = 0.82 \sin 2\pi \left(\dfrac{x - 8.24}{12.36} \right) + 1.97$$

Amplitude   Period   Mean value of y

3. Use the equation above to predict the tide level at 10:30 A.M.

4. Refer to the graph of sunset times at Saskatoon on page 158. On March 21 and September 21, the sun sets at 19.2 h. The earliest sunset time is 16.9 h and the latest is 21.5 h. Use an equation similar to the one above to predict the times of the sunset on May 24 and October 10.

5. Explain how the cosine function could have been used instead of the sine function in the above analysis.

3.7 The Tangent Function of an Angle

Another function associated with an angle in standard position is called the tangent function. This name is used because the function involves the tangent line to the unit circle at A(1, 0).

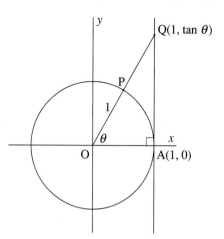

Let P be any point on the unit circle. Then P is on the terminal arm of an angle θ in standard position. Let Q be the point where the line OP intersects the tangent line at A(1, 0). We define the *tangent* of θ as the y-coordinate of Q, and we write it as tan θ.

Therefore, $(1, \tan \theta)$ are the coordinates of the point where the line OP intersects the tangent line to the circle passing through A. As P rotates around the circle, the value of tan θ changes periodically. Some values of tan θ are illustrated in these diagrams.

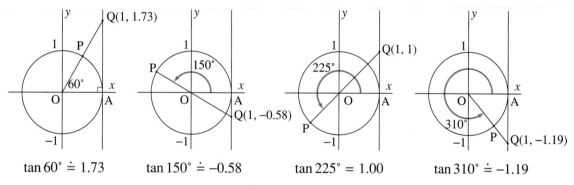

$$\tan 60° \doteq 1.73 \qquad \tan 150° \doteq -0.58 \qquad \tan 225° = 1.00 \qquad \tan 310° \doteq -1.19$$

We can use this definition to determine tan θ for certain multiples of 90°, or $\frac{\pi}{2}$.

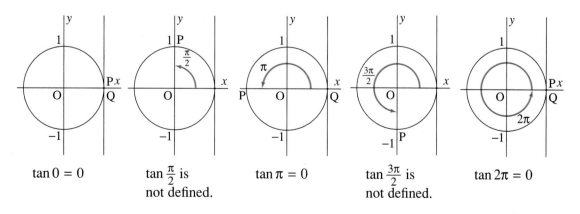

$$\tan 0 = 0 \qquad \tan \frac{\pi}{2} \text{ is not defined.} \qquad \tan \pi = 0 \qquad \tan \frac{3\pi}{2} \text{ is not defined.} \qquad \tan 2\pi = 0$$

$P(x, y)$ represents any point on the unit circle, forming an angle θ in standard position. The line through O and P intersects the tangent line through $A(1, 0)$ at Q. We define:

- The *tangent* of θ is the y-coordinate of Q.

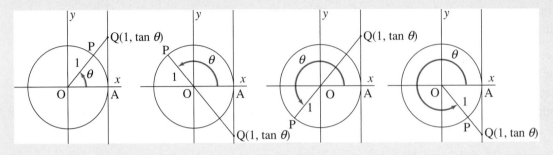

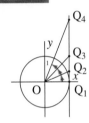

As P rotates around the unit circle starting at the positive x-axis, visualize how the y-coordinate of Q:

- changes from 0 to numbers becoming arbitrarily large and positive
- then becomes arbitrarily large and negative, and increases to 0

The cycle repeats every 180°, or π radians.

Example 1

Determine each value to 4 decimal places.

a) $\tan 162°$ **b)** $\tan 3.8$

Solution

a) Using degree mode on a calculator, $\tan 162° \doteq -0.3249$

b) Using radian mode on a calculator, $\tan 3.8 \doteq 0.7736$

As P rotates around the circle, past 180° or π, the same values of y are encountered as before. Hence, in *Example 1*, there are infinitely many other angles that have the same tangent as 162°, or the same tangent as 3.8 radians.

These are also equal to $\tan 162°$.
$\tan(162° + 180°)$, or $\tan 342°$
$\tan(162° - 180°)$, or $\tan(-18°)$

These are also equal to $\tan 3.8$.
$\tan(3.8 + \pi)$
$\tan(3.8 - \pi)$

Relating tan θ with sin θ and cos θ

In the diagram, since $\triangle PON$ and $\triangle QOA$ are similar triangles, $\dfrac{QA}{OA} = \dfrac{PN}{ON}$.

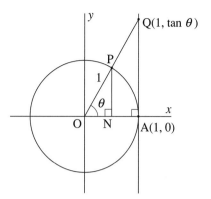

Therefore, $\dfrac{\tan \theta}{1} = \dfrac{\sin \theta}{\cos \theta}$

$\qquad\qquad \tan \theta = \dfrac{\sin \theta}{\cos \theta}$

The diagram shows θ in the first quadrant. When we use similar diagrams for θ in the other quadrants, we can show that this equation holds for any value of θ, except those for which $\cos \theta = 0$.

For any value of θ, $\qquad \tan \theta = \dfrac{\sin \theta}{\cos \theta}, \quad \cos \theta \ne 0$

In Section 3.5, we found exact values for the sines and cosines of certain special angles and their multiples. We can use the above equation to determine the tangents of the same angles. Here are three examples.

$$\tan \frac{\pi}{6} = \frac{\sin \frac{\pi}{6}}{\cos \frac{\pi}{6}}$$
$$= \frac{\frac{1}{2}}{\frac{\sqrt{3}}{2}}$$
$$= \frac{1}{2} \times \frac{2}{\sqrt{3}}$$
$$= \frac{1}{\sqrt{3}}, \text{ or } \frac{\sqrt{3}}{3}$$

$$\tan \frac{\pi}{4} = \frac{\sin \frac{\pi}{4}}{\cos \frac{\pi}{4}}$$
$$= \frac{\frac{1}{\sqrt{2}}}{\frac{1}{\sqrt{2}}}$$
$$= 1$$

$$\tan \frac{\pi}{3} = \frac{\sin \frac{\pi}{3}}{\cos \frac{\pi}{3}}$$
$$= \frac{\frac{\sqrt{3}}{2}}{\frac{1}{2}}$$
$$= \frac{\sqrt{3}}{2} \times \frac{2}{1}$$
$$= \sqrt{3}$$

$\tan \dfrac{\pi}{6} = \dfrac{1}{\sqrt{3}}, \text{ or } \dfrac{\sqrt{3}}{3}$ $\qquad\qquad$ $\tan \dfrac{\pi}{4} = 1$ $\qquad\qquad$ $\tan \dfrac{\pi}{3} = \sqrt{3}$

Example 2

Determine $\tan \dfrac{11\pi}{6}$.

Solution

Draw a diagram. The reference angle is $\dfrac{\pi}{6}$.
Label the triangle with the measures of its sides.

Since P is in the fourth quadrant, $\cos \dfrac{11\pi}{6}$ is
positive and $\sin \dfrac{11\pi}{6}$ is negative.

Hence, $\cos \dfrac{11\pi}{6} = \dfrac{\sqrt{3}}{2}$ and $\sin \dfrac{11\pi}{6} = -\dfrac{1}{2}$

Therefore, $\tan \dfrac{11\pi}{6} = \dfrac{-\dfrac{1}{2}}{\dfrac{\sqrt{3}}{2}}$

$= -\dfrac{1}{\sqrt{3}}$, or $-\dfrac{\sqrt{3}}{3}$

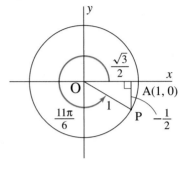

DISCUSSING THE IDEAS

1. In *Example 1*, why is tan 162° negative and tan 3.8 positive?

2. The tangent function is positive in some quadrants and negative in others. Explain how you can remember the quadrants in which the tangent function is positive and the quadrants in which it is negative.

3. Are there any maximum or minimum values for tan θ? Explain.

4. Explain why different angles can have the same tangent.

5. In *Visualizing*, page 206, what is meant by "arbitrarily large"?

A **1.** In each diagram, determine tan θ.

a)

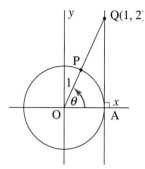

b)

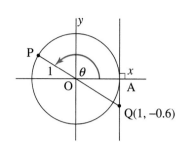

c)

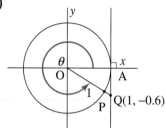

d)

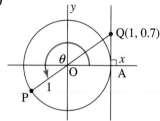

2. For each angle θ, calculate $\sin \theta$, $\cos \theta$, then $\tan \theta$. Do not use a calculator.

a) $\theta = 30°$

b) $\theta = -120°$

c) $\theta = 330°$

d) $\theta = 0$

e) $\theta = \frac{5\pi}{3}$

f) $\theta = -\frac{7\pi}{4}$

g) $\theta = \frac{\pi}{4}$

h) $\theta = \frac{3\pi}{4}$

i) $\theta = -\frac{3\pi}{2}$

3. State the quadrant in which each angle is found.

a) $\sin \theta < 0$, $\tan \theta < 0$

b) $\cos \theta < 0$, $\tan \theta > 0$

c) $\sin \theta < 0$, $\cos \theta < 0$

d) $\tan \theta < 0$, $\sin \theta > 0$

B **4. a)** Determine $\tan 25°$ to 4 decimal places.

b) Determine three other angles with the same tangent as 25°, and verify with a calculator.

5. a) Determine $\tan 1.92$ to 4 decimal places.

b) Determine three other angles with the same tangent as 1.92 radians, and verify with a calculator.

6. Recall exercise 5, page 188, where you investigated a pattern in the sines and cosines of certain angles. Find out if there is a similar pattern for tangents.

7. Point $P(x, y)$ is on the unit circle and also on the terminal arm of an angle θ in standard position. The tangent line at $A(1, 0)$ intersects line OP at Q.

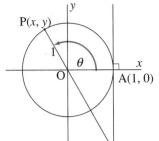

a) Write expressions for $\sin \theta$ and $\cos \theta$.

b) Substitute the results from part a into the expression $\tan \theta = \dfrac{\sin \theta}{\cos \theta}$.

c) Use the result of part b. Explain how to determine $\tan \theta$ when the coordinates of P are known.

8. The angle θ is in the second quadrant, and $\tan \theta = -\dfrac{3}{7}$.

a) Draw a diagram to show the angle in standard position and a point P on its terminal arm.

b) Determine possible coordinates for P.

9. Repeat exercise 8 for θ in the fourth quadrant, and $\tan \theta = -\dfrac{3}{2}$.

10. Repeat exercise 8 for θ in the third quadrant, and $\tan \theta = \dfrac{2}{5}$.

11. a) Graph $y = \dfrac{2}{3}x$.

b) Let θ represent the angle that the line $y = \dfrac{2}{3}x$ makes with the positive x-axis. Locate a point on the line, in the first quadrant, with integral coordinates. Use these coordinates to determine $\tan \theta$.

c) Compare the equation of the line with the result of part b. Make a conclusion based on your findings.

C 12. a) Suppose you know the tangent of an angle θ in standard position in the first quadrant. Write to explain how you would determine $\sin \theta$ and $\cos \theta$.

b) Suppose you know the sine of an angle θ in standard position in the second quadrant. Write to explain how you would determine $\cos \theta$ and $\tan \theta$.

c) Suppose you know the cosine of an angle θ in standard position in the third quadrant. Write to explain how you would determine $\sin \theta$ and $\tan \theta$.

COMMUNICATING THE IDEAS

Use the results of *Examples 1* and *2*, and the exercises in this section. Write to describe the different ways that $\tan \theta$ can be determined.

Graphing y = tan θ

1. Change to degree mode. To graph $y = \tan \theta$ for $-360° \le \theta \le 360°$, use these numbers in the window menu, and enter Y1=tan(X). Use the trace key to trace along the curve. Enter 60 to display "X=60," then press [ENTER].

   a) Is the value of y what you expected? Explain.

   b) Predict what other values of x should produce the same value of y and the opposite value of y. Enter these values to check your predictions.

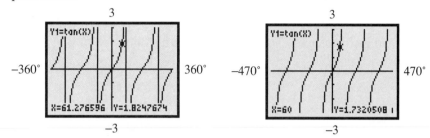

2. Change the settings in the window menu to $-470° \le \theta \le 470°$. Trace to the point where $x = 60$, and observe the value of y. Predict what other values of x should produce the same value of y and the opposite value of y. Check your predictions.

3. Change to radian mode. Change the window settings to graph $y = \tan \theta$ for $-2\pi \le \theta \le 2\pi$. Use the trace key to trace along the curve. Press [π] [÷] 3 to display "X=π/3," then press [ENTER]. Repeat exercise 1a and b.

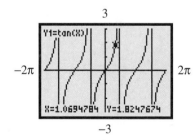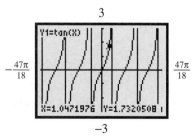

4. Change the window settings to those shown (above right). Trace to the point where $x = \frac{\pi}{3}$, and observe the value of y. Predict what other values of x should produce the same value of y and the opposite value of y. Check your predictions. Trace to the points where $x = \pm\frac{\pi}{2}$, $x = \pm\frac{3\pi}{2}$, and $x = \pm\frac{5\pi}{2}$. Explain the results.

3.8 Graphing the Tangent Function

To graph the function $y = \tan \theta$, recall the definition. Let P be a point on the unit circle, and also on the terminal arm of an angle θ in standard position. Then $\tan \theta$ is the y-coordinate of the point Q where the line OP intersects the tangent line to the circle at A(1, 0).

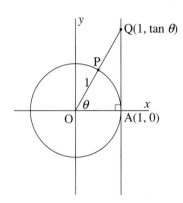

Visualize P rotating counterclockwise around the circle starting at A(1, 0). As θ increases, the position of Q on the tangent line changes periodically. This causes a periodic change in the values of $\tan \theta$.

Suppose θ starts at 0 and increases to $\frac{\pi}{2}$. Then $\tan \theta$ changes as follows.

$\theta = 0$

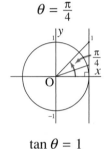

As θ increases from 0 to $\frac{\pi}{4}$, $\tan \theta$ increases from 0 to 1.

$\tan \theta = 0$

$\theta = \frac{\pi}{4}$

As θ increases from $\frac{\pi}{4}$ to $\frac{\pi}{2}$, $\tan \theta$ continues to increase.

$\tan \theta = 1$

Suppose θ continues from $\frac{\pi}{2}$ to π.

$\theta = \frac{\pi}{2}$

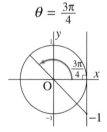

As θ increases from $\frac{\pi}{2}$ to $\frac{3\pi}{4}$, $\tan \theta$ increases to -1.

$\tan \theta$ is not defined.

$\theta = \frac{3\pi}{4}$

As θ increases from $\frac{3\pi}{4}$ to π, $\tan \theta$ increases from -1 to 0.

$\tan \theta = -1$

$\theta = \pi$

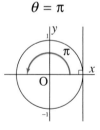

$\tan \theta = 0$

We use these results to sketch the graph of $y = \tan \theta$ for $0 < \theta \le \pi$. No value is plotted for $\tan \theta$ when $\theta = \frac{\pi}{2}$.

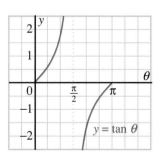

As θ continues beyond π, the line OP intersects the tangent line at A at the same locations as before, and the same values of $\tan \theta$ are encountered. Hence, the graph can be continued to the right. Similarly, the graph could be continued to the left, corresponding to a rotation in the counterclockwise direction.

When $\theta < \frac{\pi}{2}$, as θ increases and comes closer and closer to $\frac{\pi}{2}$, the graph comes closer and closer to the line $x = \frac{\pi}{2}$, but never reaches this line. Similarly, when $\theta > \frac{\pi}{2}$, as θ decreases and comes closer and closer to $\frac{\pi}{2}$, the graph comes closer and closer to the line $x = \frac{\pi}{2}$. For these reasons, the line $x = \frac{\pi}{2}$ is called an *asymptote*. The graph of $y = \tan \theta$ has asymptotes $x = \pm\frac{\pi}{2}$, $x = \pm\frac{3\pi}{2}$, $x = \pm\frac{5\pi}{2}$, ...

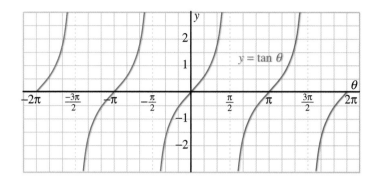

The graph above shows four cycles of the function $y = \tan \theta$.
When θ is in radians, the period of this function is π.

DISCUSSING THE IDEAS

1. Explain why the period of the function $y = \tan \theta$ is not the same as the period of the functions $y = \sin \theta$ or $y = \cos \theta$.

2. Describe and account for any symmetry in the graph of $y = \tan \theta$.

3. Explain how the diagrams on page 212 account for the asymptotes of the graph of the function $y = \tan \theta$.

3.8 EXERCISES

(A) 1. In these diagrams, graphs of $y = \tan \theta$ were started using different scales.
Copy each graph onto grid paper, then extend it for at least two cycles.

a)
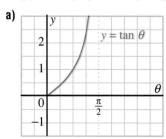

b)
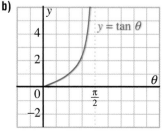

c)
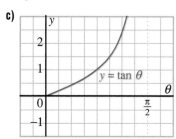

2. This graphing calculator screen shows the coordinates of one point on the graph of $y = \tan x$. Use only the information on the screen. Write the x-coordinates of four other points on the graph that have the same y-coordinate as this one.

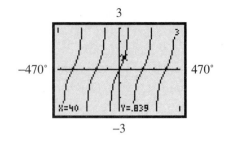

3. Write to explain how you determined the x-coordinates in exercise 2.

4. Compare the y-coordinate in the screen in exercise 2 with the y-coordinates in the screens on page 197. How are the three y-coordinates related? Explain.

5. Graph $y = \tan \theta$ for $-2\pi \le \theta \le 2\pi$.

B 6. For the graph of $y = \tan \theta$, determine:

   a) the domain   **b)** the range   **c)** the y-intercept   **d)** the θ-intercepts

7. A clinometer can determine the maximum height reached by a model rocket. The greatest angle of elevation, θ, for the rocket's trajectory and the distance, d metres, of the observer from the launch site can be measured.

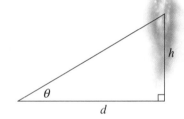

   a) Write h as a function of θ.

   b) Suppose you are 10 m from the launch site. Plot a graph of the function in part a for $0 \le \theta \le \frac{\pi}{2}$.

   c) Write to explain why the graph you constructed in part b is only useful for heights up to about 150 m.

   d) Use your graph. Explain why it is impossible to measure θ correctly as $\frac{\pi}{2}$.

   e) Write to explain how you could use the clinometer to determine the maximum height of any model rocket in flight.

   f) Why does the method described above require the rocket to be launched vertically upward, and with no wind?

8. Graph $y = \sin \theta$ and $y = \tan \theta$ on the same axes in an appropriate graphing window. Use either degree mode or radian mode.

   a) Describe some ways in which the two graphs appear to be related.

   b) When $0° < \theta < 90°$ or $0 < \theta < \frac{\pi}{2}$, which is greater, $\sin \theta$ or $\tan \theta$? Write to explain, using both the graph and the definitions of $\sin \theta$ and $\tan \theta$.

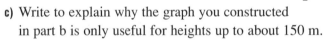

COMMUNICATING THE IDEAS

Compare the graph of the tangent function with the graph of either the sine or cosine function. In what ways are they similar? In what ways are they different? Write to explain the similarities and differences. Use examples to illustrate your explanations.

Trigonometric-Like Functions

Problem solving has been defined as the process of applying previously acquired knowledge to new and unfamiliar situations. The problems on this page illustrate examples of this definition.

In this case, your previously acquired knowledge comprise the definitions of the trigonometric functions of an angle in standard position. Recall that these were defined in terms of a point P rotating around a unit circle with centre O(0, 0).

You can apply this knowledge to the following new and unfamiliar situations. Suppose we replace the circle in the definition of the trigonometric functions with some other figure, such as a square. Could we define and graph functions that correspond to the trigonometric functions, using a square instead of a circle? Here are two possibilities.

The square could have vertical and horizontal sides:

The square could have vertical and horizontal diagonals:

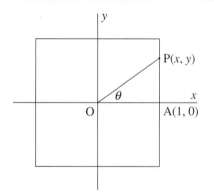

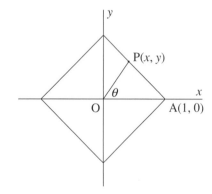

For exercises 1, 2, and 3, choose one situation above.

1. Define two functions that represent the horizontal and vertical coordinates of P (make up your own names for these functions). Draw a diagram. Use it to determine values of your functions for a few values of θ degrees.

2. Draw graphs of your functions for $0° \leq \theta \leq 360°$. You could do this with a spreadsheet or on grid paper.

3. List as many properties of the graphs as you can, including how they compare with the graphs of $y = \sin \theta$ and $y = \cos \theta$.

4. Another new and unfamiliar situation would be to use a unit circle, but with some point other than O(0, 0) as its centre. Draw such a circle and repeat exercises 1, 2, and 3 using this situation.

3.9 Reciprocal Trigonometric Functions

The functions $y = \sin \theta$, $y = \cos \theta$, and $y = \tan \theta$ are called the primary trigonometric functions. There are three other trigonometric functions, which are defined as the reciprocals of these functions.

Defining the reciprocal functions

The *cosecant*, *secant*, and *cotangent* functions are defined as the reciprocals of the sine, cosine, and tangent functions respectively.

$$\csc \theta = \frac{1}{\sin \theta}, \qquad \text{provided } \sin \theta \neq 0$$

$$\sec \theta = \frac{1}{\cos \theta}, \qquad \text{provided } \cos \theta \neq 0$$

$$\cot \theta = \frac{1}{\tan \theta}, \qquad \text{provided } \tan \theta \neq 0$$

There are no cosecant, secant, or cotangent keys on a calculator. These are not necessary, since you can find the sine or cosine of a number, then take its reciprocal.

Example 1

Determine each value to 4 decimal places.

a) $\csc 65°$ **b)** $\sec 4.38$ **c)** $\cot 295°$

Solution

> **Think...**
> Use a calculator to determine $\sin 65°$, $\cos 4.38$, and $\tan 295°$.
> Determine the reciprocal of each value.

a) Using degree mode: $\sin 65° \doteq 0.906\ 307\ 787$

    Take the reciprocal: $\csc 65° \doteq 1.1034$

b) Using radian mode: $\cos 4.38 \doteq -0.326\ 302\ 178$

    Take the reciprocal: $\sec 4.38 \doteq -3.0646$

c) Using degree mode: $\tan 295° \doteq -2.144\ 506\ 291$

    Take the reciprocal: $\cot 295° \doteq -0.4663$

Graphing the reciprocal functions

We can sketch the graphs of the reciprocal functions by using the properties of the graphs of the corresponding primary functions, along with the reciprocal tool introduced in Chapter 1.

Example 2

Graph the function $y = \csc \theta$ over three cycles.

Solution

Step 1.

Graph the function $y = \sin \theta$. Draw vertical dotted lines through the points where the graph intersects the θ-axis. At these points, $\sin \theta = 0$, and $\csc \theta$ is not defined.

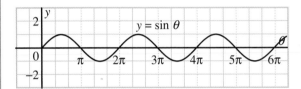

Step 2. Consider values of θ from 0 to π.

Since $\sin 0 = 0$, $\csc 0$ is undefined.

Since $\sin \pi = 0$, $\csc \pi$ is undefined.

As θ increases from 0 to $\frac{\pi}{2}$ and then to π, $\sin \theta$ increases from 0 to 1, then decreases to 0. According to the reciprocal tool, $\csc \theta$ decreases to 1, then increases.

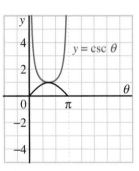

Step 3. Consider values of θ from π to 2π.

Since $\sin \pi = 0$, $\csc \pi$ is undefined.

Since $\sin 2\pi = 0$, $\csc 2\pi$ is undefined.

As θ increases from π to $\frac{3\pi}{2}$ and then to 2π, $\sin \theta$ decreases from 0 to -1, then increases to 0. According to the reciprocal tool, $\csc \theta$ increases to -1, then decreases.

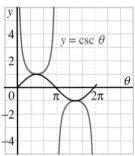

Step 4.

Since the sine function is periodic with period 2π, the cosecant function is also periodic with period 2π. Hence, Steps 2 and 3 complete one cycle. Other cycles of the graph of $y = \csc \theta$ can be completed by repeating the first cycle.

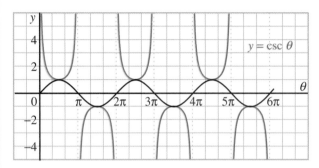

The functions $y = \sec \theta$ and $y = \cot \theta$ can be graphed in a similar way.

DISCUSSING THE IDEAS

1. Does the cosecant function have any maximum or minimum values? Explain.

2. The cosecant function is positive in some quadrants and negative in others. Explain how you can remember the quadrants in which the cosecant function is positive and the quadrants in which it is negative.

3.9 EXERCISES

 1. Compare the graphs of $y = \sin \theta$ and $y = \csc \theta$. In what ways are they similar? In what ways are they different?

2. Determine each value to 4 decimal places.

   a) $\csc 110°$ b) $\sec 256°$ c) $\cot (-95°)$ d) $\sec 272°$

   e) $\cot 123°$ f) $\csc (-269°)$ g) $\sec 184°$ h) $\cot 355°$

3. Determine each value to 4 decimal places.

   a) $\sec 1.75$ b) $\csc 3.22$ c) $\cot 2.16$ d) $\sec 6.75$

   e) $\csc 2.5$ f) $\cot (-1.65)$ g) $\sec 6.25$ h) $\csc (-2.47)$

4. This graphing calculator screen shows the coordinates of one point on the graph of $y = \csc \theta$. Write the x-coordinates of four other points on the graph that have the same y-coordinate as this point.

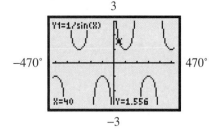

5. Write to explain how you determined the x-coordinates in exercise 4.

6. Compare the y-coordinate in the screen in exercise 4 with the y-coordinate in the first screen on page 197. How are the two y-coordinates related?

B 7. Graph each function for $-2\pi \leq \theta \leq 2\pi$.

    a) $y = \csc \theta$ **b)** $y = \sec \theta$ **c)** $y = \cot \theta$

8. The graphing calculator screen (below left) shows the coordinates of one point on the graph of $y = \sec x$. Write the x-coordinates of four other points on the graph that have the same y-coordinate as this point.

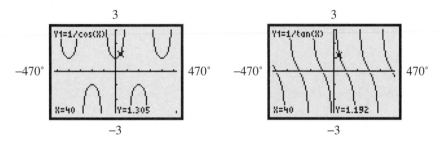

9. The graphing calculator screen (above right) shows the coordinates of one point on the graph of $y = \cot x$. Write the x-coordinates of four other points on the graph that have the same y-coordinate as this point.

10. Choose either exercise 8 or 9. Write to explain how you determined the x-coordinates.

11. Compare the y-coordinate in the screen in exercise 8 with the y-coordinate in the second screen on page 197. How are the two y-coordinates related?

12. Compare the y-coordinate in the screen in exercise 9 with the y-coordinate in the screen on page 214. How are the two y-coordinates related?

COMMUNICATING THE IDEAS

Compare the graphs of the cosecant, secant, and cotangent functions. In what ways are they similar? In what ways are they different? Write to explain the similarities and differences. Use examples to illustrate your explanations.

1. Earth travels in a nearly circular orbit around the sun. The radius of the orbit is about 149 000 000 km.

   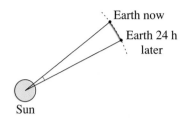

   Earth now
   Earth 24 h later
   Sun

   a) What is the measure, in radians, of the angle subtended at the sun by the positions of Earth at two different times 24 h apart?

   b) About how far does Earth travel in one day in its orbit around the sun?

   c) How fast is Earth travelling around the sun?

2. Refer to page 169. Students conducted an experiment to verify Snell's Law. They measured different angles of incidence and the corresponding angles of refraction. The students input their data on a spreadsheet.

Angle of incidence, i (degrees)	Angle of refraction, r (degrees)	sin i	sin r	$\dfrac{\sin i}{\sin r}$
10	7	−0.544	0.657	−0.83
20	13	0.913	0.420	2.17
30	19	−0.988	0.150	−6.59
40	25	0.745	−0.132	−5.63
50	30	−0.262	−0.988	0.27
60	35	−0.305	−0.428	0.71
70	38	0.774	0.296	2.61
80	40	−0.994	0.745	−1.33

   a) Look at the spreadsheet. Explain what is wrong with it.

   b) Correct the spreadsheet.

   c) Use your modified spreadsheet to estimate the index of refraction.

3. For each angle in standard position, determine θ in radians and degrees.

   a)
   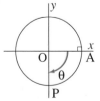

   b)
   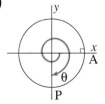

   c)

4. When you turn the handlebars of a bicycle through an angle θ, the front wheel turns through the same angle. As you pedal forward, the front wheel turns in a circle with radius r_1 and the rear wheel turns in a circle with radius r_2. For the bicycle shown, the distance between the axles of the two wheels is 1 m.

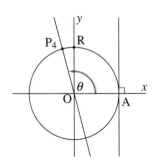

a) On this diagram, $r_1 > r_2$. Is this always true, or are there some values of θ for which $r_1 < r_2$ or $r_1 = r_2$? Explain.

b) What values of θ are possible? Explain.

c) Describe how r_1 and r_2 change in each case.
   i) θ becomes smaller and smaller.
   ii) θ becomes larger and larger.

d) Explain why $\angle BOF = \theta$.

e) Express r_1 and r_2 as functions of θ. State the domain of each function.

f) Graph the functions on the same grid for values of θ in the domain.

g) Explain how the graph supports your answers to parts a, b, and c.

5. a) To help you understand why $\tan 90°$ is undefined, visualize what happens to Q as the terminal arm of θ moves from P_3 to R (below left). Write to describe what happens.

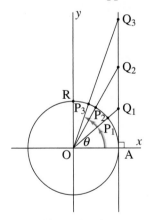

b) Visualize what happens to the position of Q as the terminal arm moves from R to P_4 (above right). Write to describe what happens.

6. Suppose the angle θ is in degrees instead of radians.

a) How would the graph of $y = \tan \theta$ on page 213 change?

b) Is the graph of $y = \tan \theta$ when θ is in degrees the same graph as the graph of $y = \tan \theta$ when θ is in radians? Explain.

1. **a)** Graph the function $f(x) = \frac{1}{x^3}$.

   b) Explain the appearance of the graph.

   c) Use the graph of this function as a guide. Sketch the graphs of $y = f(-x)$, $y = -f(x)$, and $x = f(y)$.

   d) Which of the graphs in part c represent functions? Explain.

2. Describe what happens to the graph of a function when you make each change to its equation.

   a) Replace x with $2x$.

   b) Replace y with $-4y$.

   c) Replace x with $\frac{1}{3}x$.

   d) Replace y with $\frac{1}{4}y$.

   e) Replace x with $3x$ and y with $3y$.

   f) Replace x with $-x$ and y with $\frac{1}{2}y$.

3. An object with mass 6 kg and kinetic energy, E joules, is travelling at v metres per second, where $v = \sqrt{\frac{E}{3}}$.

   a) Sketch a graph of the function for $0 \le E \le 100$.

   b) What are the domain and range of the function in part a?

4. Refer to the frequencies of a piano's A note on page 116, exercise 6b.

   a) Determine the base-10 logarithm of each frequency.

   b) There is a pattern in the frequencies. What pattern can you find in the logarithms in part a?

5. **a)** Evaluate each logarithm.

   i) $\log_2 8$ and $\log_8 2$ **ii)** $\log_5 25$ and $\log_{25} 5$

   b) On the basis of the results of part a, make a conjecture about how $\log_a b$ and $\log_b a$ are related, where $a, b > 0$. Prove your conjecture.

6. Using only the definition of a logarithm, simplify the expression $a^{\log_a x}$, and explain your reasoning.

7. Simplify the expressions in each list. Explain the results.

   a) $\log 100 - \log 10$
   $\log 200 - \log 20$
   $\log 300 - \log 30$
   $\log 400 - \log 40$

   b) $\log 6 + \log 10$
   $\log 6 + \log 100$
   $\log 6 + \log 1000$
   $\log 6 + \log 10\ 000$

8. These equations were solved graphically on pages 69–70. Solve each equation using logarithms.

a) exercise 1b: $1000(1.08)^n = 2500$

b) exercise 3c: $2.76(1.022)^n = 6$

c) exercise 4a: $100(0.95)^n = 50$

d) exercise 6b: $100(0.87)^n = 20$

e) exercise 7c: $1000(1.04)^{2n} = 2500$

f) exercise 9c: $130\ 000 \times 1.45^n = 10\ 000\ 000$

9. Choose one equation from exercise 8. Write to explain how you solved it.

10. Acid rain is a major environmental problem in some areas. Fish are affected when the pH value of a lake drops below 6.0. Trees and plants are affected when the pH value of rainwater drops below 3.5. How many times as acidic is rainwater with a pH value of 3.5 than a lake with a pH value of 6.0?

11. Recall the equation $c = 120 \times 1.12^n$ from page 68. It models the mass, c tonnes, of all the satellites in orbit as a function of n, the number of years since 1960.

a) Change the equation so that n represents the number of years since 1991.

b) Change the equation so that n represents the year.

c) What do the vertical intercepts of the graphs of the equations in parts a and b represent?

12. These sequences start with the same two terms.

Arithmetic sequence: 3, 12, 21, ...   ①

Geometric sequence: 3, 12, 48, ...   ②

a) Show that t_3 of ② is the same as t_6 of ①.

b) Which term in ① is the same as t_4 in ②?

c) Show that every term in the geometric sequence is also a term in the arithmetic sequence.

13. Solve and check.

a) $4^x + 4^{x+1} = 40$ b) $3^x - 3^{x-1} = \frac{2}{27}$ c) $5(2)^x - 3(2)^{x-1} = 224$

14. a) Graph $y = \tan x$ using a graphing window defined by $-470° \le x \le 470°$ and $-3 \le y \le 3$. Trace to the point where $x = 60°$. Observe the corresponding value of y.

b) Predict some other values of x that produce the same value of y. Trace to these points to verify your predictions.

15. Compare the graphs of $y = \tan \theta$ and $y = \cot \theta$. In what ways are they similar? In what ways are they different?

4 TRIGONOMETRIC FUNCTIONS OF REAL NUMBERS

Predicting Hours of Daylight

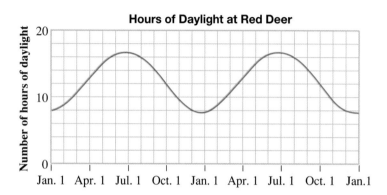

Hours of Daylight at Red Deer

CONSIDER THIS SITUATION

This graph shows how the number of hours of daylight at Red Deer varies during a two-year period.

• Why are there more daylight hours in summer than in winter?

• Estimate the number of hours of daylight at Red Deer on each date.
   a) March 1   **b)** July 15   **c)** November 11

• How was the graph obtained?

• Suppose a similar graph were drawn for each city. How would it differ from the graph above? Explain.
   a) Yellowknife   **b)** Mexico City

The graph is a model for predicting the number of hours of daylight. On pages 270 and 271, you will develop an improved model. It will be the equation of the curve above.

 FYI Visit www.awl.com/canada/school/connections

For information related to the above problem, click on <u>MATHLINKS</u> followed by <u>AWMath</u>. Then select a topic under Predicting Hours of Daylight.

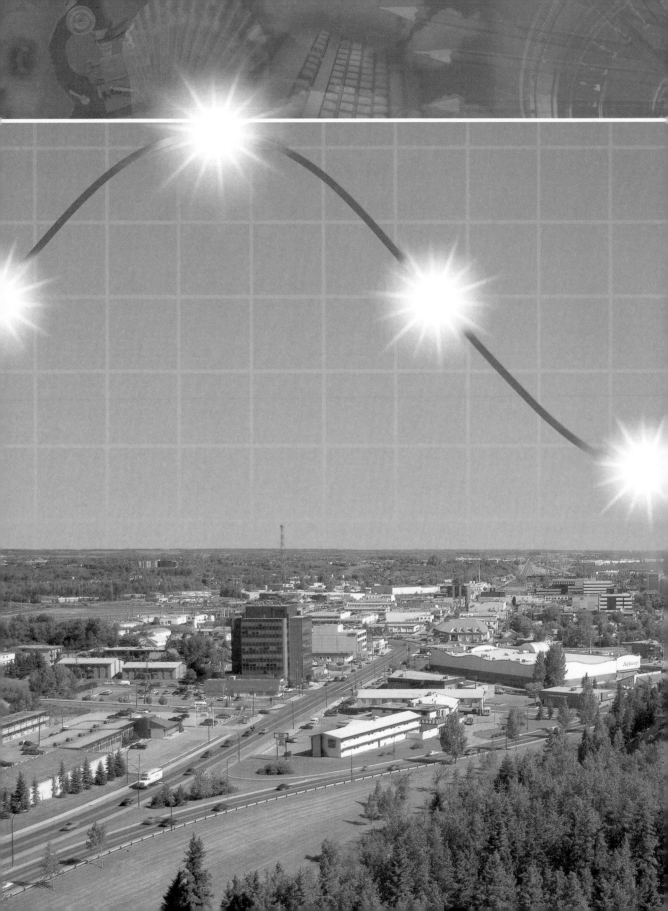

INVESTIGATE Defining and Graphing $y = \sin x$ and $y = \cos x$

The diagram below shows a unit circle with centre (0, 0). Ask your teacher for a large scale version of this diagram from the Teacher's Resource Book. The scale on the circle is the same as the scale on the coordinate axes. This scale indicates the distance from A(1, 0) measured counterclockwise. In principle, this scale continues indefinitely in both directions, but only the part from 0 to 2π is shown to avoid overlapping.

The functions $y = \sin x$ and $y = \cos x$ are defined as follows. Let x represent the arc length measured along the circle from A(1, 0) to any point P on the circle. The first coordinate of P is the *cosine* of x. The second coordinate of P is the *sine* of x.

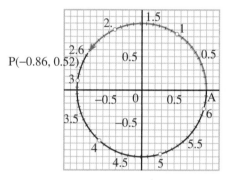

When $x = 2.6$, the coordinates of P are approximately $(-0.86, 0.52)$.
Hence, $\cos 2.6 \doteq -0.86$, and $\sin 2.6 \doteq 0.52$

Although estimates accurate to 2 decimal places are possible using the large scale diagram, you may not always be able to determine the hundredths digit for certain. Your best estimate is sufficient for this investigation.

1. Use the large scale diagram. Check that $\cos 2.6 \doteq -0.86$ and $\sin 2.6 \doteq 0.52$.

2. Use this method to make tables of values for both functions $y = \cos x$ and $y = \sin x$. Use values of x corresponding to points spread around the circumference of the circle.

3. Sketch the graphs of the functions of $y = \sin x$ and $y = \cos x$.

4. Use the large scale diagram to explain your answer to each question.

    a) What are the domain and the range of $y = \cos x$?

    b) What is the period of $y = \cos x$?

    c) What are the x-intercepts of the graph of $y = \cos x$?

5. Repeat exercise 4, using the function $y = \sin x$.

In the first section of Chapter 3, several graphs illustrated these examples of periodic functions.

- time of the sunset
- lengths of shadows
- volume and pressure of blood in the heart
- volume of air in the lungs

These graphs differ from the other trigonometric graphs in Chapter 3 in one major way. They show periodic functions without the use of angles.

Up to now, the horizontal axis has been scaled in angle measures of degrees or radians. These scales are not useful in applications such as those above, where the horizontal axis is usually marked in time intervals. To use trigonometric functions in applications, we will redefine them by using variables such as x or t in their equations instead of θ. The new definitions will be consistent with the previous definitions. The key to the new definitions is the radian measure that was introduced in Chapter 3.

One radian is the measure of the angle subtended at the centre of a circle by an arc whose length equals the radius. If the circle is a unit circle, the arc has length 1. For a point rotating around the circle, the number of units of distance it has travelled is the same as the number of radians.

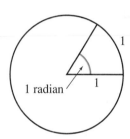

A length can be measured along the circumference of the unit circle, just as an angle is measured at the centre. When the angle is in radians, the number of length units is the same as the number of radians.

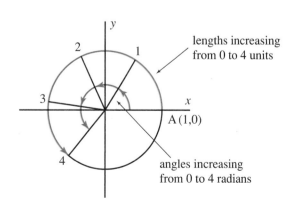

We will use the arc length instead of the angle at the centre as the variable in the definition of the trigonometric functions. Visualize a point P rotating around a unit circle with centre O(0, 0). Let x represent the length of the arc from A(1, 0) to P. When the rotation is counterclockwise, x is positive; when the rotation is clockwise, x is negative. For any position of P on the circle, we define the first coordinate of P to be cos x, and the second coordinate of P to be sin x.

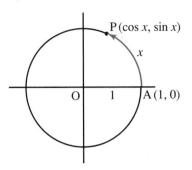

As P rotates around the circle, the values of cos x and sin x change periodically. You can use a calculator in radian mode to determine these values for any value of x. Some typical values are illustrated below. For each example, the distance from A to P measured along the circle is indicated. Observe how this distance is used to determine the coordinates of P.

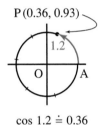

P(0.36, 0.93)

cos 1.2 ≐ 0.36
sin 1.2 ≐ 0.93

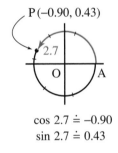

P(−0.90, 0.43)

cos 2.7 ≐ −0.90
sin 2.7 ≐ 0.43

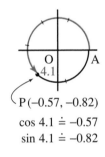

P(−0.57, −0.82)

cos 4.1 ≐ −0.57
sin 4.1 ≐ −0.82

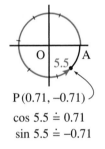

P(0.71, −0.71)

cos 5.5 ≐ 0.71
sin 5.5 ≐ −0.71

The variable x represents the length of the arc from A(1, 0) to any point P on the unit circle.

- The *cosine* of x is the horizontal coordinate of P.
- The *sine* of x is the vertical coordinate of P.

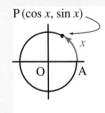

P(cos x, sin x)

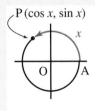

P(cos x, sin x)

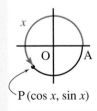

P(cos x, sin x)

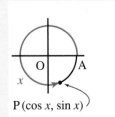

P(cos x, sin x)

Example

Determine sin 3.7 and cos 3.7. Draw a diagram to explain the meaning of the results.

Solution

Be sure your calculator is in radian mode.

$\sin 3.7 \doteq -0.5298 \qquad \cos 3.7 \doteq -0.8481$

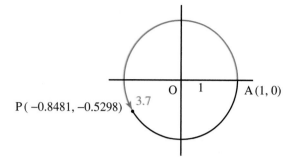

$\sin 3.7 = -0.5298$ and $\cos 3.7 = -0.8481$ mean that when the length of arc AP in a unit circle is 3.7 units, the coordinates of P are approximately $(-0.8481, -0.5298)$.

Previously, in your study of mathematics, you encountered these functions of real numbers:

Function type	Example
linear functions	$y = 2x + 3$
quadratic functions	$y = x^2 + 4x - 7$
polynomial functions	$y = x^3 - 2x^2 + x$
reciprocal functions	$y = \dfrac{1}{x^2 - 4}$
rational functions	$y = \dfrac{x}{x - 2}$
exponential functions	$y = 2^x$
logarithmic functions	$y = \log_2 x$

The functions $y = \sin x$ and $y = \cos x$ are new functions. Just as with other functions, the variables x and y represent real numbers. The variable x represents the length of the arc from A(1, 0) to any point P on the unit circle. The unit of length for this measurement is the same as the unit of length measured along the coordinate axes. Therefore, when $x = 1$, $\angle POA = 1$ radian. This is the reason why radians are a more fundamental unit of angle measure than degrees.

Graphing the function $y = \sin x$

The graph of $y = \sin x$ corresponds to that of $y = \sin \theta$, but the horizontal axis is marked as the x-axis. Think of the values of x along this axis as representing real numbers, not angles.

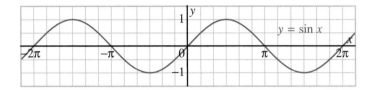

On a graph of a trigonometric function, it is customary to mark the horizontal axis with the number π and its multiples, as on the graph above. The advantage of this arrangement is that the x-intercepts are shown exactly.

However, on graphs of other functions, such as those in the list on page 229, it is customary to mark the horizontal axis with integers. We can also do this with trigonometric functions. The graphs below and above are identical; but the horizontal axis of the graph below is labelled with integers instead of multiples of π. This arrangement facilitates working with coordinates of points on the graph.

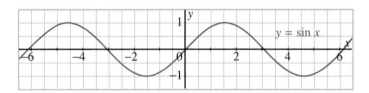

In this chapter and the next, we will use both types of horizontal scales.

Properties of the function $y = \sin x$

Period: 2π, or approximately 6.28

Maximum value of y: 1   Minimum value of y: -1

Domain: all real numbers   Range: $-1 \leq y \leq 1$

x-intercepts: $0, \pm\pi, \pm 2\pi, \ldots$; or 0, and approximately $\pm 3.14, \pm 6.28, \ldots$

y-intercept: 0

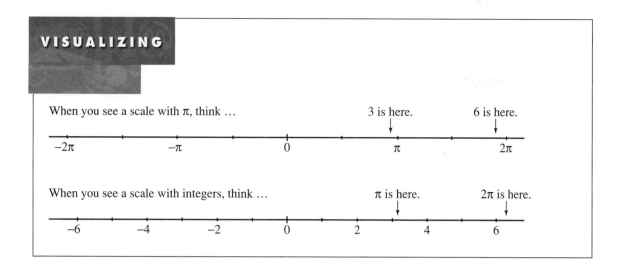

When you see a scale with π, think ...   3 is here.   6 is here.

-2π $-\pi$ 0 π 2π

When you see a scale with integers, think ... π is here. 2π is here.

-6 -4 -2 0   2   4   6

Graphing the function $y = \cos x$

Corresponding graphs of the function $y = \cos x$ are shown below. Both graphs show the same function drawn to the same scale; the only difference is the numbers that are marked on the horizontal axes.

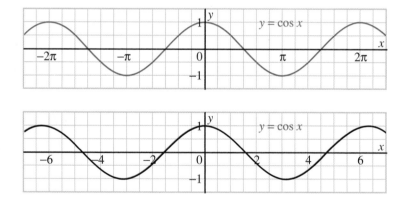

The second graph is more useful for applied problems. Methods of adapting the sine and cosine functions to solve applied problems will be developed in the following sections. The graphs in the solutions of these problems will look like the second graph above (or the second graph of $y = \sin x$ on page 230), but they will be expanded or compressed vertically and horizontally, and their positions relative to the axes will be different. The curves look like waves, and are called *sinusoids*. The corresponding functions are *sinusoidal functions*.

Properties of the function $y = \cos x$

Period: 2π, or approximately 6.28

Maximum value of y: 1   Minimum value of y: -1

Domain: all real numbers   Range: $-1 \le y \le 1$

x-intercepts: $\pm\frac{\pi}{2}, \pm\frac{3\pi}{2}, \pm\frac{5\pi}{2}, \ldots$; or approximately $\pm1.57, \pm4.71, \pm7.85, \ldots$

y-intercept: 1

DISCUSSING THE IDEAS

1. Refer to the diagram in *Visualizing*, page 227. What numbers on the arc length scale correspond to the points $(-1, 0)$ and $(1, 0)$? Are these numbers unique? Explain.

2. Explain why radians are a more fundamental unit of angle measure than degrees.

3. Explain why the two graphs on page 230 are identical.

4.1 EXERCISES

A 1. Use the graph (below left) to estimate each value.

   a) $\sin 0.4$ **b)** $\cos 1.1$ **c)** $\cos 2.7$ **d)** $\sin 5.5$

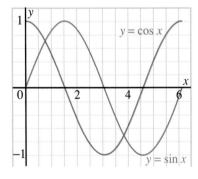
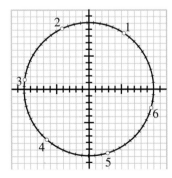

2. Use the unit circle (above right) to estimate each value in exercise 1.

3. Use a calculator to determine each value in exercise 1.

4. Compare the answers in exercise 3 with your estimates in exercises 1 and 2.

    a) Did exercise 1 or exercise 2 give the more accurate estimates? Which method do you prefer?

    b) Write to explain why the values obtained from the graph in exercise 1 should be the same as those obtained from the circle in exercise 2.

B **5.** Determine each pair of values. Draw a diagram to explain the meaning of the results.

    a) $\sin 0.5$ and $\cos 0.5$ **b)** $\sin 1$ and $\cos 1$

    c) $\sin 2$ and $\cos 2$ **d)** $\sin 2.8$ and $\cos 2.8$

    e) $\sin 4.3$ and $\cos 4.3$ **f)** $\sin 10$ and $\cos 10$

    g) $\sin (-0.5)$ and $\cos (-0.5)$ **h)** $\sin (-3.7)$ and $\cos (-3.7)$

6. a) Graph the functions $y = \sin x$ and $y = \cos x$ on the same axes for $-2\pi \le x \le 2\pi$.

    b) How do the two graphs appear to be related? Explain.

    c) What is the period of each function?

    d) What are the coordinates of the points of intersection?

7. Suppose you do not realize that a calculator is in degree mode, and you use it to graph $y = \sin x$ for $-2\pi \le x \le 2\pi$.

    a) What would the resulting graph look like? Explain.

    b) Use a graphing calculator to check your prediction.

8. The square in the first screen on the next page has vertices $(\pm 1, 0)$ and $(0, \pm 1)$. Visualize the square rotating counterclockwise about $O(0, 0)$ to the positions in the three image screens. On each screen, the distance travelled by the marked vertex is shown in the second-last line, and indicated by the solid arc on the unit circle. The last line shows the coordinates of this vertex. Choose one image screen.

    a) Write the angle in standard position, in radians, corresponding to each vertex of the square.

    b) Determine the coordinates of each vertex.

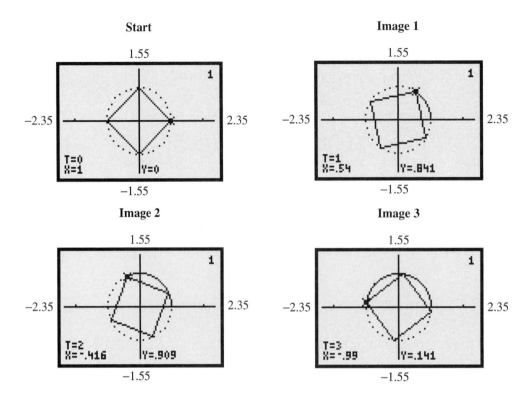

Start

1.55
1
−2.35   2.35
−1.55

T=0
X=1   Y=0

Image 1

1.55
1
−2.35   2.35
−1.55

T=1
X=.54   Y=.841

Image 2

1.55
1
−2.35   2.35
−1.55

T=2
X=−.416   Y=.909

Image 3

1.55
1
−2.35   2.35
−1.55

T=3
X=−.99   Y=.141

9. Choose one image screen from exercise 8. Write to explain how you determined the coordinates of the other three vertices.

10. The graphing calculator screen (below left) shows the coordinates of one point on the graph of $y = \sin x$, to the nearest thousandth. Use only the information on the screen. Write the x-coordinates of four other points on the graph that have the same y-coordinate as this point.

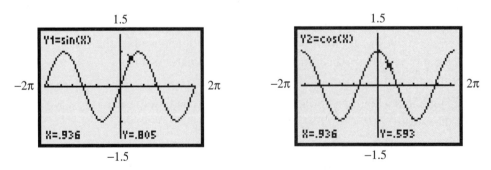

1.5
Y1=sin(X)
−2π   2π
−1.5
X=.936   Y=.805

1.5
Y2=cos(X)
−2π   2π
−1.5
X=.936   Y=.593

11. The graphing calculator screen (above right) shows the coordinates of one point on the graph of $y = \cos x$, to the nearest thousandth. Repeat exercise 10 for this screen.

12. Choose either exercise 10 or 11. Write to explain how you determined the x-coordinates.

13. a) Use the mode menu of a graphing calculator to display numbers to 3 decimal places. Graph $y = \sin x$ using a graphing window defined by $-2\pi \leq x \leq 2\pi, -1.5 \leq y \leq 1.5$. Trace to the point where $x = 1.069$. Observe the corresponding value of y.

b) Mentally predict some other values of x that produce the same value of y. Trace to these points to verify your predictions.

14. a) Graph $y = \cos x$ using the same graphing window as in exercise 13. Trace to the point where $x = 1.069$. Observe the corresponding value of y.

b) Mentally predict some other values of x that produce the same value of y. Trace to these points to verify your predictions.

c) Compare your answers in part b and exercise 13b. Explain why some x-coordinates are the same and why some are different.

15. Compare the two graphs in exercises 10 and 11. The x-coordinates of the indicated points are the same.

a) How are the y-coordinates of these points related? Explain.

b) Is the relationship you identified in part a true for all points with the same x-coordinate on the two graphs, or only for the points indicated? Explain.

16. a) Graph $y = \sin x$ using a graphing window defined by $-4.7 \leq x \leq 4.7$, $-3.1 \leq y \leq 3.1$. Use the zoom feature to enlarge the part of the graph containing the origin. Zoom in enough times until the visible part of the graph looks like a straight line.

b) Use the trace and arrow keys to move the cursor along the graph. What do you notice about the coordinates of the points on the graph? Use the definition of $\sin x$ to explain.

C **17. a)** In exercise 8, explain why the vertices in Image 3 lie almost on the coordinate axes.

b) Suppose the distance travelled by the marked vertex is an integer. Would it be possible for the vertices to lie on the axes? Explain.

18. A function $y = f(x)$ is defined to be *periodic* if there is a number p such that $f(x + p) = f(x)$ for all values of x in the domain. Use this definition to prove that the functions, $y = \sin x$ and $y = \cos x$, are periodic.

COMMUNICATING THE IDEAS

Your friend missed today's mathematics lesson and phones you to ask about it. How would you explain the difference between the functions $y = \sin \theta$ and $y = \sin x$?

Graphing $y = \sin x$ and $y = \cos x$

In *Exploring with a Graphing Calculator,* Chapter 3, page 191, you graphed $y = \sin \theta$ and $y = \cos \theta$ where θ is in radians. The graphs of $y = \sin x$ and $y = \cos x$ are the same as these. In applied problems, however, it is inconvenient to use π in the window settings and horizontal scale. We prefer to use rational numbers, as we do with other functions.

In the exercises on this page, your calculator must be in radian mode. You will be working with these screens. They show the graph of $y = \sin x$ using two different window settings, shown below each screen.

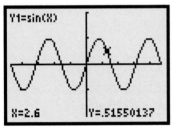

$$-9.4 \leq x \leq 9.4$$
$$-2 \leq y \leq 2$$

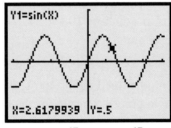

$$-\frac{47\pi}{18} \leq x \leq \frac{47\pi}{18}$$
$$-2 \leq y \leq 2$$

1. For the first window setting:

   a) Graph the function $y = \sin x$. Use the trace key to trace along the curve.

   b) Predict the five x-intercepts that are less than 7. Use the zero feature in the calculate menu to check your predictions.

   c) Predict the coordinates of the three maximum points on the screen. Use the maximum feature in the calculate menu to check.

   d) Predict the coordinates of the minimum points on the screen. Check using the minimum feature.

   e) Use the results of parts a to c to determine the period of the function. Try to do this in more than one way.

   f) Can the period of the function be determined by tracing? Explain.

2. Repeat exercise 1 using the second window setting.

3. Write to explain the advantage of each window setting.

To solve applied problems, we require a graph that combines the advantages of both window settings. This will be done in Section 4.4.

4. Explain why the graphs of $y = \sin x$ and $y = \cos x$ are the same as the graphs of $y = \sin \theta$ and $y = \cos \theta$ in Chapter 3, page 191.

4.2 Graphing $y = a \sin (x - c) + d$ and $y = a \cos (x - c) + d$

To use the functions $y = \sin x$ and $y = \cos x$ in an applied situation, we change the graphs and the equations so that the maximum and minimum values, and the periods, correspond to the quantities in the situation. To do this, we relate the changes in the equations to the changes in the graphs.

INVESTIGATE Transforming Sinusoidal Graphs Part I

The graphs of $y = \sin x$ and $y = \cos x$ are shown below. In this investigation, you will graph other related equations on the same screen. You will use the tools from the functions toolkit in Chapter 1 to explain how the changes in the graphs are related to the changes in the equations.

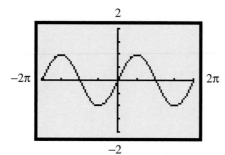

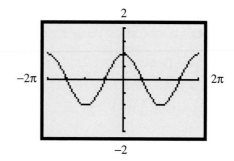

Comparing the graphs of $y = a \sin x + d$ with $y = \sin x$ and $y = a \cos x + d$ with $y = \cos x$

1. a) Graph the function $y = 2 \sin x + 1$ using the same window settings as above.

   b) Describe how the graph of $y = 2 \sin x + 1$ is related to the graph of $y = \sin x$. Explain this relationship.

   c) Repeat part a, using numbers other than 2 and 1 in the equation. Use both positive and negative numbers, including those between -1 and 1. Are the results what you expected? Explain.

2. a) What information about the graph of $y = a \sin x + d$ do a and d provide? Explain.

   b) What happens to the graph of $y = a \sin x + d$ when a is changed? What happens when d is changed?

3. Repeat exercises 1 and 2 using cosine functions.

4. Graph these functions for $-2\pi \le x \le 2\pi$.

    a) $y = 2\sin x + 1$
    b) $y = 3\sin x - 4$
    c) $y = -2\sin x + 2$

    d) $y = 3\cos x + 2$
    e) $y = 2\cos x - 1$
    f) $y = -2\cos x + 4$

Comparing the graphs of $y = \sin(x - c)$ with $y = \sin x$ and $y = \cos(x - c)$ with $y = \cos x$

5. a) Graph the function $y = \sin\left(x - \frac{\pi}{3}\right)$. Describe how the graph of $y = \sin\left(x - \frac{\pi}{3}\right)$ is related to the graph of $y = \sin x$.

    b) Repeat part a, using numbers other than $\frac{\pi}{3}$ in the equation. Use both positive and negative numbers. Are the results what you expected? Explain.

6. a) What information about the graph of $y = \sin(x - c)$ does c provide? Explain.

    b) What happens to the graph of $y = \sin(x - c)$ when c is changed?

7. Repeat exercises 5 and 6 using cosine functions.

8. Graph these functions for $-2\pi \le x \le 2\pi$.

    a) $y = 2\sin\left(x - \frac{\pi}{4}\right) + 2$
    b) $y = 3\sin\left(x + \frac{\pi}{4}\right) - 1$

    c) $y = 2\cos\left(x - \frac{\pi}{3}\right) + 2$
    d) $y = -2\cos\left(x + \frac{\pi}{3}\right) + 3$

In *Investigate*, you made certain changes to the equations $y = \sin x$ and $y = \cos x$, and used the tools from the functions toolkit in Chapter 1 to explain what happened to the graphs. You can use these tools to sketch the graphs of many sinusoidal functions without using a graphing calculator or graphing software.

Stretching and reflection tools   *Translation tools*

Vertical expansion or compression, and/or reflection

Translate left or right.

Translate up or down.

$$y = a\sin(x - c) + d$$
$$y = a\cos(x - c) + d$$

The vertical displacement of a sinusoidal function

According to the translation tool, the constant term d in $y = \sin x + d$ or $y = \cos x + d$ has the effect shown below. The number d is the *vertical displacement*.

The graphs of $y = \sin x + d$ and $y = \cos x + d$ are the images of the graphs of $y = \sin x$ and $y = \cos x$, respectively, under a vertical translation of d units.

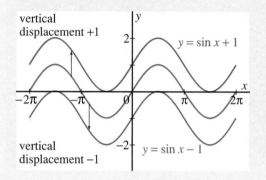

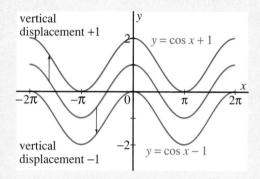

Pedalling a bicycle illustrates vertical displacement. The heights of the pedals above the ground change periodically. A graph of the height of a pedal against time is sinusoidal with vertical displacement equal to the mean height of the pedal above the ground.

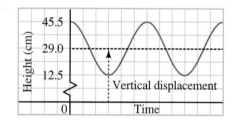

The amplitude of a sinusoidal function

According to the stretching and reflection tools, the coefficient a in $y = a \sin x$ or $y = a \cos x$ has the effect shown below. The number $|a|$ is the *amplitude*. It represents the distance from any maximum or minimum point to the x-axis.

The graphs of $y = a \sin x$ and $y = a \cos x$ are the images of the graphs of $y = \sin x$ and $y = \cos x$, respectively, under a vertical expansion or compression. If $a < 0$, there is also a reflection in the x-axis.

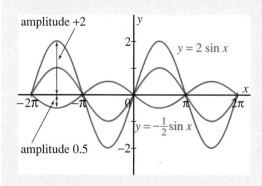

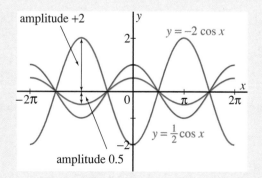

The movement of a tower or skyscraper illustrates amplitude. It must be built so that its top can sway with the wind. This graph shows how the distance of the Space Deck of the CN Tower from its rest position might change with time in a high wind. The distance the tower sways from its rest position is the amplitude of the vibration.

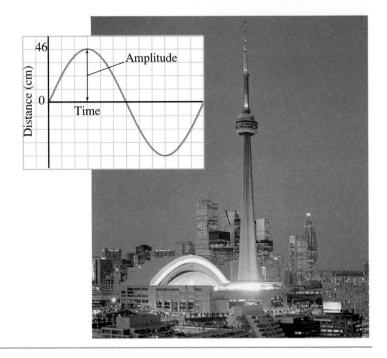

Example 1

a) Graph the function $f(t) = 2 \sin t + 3$ over two cycles.

b) State the amplitude, the vertical displacement, and the period of the function.

Solution

a) Graph $y = \sin t$, and expand it vertically by a factor of 2. This produces the image $y = 2 \sin t$. Then translate the image 3 units up.

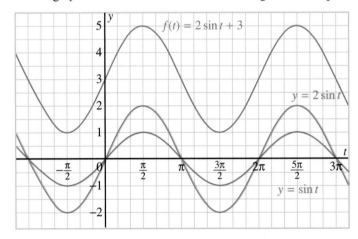

b) The amplitude is 2. The vertical displacement is 3. The period is 2π.

In *Example 1*, we can use the graph of $f(t) = 2 \sin t + 3$ to obtain a definition for the amplitude of a periodic function. In this graph, the maximum value of the function is 5, and the minimum value is 1. The amplitude is one-half the distance from the minimum to the maximum, measured in the vertical direction. For this function, the amplitude is $\frac{1}{2}(5 - 1)$, or 2.

When M represents the maximum value of a sinusoidal function in any cycle, and m represents the minimum value in that cycle, then the *amplitude A* of the function is:

$$A = \frac{M - m}{2}$$

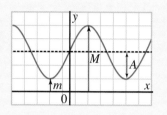

The phase shift of a sinusoidal function

According to the translation tool, the term c in $y = \sin(x - c)$ has the effect shown below. The number c is the *phase shift*. It represents the distance the graph is translated horizontally.

The graphs of $y = \sin(x - c)$ and $y = \cos(x - c)$ are the images of the graphs of $y = \sin x$ and $y = \cos x$, respectively, under a horizontal translation of c units.

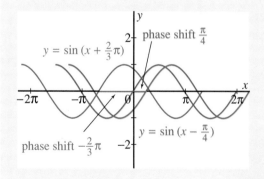

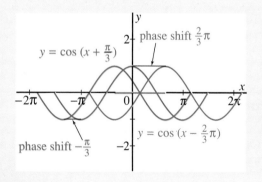

Changing wildlife populations can illustrate phase shift. In a certain region, the number of rabbits and the number of wolves increases and decreases periodically. The population graph for the wolves shifts horizontally relative to the population graph for the rabbits.

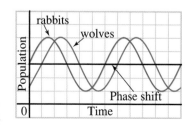

Phase Shift of $y = \sin(x - c)$ and $y = \cos(x - c)$

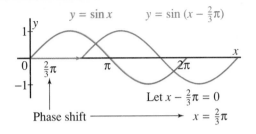

$y = \sin x$ $y = \sin(x - \frac{2}{3}\pi)$

Let $x - \frac{2}{3}\pi = 0$

Phase shift ⟶ $x = \frac{2}{3}\pi$

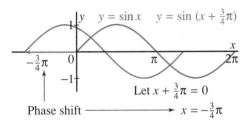

$y = \sin x$ $y = \sin(x + \frac{3}{4}\pi)$

Let $x + \frac{3}{4}\pi = 0$

Phase shift ⟶ $x = -\frac{3}{4}\pi$

To determine the phase shift of $y = \sin(x - c)$, let $x - c = 0$, then solve for x.
The phase shift of a sine function is the x-coordinate of a point where the sine cycle begins.

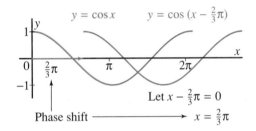

$y = \cos x$ $y = \cos(x - \frac{2}{3}\pi)$

Let $x - \frac{2}{3}\pi = 0$

Phase shift ⟶ $x = \frac{2}{3}\pi$

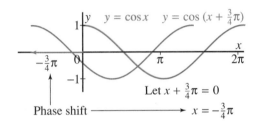

$y = \cos x$ $y = \cos(x + \frac{3}{4}\pi)$

Let $x + \frac{3}{4}\pi = 0$

Phase shift ⟶ $x = -\frac{3}{4}\pi$

To determine the phase shift of $y = \cos(x - c)$, let $x - c = 0$, then solve for x.
The phase shift of a cosine function is the x-coordinate of a maximum point.

Example 2

Graph the function $f(x) = 3\sin(x - \frac{2\pi}{3}) + 2$ over two cycles. State the
vertical displacement, the amplitude, and the phase shift.

Solution

Graph $y = \sin x$, and expand it vertically by a factor of 3 to obtain the image
$y = 3\sin x$. Then translate the image $\frac{2\pi}{3}$ units right and 2 units up.

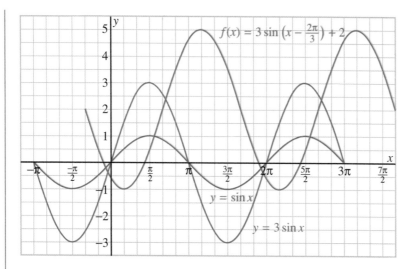

The vertical displacement is 2, the amplitude is 3, and the phase shift is $\frac{2\pi}{3}$.

Example 3

A sinusoidal curve is shown. For the function defined by this curve:

a) Write two different cosine equations.

b) Write two different sine equations.

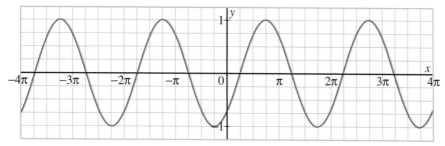

Solution

Think...

The graph could be a cosine function, with its phase shift from the *y*-axis to any maximum point. The graph could also be a sine function, with its phase shift from the origin to any point where the sine cycle begins on the *x*-axis.

a) The horizontal distances from the y-axis to the maximum points on either side of the y-axis are $\frac{3\pi}{4}$ and $-\frac{5\pi}{4}$. Hence, these are two possible phase shifts for a cosine function. The corresponding equations are $y = \cos\left(x - \frac{3\pi}{4}\right)$ and $y = \cos\left(x + \frac{5\pi}{4}\right)$.

b) Two points where the sine cycle begins on the x-axis are $\left(\frac{\pi}{4}, 0\right)$ and $\left(-\frac{7\pi}{4}, 0\right)$. Hence, two possible phase shifts for a sine function are $\frac{\pi}{4}$ and $-\frac{7\pi}{4}$. The corresponding equations are $y = \sin\left(x - \frac{\pi}{4}\right)$ and $y = \sin\left(x + \frac{7\pi}{4}\right)$, respectively.

DISCUSSING THE IDEAS

1. Explain the difference between amplitude and vertical displacement.

2. In the graphs in the display on page 239, some values of a are negative. Why aren't the corresponding amplitudes negative?

3. In *Example 2*, explain why the graph is shifted right and up, when there is a negative sign inside the brackets and a positive sign outside the brackets.

4. In *Example 3*, how can you determine the points where the sine cycle begins on the x-axis?

5. Refer to the wildlife population graph on page 241. Explain why the graph for the wolves is shifted horizontally relative to the graph for the rabbits.

6. Refer to the graph of $y = \cos\left(x - \frac{2}{3}\pi\right)$ in *Visualizing*, page 242. Explain why you can determine the phase shift by letting $x - \frac{2}{3}\pi = 0$.

7. In the solution of *Example 2*, the vertical expansion was applied before the translations. Could the translations be applied before the vertical expansion? Explain.

8. Explain why the graph of a sinusoidal function can have more than one equation. Use an example to illustrate your explanation.

9. **a)** Can the x-coordinate of any maximum point on the graph of a cosine function be taken as the phase shift? Explain.

   b) Can the x-coordinate of any point where the graph of a sine function intersects the x-axis be taken as the phase shift? Explain.

A **1.** The graph of $y = \cos x$ is transformed as described. The equation of its image has the form $y = a\cos(x - c) + d$. Determine a, c, and d for each transformation.

 a) Translate the graph 2 units left.

 b) Expand the graph vertically by a factor of 3, then translate it 2 units down.

 c) Compress the graph vertically by a factor of $\frac{1}{2}$, then translate it 3 units down.

 d) Expand the graph vertically by a factor of 2, then translate it $\frac{\pi}{3}$ units left and 2 units down.

2. Write to explain the difference between translating a graph vertically up and expanding it vertically.

3. Consider a sinusoidal function of the form $y = a\sin(x - c) + d$. Write to explain how its graph will change for each change described.

 a) The sign of a is changed and d is increased by 2.

 b) c is increased by $\frac{3\pi}{4}$ and d is increased by 3.

 c) a is doubled, c is decreased by 3, and d is increased by 1.

 d) a is halved, c is decreased by 2, and d is decreased by 4.

4. The screen (below left) shows the graphs of $y = \cos x + d$ for $d = -3, -2, -1, 0, 1, 2,$ and 3. The coordinates of one point on one graph are shown on the screen. Write the coordinates of the points on the other graphs that have the same x-coordinate as this point.

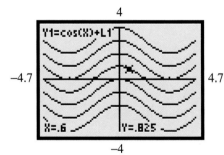

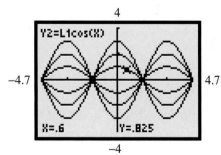

5. The screen (above right) shows the graphs of $y = a\cos x$ for $a = -3, -2, -1, 0, 1, 2,$ and 3. The coordinates of one point on one graph are shown on the screen. Write the coordinates of the points on the other graphs that have the same x-coordinate as this point.

6. a) Use the graphing window in exercise 4. Graph $y = \sin x + d$ for $d = -3, -2, -1, 0, 1, 2,$ and 3. Trace to the point on the graph of $y = \sin x$ where $x = 0.6$. Observe the corresponding value of y.

 b) Predict the coordinates of a point on each of the other graphs that has the same value of x. Trace to these points to verify your predictions.

7. Repeat exercise 6, using $y = a \sin x$ for $a = -3, -2, -1, 0, 1, 2,$ and 3.

8. The screen shows the graphs of $y = \cos(x - c)$ for $c = -2, -1, 0, 1,$ and 2. The coordinates of one point on one graph are shown. Write the coordinates of one point on each of the other graphs that has the same y-coordinate as this point.

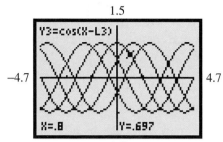

1.5

Y3=cos(X-L3)

−4.7   4.7

X=.8   Y=.697

−1.5

9. a) Use the graphing window in exercise 8.
   Graph $y = \sin(x - c)$ for $c = -2, -1, 0, 1,$ and 2. Trace to the point on the graph of $y = \sin x$ where $x = 0.8$. Observe the corresponding value of y.

   b) Predict the coordinates of a point on each of the other graphs that has the same value of x. Trace to these points to verify your predictions.

10. Graph each set of functions on the same grid for $-\pi \leq x \leq \pi$.

   a) $y = \sin x$ $y = \sin x + 1.5$ $y = \sin x - 2$

   b) $y = \sin x$ $y = \sin(x + 1.5)$ $y = \sin(x - 2)$

   c) $y = \cos x$ $y = \cos x - 3$ $y = \cos x + 4$

   d) $y = \cos x$ $y = \cos(x - 3)$ $y = \cos(x + 4)$

11. For each list, predict what the graphs of the equations would look like. Use a graphing calculator to check your predictions.

   a) $y = 3\cos x + 3$
       $y = 2\cos x + 2$
       $y = \cos x + 1$
       $y = -\cos x - 1$
       $y = -2\cos x - 2$
       $y = -3\cos x - 3$

   b) $y = 3\sin x - 3$
       $y = 2\sin x - 2$
       $y = \sin x - 1$
       $y = -\sin x + 1$
       $y = -2\sin x + 2$
       $y = -3\sin x + 3$

12. For each list, predict what the graphs of the functions would look like. Use a graphing calculator to check your predictions.

   a) $y = \sin x$
       $y = \sin(x - 2\pi)$
       $y = \sin(x - 4\pi)$
       $y = \sin(x - 6\pi)$

   b) $y = \sin x$
       $y = \sin(x - \pi)$
       $y = \sin(x - 2\pi)$
       $y = \sin(x - 3\pi)$

   c) $y = \sin x$
       $y = \sin(x - \frac{\pi}{2})$
       $y = \sin(x - \pi)$
       $y = \sin(x - \frac{3\pi}{2})$

13. Suppose the sine functions in exercise 12 were replaced with cosine functions. Would you get the same results? Check using a graphing calculator.

14. The function below can be considered as a sine function. Determine two possible values for the phase shift. What is the equation of the function for each phase shift?

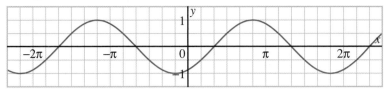

15. The function in exercise 14 can also be considered as a cosine function. Determine two possible values for the phase shift. What is the equation of the function for each phase shift?

16. Write to explain why the graph in exercise 14 defines more than one function.

17. Construct axes for $0 \leq x \leq 10$ in increments of 1, and for $-1.5 \leq y \leq 1.5$ in increments of 0.5. Identify each number on the x-axis.

a) π b) $\frac{\pi}{2}$ c) $\frac{\pi}{3}$ d) $\frac{\pi}{4}$ e) $\frac{\pi}{6}$

18. Each function below has the form $y = a \sin x + d$ or $y = a \cos x + d$. Write the equation for each graph.

a)

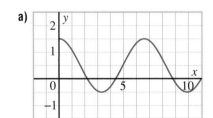

b)

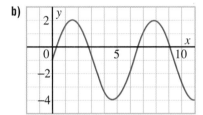

c)

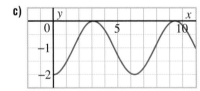

d)

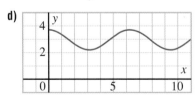

Ⓑ 19. Graph each sinusoidal function over two cycles. Determine its domain and range, its phase shift, its amplitude, and its maximum and minimum values.

a) $y = 5 \sin x$ b) $y = 3 \cos x$ c) $h = 0.25 \sin d + 4$

d) $f(x) = 2 \cos x - 3$ e) $f(x) = 4 \sin x - 2$ f) $f(x) = 1.5 \cos x + 3$

g) $f(x) = \frac{1}{2} \sin x - 1$ h) $h(t) = 2 + 2 \sin t$ i) $g(a) = 1 + 3.5 \cos a$

j) $y = \sin(x - \frac{\pi}{4})$ k) $y = \sin(x - \frac{4\pi}{3})$ l) $y = 2 \sin(x + \frac{5\pi}{6})$

m) $y = 3 \cos(x - \frac{\pi}{6}) + 3$ n) $y = 2 \cos(x + \frac{5\pi}{3}) - 2$ o) $y = 5 \cos(x - \frac{7\pi}{6}) + 2$

20. In the display on page 241, a formula for the amplitude of a sinusoidal function is given in terms of the maximum and minimum values. Write a similar formula for the vertical displacement.

21. For the function $f(x) = a \sin x + d$, write to explain how you can determine its maximum and minimum values, and the corresponding values of x.

22. Repeat exercise 21 for the function $f(x) = a \cos x + d$.

23. a) Graph the function $f(x) = \cos(x - \frac{\pi}{2})$. What conclusion can you make?

   b) Determine values of c for which the graph of $y = \cos(x - c)$ coincides with the graph of $y = \sin x$.

24. a) Graph the function $f(x) = \sin(x + \frac{\pi}{2})$. What conclusion can you make?

   b) Determine values of c for which the graph of $y = \sin(x - c)$ coincides with the graph of $y = \cos x$.

25. Graph each function over two cycles, and state its amplitude.

   a) $y = -4 \sin x$ b) $y = -0.2 \sin x$ c) $n = -9 \cos s$

   d) $f(x) = -3 \sin x + 2$ e) $f(x) = 4 - 2 \cos x$ f) $h(t) = 5 - 10 \sin t$

26. A function is given by the equation $y = 2 \sin(x + \frac{\pi}{2})$. Explain why you might prefer to rewrite it using cosine instead of sine.

Ⓒ 27. Determine the equation of a function of the form $f(x) = \sin x + p$ whose graph just touches the x-axis. How many such functions are there? Explain.

28. Consider the function $f(x) = a \sin(x - c) + d$, where $a > 0$.

   a) What is the maximum value of $f(x)$? For what values of x does this occur?

   b) What is the minimum value of $f(x)$? For what values of x does this occur?

29. a) Determine an equation of a function of the form $f(x) = \sin(x - c) + d$ that has a maximum value of 3 when $x = 0$.

   b) Is your answer in part a unique? Explain.

COMMUNICATING THE IDEAS

Write to explain how changing a, c, and d in the equations of the sinusoidal functions $y = a \sin(x - c) + d$ and $y = a \cos(x - c) + d$ affect their graphs. Include graphs as part of your explanation.

4.3 Graphing $y = a \sin b(x - c) + d$ and $y = a \cos b(x - c) + d$

INVESTIGATE Transforming Sinusoidal Graphs Part II

You will continue the investigations of the previous section, making other changes to the equations $y = \sin x$ and $y = \cos x$.

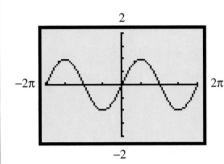

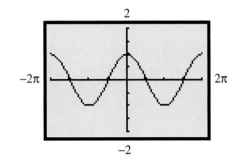

Comparing the graphs of $y = \sin bx$ with $y = \sin x$ and $y = \cos bx$ with $y = \cos x$

1. a) Graph the functions $y = \sin x$ and $y = \sin 2x$ on the same axes.

   b) Describe how the graph of $y = \sin 2x$ is related to the graph of $y = \sin x$. Explain this relationship.

   c) What is the period of the function $y = \sin 2x$? How is it related to the period of $y = \sin x$?

2. Repeat exercise 1, using the equation $y = \sin \frac{1}{2}x$.

3. Repeat exercise 1, using other coefficients of x. Use coefficients that are both greater than 1 and less than 1. Are the results what you expected?

4. a) What information about the graph of $y = \sin bx$ does b provide? Explain.

   b) What happens to the graph of $y = \sin bx$ when b is changed?

5. Repeat exercises 1 to 4, using cosine functions.

6. Sketch these functions for $-2\pi \leq x \leq 2\pi$.

   a) $y = 3 \sin 2x$ b) $y = 0.5 \cos 2x + 2$

   c) $y = -\sin 3x + 4$ d) $y = -2 \cos 0.5x$

In *Investigate*, you made certain changes to the equations $y = \sin x$ and $y = \cos x$, and used the tools from the functions toolkit in Chapter 1 to explain what happened to the graphs. You can use these tools to sketch the graphs of many sinusoidal functions without using graphing technology.

The period of a sinusoidal function

According to the stretching tool, the coefficient b in the equation $y = \sin bx$ or $y = \cos bx$ has the effect shown below. Recall that the *period* of a periodic function is the length of the shortest part of the graph that repeats, measured along the x-axis.

The graphs of $y = \sin bx$ and $y = \cos bx$ are the images of the graphs of $y = \sin x$ and $y = \cos x$, respectively, under a horizontal compression or expansion.

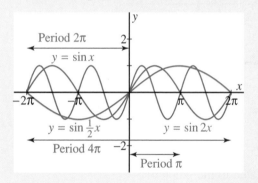
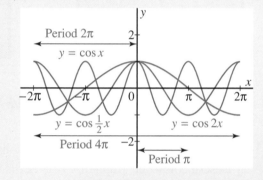

Applications of the period occur in astronomy. In 1968, scientists were astonished when two astronomers detected extremely massive stars that spin on their axes in a fraction of a second. Since a pulse of radio energy is sent out on each rotation, these stars are called pulsating stars, or pulsars.

One pulsar, in the Crab nebula, pulses every 0.033 s. This time is its period.

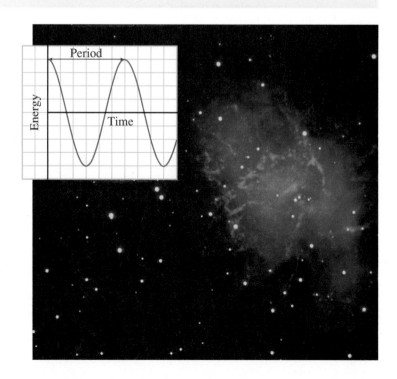

Period of $y = \sin bx$ and $y = \cos bx$

When $b > 1$, there is a horizontal compression. The period is less than 2π.

For example, when $b = 3$, there is a compression by a factor of $\frac{1}{3}$. The period of $y = \sin 3x$ and $y = \cos 3x$ is $\frac{1}{3}$ the period of $y = \sin x$ and $y = \cos x$.

When $0 < b < 1$, there is a horizontal expansion. The period is greater than 2π.

For example, when $b = \frac{1}{2}$, there is an expansion by a factor of 2. The period of $y = \sin \frac{1}{2}x$ and $y = \cos \frac{1}{2}x$ is twice the period of $y = \sin x$ and $y = \cos x$.

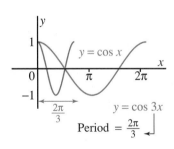

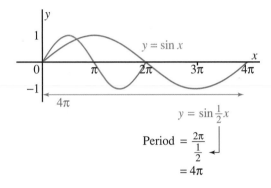

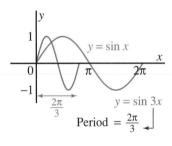

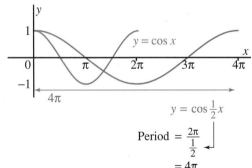

The period of $y = \sin bx$ and $y = \cos bx$ is $\frac{2\pi}{b}$, $b > 0$.

Example 1

a) Graph the function $f(x) = 2\cos 3(x - \frac{\pi}{2})$ over two cycles.

b) State the vertical displacement, the phase shift, the period, and the amplitude.

c) State the domain and the range.

Solution

a) Graph $y = 2\cos x$, then compress it horizontally by a factor of $\frac{1}{3}$. At this
point, the equation of the curve is $y = 2\cos 3x$.
Then translate the image $\frac{\pi}{2}$ units right.

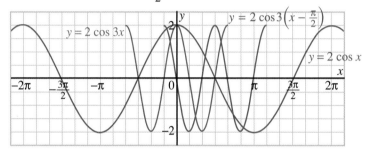

b) The vertical displacement is 0, the phase shift is $\frac{\pi}{2}$, the period is $\frac{2\pi}{3}$, and
the amplitude is 2.

c) The domain is the set of all real numbers. The range is the set of real
numbers between -2 and 2, inclusive.

When we graph a function, we customarily draw the axes before we draw the curve.
Then we make the curve fit the scales on the axes. We used this method in *Example 1*.
However, with sinusoidal functions, it is easier to draw the curve first, then the axes.
That is, we make the axes fit the curve. We use this method in *Example 2*.

Example 2

Graph the function $y = 3\cos 2(x - \frac{\pi}{3}) + 4$ over two complete cycles.

Solution

Step 1.   Draw a sinusoidal curve, without axes. Determine the phase shift
and the period, and use them to establish a horizontal scale.

The phase shift is $\frac{\pi}{3}$, and the period is $\frac{2\pi}{2}$, or π.

Since the function is a cosine function, the phase shift $\frac{\pi}{3}$ is the
x-coordinate of a maximum point. Since the period is π, the

x-coordinate of the next maximum point is $\frac{\pi}{3} + \pi$, or $\frac{4\pi}{3}$.
Label these points.

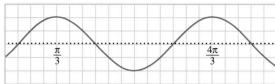

Step 2. Complete the graph by drawing the axes and their scales.

The graph was drawn so that the period, π, corresponds to 12 squares. Hence, 4 squares correspond to $\frac{\pi}{3}$. Then, the origin, O, is 4 squares to the left of the first maximum point. Draw the *y*-axis through this point.

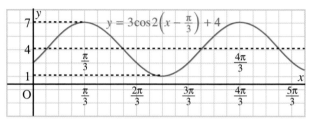

The vertical displacement is 4. Since the amplitude is 3, the maximum value of *y* is 7 and the minimum value is 1. Use these values to mark the vertical scale. Draw the *x*-axis and mark its scale.

DISCUSSING THE IDEAS

1. Could the period of a periodic function be defined as the distance between two consecutive maximum points? Explain.

2. The functions in *Examples 1* and *2* could have been graphed with a graphing calculator or a computer. What are the advantages of not using graphing technology to graph these functions?

3. In *Visualizing*, page 251, why is the period less than 2π when $b > 1$ and greater than 2π when $0 < b < 1$?

4. A disadvantage of the method used in *Example 2* is that if you always start with the same sinusoidal curve, then all your graphs will have the same size and shape. How might this disadvantage be overcome?

5. a) What is the period of $y = \sin bx$ and $y = \cos bx$ if $b < 0$? Explain.

   b) What happens when $b = 0$?

A **1.** These graphs are sinusoidal. Identify the period and write the equation of each graph.

a)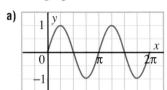

b)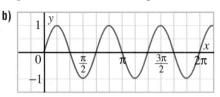

c)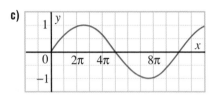

d)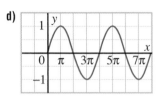

2. The graph of $y = \cos x$ is transformed as described below. The equation of its image has the form $y = a \cos b(x - c) + d$. Determine a, b, c, and d for each transformation.

a) Compress the graph horizontally by a factor of $\frac{1}{2}$.

b) Expand the graph horizontally by a factor of 5.

c) Compress the graph horizontally by a factor of $\frac{1}{3}$ and expand it vertically by a factor of 4.

d) Expand the graph horizontally by a factor of 2, expand it vertically by a factor of 3, then translate it $\frac{\pi}{3}$ units right.

3. Consider a sinusoidal function of the form $y = a \sin b(x - c) + d$. Explain how its graph will change for each change described.

a) b is doubled.

b) b is halved.

c) a and b are both doubled, and c is decreased by 2.

d) a is tripled, b is halved, c is increased by 4, and d is increased by 3.

4. The screen shows the graphs of $y = \cos bx$ for $b = 2, 1$, and 0.5. The coordinates of one point on one graph are shown. Write the coordinates of one point on each of the other graphs that has the same y-coordinate as this point.

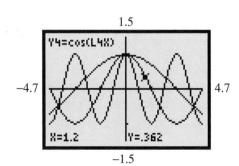

5. a) Use the same graphing window as in exercise 4. Graph $y = \sin bx$ for $b = 2, 1$, and 0.5. Trace to the point on the graph of $y = \sin x$ where $x = 0.8$. Observe the corresponding value of y.

b) Predict the coordinates of a point on each of the other graphs that has the same value of y. Trace to these points to verify your predictions.

6. a) Graph each set of functions on the same grid for $-\pi \le x \le \pi$.

  i) $y = \sin 2x$ $\qquad y = \sin x$ $\qquad y = \sin \frac{1}{2}x$

  ii) $y = \cos 3x$ $\qquad y = \cos x$ $\qquad y = \cos \frac{1}{3}x$

b) Describe the effect on the graphs of $y = \sin bx$ and $y = \cos bx$ as b varies.

7. Each function below is sinusoidal. For each graph, state:

  i) the amplitude $\qquad$ **ii)** the period $\qquad$ **iii)** a possible phase shift

  iv) the maximum value of y, and the values of x for which it occurs

  v) the minimum value of y, and the values of x for which it occurs

  vi) the domain and the range $\qquad$ **vii)** the vertical displacement

a)

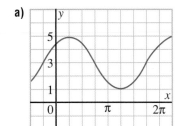

b)

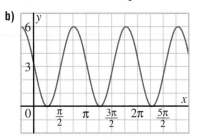

c)

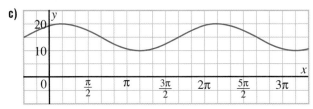

8. Write an equation to represent each function in exercise 7.

9. Graph each sinusoidal function, and state its amplitude and period.

  a) $y = 2 \sin 2x$ $\qquad$ **b)** $y = 3 \sin \frac{1}{2}x$ $\qquad$ **c)** $y = 4 \sin 2x$

  d) $y = 4 \cos \frac{1}{2}x$ $\qquad$ **e)** $y = 5 \cos 2x$ $\qquad$ **f)** $y = 3 \cos 3x$

10. Write to explain how to sketch the graph of the function $y = a \sin bx$.

11. State the amplitude, the period, and the phase shift for each function.

  a) $f(x) = 5 \cos 3(x - \pi)$ $\qquad$ **b)** $f(x) = 2 \cos 3(x - \frac{\pi}{2})$

  c) $f(x) = 2.5 \sin 6(x - \frac{2\pi}{3})$ $\qquad$ **d)** $f(x) = 0.5 \cos 5(x + \frac{5\pi}{4})$

12. For each graph, write an equation in the form $y = \cos bx$.

a)

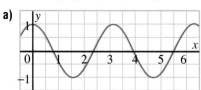

b)

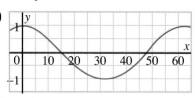

c)

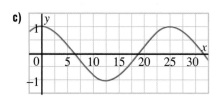

d)

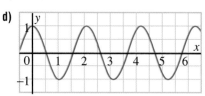

13. Describe what happens to each graph.

a) $y = a \sin 2(x - \frac{\pi}{6}) + 5$, as a varies

b) $y = 3 \sin b(x - \frac{\pi}{6}) + 5$, as b varies

c) $f(x) = 3 \sin 2(x - c) + 5$, as c varies

d) $g(x) = 3 \sin 2(x - \frac{\pi}{3}) + d$, as d varies

B **14.** Consider the function $y = 2 \cos 2(x - \frac{\pi}{3}) + 4$.

a) Find the phase shift and the period.

b) Determine the vertical displacement and the amplitude.

c) Graph the function by graphing a sinusoidal curve first, then make the axes fit the curve.

15. Consider the function $f(x) = 3 \sin 2(x - \frac{\pi}{4}) + 3$.

a) Find the phase shift and the period.

b) Determine the vertical displacement and the amplitude.

c) Graph the function by graphing a sinusoidal curve first, then make the axes fit the curve.

16. Graph each function over two cycles, then state its amplitude, period, and phase shift.

a) $f(x) = 2 \sin (2x + \frac{\pi}{3})$

b) $f(x) = 5 \cos (2x - \frac{\pi}{2})$

c) $f(x) = 3 \cos (2x - \frac{\pi}{2})$

d) $f(x) = 5 \sin (2x + \frac{\pi}{3})$

17. Determine the phase shift and the period of each function, then draw its graph.

a) $y = \sin (2x - \pi)$

b) $y = 2 \cos (3x - \pi) + 1$

c) $y = 2 \cos (3x - \pi) + 4$

d) $y = 5 \sin (4x + \pi) - 3$

18. Write to explain how to sketch the graph of the function $y = a \sin b(x - c) + d$.

19. Graph each function.

a) $y = 4 \cos 3(x - \frac{\pi}{2}) + 4$

b) $y = 2 \cos 4(x + \pi) - 3$

c) $f(x) = 3 \sin 2(x + \frac{\pi}{6}) - 6$

d) $f(x) = 4 \sin 3(x - \frac{\pi}{6}) + 2$

20. These graphs were produced by *Graphmatica*. Each pattern shows the graphs of ten sinusoidal functions. For each pattern, write a set of equations that could be used to make it.

a)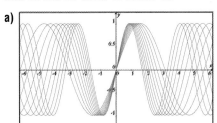

b)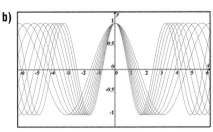

21. Make patterns like those in exercise 20.

22. Write to explain how the graph of $y = \sin(4x + 1)$ differs from that of $y = \sin 4x + 1$.

C 23. Consider the function $f(x) = a \sin b(x - c) + d$, where $a > 0$.

   a) Write an expression for the maximum value of $f(x)$. For what values of x does this occur?

   b) Write an expression for the minimum value of $f(x)$. For what values of x does this occur?

24. Repeat exercise 23 for the function $f(x) = a \cos b(x - c) + d$, where $a > 0$.

25. Two of these equations represent the same function. Which two are they? Explain.

   a) $y = 3 \sin 2(x + \frac{\pi}{2})$ **b)** $y = 3 \cos 2x$

   c) $y = 3 \cos 2(x + \frac{\pi}{4})$ **d)** $y = 3 \sin 2(x + \pi)$

26. Two of these equations represent the same function. Which two are they? Explain.

   a) $f(x) = 3 \sin(2x + \pi)$ **b)** $f(x) = 3 \cos(2x + \frac{\pi}{2})$

   c) $f(x) = 3 \cos 2x$ **d)** $f(x) = 3 \sin(2x + 2\pi)$

COMMUNICATING THE IDEAS

Write to explain why increasing the value of a in the equation $y = a \sin bx$ expands the graph vertically, but increasing the value of b compresses the graph horizontally, where both $a > 0$ and $b > 0$. Include some graphs in your explanation.

Dynamic Graphs

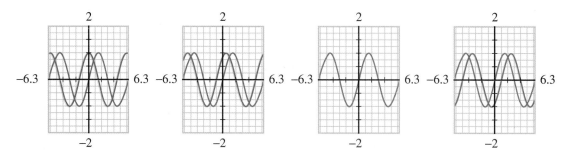

The graphs on these two pages were produced using software called *Graphs! Graphs! Graphs!* This program contains a feature called a "slider" that lets you change a coefficient in an equation gradually from one value to another. The graph moves smoothly in response to the change. This was done to produce the sequence of graphs above, and at the top of the next page. They show how the graph of $y = \cos(x - c)$ changes as c increases from 0 to $\frac{7\pi}{4}$ in steps of $\frac{\pi}{4}$. The blue curve representing $y = \sin x$ is shown for comparison.

These graphs were obtained from the graph of $y = \cos(x - c)$, which is built into the program. The variable c is called a "slider variable." The slider at the right is set with $c = 0$, and corresponds to the first graph above. The dot beside c indicates that c is active. When the slider box is dragged up, the values of c increase gradually from 0 to 6.3 (that is, from 0 to 2π), producing graphs like those above.

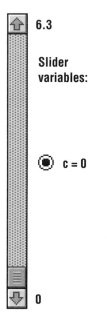

Slider variables:

$c = 0$

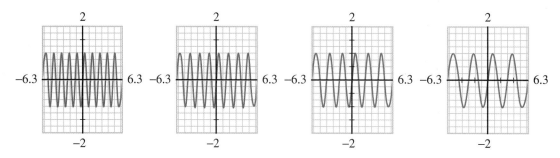

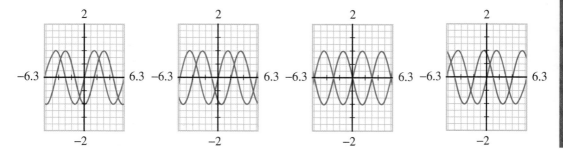

1. **a)** Scan the eight graphs at the top of these two pages, from left to right. Visualize how the cosine curve gradually changes as c increases.

   b) Determine the value of c for each graph in terms of π, and in decimal form.

   c) For what values of c does $\cos(x - c) = \sin x$?

   d) For what values of c does $\cos(x - c) = -\sin x$?

2. The slider at the right corresponds to one graph at the bottom of these two pages. This sequence of graphs shows how $y = \sin ax$ changes as a decreases from 5 to $\frac{1}{10}$.

   a) Identify the graph below that is indicated by the slider.

   b) Scan the graphs back and forth across the two pages. Visualize how the sine curve changes as a changes.

   c) Write the equation of each graph, from left to right.

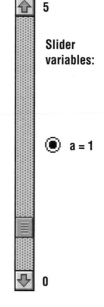

5

Slider variables:

⦿ $a = 1$

0

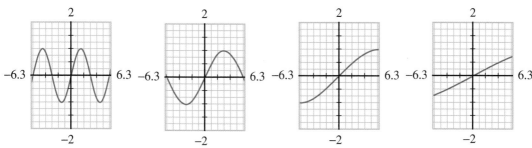

Mathematics & Technology

4.4 Graphing $y = \sin \frac{2\pi}{p}x$ and $y = \cos \frac{2\pi}{p}x$

In the preceding sections, the sinusoidal functions had equations of the form $y = a \sin b(x - c) + d$. In all the examples and exercises, b was a rational number. This meant that the period, $\frac{2\pi}{b}$, was an irrational number. In an applied problem, the period is usually a rational number. Hence, it is useful to write the equations of sinusoidal functions in a form that ensures the period is a rational number.

INVESTIGATE | **Transforming Sinusoidal Graphs Part III**

In this investigation, you will adjust the periods of sinusoidal functions to make them rational numbers.

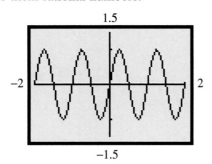

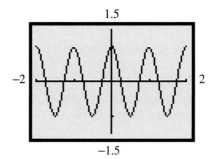

Sinusoidal functions with period that is a whole number

1. Both graphs above were obtained from a graphing calculator using the window settings $-2 \le x \le 2$ and $-1.5 \le y \le 1.5$.

   a) Use the graphs to estimate the period of each function.

   b) Recall that the period of the function $y = \sin bx$ is $\frac{2\pi}{b}$. What value of b is needed to produce each period in part a?

   c) What are the equations of the two functions graphed above?

   d) Check your answer to part c by graphing your equations.

2. Predict the period of each function. Graph the function to check.

   a) $y = \sin \frac{2\pi}{2}x$ **b)** $y = \sin \frac{2\pi}{3}x$ **c)** $y = \sin \frac{2\pi}{4}x$ **d)** $y = \sin \frac{2\pi}{5}x$

3. Write the equation of a cosine function with each period.

   a) 2   **b)** 3   **c)** 4   **d)** 5

4. What is the period of the function $y = \sin \frac{2\pi}{p}x$? Explain.

The results of *Investigate* are summarized below. You can use these results to sketch the graphs of sinusoidal functions whose periods are rational numbers.

The graphs of $y = \sin 2\pi x$ and $y = \cos 2\pi x$

Recall that the period of $y = \sin bx$ and $y = \cos bx$ is $\frac{2\pi}{b}$.

For the functions $y = \sin 2\pi x$ and $y = \cos 2\pi x$, the value of b is 2π.

Hence, their periods are $\frac{2\pi}{2\pi}$, or 1.

VISUALIZING   **Graphs of $y = \sin 2\pi x$ and $y = \cos 2\pi x$**

Visualize four squares of the coordinate grid, with the origin at the centre. Two cycles of the function $y = \sin 2\pi x$ fit in these squares.

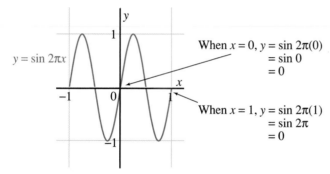

When $x = 0$, $y = \sin 2\pi(0)$
$= \sin 0$
$= 0$

When $x = 1$, $y = \sin 2\pi(1)$
$= \sin 2\pi$
$= 0$

Two cycles of the function $y = \cos 2\pi x$ fit in the same squares.

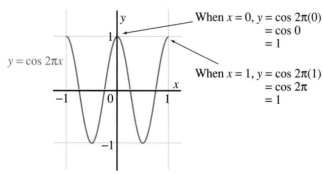

When $x = 0$, $y = \cos 2\pi(0)$
$= \cos 0$
$= 1$

When $x = 1$, $y = \cos 2\pi(1)$
$= \cos 2\pi$
$= 1$

The graphs of $y = \sin \dfrac{2\pi}{p}x$ and $y = \cos \dfrac{2\pi}{p}x$

The period of $y = \sin bx$ and $y = \cos bx$ is $\dfrac{2\pi}{b}$.

For the functions $y = \sin \dfrac{2\pi}{p}x$ and $y = \cos \dfrac{2\pi}{p}x$, the value of b is $\dfrac{2\pi}{p}$.

Hence, their periods are $\dfrac{2\pi}{\frac{2\pi}{p}} = \dfrac{2\pi}{1} \times \dfrac{p}{2\pi}$

$$= p$$

VISUALIZING   **Graphs of $y = \sin \dfrac{2\pi}{p}x$ and $y = \cos \dfrac{2\pi}{p}x$**

When $p > 1$, there is a horizontal expansion relative to $y = \sin 2\pi x$, or $y = \cos 2\pi x$.
When $0 < p < 1$, there is a compression.

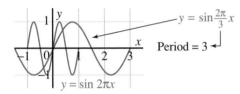

$y = \sin \dfrac{2\pi}{3}x$ Period = 3

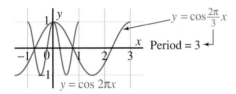

$y = \cos \dfrac{2\pi}{3}x$ Period = 3

The periods of $y = \sin \dfrac{2\pi}{p}x$ and $y = \cos \dfrac{2\pi}{p}x$ are p.

Example 1

Determine the period of this function, then sketch its graph.

$$y = \sin \dfrac{2\pi x}{5}$$

Solution

Compare $y = \sin \dfrac{2\pi x}{5}$ with the general equation $y = \sin \dfrac{2\pi x}{p}$.

The period of the function is 5.
Sketch a sine curve, then label the x-axis so the period is 5.

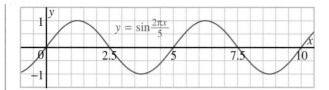

When π occurs in the equation as a factor of the quantity whose sine or cosine is to be found, then it does not appear as part of the scale on the horizontal axis. The fact that π does not occur on the axis means that it must occur in the equation.

Example 2

Graph the function $y = 3 \sin 2\pi \dfrac{(x-2)}{4} + 6$.

Solution

Step 1. Draw a sinusoidal curve. Find the phase shift and the period, and use them to establish a horizontal scale.
The phase shift is 2. The period is 4.

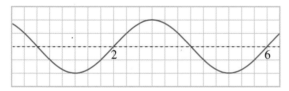

Step 2. Find the amplitude and vertical displacement.
The amplitude is 3 and the vertical displacement is 6.
Complete the graph by drawing the axes and their scales.

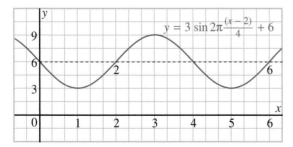

In *Example 2*, the period, amplitude, phase shift, and vertical displacement are represented by numbers in the equation.

Phase shift

$$y = 3 \sin 2\pi \frac{(x-2)}{4} + 6$$

Amplitude   Period   Vertical displacement

You can use this pattern to write the equation when these data are given or when they can be read from a graph. Notice that the coefficient of x is $\frac{2\pi}{4}$. Hence, the period is 4.

Example 3

The volume of air in the lungs is a sinusoidal function of time. A graph illustrating this variation for normal breathing is shown on page 160. Write an equation for this function.

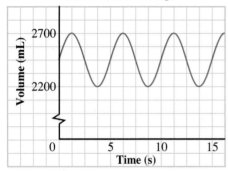

Volume of air in lungs

Solution

The vertical displacement is $\frac{2700 + 2200}{2}$ mL = 2450 mL.

The amplitude is $\frac{2700 - 2200}{2}$ mL = 250 mL.

The period is 5 s.
Let V millilitres represent the volume of air in the lungs at time t seconds.

Substitute the known values in the general equation $y = a \sin \frac{2\pi x}{p} + d$.

An equation for the function is $V = 250 \sin \frac{2\pi t}{5} + 2450$.

DISCUSSING THE IDEAS

1. Give two reasons why the period of $y = \sin \frac{2\pi}{p} x$ is p.

2. For what values of p in $y = \sin \frac{2\pi}{p} x$ is there a horizontal expansion? How do these compare with the values of b in $y = \sin bx$ for which there is an expansion? Explain.

3. What are some other possible equations for the function in *Example 3*? Explain each equation.

4.4 EXERCISES

(A) 1. State the amplitude, period, phase shift, and vertical displacement for each function.

a) $y = 3 \sin 2\pi \frac{(t - 1)}{5} + 4$

b) $y = 2 \cos 2\pi \frac{(t - 5)}{4} - 6$

2. Identify the period of each function, then write its equation in the form $y = \sin \frac{2\pi}{p} x$.

a)
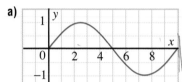

b)
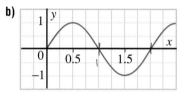

c)
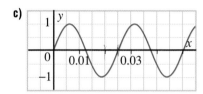

d)
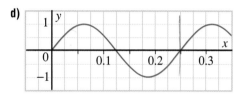

3. Each function below is sinusoidal. For each graph, state:

   i) the amplitude   **ii)** the period   **iii)** a possible phase shift

   iv) the maximum value of y, and the values of x for which it occurs

   v) the minimum value of y, and the values of x for which it occurs

   vi) the domain and the range   **vii)** the vertical displacement

a)
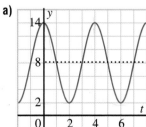

b)
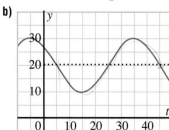

4. Write an equation to represent each function in exercise 3.

5. Write the equation of each function in the form $y = a \cos \frac{2\pi}{p}x + b$.

a)

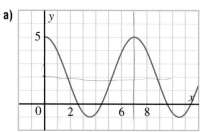

b)

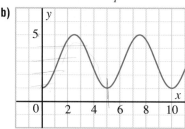

c)

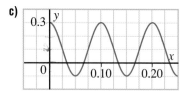

d)

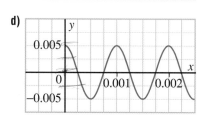

6. Write to explain how you can determine the period of a function written in the form $y = \cos 2k\pi x$, where k is an integer.

7. Write an equation for a cosine function with the following properties.

   a) amplitude: 5   period: 1   phase shift: 9   vertical displacement: −4

   b) amplitude: 12   period: 0.5   phase shift: −3   vertical displacement: 1.5

   c) amplitude: 2.4   period: 27   phase shift: 19   vertical displacement: 15.1

8. The graphing calculator screen (below left) shows the coordinates of one point on the graph of $y = \sin 2\pi x$. Use only the information on the screen. Write the x-coordinates of four other points on the graph that have the same y-coordinate as this point.

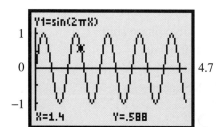

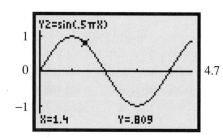

9. The graphing calculator screen (above right) shows the coordinates of one point on the graph of $y = \sin \frac{\pi}{2}x$. Use only the information on the screen. Write the x-coordinates of four other points on the graph that have the same y-coordinate as this point.

10. Choose either exercise 8 or exercise 9. Write to explain how you determined the x-coordinates.

11. Compare the two graphs in exercises 8 and 9. Explain why the indicated points have different y-coordinates when they have the same x-coordinate.

12. a) Graph $y = \sin 2\pi x$ using a graphing window defined by $0 \le x \le 4.7, -1.5 \le y \le 1.5$. Trace to the point where $x = 0.75$. Observe the corresponding value of y.

   b) Predict some other values of x that should produce the same value of y. Trace to these points to verify your predictions.

13. Repeat exercise 12, using $y = \sin \frac{\pi}{2}x$.

B 14. Sketch a graph of each function. What is its domain and range?

   a) $y = 2 \cos 2\pi \frac{(x-1)}{3} + 4$

   b) $y = 3 \cos 2\pi \frac{(x-4)}{2} - 3$

   c) $f(x) = 2.4 \cos 2\pi \frac{(x+3)}{12} - 3.6$

   d) $f(t) = 3.5 \sin 2\pi \frac{(t-8.4)}{9.2} + 10$

15. Sketch a graph of each function for $-5 \le t \le 5$. State the maximum and minimum values of y, and the values of t for which they occur.

   a) $y = 2 \cos 2\pi \frac{(t-1)}{3} - 3$

   b) $y = 4 \sin 2\pi \frac{(t+2)}{5} - 4$

   c) $y = 2 \sin 2\pi \frac{(t-1)}{3} + 6$

   d) $y = 5 \cos 2\pi \frac{(t+3)}{6} + 2$

16. Write an equation to represent a sine function with the following properties.

   a) maximum: 23   minimum: 11   period: 5   phase shift: 9

   b) maximum: 17.2   minimum: 8.6   period: 3.9   phase shift: 4.7

17. Write typical values for the amplitude, vertical displacement, phase shift, and period of a sine function. Use these values to write an equation of the function. Exchange equations with a classmate. Determine the amplitude, vertical displacement, phase shift, and period of your classmate's function.

18. Write the equations of the sinusoidal functions in each pattern.

a)

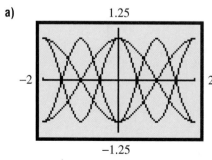

b)

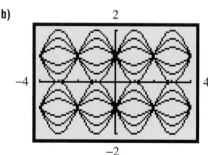

19. Make the patterns in exercise 18. Then make a different pattern.

20. Write an equation for the volume of air in the lungs during deep breathing, when the variation is from 1000 mL to 5000 mL. Assume the period is 10 s.

In *Example 3* and exercise 20, you used a sinusoidal function to model the volume of air in the lungs.

- What are some reasons why the volume of air in the lungs might be different from the volume predicted by the models?
- For each reason, how would the graph of volume against time be affected?

21. The twin towers of the World Trade Center in New York were once the tallest buildings in the world. During a strong wind, the top of each tower can swing back and forth as much as 80 cm, with a period of 10 s.

   a) Draw a graph to show the departure of the top of one of the buildings from the normal position as a function of time, for 20 s.

   b) Write an equation for the function in part a.

22. A piston in an engine moves up and down in the cylinder, as shown in the diagram. The height, h centimetres, of the piston at time, t seconds, is given by this formula. $h = 20 \sin \frac{2\pi t}{0.05} + 20$

   a) State the piston's maximum height, minimum height, and period.

   b) Suppose the piston operates for one hour. How many complete cycles does it make?

23. The graph shows how the height varies with time for the pistons in a six-cylinder engine.

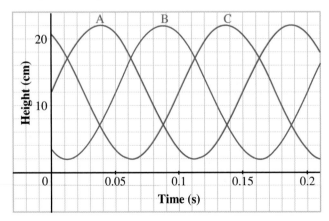

   a) Determine an equation for the graph labelled A.

   b) Determine a possible phase shift for graph B and for graph C.

c) Write an equation for graph B and for graph C.

d) The graph describes the motion of all 6 cylinders. Suggest why only 3 traces are visible.

e) Determine the maximum speed of the pistons.

24. Write to explain why it is useful to write an equation in the form $y = \sin \frac{2\pi}{p}x$ instead of $y = \sin kx$.

25. The fundamental tone of a guitar string with length L is associated with a sinusoidal function with period $2L$. The period of the first overtone is $\frac{2L}{2}$; the period of the second overtone is $\frac{2L}{3}$; and so on.

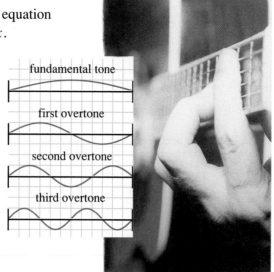

fundamental tone

first overtone

second overtone

third overtone

a) Assume the string is 50 cm long, and the amplitude of the vibration is 0.5 cm. Write the equations of the functions associated with the fundamental tone and the first three overtones.

b) Draw the graphs of the functions in part a on the same axes.

MODELLING a Vibrating Guitar String

In exercise 25, you used sinusoidal functions to model the vibration of a guitar string.

- What do the variables x and y in the equations of the functions represent?
- How would the motion of the guitar string be different from that predicted by the model?
- How would this affect the graphs?

C **26.** Two of these equations represent the same function. Which two are they? Explain.

a) $y = 3 \sin 2\pi \frac{(t-1)}{8} + 2$

b) $y = 3 \cos 2\pi \frac{(t-5)}{8} + 2$

c) $y = 3 \sin 2\pi \frac{(t-3)}{8} + 2$

d) $y = 3 \cos 2\pi \frac{(t+1)}{8} + 2$

COMMUNICATING THE IDEAS

Write to explain the similarities and the differences between graphing a function whose equation has the form $y = a \cos b(x - c) + d$ and one whose equation has the form $y = a \cos 2\pi \frac{(x-c)}{p} + d$.

Predicting Hours of Daylight

On page 224, we considered the variation in the number of hours of daylight at Red Deer during a two-year period. The following data, obtained from the Internet, were used to produce the graph on page 224. The times are in hours and minutes for a 24-h clock.

Sunset and Sunrise Times at Red Deer (h:min)

Date	Dec. 21	Mar. 21	June 21	Sept. 21	Dec. 21
Sunset time	16:24	18:50	21:01	18:37	16:24
Sunrise time	8:43	6:36	4:13	6:19	8:43

From these data, the following information can be calculated. The dates are given as the number of the day in the year, with January 1 being day 1 and December 31 as day 365. The times are hours in decimal form.

Hours of Daylight at Red Deer

Day of year	−10	80	172	264	355
Hours	7.68	12.23	16.80	12.30	7.68

Check that these numbers are correct. You can use this information to determine a sinusoidal equation that models the number of hours of daylight at Red Deer.

 DEVELOP A MODEL

Let h hours represent the number of hours of daylight on the nth day of the year. Visualize a cosine function whose equation has this form.

$$h = a \cos 2\pi \frac{(n-c)}{p} + d$$

a, c, p, and d represent the amplitude, phase shift, period, and vertical displacement, respectively.

1. For the data in the second table above, what are the values of a, c, p, and d?

2. Write an equation that models the number of hours of daylight at Red Deer.

3. Graph the equation.

You can use your equation to estimate the number of hours of daylight on any day of the year. To determine the number of the day of the year, use this table.

Cumulative Days in the Year (non-leap years)

Jan.	Feb.	Mar.	Apr.	May	Jun.	Jul.	Aug.	Sep.	Oct.	Nov.	Dec.
31	59	90	120	151	181	212	243	273	304	334	365

4. Use the equation from exercise 2. Estimate the number of hours of daylight on each date.

a) March 1   **b)** July 15   **c)** November 11

5. a) Estimate the numbers of hours of daylight on March 21, June 21, and September 21.

b) Compare your answers in part a with the data on page 270. Account for the discrepancies in the results.

6. The equation from exercise 2 was entered into a graphing calculator. The screen at the right was obtained.

a) What information is shown?

b) Use the trace feature on a graphing calculator. Estimate the number of hours of daylight on each date in exercise 4.

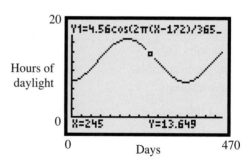

7. a) Use the data on page 270. Determine equations that model:

   i) the sunset times at Red Deer   **ii)** the sunrise times at Red Deer

b) Use the equations in part a. Estimate the times of the sunset and the sunrise at Red Deer on March 1, July 15, and November 11. Compare the results with your answers to exercise 4.

8. The graphs show standard times. How would the graphs change for daylight-saving time?

4.5 Modelling Real Situations Using Trigonometric Functions

The tides are the periodic rise and fall of the water in the oceans, caused almost entirely by the gravitational attraction of the moon and the sun. An equation that expresses the depth of the water as a function of time is complicated, since the distances and relative positions of the moon and sun are constantly changing. However, the depth can be approximated by a sinusoidal function. The amplitude of the function depends on the location and, at any particular location, it varies considerably at different times of the year.

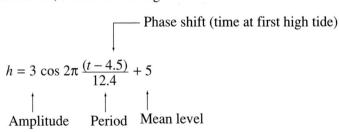

The world's highest tides occur in the Bay of Fundy, where 100 billion tonnes of water pulse in and out twice a day. This graph shows how the depth of the water at the Annapolis Tidal Generating Station varies during a typical day.

We can find the vertical displacement, the phase shift, the period, and the amplitude of the function from the graph, and use these to write an equation of the function.

Tides at Annapolis Tidal Generating Station

Since the mean level is 5 m, the vertical displacement is 5 m. The amplitude is the difference between high and mean tide levels, 3 m. The first high tide occurs at 4.5 h. If we think of the function as a cosine function, the phase shift is 4.5 h. The period is the time between two high tides, 12.4 h. Hence, for a depth of h metres, and a time of t hours, an equation of the function is:

Phase shift (time at first high tide)

$$h = 3 \cos 2\pi \frac{(t - 4.5)}{12.4} + 5$$

Amplitude   Period   Mean level

Example 1

a) Estimate, to the nearest tenth of a metre, the depth of the water at 2:45 P.M.

b) Estimate, to the nearest minute, one of the times when the water is 2.5 m deep on the day represented by the equation.

Solution

> **Think...**
>
> Express times in decimals of hours, on a 24-h clock: 2.45 P.M. = (12 + 2.75) h
>
> $$= 14.75 \text{ h}$$

Using a graphing calculator

a) Graph the function above.
Press TRACE 14.75 ENTER.
At 2:45 P.M., the depth of the water is approximately 6.4 m.

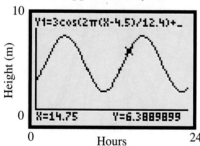

b) Graph $y = 2.5$ on the same graph.
Activate the intersect feature.
A depth of 2.5 m occurs at approximately 11.86 h, or 11:51 A.M.

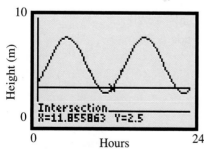

Using a scientific calculator

a) Use the equation preceding *Example 1*.

$$h = 3 \cos 2\pi \frac{(t - 4.5)}{12.4} + 5$$

Substitute $t = 14.75$ in the equation.

$$h = 3 \cos 2\pi \frac{(14.75 - 4.5)}{12.4} + 5$$

$$\doteq 6.388\ 989\ 932$$

At 2:45 P.M., the depth of the water is approximately 6.4 m.

b) Substitute $h = 2.5$ in the equation, then solve for t.

$$3 \cos 2\pi \frac{(t - 4.5)}{12.4} + 5 = 2.5$$

$$3 \cos 2\pi \frac{(t - 4.5)}{12.4} = -2.5$$

$$\cos 2\pi \frac{(t - 4.5)}{12.4} = \frac{-2.5}{3}$$

Find a number whose cosine is $\frac{-2.5}{3}$.
Use the $\boxed{\cos^{-1}}$ key. One number is approximately 2.555 907 11.

Hence,

$$2\pi \frac{(t - 4.5)}{12.4} \doteq 2.555\ 907\ 11$$

$$t - 4.5 \doteq \frac{2.555\ 907\ 11 \times 12.4}{2\pi}$$

$$t \doteq 9.544\ 137\ 108$$

A depth of 2.5 m occurs at approximately 9.54 h, or 9:33 A.M.

We can use the pattern suggested by *Example 1* to solve other problems involving quantities that change periodically. In each case, use a sinusoidal function to represent the data. This is the general pattern:

Amplitude ┌─ Phase shift (value of *t* at first maximum)

$$y = A \cos 2\pi \frac{(t-S)}{P} + M$$

Period Mean value or vertical displacement

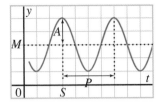

Example 2

A Ferris wheel has a radius of 20 m. It rotates once every 40 s. Passengers get on at point S, which is 1 m above ground level. Suppose you get on at S and the wheel starts to rotate.

a) Graph how your height above the ground varies during the first two cycles.

b) Write an equation that expresses your height as a function of the elapsed time.

c) Estimate your height above the ground after 45 s.

d) Estimate one of the times when your height is 35 m above the ground.

Solution

a) *Step 1.* Draw a sinusoidal curve. Indicate the phase shift, the period, the vertical displacement, and the amplitude.

For a cosine function, the phase shift is the *t*-coordinate of the first maximum, point A. Since the Ferris wheel rotates once every 40 s, the period is 40 s. Since you take 20 s to reach A, the phase shift is 20 s. Hence, the *t*-coordinates of two consecutive maximum points are 20 and 60. The vertical displacement is 21 m, and the amplitude is 20 m.

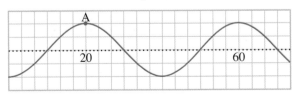

Step 2. Complete the graph by drawing the axes and their scales.

The numbers 20 and 60 on the graph above establish a horizontal scale, where 3 squares represent 10 s. The vertical axis is 6 squares to the left of the first maximum point. The minimum and maximum heights of 1 m and 41 m establish a vertical scale, where 1 square represents 10 m. The horizontal axis is one-tenth of a square below the minimum points.

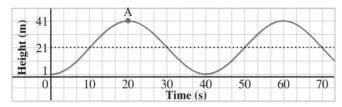

b) Substitute from part a into the general equation $y = A \cos 2\pi \frac{(t - S)}{P} + M$.
An equation that expresses height as a function of time is $h = 20 \cos 2\pi \frac{(t - 20)}{40} + 21$.

Using a graphing calculator

c) Graph the equation. Press [TRACE] 45 [ENTER]. After 45 s, you will be about 6.9 m above the ground.

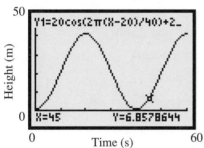

d) Graph $y = 35$ on the same graph. Activate the intersect feature. After about 14.9 s, you will be 35 m above the ground.

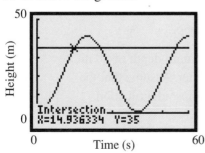

Using a scientific calculator

c) Substitute $t = 45$ in the equation.
$$h = 20 \cos 2\pi \frac{(45 - 20)}{40} + 21$$
$$\doteq 6.857\ 864\ 376$$
After 45 s, you will be about 6.9 m above the ground.

d) Substitute $h = 35$, then solve for t.
$$20 \cos 2\pi \frac{(t - 20)}{40} + 21 = 35$$
$$\cos 2\pi \frac{(t - 20)}{40} = \frac{14}{20}$$
Use the [cos⁻¹] key. One number whose cosine is $\frac{14}{20}$ is approximately 0.795 398 83.

Hence,
$$2\pi \frac{(t - 20)}{40} = 0.795\ 398\ 83$$
$$t - 20 = \frac{0.795\ 398\ 83 \times 40}{2\pi}$$
$$t = 25.063\ 666\ 22$$

After about 25.1 s, you will be 35 m above the ground.

1. On page 272, an equation is shown for the function defined by the graph. What are some other equations for this function? Explain.

2. Suppose the equation next to the graph on page 274 were written in terms of a sine function instead of a cosine function. Would any change be needed? Explain.

3. The two solutions to *Example 1b* have different answers. Explain how this is possible.

4. Choose one solution from *Example 1b*. Explain how the time is converted in the last step of the solution.

5. What are some other equations for the function in *Example 2*? Explain.

4.5 EXERCISES

A 1. At a seaport, the depth of the water, h metres, at time, t hours, during a certain day is given by this formula. $h = 1.8 \sin 2\pi \frac{(t - 4.00)}{12.4} + 3.1$

 a) State.

 i) the period   ii) the amplitude   iii) the phase shift

 b) What is the maximum depth of the water? When does it occur?

 c) Estimate the depth of the water at 5:00 A.M. and at 12 noon.

 d) Estimate one of the times when the water is 2.25 m deep.

2. This equation gives the depth of the water, h metres, at an ocean port at any time, t hours, during a certain day. $h = 2.5 \sin 2\pi \frac{(t - 1.5)}{12.4} + 4.3$

 a) Explain the significance of each number in the equation.

 i) 2.5   ii) 12.4   iii) 1.5   iv) 4.3

 b) What is the minimum depth of the water? When does it occur?

 c) Estimate the approximate depth of the water at 9:30 A.M.

 d) Estimate one of the times when the water is 4.0 m deep.

B 3. On a typical day at an ocean port, the water has a maximum depth of 20 m at 8:00 A.M. The minimum depth of 8 m occurs 6.2 h later. Assume that the relation between the depth of the water and time is a sinusoidal function.

 a) What is the period of the function?

 b) Write an equation for the depth of the water at any time, t hours.

 c) Estimate the depth of the water at 10:00 A.M.

 d) Estimate one of the times when the water is 10 m deep.

4. Tidal forces are greatest when Earth, the sun, and the moon are in line. When this occurs at the Annapolis Tidal Generating Station, the water has a maximum depth of 9.6 m at 4:30 A.M. and a minimum depth of 0.4 m 6.2 h later.

a) Write an equation for the depth of the water at any time, t hours.

b) Estimate the depth of the water at 2:45 P.M.

c) Compare the results of part b with *Example 1a*.

5. Repeat exercise 4 when the tidal forces are weakest. The maximum depth of 6.4 m occurs at 7:45 A.M. and the minimum depth of 3.6 m occurs 6.2 h later.

6. A mass is supported by a spring so that it rests 0.5 m above a table top. The mass is pulled down 0.4 m and released at time $t = 0$. This creates a periodic up-and-down motion, called *simple harmonic motion*. It takes 1.2 s for the mass to return to the low position each time.

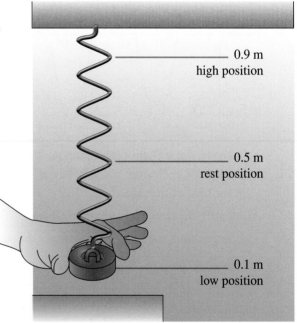

0.9 m
high position

0.5 m
rest position

0.1 m
low position

a) Graph the height of the mass above the table top as a function of time for the first 2.0 s.

b) Write an equation of a sinusoidal function that describes the graph in part a.

c) Use your equation to estimate the height of the mass above the table top after each time.
   i) 0.3 s   ii) 1.2 s

d) Estimate one of the times when the height of the mass is 0.75 m.

MODELLING the Motion of a Spring

In exercise 6, you used sinusoidal functions to model the motion of a mass on a spring.

- How would the motion of the spring be different from that predicted by the model?
- How would this affect the graph?
- Suggest how the equation of the function could be modified to improve the model.

7. A Ferris wheel has a radius of 25 m. Its centre is 26 m above the ground. It rotates once every 50 s. Suppose you get on at the bottom at $t = 0$.

a) Graph how your height above the ground changes during the first 2 min.

b) Write an equation for the function in part a.

c) Estimate how high you will be above the ground after each time.
   i) 10 s   **ii)** 20 s   **iii)** 40 s   **iv)** 60 s

d) Estimate one of the times when you are 50 m above the ground.

8. This graph shows how voltage, V volts, varies with time, t seconds, for the electricity provided to an electrical appliance.

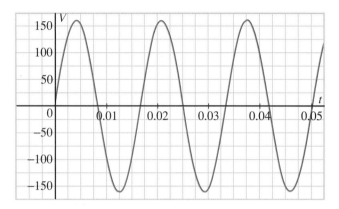

a) Identify the period of the function. Use the period to determine the number of cycles per second. This quantity, the frequency, is the reciprocal of the period.

b) Identify the peak (maximum) voltage.

c) Use the amplitude and period. Write an equation that describes this graph.

9. Technicians often use oscilloscopes to display graphs of voltage against time. The display may be sinusoidal. This graph shows the display that might result if a microphone was connected to the oscilloscope, and the voltage, V volts, was recorded as a tuning fork was sounded for t seconds.

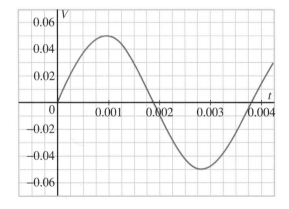

a) Estimate the period. Use it to determine the frequency of the tuning fork.

b) Write an equation that describes this graph.

10. The pedals of a bicycle are mounted
on a bracket whose centre is 29.0 cm
above the ground. Each pedal is 16.5 cm
from the centre of the bracket.
Assume that the bicycle is pedalled
at 12 cycles per minute.

a) Graph the height of one pedal above
the ground for the first few cycles.
Assume the pedal starts at
the topmost position at $t = 0$.

b) Write an equation of a sinusoidal function
that describes the graph in part a.

c) Estimate the height of the pedal after each time.
   i) 5 s   ii) 12 s   iii) 18 s

d) Estimate one of the times when the pedal is 40 cm above the ground.

11. On December 21, the sun is closest to
Earth, at approximately 147.2 million
kilometres. On June 21, the sun is at
its greatest distance, approximately
152.2 million kilometres.

a) Express the distance, d million
kilometres, from Earth to the sun
as a sinusoidal function of the number of days of the year.

b) Use the function in part a. Estimate the distance from Earth to the sun on
each date.
   i) March 1   ii) April 30   iii) September 2

c) On which dates is the sun about 150.0 million kilometres from Earth?

C 12. Refer to *Example 1*. Suppose your calculator is in degree mode. Would you
be able to solve this problem? If so, explain how. If not, explain why not.

13. In the solution of *Example 1*, a cosine function was used. Could you solve
Example 1 using a sine function? If so, explain how. If not, explain why not.

COMMUNICATING THE IDEAS

Write to explain how a sinusoidal function can be used to model a situation where a
quantity increases and decreases periodically. To illustrate your explanation, use an example
with equations and graphs.

Visualizing Graphs of Composite Functions

Recall that when two functions are applied in succession, the resulting function is called the composite of the two given functions. For example, consider the functions $y = \sin x$ and $y = 2x$. These functions can be applied in succession in two different ways.

Take the sine of x and multiply the result by 2: $y = 2\sin x$

Multiply x by 2 and take the sine of the result: $y = \sin 2x$

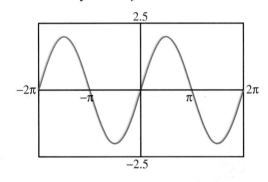

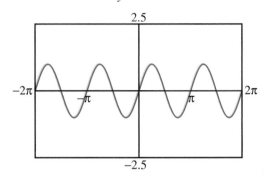

In the problems below, you will explore other composite functions obtained by applying a sine or cosine function with another function. Use a graphing calculator or computer graphing software. The graphs shown here were produced using *Graphmatica*.

Composite functions involving absolute value

1. One graph below represents $y = \sin |x|$, and the other graph represents $y = |\sin x|$. Visualize how the values of y would be calculated for each function. Then identify the equation of each graph. Explain your choices.

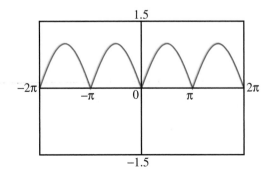

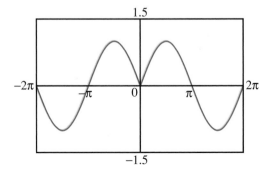

2. Predict what the graph of each equation would look like. Use a graphing calculator or graphing software to verify your predictions.

a) $y = \cos|x|$ **b)** $y = |\cos x|$

Composite functions involving powers

3. One graph below represents $y = \sin x^2$, and the other graph represents $y = (\sin x)^2$. Visualize how the values of y would be calculated for each function. Then identify the equation of each graph. Explain your choices.

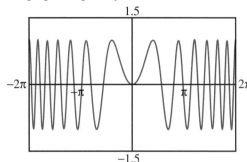 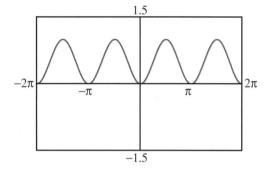

4. Predict what the graph of each equation would look like. Use a graphing calculator or graphing software to verify your predictions.

a) $y = \cos x^2$ **b)** $y = (\cos x)^2$

5. Predict what the graph of $y = (\sin x)^n$ and the graph of $y = (\cos x)^n$ would look like for $n = 4, 6, 8, \ldots$ (including much larger even numbers, such as 50). Verify your predictions.

6. One graph below represents $y = \sin x^3$, and the other graph represents $y = (\sin x)^3$. Visualize how the values of y would be calculated for each function. Then identify the equation of each graph. Explain your choices.

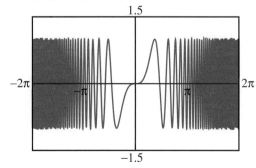 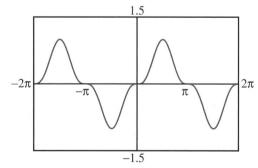

7. Predict what the graph of each equation would look like. Use a graphing calculator or graphing software to verify your predictions.

a) $y = \cos x^3$ **b)** $y = (\cos x)^3$

8. Predict what the graph of $y = (\sin x)^n$ and the graph of $y = (\cos x)^n$ would look like for $n = 5, 7, 9, \ldots$ (including much larger odd numbers, such as 51). Verify your predictions.

Composite functions involving exponentials

9. One graph below represents $y = \sin 2^x$, and the other graph represents $y = 2^{\sin x}$. Visualize how the values of y would be calculated for each function. Then identify the equation of each graph. Explain your choices.

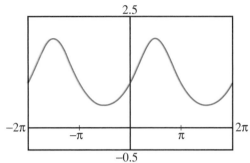

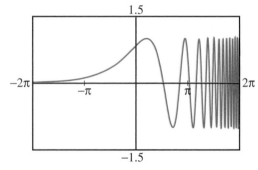

10. Predict what the graph of each equation would look like. Use a graphing calculator or graphing software to verify your predictions.

   a) $y = \cos 2^x$
   b) $y = 2^{\cos x}$

Relating pairs of graphs

11. The two graphs from exercise 3 are reproduced below. The graph of $y = x^2$ is shown in blue. For small values of x, the graphs of $y = \sin x^2$ and $y = (\sin x)^2$ are practically identical to the graph of $y = x^2$.

   a) Explain why this is so.

   b) Determine if a similar property occurs with the other pairs of graphs in the above exercises. For each pair of graphs, explain why the property occurs or does not occur.

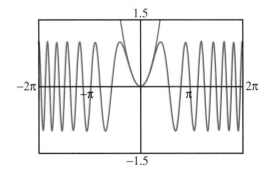

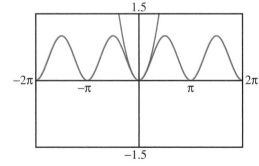

The Sinusoid of Best Fit

If you have a TI-83 calculator, you can use it to determine the equation of the sinusoid of best fit for given data. This type of calculation is called a *sinusoidal regression*. Your calculator must be in radian mode, and you must have at least five data points (better results are obtained with more points). The screens below show results obtained using data from page 270. To carry out a sinusoidal regression, follow these steps.

- Clear the Y= list.
- Define an appropriate graphing window.
- Press [STAT] **1**, and enter the data to be plotted in the first two columns.
- Press [STAT PLOT], and set up the first plot.
- Press [GRAPH] to graph the data.
- Press [STAT] [▶]. Scroll down and select SinReg.
- Press [VARS] [▶] **1 1** [ENTER]. The calculator determines the equation of the sinusoid of best fit and copies it into the Y= list beside Y1.
- Press [GRAPH] to see the plotted points and the graph of the sinusoid of best fit.

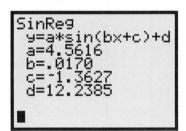

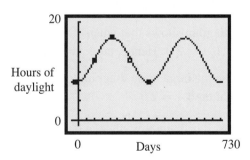

1. **a)** Use the data on page 270 to produce the results above.

   b) Determine the amplitude, phase shift, and period of the function in the screens above. Explain the results.

2. Complete some of 4.5 Exercises using a sinusoidal regression.

3. **a)** Use sinusoidal regression to determine an equation of a sine curve that passes through the points $(0, 0), (1, 1), (2, 0), (3, -1)$, and $(4, 0)$.

   b) Explain the result.

4.6 The Function $y = \tan x$

Defining and Graphing $y = \tan x$

The diagram below shows a unit circle with centre $(0,0)$ and the tangent line at $A(1,0)$. Ask your teacher for a large scale version of this diagram from the Teacher's Resource Book. The scale on the tangent line is the same as the one that could have been drawn on the coordinate axes. The scale on the circle indicates the distance from A measured counterclockwise. In principle, this scale continues indefinitely in both directions, but only the part from 0 to 2π is shown to avoid overlapping.

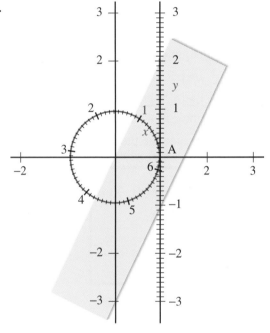

The function $y = \tan x$ is defined as follows. Place a straightedge at the origin so that it intersects the circle and tangent line, as shown. Let x represent the arc length measured along the circle from A to the straightedge. Let y represent the distance measured along the tangent line from A to the straightedge. This value of y is the *tangent* of x, and we write $y = \tan x$.

The diagram shows that when $x = 1.1$, $y \doteq 2.0$. Hence, $\tan 1.1 \doteq 2.0$

1. Use the large scale diagram to check that $\tan 1.1 \doteq 2.0$

2. Use this method to make a table of values for the function $y = \tan x$. Use values of x corresponding to points around the circumference of the circle.

3. Describe what happens to the values of $\tan x$ as x increases, starting at 0.

4. Graph the function $y = \tan x$.

5. Use the large scale diagram to explain your answer to each question.

   a) For what values of x is $\tan x$ undefined?

   b) What is the domain of the function $y = \tan x$? What is the range?

   c) What is the period of the function $y = \tan x$?

   d) What are the x-intercepts of the graph of the function $y = \tan x$?

   e) What are the equations of the asymptotes?

In Section 3.8, you used a figure similar to this to define tan θ and to construct the graph of $y = \tan \theta$, where θ is an angle in standard position. In this section, we will use arc length as the variable in the definition of the tangent function.

Visualize a point P rotating around a unit circle with centre $O(0,0)$. Let x represent the length of the arc from $A(1,0)$ to P. Let Q be the point where the line OP intersects the tangent line at $A(1,0)$. We define the second coordinate of Q to be $\tan x$.

Hence, the coordinates of Q are $(1, \tan x)$.

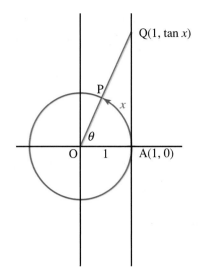

Example

Determine tan 2.1. Draw a diagram to explain the meaning of the result.

Solution

Using radian mode, $\tan 2.1 \doteq -1.7098$

This means that when the length of arc AP in a unit circle is 2.1 units, the line through O and P intersects the tangent line at $A(1,0)$ at the point with coordinates $(1, -1.7098)$.

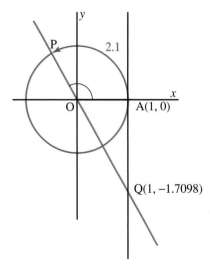

Graphing the function $y = \tan x$

The graph of $y = \tan x$ corresponds to that of $y = \tan \theta$, with the horizontal axis marked as the x-axis.

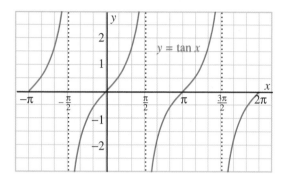

Properties of the function $y = \tan x$

Period: π

Domain: all real numbers except $\ldots, -\dfrac{3\pi}{2}, -\dfrac{\pi}{2}, \dfrac{\pi}{2}, \dfrac{3\pi}{2}, \ldots$

Range: all real numbers

x-intercepts: $0, \pm\pi, \pm 2\pi, \ldots$ y-intercept: 0

Asymptotes: $x = \pm\dfrac{\pi}{2}, \pm\dfrac{3\pi}{2}, \pm\dfrac{5\pi}{2}, \ldots$

DISCUSSING THE IDEAS

1. Explain why the function $y = \tan x$ is not defined for all values of x.

2. Explain why the function $y = \tan x$ is periodic.

3. Explain why the period of the function $y = \tan x$ is not the same as the period of $y = \sin x$ or $y = \cos x$.

4.6 EXERCISES

Ⓐ Graphing $y = \tan x + d$

1. a) Predict how the graph of the function $y = \tan x + 1$ compares with the graph of $y = \tan x$.

   b) Use graphing technology or grid paper to verify your prediction in part a.

   c) Write to explain how you could use transformation techniques to sketch the graph of $y = \tan x + d$ if you know the value of d.

2. Graph each function over two cycles. For each function, determine the domain, range, period, and the equations of the asymptotes.

   a) $y = \tan x - 2$ b) $y = \tan x + 3$

Graphing $y = a \tan x$

3. a) Predict how the graph of the function $y = 3 \tan x$ compares with the graph of $y = \tan x$.

   b) Use graphing technology or grid paper to verify your prediction in part a.

   c) Write to explain how you could use transformation techniques to sketch the graph of $y = a \tan x$ if you know the value of a.

4. Repeat exercise 2 for these functions.

   a) $y = 5 \tan x$ b) $y = 0.25 \tan x$

Graphing $y = \tan (x - c)$

5. a) Predict how the graph of the function $y = \tan (x - \frac{\pi}{4})$ compares with the graph of $y = \tan x$.

   b) Use graphing technology or grid paper to verify your prediction in part a.

   c) Write to explain how you could use transformation techniques to sketch the graph of $y = \tan (x - c)$ if you know the value of c.

6. Repeat exercise 2 for these functions.

   a) $y = \tan (x + \frac{\pi}{4})$ b) $y = \tan (x - \frac{\pi}{2})$

Graphing $y = \tan bx$

7. a) Predict how the graph of the function $y = \tan 2x$ compares with the graph of $y = \tan x$.

   b) Use graphing technology or grid paper to verify your prediction in part a.

   c) Write to explain how you could use transformation techniques to sketch the graph of $y = \tan bx$ if you know the value of b.

8. Repeat exercise 2 for these functions.

   a) $y = \tan 4x$ b) $y = \tan \frac{x}{2}$

Graphing $y = \tan\frac{\pi}{p}x$

9. a) Predict how the graph of the function $y = \tan\frac{\pi}{2}x$ compares with the graph of $y = \tan x$.

b) Use graphing technology or grid paper to verify your prediction in part a.

c) Write to explain how you could use transformation techniques to sketch the graph of $y = \tan\frac{\pi}{p}x$ if you know the value of p.

10. Graph each function over two cycles. For each function, determine the domain, range, period, and the equations of the asymptotes.

a) $y = \tan\frac{\pi}{4}x$
                                                    b) $y = \tan 2\pi x$

B **11.** Determine each value. Sketch a diagram to explain the meaning of the result.

a) $\tan 2.5$ **b)** $\tan 3.4$ **c)** $\tan(-1.5)$ **d)** $\tan(-5.0)$

e) $\tan 2\pi$ **f)** $\tan\frac{3\pi}{4}$ **g)** $\tan\frac{5\pi}{4}$ **h)** $\tan\left(-\frac{\pi}{2}\right)$

12. The graphing calculator screen (below left) shows the coordinates of one point on the graph of $y = \tan x$. Use only the information on the screen. Write the x-coordinates of four other points on the graph that have the same y-coordinate as this point.

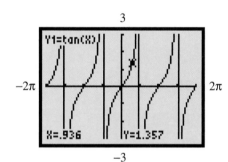
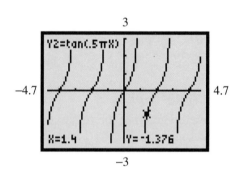

13. The graphing calculator screen (above right) shows the coordinates of one point on the graph of $y = \tan\frac{\pi}{2}x$. Use only the information on the screen. Write the x-coordinates of four other points on the graph that have the same y-coordinate as this point.

14. Choose either exercise 12 or 13. Write to explain how you determined the x-coordinates.

15. a) Use the mode menu of a graphing calculator to display numbers to three decimal places. Graph $y = \tan x$ using a graphing window defined by $-2\pi \le x \le 2\pi$, $-3 \le y \le 3$. Trace to the point where $x = 1.069$. Observe the corresponding value of y.

b) Predict some other values of x that produce the same value of y. Trace to these points to verify your predictions.

16. a) Graph $y = \tan \frac{\pi}{2}x$ using a graphing window defined by $-4.7 \le x \le 4.7$, $-4 \le y \le 4$. Trace to the point where $x = 0.8$. Observe the corresponding value of y.

b) Predict some other values of x that produce the same value of y. Trace to these points to verify your predictions.

17. A police cruiser is parked so that the beacon on its roof is 3 m from a brick wall. As the beacon rotates, a spot of light moves along the wall. The beacon makes one complete rotation in 2 s.

a) Determine an equation that expresses the distance, d metres, as a function of time, t seconds.

b) Graph the function in part a.

18. a) Graph $y = \tan x$ using a graphing window defined by $-4.7 \le x \le 4.7$, $-3.1 \le y \le 3.1$. Use the zoom feature to enlarge the part of the graph containing the origin. Zoom in enough times until the visible part of the graph looks like a straight line.

b) Use the trace and arrow keys to move the cursor along the graph. What do you notice about the coordinates of the points on the graph? Use the definition of $\tan x$ to explain.

c) Compare the results of parts a and b with the results of exercise 16 on page 235. Which is greater, $\sin x$ or $\tan x$, when x is close to 0? Explain.

19. Write the equations of the tangent functions in each pattern.

a)

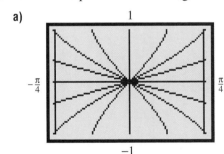

b)

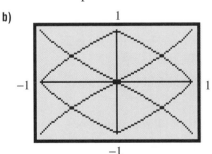

Prepare a table that compares and contrasts the functions $y = \sin x$ and $y = \tan x$.

4.7 The Functions $y = \csc x$, $y = \sec x$, and $y = \cot x$

The *cosecant*, *secant*, and *cotangent* functions of a real number are defined as the reciprocals of the sine, cosine, and tangent functions, respectively.

$\csc x = \dfrac{1}{\sin x}$, provided $\sin x \neq 0$

$\sec x = \dfrac{1}{\cos x}$, provided $\cos x \neq 0$

$\cot x = \dfrac{1}{\tan x}$, provided $\tan x \neq 0$

To draw and analyze the graphs of reciprocal functions, we apply the same methods as with the primary functions.

Example

a) Graph the function $y = 2 \sec 3x$ over two cycles.

b) What is the period of the function?

c) State the domain and range of the function.

d) What are the equations of the asymptotes of the graph of the function?

Solution

a) Graph the corresponding cosine function $y = 2 \cos 3x$. The graph has an amplitude of 2 and a period of $\frac{2\pi}{3}$. Draw vertical broken lines through the points where the graph of $y = 2 \cos 3x$ intersects the x-axis. At these points, $\cos 3x = 0$ and $\sec 3x$ is not defined. Use these lines as guides to sketch the graph of $y = 2 \sec 3x$.

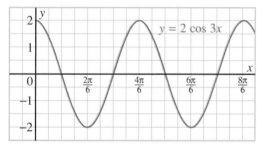

b) The period of $y = 2 \sec 3x$ is the same as that of $y = 2 \cos 3x$, that is, $\frac{2\pi}{3}$.

c) The domain consists of all real numbers except $\pm\frac{\pi}{6}, \pm\frac{3\pi}{6}, \pm\frac{5\pi}{6}, \ldots$.

The range consists of all real numbers greater than or equal to 2, or less than or equal to -2.

d) The equations of the asymptotes are $x = \pm\frac{\pi}{6}, x = \pm\frac{3\pi}{6}, x = \pm\frac{5\pi}{6}, \ldots$.

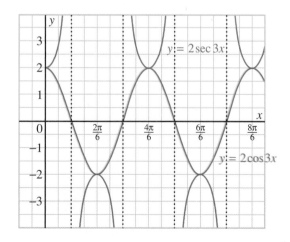

1. What are the values of x for which each function is not defined? Explain.

   a) $y = \csc x$ **b)** $y = \sec x$ **c)** $y = \cot x$

2. In the solution of the *Example*, the scale on the x-axis is marked $\frac{2\pi}{6}, \frac{4\pi}{6}, \frac{6\pi}{6}, \ldots$

   a) Why were these numbers not expressed in lowest terms?

   b) Would it be acceptable to express the numbers in lowest terms? Explain.

4.7 EXERCISES

A

1. Graph the function $y = \csc x$ for $-2\pi \leq x \leq 2\pi$.

2. Look at the list of properties on page 230 for $y = \sin x$ and on page 286 for $y = \tan x$. Decide which items should appear in a list of properties for the function $y = \csc x$. Use your graph in exercise 1. Make a list of the properties for the function $y = \csc x$.

3. Graph the function $y = \sec x$ for $-2\pi \leq x \leq 2\pi$.

4. Look at the list of properties on page 232 for $y = \cos x$ and on page 286 for $y = \tan x$. Decide which items should appear in a list of properties for the function $y = \sec x$. Use your graph in exercise 3. Make a list of the properties for the function $y = \sec x$.

5. Graph the function $y = \cot x$ for $-2\pi \leq x \leq 2\pi$.

6. Look at the list of properties on page 286 for $y = \tan x$. Use your graph in exercise 5. Make a similar list of the properties for the function $y = \cot x$.

B

7. Graph each function over two cycles. For each function, determine the domain, range, period, and the equations of the asymptotes.

   a) $y = \csc 2x$ **b)** $y = 3 \csc 4x$ **c)** $y = 3 \csc 4(x - \frac{\pi}{6})$

   d) $y = \sec 3x$ **e)** $y = 2 \sec 0.5x$ **f)** $y = 4 \sec 2(x + \frac{\pi}{3})$

   g) $y = \cot 2x$ **h)** $y = \cot 0.5x$ **i)** $y = 0.5 \cot 3x$

8. Graph each function over two cycles. For each function, determine the domain, range, period, and the equations of the asymptotes.

   a) $y = \csc 2\pi x$ **b)** $y = 2 \csc \pi x$ **c)** $y = 3 \csc \frac{2\pi}{3} x$

   d) $y = \sec 4\pi x$ **e)** $y = 3 \sec 3\pi x$ **f)** $y = 2 \sec \frac{2\pi}{5} x$

   g) $y = \cot \pi x$ **h)** $y = 2 \cot 2\pi x$ **i)** $y = 10 \cot \frac{\pi}{10} x$

9. Choose one part of exercise 7 or 8. Write to explain how you graphed the function, and how you determined its properties.

10. A graphing calculator was used to produce the two graphs below. Each function is a reciprocal trigonometric function. Determine a possible equation for each graph.

a)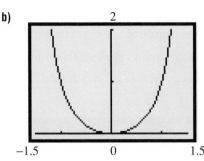

b)

11. Choose one part of exercise 10. Write to explain how you determined a possible equation for the graph.

C 12. A 100-m building is on the equator. On June 21 each year the sun rises at 6:00 A.M. in the east and sets at 6:00 P.M. in the west.

   a) Determine an equation that could be used to calculate the length of the building's shadow at any time of the day.

   b) Draw a graph to show the length of the shadow as a function of the time of day.

COMMUNICATING THE IDEAS

Write to explain how the six trigonometric functions of a real number x are defined. In your explanation, describe some of the basic properties of these six trigonometric functions. Use diagrams and graphs to illustrate your explanation.

1. Each function below is sinusoidal. Write an equation for each function. State the vertical displacement, the amplitude, the maximum value of y, the minimum value of y, the domain and range, and the y-intercept.

a)

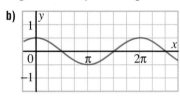

b)

c)

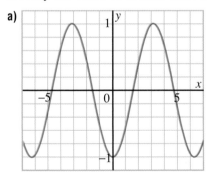

d)

2. These graphs have equations of the form $y = \cos(x - c)$. For each graph, identify the value of c, then write the equation.

a)

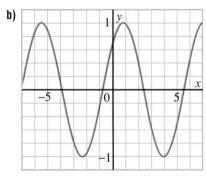

b)

3. Write to explain how you can write the equation $y = \sin x$ in the form $y = \cos(x - c)$.

4. Write to explain how you can write the equation $y = \cos x$ in the form $y = \sin(x - c)$.

5. Graph each function.

  a) $y = \sin 2(x - \frac{\pi}{2})$ **b)** $y = 2\sin 3(x - \frac{\pi}{3}) - 5$

  c) $g(x) = 3\cos 2(x + \frac{\pi}{4}) + 1$ **d)** $g(x) = 3\cos 3(x - \frac{2\pi}{3}) + 4$

6. Refer to exercise 10, page 279.

  a) Graph the height of the other pedal above the ground for the first few cycles.

  b) Write an equation of a sinusoidal function that describes the graph in part a.

1. The function $t = \frac{300}{v}$ models the time, t hours, required to drive 300 km at v kilometres per hour.

   a) Graph the function for reasonable values of v.

   b) Calculate the times required to drive 300 km at 50 km/h, 60 km/h, and 100 km/h.

   c) Suppose the speed is doubled. What happens to the time required to drive 300 km?

2. Describe how the graph of $y = 2x^2$ compares to the graph of each function.

   a) $y = 2x^2 + 5$ b) $y = 2(x - 5)^2$ c) $y = 2(x + 5)^2$

3. A bacterial culture doubles in size every 8 h. Suppose there are 1000 bacteria in the culture. Calculate the number of bacteria in the culture after each time.

   a) 16 h   b) 44 h

4. For the bacterial culture in exercise 3, calculate the amount of time to obtain each number of bacteria.

   a) 2500   b) 6000

5. Write as a single logarithm.

   a) $\log_2 x + \log_2 y - \log_2 z$ b) $2 \log_4 x - \log_4 y$

   c) $3 \log_7 x + 5 \log_7 y$ d) $\frac{1}{2} \log_5 x + 3 \log_5 y$

   e) $\log (2x - 3) + \log (y + 5)$ f) $3 \log_8 (x + y) - \log_8 (x - y)$

6. Find the sum of each infinite geometric series.

   a) $63 - 21 + 7 - \frac{7}{3} + \ldots$ b) $1 + \frac{1}{9} + \frac{1}{81} + \frac{1}{729} + \ldots$

7. A ball is dropped from a height of 2 m to a floor. On each bounce the ball rises to 50% of the height from which it fell. Calculate the total distance the ball travels before coming to rest. Identify any assumptions you made.

8. Sketch each angle in standard position.

   a) $\theta = -\frac{7\pi}{2}$ radians   b) $\theta = \frac{12\pi}{5}$ radians

   c) $\theta = -\frac{9\pi}{2}$ radians   d) $\theta = \frac{7\pi}{3}$ radians

9. Without using a calculator, determine if each quantity is positive or negative. Justify your answer.

   a) $\sin 220°$ b) $\cos 195°$ c) $\tan 20°$ d) $\sin 350°$

   e) $\cos 170°$ f) $\sin 90°$ g) $\cos 75°$ h) $\tan 185°$

10. Match each graph below with its equation.

   i) $y = \sin(x + \frac{\pi}{3})$ **ii)** $y = \sin(x - \frac{\pi}{2})$ **iii)** $y = \sin(x - \frac{\pi}{6})$ **iv)** $y = \sin(x + \frac{\pi}{4})$

a)

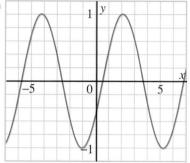

b)

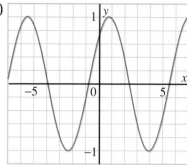

c)

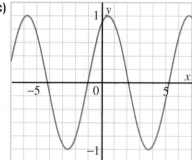

d)

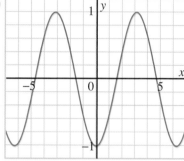

11. For each graph in exercise 10, write a second equation to describe it.

12. One energy-saving idea is to design houses with recessed windows on the side facing the sun. Sunlight enters the window in winter when the sun's angle is low, but it is blocked in the summer when its angle is much higher. The highest angle of elevation of the sun occurs at noon in mid-June, when the angle of elevation, in degrees, is $l + 23.5$, where l degrees is the latitude of the location.

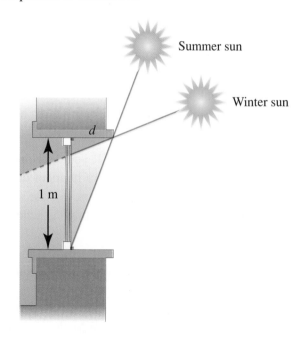

a) Assume that the window shown is designed to block the sun's rays when it is at its highest elevation in the summer. Express the depth, d metres, of the recess as a function of the latitude l.

b) Graph d against l, then determine the value of d for your latitude.

5 TRIGONOMETRIC EQUATIONS AND IDENTITIES

When Will They be in the Dark?

School starting and closing times affect the routes and times of school buses. Since the sun rises later and sets earlier in the winter than it does in the summer, some students may be riding in darkness on certain days.

 CONSIDER THIS SITUATION

At a high school in Saskatchewan, classes start at 8:30 A.M. and end at 3:15 P.M. each day. Some students must catch the school bus as early as 7:15 A.M. and they do not get home until 4:30 P.M.

The least amount of daylight occurs on December 21 when the sun rises at 8:45 A.M and sets at 4:57 P.M. The greatest amount of daylight occurs on June 21 when the sun rises at 4:47 A.M. and sets at 9:01 P.M.

- Would there be some days when students would be getting on the school bus in darkness, or arriving home in darkness? Explain.

- Do you think school buses should be allowed to pick up students in darkness in the morning, or return them home in darkness in the afternoon? Why?

Recall that on pages 270 and 271, you developed a sinusoidal model for predicting the number of hours of daylight on any day of the year. On pages 306 and 307, you will use similar models to predict when students would be riding the school bus before sunrise or after sunset.

 FYI Visit www.awl.com/canada/school/connections

> For information related to the above problem, click on MATHLINKS, followed by AWMath. Then select a topic under When Will They be in the Dark?

5.1 Solving Trigonometric Equations Using Graphing Technology

Recall that in *Examples 1* and *2* of Section 4.5, we solved the equations $3\cos 2\pi\frac{(t-4.5)}{12.4} + 5 = 2.5$ and $20\cos 2\pi\frac{(t-20)}{40} + 21 = 35$. These equations are examples of trigonometric equations.

An equation involving one or more trigonometric functions of a variable is called a *trigonometric equation*.

These are trigonometric equations:

$3\sin 2x - \cos x = 0$

$x + \sin x = 2$

These are not trigonometric equations:

$2x + 5\tan 4 = 1.5$

$x^3 - \cos\pi = 0$

All trigonometric equations can be solved graphically using a graphing calculator or computer graphing software. In Section 5.2, you will learn how some trigonometric equations can be solved algebraically.

Example 1

a) Solve the equation $\cos x = 0.35$ to 4 decimal places for the domain $0 \le x < 2\pi$.

b) Use a diagram to explain the meaning of the results in part a.

c) Suppose the domain is the set of all real numbers. Write expressions to represent all the roots of the equation $\cos x = 0.35$.

Solution

a) Graph the functions $y = \cos x$ and $y = 0.35$ on the same screen. There are two roots in the interval $0 \le x < 2\pi$.

Each root of the equation $\cos x = 0.35$ is the x-coordinate of the point of intersection of the graphs of the two functions. Activate the intersect feature.

One root is $x \doteq 1.2132$.   Another root is $x \doteq 5.0700$.

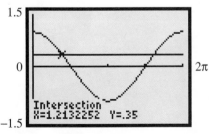

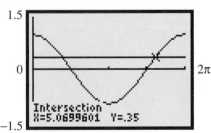

Check.

$\cos x = 0.35$

When $x \doteq 1.2132$, L.S. $= \cos 1.2132$ R.S. $= 0.35$
$\doteq 0.35$

When $x \doteq 5.0700$, L.S. $= \cos 5.0700$ R.S. $= 0.35$
$\doteq 0.35$

The roots are correct.

b)

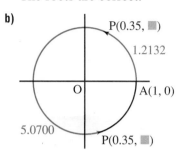

Draw a unit circle.
When the first coordinate of P is 0.35, the arc length from A(1, 0) could be 1.2132 units or 5.0700 units.

c) The cosine function has period 2π. Hence, if there is no restriction on x, all other roots are obtained by adding or subtracting a multiple of 2π to the two roots in part a. Expressions representing all roots of the equation are $x \doteq 1.2132 + 2n\pi$ and $x \doteq 5.0700 + 2n\pi$, where n is any integer.

In *Example 1c*, the expressions $x \doteq 1.2132 + 2n\pi$ and $x \doteq 5.0700 + 2n\pi$ are called the *general solution* of the equation $\cos x = 0.35$. The expressions represent all the roots of the equation. We can determine particular roots by substituting different integers for n. For example, if $n = 10$, the above expressions become

$\quad\quad x \doteq 1.2132 + 20\pi \quad\quad x \doteq 5.0700 + 20\pi$
$\quad\quad \doteq 64.0451 \quad\quad\quad\quad\quad \doteq 67.9019$

These roots correspond to arc lengths obtained by moving 10 times counterclockwise around the circle beyond the roots obtained in *Example 1*. To check, determine the cosine of each number. We obtain $\cos 64.0451 \doteq 0.35$ and $\cos 67.9019 \doteq 0.35$.

Example 2

a) Solve the equation $3 + \sin 2x = 1 - 5\sin 2x$, over the domain $0 \le x < 2\pi$.

b) Determine the general solution of the equation where the domain is the set of real numbers.

Solution

a) Graph the functions $y = 3 + \sin 2x$ and $y = 1 - 5\sin 2x$ on the same screen, for $0 \le x < 2\pi$. From the graph, there are four roots in the given domain. Activate the intersect feature. Visualize numbering the roots from left to right. The first root is $x \doteq 1.7407$. The second root is $x \doteq 2.9717$. Use the intersect feature to determine the other two roots, but also use the pattern on the graphs. Since the period of each function is π, determine the other two roots by adding π to the known roots.

Hence, the third root is $x \doteq 1.7407 + \pi$, or approximately 4.8823.

The fourth root is $x \doteq 2.9717 + \pi$, or approximately 6.1133.

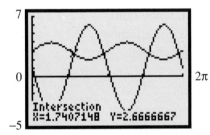 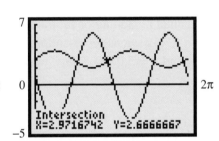

b) All other roots can be obtained by adding or subtracting a multiple of π to the first two roots. Hence, the solution is $x \doteq 1.7407 + n\pi$ and $x \doteq 2.9717 + n\pi$, where n is an integer.

Recall that a quadratic equation has the form $ax^2 + bx + c = 0$. This is also called a second-degree equation. A second-degree trigonometric equation has a similar form, except that x is replaced with a trigonometric function, such as $\sin x$, $\cos x$, or $\tan x$. These equations contain expressions such as $(\sin x)^2$. These expressions occur so often that they are abbreviated as $\sin^2 x$. For example, an equation such as $2(\sin x)^2 + (\sin x) - 2 = 0$ is written $2\sin^2 x + \sin x - 2 = 0$.

Example 3

Solve $2\sin^2 x + \sin x - 2 = 0$, then write the general solution.

Solution

Graph the function $y = 2\sin^2 x + \sin x - 2$. The period of the function is 2π. From the graph, there are two roots in the interval $0 \le x < 2\pi$. Activate the zero feature. One root is $x \doteq 0.8959$. The other root is $x \doteq 2.2457$.

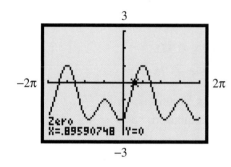
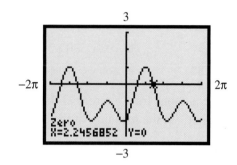

The general solution is $x \doteq 0.8959 + 2n\pi$ and $x \doteq 2.2457 + 2n\pi$.

DISCUSSING THE IDEAS

1. In *Example 1a*, explain why the interval for the roots is written as $0 \le x < 2\pi$ and not as $0 \le x \le 2\pi$.

2. What other way could you solve *Example 2*?

3. Explain why the period of the function in *Example 3* is 2π.

4. In *Example 3*, suppose the constant term in the equation was changed from -2 to other numbers. What possible numbers of solutions could the equation have? Use examples to explain your answers.

5. What other way could you solve *Example 3*?

Use a graphing calculator or computer graphing software for each exercise.

1. Solve each equation to 4 decimal places for $0 \le x < 2\pi$, and check.

   a) $\sin x = 0.8$ **b)** $\sin x = -0.45$ **c)** $\cos x = 0.75$

   A **d)** $\cos x = -0.25$ **e)** $\tan x = 1.5$ **f)** $\tan x = -3.0$

2. Choose two equations from exercise 1 that involve different functions. Draw a diagram to explain the meaning of the solutions for each equation.

3. Solve each equation to 4 decimal places for $-\pi \le x < \pi$.

   a) $\sin x = 0.55$ **b)** $\cos x = 0.82$ **c)** $\tan x = 0.375$

   d) $\sin x = -0.88$ **e)** $\cos x = -0.66$ **f)** $\tan x = -5.75$

4. Choose two equations from exercise 3 that involve different functions. Draw a diagram to explain the meaning of the solutions for each equation.

5. Solve each equation to 4 decimal places for $-\pi \le x < \pi$.

   a) $2 + 3\sin x = 5 - 2\sin x$ **b)** $1 + \cos x = 2 - 2\cos x$

   c) $4 - 2\tan x = 6 - 3\tan x$ **d)** $3 + 2\cos x = 1 + \cos x$

   e) $2 + 3\sin x = 3 + 2\sin x$ **f)** $3 + 3\tan x = 2 + 2\tan x$

   g) $1 + 3\sin x = 3 + 2\sin x$ **h)** $3 + 3\sin x = 1 + 2\sin x$

B 6. Solve each equation for $0 \le x < 2\pi$.

   a) $\sin 2x = 0$ **b)** $\sin 3x = 0$ **c)** $\tan 4x = 0$

   d) $\sin \frac{1}{2}x = 0$ **e)** $\sin \frac{1}{3}x = 0$ **f)** $\sin \frac{1}{4}x = 0$

   g) $\cos 2x = \frac{1}{2}$ **h)** $\tan 3x = \frac{1}{2}$ **i)** $\cos 4x = \frac{1}{2}$

   j) $\sin 2x = 0.75$ **k)** $\sin 3x = 0.75$ **l)** $\sin 4x = 0.75$

7. Solve each equation, then write the general solution.

   a) $\cos \left(x + \frac{\pi}{2}\right) = \frac{1}{2}$ **b)** $\cos (x + \pi) = \frac{1}{2}$ **c)** $\cos (x + 2\pi) = \frac{1}{2}$

   d) $\sin \left(x + \frac{\pi}{2}\right) = 0.75$ **e)** $\sin (x + \pi) = 0.75$ **f)** $\sin (x + 2\pi) = 0.75$

   g) $\cos \left(x + \frac{\pi}{3}\right) = \cos x$ **h)** $\cos \left(x + \frac{\pi}{2}\right) = \cos x$

   i) $\cos (x + \pi) = \cos x$ **j)** $\cos (x + 2\pi) = \cos x$

8. Solve each equation, then write the general solution.

   a) $\cos (x + 1) = \cos x$ **b)** $\cos \frac{\pi}{2}(x + 1) = \cos \frac{\pi}{2}x$

   c) $\cos \pi(x + 1) = \cos \pi x$ **d)** $\cos 2\pi(x + 1) = \cos 2\pi x$

   e) $\sin x = \cos x$ **f)** $\sin 2x = \cos 2x$

   g) $\sin x = \cos 2x$ **h)** $\sin 2x = \cos x$

9. Write to explain how the roots of the equations in exercises 8a and b are related. Include in your writing the reasons the roots are related this way.

10. Solve the equations in each list, then write the general solutions.

a) $\sin^2 x + 3 \sin x - 1 = 0$

$\sin^2 x + 2 \sin x - 1 = 0$

$\sin^2 x + 1 \sin x - 1 = 0$

$\sin^2 x + 0 \sin x - 1 = 0$

b) $3 \sin^2 x + \sin x - 1 = 0$

$2 \sin^2 x + \sin x - 1 = 0$

$1 \sin^2 x + \sin x - 1 = 0$

$0 \sin^2 x + \sin x - 1 = 0$

11. a) Graph $y = x - 2 \sin x$.

b) Use the graph. Find all the solutions of the equation $2 \sin x = x$, to 3 decimal places.

12. Repeat exercise 11, replacing $\sin x$ with each function.

a) $\cos x$

b) $\tan x$

13. The screen (below left) shows the graphs of the functions $y = 2 \cos x$ and $y = \frac{1}{x}$ in the interval $-2\pi \le x \le 2\pi$.

a) How many roots of the equation $2 \cos x = \frac{1}{x}$ appear to be in this interval? Explain.

b) Determine all the roots of the equation in this interval. How many roots are there?

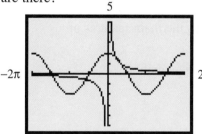

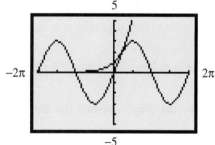

14. The screen (above right) shows the graphs of the functions $y = 3 \sin x$ and $y = 3^x$ in the interval $-2\pi \le x \le 2\pi$.

a) How many roots of the equation $3 \sin x = 3^x$ appear to be in this interval? Explain.

b) Determine all the roots of the equation in this interval. How many roots are there?

15. Use the graph in exercise 14. How many roots of the equation $3 \cos x = 3^x$ are there in the interval $-2\pi \le x \le 2\pi$? Explain.

16. The screen shows the graphs of the functions $y = \cos 2x$, $y = \cos x$, and $y = \cos \frac{1}{2}x$. The three graphs appear to intersect at common points near the right and left sides of the screen, and at the centre.

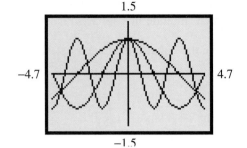

a) Use a graphing calculator. Confirm that the three graphs do intersect at common points.

b) Determine the exact coordinates of the three common points shown on the screen.

c) Verify that the coordinates of these points satisfy the equations of the three functions.

17. Solve each equation graphically. How many roots are there? Explain using both the graph and the definition in terms of a unit circle.

a) $\sin x = x$ b) $\cos x = x$ c) $\tan x = x$

C 18. In *Example 1b*, an expression such as $1.2132 + 2n\pi$ for a general solution may seem strange, since the first term has been rounded to 4 decimal places, while the second term is an exact expression. Would it be correct to replace π with its decimal approximation to 4 decimal places and write the expression as $1.2132 + 2n(3.1416)$, or $1.2132 + 6.2832n$? Use an example to illustrate your explanation.

19. a) Graph $y = x \sin x$. Use a graphing window that shows the main features of the graph.

b) Use the graph to determine the general solutions of the equation $x \sin x = 0$.

c) Explain why the graph appears the way it does.

20. a) Graph $y = x \cos x$. Use a graphing window that shows the main features of the graph.

b) Use the graph to determine the general solutions of the equation $x \cos x = 0$.

c) Explain why the graph appears the way it does.

COMMUNICATING THE IDEAS

Write to explain what is meant by the general solution of a trigonometric equation. Use examples to illustrate your explanation.

Investigating y = x − 2 sin x and y = x − 2 cos x

Recall the functions $y = x - 2\sin x$ and $y = x - 2\cos x$ in exercises 11 and 12, page 303. You will explore the graphs of these functions by zooming out from the origin. Before you start, set your calculator as follows. These instructions are for the TI-83 calculator. Other graphing calculators can be set up in a similar way.

Set the graphing window.

Set the graphing window so that $-3 \le x \le 3$ and $-2 \le y \le 2$. This puts the origin at the centre of the screen.

Set both Xscl and Yscl to 0. This avoids having crowded tick marks on the axes when you zoom out.

Set the zoom factor to 10.

Press ⌊ZOOM⌋ ⌊▶⌋ **4** to get the zoom factors menu.

Set both XFact and YFact to 10. This ensures that you zoom in or out by a factor of 10 each time you press ⌊ZOOM⌋.

Press ⌊QUIT⌋ to return to the main screen.

1. **a)** Graph the function $y = x - 2\sin x$.

   b) Press ⌊ZOOM⌋ **3** ⌊ENTER⌋ to zoom out one step. Explain why the graph appears the way it does.

2. Predict what the graph will look like if you zoom out another step. Justify your prediction with an explanation. Use your calculator to check your prediction.

3. Predict what the graph will look like if you zoom in several steps. Use your calculator to check your prediction.

4. Repeat exercises 1, 2, and 3 using the function $y = x - 2\cos x$.

5. Determine the coordinates of the points of intersection of the graphs of the functions $y = x - 2\sin x$ and $y = x - 2\cos x$. Explain the results.

When Will They be in the Dark?

On page 296, we considered the problem of whether there would be any days when students in a school in Saskatchewan would be riding the school bus before sunrise in the morning or after sunset in the afternoon. The data from page 296 are summarized below, with times for a 24-h clock.

**Sunrise and sunset times
(h:min)**

Sunrise time	08:45	04:47
Sunset time	16:57	21:01

**School bus times
(h:min)**

	All days
Morning pickup	07:15
Afternoon drop-off	16:30

 **DEVELOP A MODEL**

You will develop sinusoidal models for the sunset and sunrise times. Use a graphing calculator or computer graphing software.

Modelling sunset times

Let s hours represent the time of the sunset on the nth day of the year. Choose either a cosine function or a sine function whose equation has the form:

$$s = a \cos 2\pi \frac{(n-c)}{p} + d \quad \text{or} \quad s = a \sin 2\pi \frac{(n-c)}{p} + d$$

where a, c, p, and d represent the amplitude, phase shift, period, and vertical displacement, respectively.

1. What are the values of a, c, p, and d for the function you chose?

2. Write an equation that models the time of the sunset above.

3. Graph the equation.

Modelling sunrise times

4. a) Determine a similar equation for the sunrise times.

   b) Graph the equation on the same screen as the sunset times.

Modelling school bus times

5. a) Write equations that model the morning and afternoon school bus times.

   b) Graph the equations on the same screen as the sunset times.

LOOK AT THE IMPLICATIONS

The graphs that model the school bus times are horizontal lines. These may or may not intersect the sinusoidal curves modelling the sunset and sunrise times.

6. a) Determine the coordinates of any points where the horizontal lines intersect the sinusoidal curves.

   b) Between which dates, if any, does the school bus pick up students before sunrise in the morning? For approximately how many days will this last?

   c) Between which dates, if any, does the school bus drop off students after sunset in the afternoon? For approximately how many days will this last?

7. What changes, if any, in the school bus times would you recommend to reduce the number of days when students will be travelling in darkness? Support your recommendations with some graphs and calculations.

REVISIT THE SITUATION

8. There is no daylight savings time in Saskatchewan. How would the graphs change if there were daylight savings time?

Model a similar situation for your school by completing exercises 9 and 10.

9. From friends or other sources of information, determine the earliest time when a school bus picks up students in the morning and the latest time when it drops them off in the afternoon.

10. a) Use the Internet, an almanac, or newspaper files. Determine the approximate time the sun rises and sets in your locality on June 21 and December 21.

    b) Determine equations that represent the time the sun rises and sets on the nth day of the year. Graph the equations on the same screen.

    c) Graph the equations representing the school bus times from exercise 9.

    d) Repeat exercise 6b.

    e) Repeat exercise 6c.

11. What changes, if any, in the school bus times would you recommend to reduce the number of days when students will be travelling in darkness? Support your recommendations with some graphs and calculations.

5.2 Solving Trigonometric Equations without Using Graphing Technology

In Section 5.1, we solved trigonometric equations by using graphing technology. Some trigonometric equations can be solved without using graphing technology. Recall from Chapter 3 that certain values of the sine, cosine, and tangent functions can be determined from geometric relationships.

- Sines and cosines of 0 and multiples of $\frac{\pi}{2}$ include 0 and ± 1.

- Sines and cosines of multiples of $\frac{\pi}{4}$ include $\pm\frac{\sqrt{2}}{2}$ as well as 0 and ± 1.

- Sines and cosines of multiples of $\frac{\pi}{6}$ include $\pm\frac{1}{2}$ and $\pm\frac{\sqrt{3}}{2}$ as well as 0 and ± 1.

Equations involving these values can be solved without using graphing technology. This means that the roots will be expressed in terms of π, without using decimal approximations. These diagrams will help you identify the roots.

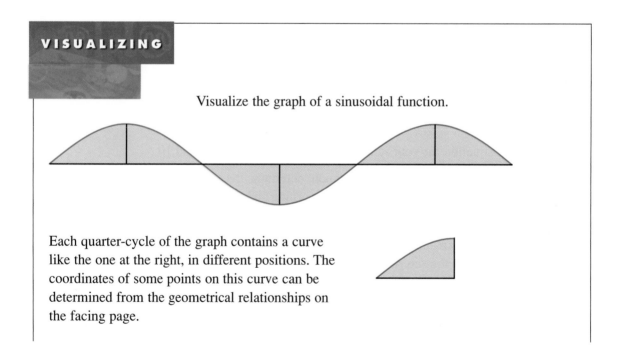

VISUALIZING

Visualize the graph of a sinusoidal function.

Each quarter-cycle of the graph contains a curve like the one at the right, in different positions. The coordinates of some points on this curve can be determined from the geometrical relationships on the facing page.

These relationships occur with the sines and cosines of $\frac{\pi}{4}$ and $\frac{\pi}{2}$, and their multiples.

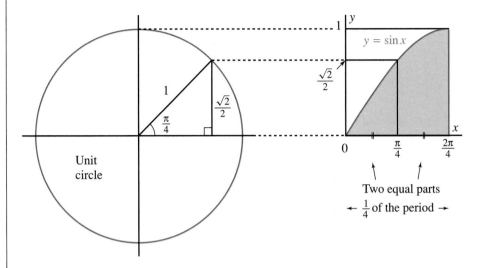

Two equal parts
← $\frac{1}{4}$ of the period →

These relationships occur with the sines and cosines of $\frac{\pi}{6}$, $\frac{\pi}{3}$, and $\frac{\pi}{2}$, and their multiples.

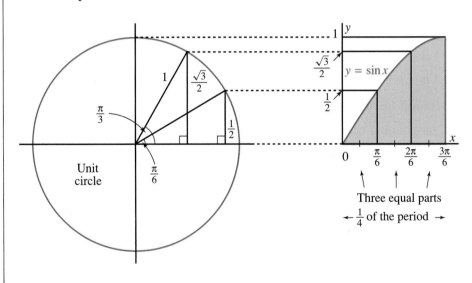

Three equal parts
← $\frac{1}{4}$ of the period →

Example 1

a) Solve the equation $\sin x = -\frac{1}{2}$ for $0 \le x < 2\pi$.

b) Write the general solution of the equation.

Solution

a)

> **Think...**
>
> The value $-\frac{1}{2}$ indicates that the solution involves multiples of $\frac{\pi}{6}$. Hence, the equation can be solved in exact form.

Sketch the graph of $y = \sin x$. Since the period of this function is 2π, each quarter cycle is $\frac{\pi}{2}$ units long.

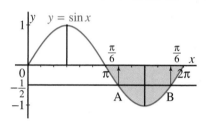

Draw the line $y = -\frac{1}{2}$. The x-coordinates of A and B are the roots of the equation $\sin x = -\frac{1}{2}$. Compare the third quarter-cycle with the second diagram in *Visualizing*, page 309. Since the y-coordinate of A is $-\frac{1}{2}$, the x-coordinate of A is one-third of the way from π to $\frac{3\pi}{2}$. Hence, the x-coordinate of A is $\pi + \frac{1}{3} \times \frac{\pi}{2} = \pi + \frac{\pi}{6}$, or $\frac{7\pi}{6}$.

Similarly, the x-coordinate of B is $2\pi - \frac{\pi}{6} = \frac{11\pi}{6}$.

For $0 \le x < 2\pi$, the solution of the equation is $x = \frac{7\pi}{6}$ and $x = \frac{11\pi}{6}$.

b) Since the period of $\sin x$ is 2π, the general solution is $x = \frac{7\pi}{6} + 2n\pi$ and $x = \frac{11\pi}{6} + 2n\pi$, where n is any integer.

Example 2

Solve the equation $\cos 3x = \frac{\sqrt{2}}{2}$ for $0 \le x < 2\pi$.

Solution

> **Think...**
>
> The value $\frac{\sqrt{2}}{2}$ indicates that the solution involves multiples of $\frac{\pi}{4}$. Hence, the equation can be solved in exact form.

Sketch the graph of $y = \cos 3x$ for $0 \le x < 2\pi$. Since the period of this function is $\frac{2\pi}{3}$, each quarter cycle is $\frac{1}{4} \times \frac{2\pi}{3} = \frac{\pi}{6}$ units long.

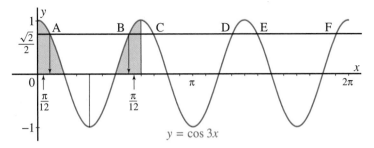

$y = \cos 3x$

Draw the line corresponding to $y = \frac{\sqrt{2}}{2}$. There are six roots when $0 \le x < 2\pi$, corresponding to the x-coordinates of A, B, C, D, E, and F.

Compare the first quarter-cycle with the first diagram in *Visualizing*, page 309. Since the y-coordinate of A is $\frac{\sqrt{2}}{2}$, the x-coordinate of A is halfway from 0 to $\frac{\pi}{6}$; that is, $\frac{\pi}{12}$.

Similarly, the x-coordinate of B is $\frac{2\pi}{3} - \frac{\pi}{12} = \frac{7\pi}{12}$.

Two roots of the equation are $\frac{\pi}{12}$ and $\frac{7\pi}{12}$.

To obtain the other roots in the interval $0 \le x < 2\pi$, add the period to each root, and repeat.

The x-coordinates of C and E are $\frac{\pi}{12} + \frac{2\pi}{3} = \frac{9\pi}{12}$ and $\frac{9\pi}{12} + \frac{2\pi}{3} = \frac{17\pi}{12}$, respectively.

The x-coordinates of D and F are $\frac{7\pi}{12} + \frac{2\pi}{3} = \frac{15\pi}{12}$ and $\frac{15\pi}{12} + \frac{2\pi}{3} = \frac{23\pi}{12}$, respectively.

When $0 \le x < 2\pi$, the solution of the equation is:

$x = \frac{\pi}{12}, x = \frac{7\pi}{12}, x = \frac{9\pi}{12}, x = \frac{15\pi}{12}, x = \frac{17\pi}{12}$, and $x = \frac{23\pi}{12}$

Example 3

a) Solve the equation $2\cos^2 x - \cos x - 1 = 0$ for $0 \le x < 2\pi$.

b) Write the general solution of the equation.

Solution

Think...

The expression $2\cos^2 x - \cos x - 1$ is like the trinomial $2a^2 - a - 1$, which can be factored as $(2a + 1)(a - 1)$. Hence, $2\cos^2 x - \cos x - 1$ can be factored similarly.

a)
$$2\cos^2 x - \cos x - 1 = 0$$
$$(2\cos x + 1)(\cos x - 1) = 0$$

Either $2\cos x + 1 = 0$ or $\cos x - 1 = 0$

$$\cos x = -\frac{1}{2} \qquad\qquad \cos x = 1$$

Sketch the graph of $y = \cos x$.

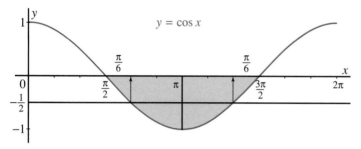

For $0 \le x < 2\pi$, the solution of the equation $\cos x = -\frac{1}{2}$ is $x = \frac{\pi}{2} + \frac{\pi}{6}$, or $\frac{2\pi}{3}$, and $x = \frac{3\pi}{2} - \frac{\pi}{6}$, or $\frac{4\pi}{3}$. The solution of the equation $\cos x = 1$ is 0.

b) The general solution of the equation is $x = \frac{2\pi}{3} + 2n\pi$, $x = \frac{4\pi}{3} + 2n\pi$, and $x = 0 + 2n\pi$, or $2n\pi$. For all values of n, all these expressions are multiples of $\frac{2\pi}{3}$. Hence, the general solution may be written as $x = \frac{2n\pi}{3}$.

DISCUSSING THE IDEAS

1. In *Visualizing*, page 309, why are the numbers on the horizontal axes not reduced to lowest terms?

2. In *Example 1*, explain how the solution would change for each equation.

  a) $\sin x = -\frac{\sqrt{3}}{2}$ **b)** $\sin x = \frac{1}{2}$ **c)** $\sin x = \frac{\sqrt{3}}{2}$

3. Why does the equation in *Example 1* have only two solutions if $0 \le x < 2\pi$, while the equation in *Example 2* has six solutions?

4. In *Example 2*, explain why the value $\frac{\sqrt{2}}{2}$ indicates that the solution involves multiples of $\frac{\pi}{4}$, but the roots of the equation are not multiples of $\frac{\pi}{4}$.

5. In *Example 3*, explain why the expressions $\frac{2\pi}{3} + 2n\pi$, $\frac{4\pi}{3} + 2n\pi$, and $2n\pi$ (which are all different) can be represented by the single expression $\frac{2n\pi}{3}$.

6. Could all the equations in *Examples 1, 2,* and *3* be solved graphically? Explain.

5.2 EXERCISES

A 1. For $0 \le x < 2\pi$, how many roots does each equation have? Explain.

a) $\sin x = \frac{1}{2}$

b) $\sin 2x = \frac{1}{2}$

c) $\sin 3x = \frac{1}{2}$

d) $\cos 2x = \frac{\sqrt{3}}{2}$

e) $\cos 4x = \frac{\sqrt{3}}{2}$

f) $\cos 8x = \frac{\sqrt{3}}{2}$

2. a) Sketch the graph of $y = \cos 2x$.

b) Use the graph. Determine all the solutions of the equation $\cos 2x = 0$ in the interval $0 \le x < 2\pi$.

c) Write an expression to represent all the solutions of the equation $\cos 2x = 0$ when there is no restriction on x.

3. Repeat exercise 2 using the function $y = \sin 2x$.

4. Repeat exercise 2 using the function $y = \tan 2x$.

B 5. Solve each equation for $0 \le x < 2\pi$.

a) $\cos x = -\frac{1}{2}$

b) $\sin x = -1$

c) $\tan x = \frac{\sqrt{3}}{3}$

d) $\cos x = \frac{\sqrt{3}}{2}$

e) $1 + \cos x = 1 - \cos x$

f) $\sin x = \sqrt{3} - \sin x$

6. Solve the equations in each list for $0 \le x < 2\pi$.

a) $\cos x = \frac{1}{2}$
$\cos 2x = \frac{1}{2}$
$\cos 3x = \frac{1}{2}$
$\cos 4x = \frac{1}{2}$

b) $\sin x = \frac{\sqrt{3}}{2}$
$\sin 2x = \frac{\sqrt{3}}{2}$
$\sin 3x = \frac{\sqrt{3}}{2}$
$\sin 4x = \frac{\sqrt{3}}{2}$

c) $\tan x = \sqrt{3}$
$\tan 2x = \sqrt{3}$
$\tan 3x = \sqrt{3}$
$\tan 4x = \sqrt{3}$

7. Factor each expression.

a) $\cos^2 x + 2\cos x$

b) $\sin^2 x + 5\sin x + 6$

c) $2\sin^2 x + \sin x - 6$

d) $3\cos^2 x - 2\cos x - 1$

e) $2\cos^2 x - 7\cos x + 3$

f) $6\sin^2 x + \sin x - 1$

8. Solve each equation for $0 \le x < 2\pi$.

a) $\cos^2 x + 2\cos x = 0$

b) $\sin^2 x + 5\sin x + 6 = 0$

c) $2\sin^2 x + \sin x - 6 = 0$

d) $3\cos^2 x - 2\cos x - 1 = 0$

e) $2\cos^2 x - 7\cos x + 3 = 0$

f) $6\sin^2 x + \sin x - 1 = 0$

9. Solve each equation for $0 \le x < 2\pi$.

a) $\sin^2 x - \sin x = 0$

b) $\cos^2 x + \cos x = 0$

c) $2 \sin^2 x + \sin x - 1 = 0$

d) $2 \sin^2 x - \sin x - 1 = 0$

e) $\cos^2 x + 3 \cos x + 2 = 0$

f) $\sin^2 x + 2 \sin x - 3 = 0$

g) $2 \cos^2 x + 3 \cos x + 1 = 0$

h) $2 \cos^2 x - 3 \cos x + 1 = 0$

i) $\sin^2 x + 3 \sin x + 2 = 0$

j) $\sin^2 x + 5 \sin x + 6 = 0$

k) $4 \cos^2 x - 4 \cos x + 1 = 0$

l) $4 \sin^2 x - 1 = 0$

10. The diagrams in *Visualizing*, page 309, were based on the sine function. Draw similar diagrams based on each function.

a) the cosine function

b) the tangent function

11. Solve each equation for $0 \le x < 2\pi$.

a) $\tan x = 1$

b) $\tan 2x = 1$

c) $\tan 4x = 1$

d) $\tan x = \sqrt{3}$

e) $\tan 4x = -\sqrt{3}$

f) $\tan 2x = \dfrac{\sqrt{3}}{3}$

12. Solve each equation for $0 \le x < 2\pi$.

a) $\tan^2 x - \tan x = 0$

b) $\tan^2 x + \tan x = 0$

c) $\tan^2 x - 1 = 0$

d) $\tan^2 x - (\sqrt{3} + 1) \tan x + \sqrt{3} = 0$

13. Write a trigonometric equation that has the given roots over the domain $0 \le x < 2\pi$.

a) 0

b) $\dfrac{\pi}{6}$

c) $\dfrac{\pi}{2}$

d) $\dfrac{\pi}{3}$

e) $\dfrac{\pi}{4}$

f) $\dfrac{3\pi}{4}$

14. Write a trigonometric equation that has the given roots over the domain $0 \le x < 2\pi$.

a) $\dfrac{\pi}{6}, \dfrac{\pi}{2}$

b) $\dfrac{\pi}{3}, \dfrac{\pi}{4}$

c) $\dfrac{\pi}{4}, \dfrac{3\pi}{4}$

C **15.** Refer to exercise 11 in 3.5 Exercises, page 189. Use the information in that exercise to solve the equation $4 \cos^2 x - 2 \cos x - 1 = 0$ for $0 \le x < 2\pi$.

16. Determine the general solution of each equation.

a) $2 \sin x = 3 + 2 \csc x$

b) $4 \tan x + \cot x = 5$

COMMUNICATING THE IDEAS

Write to explain the advantages and the disadvantages of solving trigonometric equations without using graphing technology. Use some examples to illustrate your explanations.

A graphing calculator was used to solve the equations $\sin x = \sin\left(\frac{\pi}{2} - x\right)$ and $\sin x = \sin(\pi - x)$. Here are the results.

$\sin x = \sin\left(\frac{\pi}{2} - x\right)$

The graphs of $y = \sin x$ and $y = \sin\left(\frac{\pi}{2} - x\right)$ intersect at distinct points. The first coordinate of each point of intersection is a solution of the equation $\sin x = \sin\left(\frac{\pi}{2} - x\right)$. One solution is $x \doteq 0.7854$. Since the functions are periodic, there are infinitely many solutions.

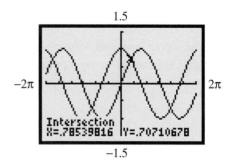

$\sin x = \sin(\pi - x)$

The graphs of $y = \sin x$ and $y = \sin(\pi - x)$ coincide. The first coordinate of every point on either graph is a solution of the equation $\sin x = \sin(\pi - x)$. Hence, every real number is a solution of the equation.

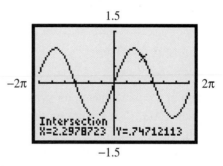

Recall that an equation that is satisfied for all values of the variable for which both sides of the equation are defined is called an identity. The equation $\sin x = \sin(\pi - x)$ is an example of a *trigonometric identity*. We can use the graph of $y = \sin x$ below to explain why this identity is true. Sin x is the y-coordinate of A and $\sin(\pi - x)$ is the y-coordinate of B. Since these coordinates are equal, $\sin x = \sin(\pi - x)$.

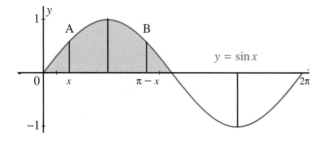

Since the graphs of the trigonometric functions are periodic, and since the curve representing a quarter cycle is repeated in different positions in each cycle, there are many trigonometric identities similar to $\sin x = \sin(\pi - x)$. In the past, these identities were important because they were used to determine values of trigonometric functions using tables.

Odd-Even Identities

You may have noticed that the graphs of $y = \sin x$ and $y = \cos x$ possess two different kinds of symmetry. These occur because the part of the curve in any quarter cycle is repeated in different positions in the other three quarters of the cycle.

The graph of $y = \cos x$ has line symmetry with respect to the y-axis. This means that if $A(x, y)$ is a point on the graph, then $B(-x, y)$ is another point on the graph. The y-coordinate of A is $\cos x$ and the y-coordinate of B is $\cos(-x)$. Since A and B have the same y-coordinate, $\cos(-x) = \cos x$.

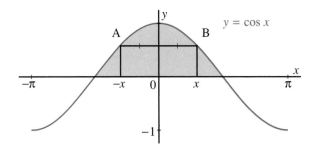

The graph of $y = \sin x$ has point symmetry with respect to the origin. This means that if $A(x, y)$ is a point on the graph, then $B(-x, -y)$ is another point on the graph. The y-coordinate of A is $\sin x$ and the y-coordinate of B is $\sin(-x)$. Since A and B have opposite y-coordinates, $\sin(-x) = -\sin x$.

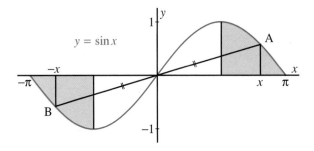

Recall from Chapter 1 that when a function $y = f(x)$ satisfies $f(-x) = f(x)$ for all values of x, it is an even function. When a function satisfies $f(-x) = -f(x)$ for all values of x, it is an odd function. Hence, $y = \cos x$ is an even function and $y = \sin x$ is an odd function. The identities $\sin(-x) = -\sin x$ and $\cos(-x) = \cos x$ are called the *odd-even identities*.

Odd-Even Identities	$\sin(-x) = -\sin x$	$\cos(-x) = \cos x$

Quotient and Pythagorean Identities

We can determine other trigonometric identities
using the definitions of the trigonometric
functions. Recall that the six trigonometric
functions of x are defined by this diagram,
where x represents the length of the arc from
$A(1, 0)$ to P on the unit circle.

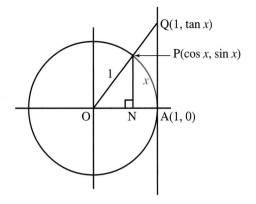

Since $\triangle QOA$ and $\triangle PON$ are similar triangles,

$$\frac{AQ}{OA} = \frac{NP}{ON}$$

Since $OA = 1$, this equation becomes
$\tan x = \frac{\sin x}{\cos x}$.

This identity is called the *quotient identity*.

Quotient Identity	$\tan x = \frac{\sin x}{\cos x} \quad (\cos x \neq 0)$

In the diagram above, the Pythagorean Theorem can be applied to $\triangle PON$.
Hence, $NP^2 + ON^2 = OP^2$, or $\sin^2 x + \cos^2 x = 1$. This identity is called the
Pythagorean identity.

Pythagorean Identity	$\sin^2 x + \cos^2 x = 1$

We can manipulate identities in different ways to obtain more identities. For
example, we can take the reciprocal of each side of the quotient identity to
obtain the identity $\cot x = \frac{\cos x}{\sin x}$. We can also divide each side of the
Pythagorean identity by $\cos^2 x$ (if $\cos^2 x \neq 0$), then use the quotient identity.

$$\sin^2 x + \cos^2 x = 1$$

$$\frac{\sin^2 x}{\cos^2 x} + \frac{\cos^2 x}{\cos^2 x} = \frac{1}{\cos^2 x}$$ Divide by $\cos^2 x$, where $\cos^2 x \neq 0$.

$$\tan^2 x + 1 = \frac{1}{\cos^2 x}$$ Using the quotient identity

$$\tan^2 x + 1 = \sec^2 x$$ Using the definition of sec x

We can verify an identity numerically by substituting values of the variable and
checking that both sides are equal. We can also verify an identity graphically by
graphing the functions defined by each side of the equation.

Example

Consider the identity $\tan x = \frac{\sin x}{\cos x}$, $\cos x \neq 0$.

a) Verify the identity numerically, when $x = \frac{\pi}{6}$ and when $x = 4$.

b) Verify the identity graphically.

Solution

a) When $x = \frac{\pi}{6}$,

$$\tan \frac{\pi}{6} = \frac{1}{\sqrt{3}}, \text{ or } \frac{\sqrt{3}}{3}$$

$$\frac{\sin \frac{\pi}{6}}{\cos \frac{\pi}{6}} = \frac{\frac{1}{2}}{\frac{\sqrt{3}}{2}}$$

$$= \frac{1}{\sqrt{3}}, \text{ or } \frac{\sqrt{3}}{3}$$

The identity is true when $x = \frac{\pi}{6}$.

Use a calculator to verify the identity when $x = 4$. The keystrokes required depend on your calculator. This screen shows the result for the TI-83. When $x = 4$, both sides of the equation have the same value.

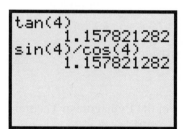

```
tan(4)
            1.157821282
sin(4)/cos(4)
            1.157821282
```

b) Graph the functions $y = \tan x$ and $y = \frac{\sin x}{\cos x}$. The graphs appear identical.

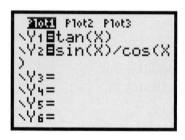

```
Plot1 Plot2 Plot3
\Y1☐tan(X)
\Y2☐sin(X)/cos(X
)
\Y3=
\Y4=
\Y5=
\Y6=
```

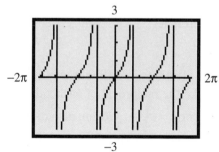

When you work with trigonometric identities, you often encounter an expression with trigonometric functions in the denominator. It is inconvenient to write the restriction(s) on the variable for every identity. When we write an identity, we mean that the expressions on both sides are equal for all values of the variable for which the expressions are defined.

1. At the beginning of this section, we used a graph to show that the equation $\sin x = \sin(\pi - x)$ is an identity. Use the unit circle definition of $\sin x$ to explain why $\sin x = \sin(\pi - x)$.

2. Explain your answer to each question.

   a) Are all equations identities?

   b) Are all identities equations?

   c) Are all equations with infinitely many solutions identities?

3. The first graphing calculator screen on page 315 shows the graphs of $y = \sin x$ and $y = \sin\left(\frac{\pi}{2} - x\right)$. Identify the graph that corresponds to $y = \sin\left(\frac{\pi}{2} - x\right)$. What other identity does this suggest? Explain.

4. Use the definitions of $\cos x$ and $\sin x$ in terms of a unit circle to explain why $\cos(-x) = \cos x$ and why $\sin(-x) = -\sin x$.

5. In the *Example* part a, why could we verify the identity for $x = \frac{\pi}{6}$ without a calculator, but we had to use a calculator to verify the identity for $x = 4$?

5.3 EXERCISES

A 1. Look at the first graph on page 316. Visualize what happens to line segment AB as x increases or decreases. Describe what this illustrates.

2. Repeat exercise 1 for the second graph on page 316.

3. Look at the diagram on page 317.

   a) Visualize what happens to $\sin x$, $\cos x$, and $\tan x$ as x increases or decreases. Describe what this illustrates.

   b) Visualize what happens to $\triangle PON$ as x increases or decreases. Describe what this illustrates.

4. Verify the Pythagorean identity numerically, for each value of x.

   a) $\frac{\pi}{6}$ b) $\frac{\pi}{4}$ c) $\frac{\pi}{2}$ d) 2.4

5. a) Verify the Pythagorean identity graphically.

   b) The Pythagorean identity can also be written as $\sin^2 x = 1 - \cos^2 x$ or as $\cos^2 x = 1 - \sin^2 x$. Choose one identity and verify it graphically.

   c) Write to explain why your graph in part b is different from your graph in part a.

B **6.** Consider the identity $\cot x = \frac{\cos x}{\sin x}$, $\sin x \neq 0$.

a) Verify the identity numerically, when $x = \frac{\pi}{3}$ and when $x = 2$.

b) Verify the identity graphically.

c) For what values of x is the identity defined?

7. Consider the identity $\tan^2 x + 1 = \sec^2 x$.

a) Verify the identity numerically, when $x = \frac{\pi}{4}$ and when $x = 5$.

b) Repeat exercise 6b and c.

8. Consider the identity $\cos(\pi - x) = -\cos x$.

a) Verify the identity numerically, when $x = \frac{\pi}{6}$ and when $x = 2.9$.

b) Repeat exercise 6b and c.

9. Determine which of these equations are identities. Justify your answer for each equation with an explanation.

a) $\tan(-x) = \tan x$

b) $\tan(-x) = -\tan x$

c) $\sin x = \sin(\pi + x)$

d) $\cos x = \cos(\pi + x)$

e) $\sin\left(\frac{\pi}{2} - x\right) = \cos x$

f) $\sin\left(\frac{\pi}{2} + x\right) = \cos x$

10. Choose two equations from exercise 9, one that is an identity and one that is not an identity. Write to explain how you determined that one equation is an identity and that the other equation is not an identity.

11. The quotient identity can also be written as $\sin x = (\cos x)(\tan x)$ or as $\cos x = \frac{\sin x}{\tan x}$, $\tan x \neq 0$.

a) Choose one identity and verify it graphically.

b) For what values of x is the identity defined?

c) Write to explain why your graph in part a is different from the graph in the *Example*.

12. Use the Pythagorean identity to show that $1 + \cot^2 x = \csc^2 x$ for any real number x, where $\sin x \neq 0$.

13. Consider the identity $\sin\left(\frac{\pi}{2} + x\right) = \sin\left(\frac{\pi}{2} - x\right)$.

a) Verify the identity numerically.   b) Verify the identity graphically.

c) How can the graph on page 315 be used to explain why the identity is true?

14. a) On the same screen, graph the functions $y = \sin^2 x$ and $y = \cos^2 x$.

b) Use the graphs in part a. Explain why $\sin^2 x + \cos^2 x = 1$.

c) On the same axes, graph the function $y = 1$ to illustrate your explanation in part b.

15. On this graph, the four line segments have the same length. Write two different identities that are illustrated by each pair of line segments.

a) ① and ②　　　　**b)** ② and ③　　　　**c)** ① and ④

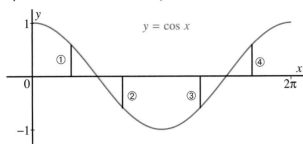

16. a) On the same screen, graph the functions $y = \sin^2 x$, $y = \cos^2 x$, and $y = \sin^2 x - \cos^2 x$.

b) Make a conjecture about the period and the amplitude of the function $y = \sin^2 x - \cos^2 x$.

c) Use the results of parts a and b to predict an identity for $\sin^2 x - \cos^2 x$.

17. Solve each equation algebraically for $0 \le x < 2\pi$. Express the roots in exact form.

a) $2 \sin x = 1 - \cos^2 x$　　　　　**b)** $2 \cos x = 1 - \sin^2 x$

c) $\cos^2 x - \sin^2 x = 1$　　　　　**d)** $\sin^2 x - \cos^2 x = 1$

C **18.** On page 317 and in exercise 11, the quotient identity was written in three different forms, with $\sin x$, $\cos x$, or $\tan x$ on the left side. Use the definitions of the reciprocal functions. Write the quotient identity in three other forms, with $\csc x$, $\sec x$, or $\cot x$ on the left side.

19. The quotient identity expresses one trigonometric function, $\tan x$, as a quotient of two other trigonometric functions, $\sin x$ and $\cos x$.

a) Would it be possible to express each of the six trigonometric functions as a quotient of two other trigonometric functions? Explain.

b) Would it be possible to express each of the six trigonometric functions as a product of two other trigonometric functions? Explain.

COMMUNICATING THE IDEAS

How would you explain trigonometric identities to a friend who missed today's class? Write down your ideas, with some examples to illustrate your explanation. Use examples that are different from the examples in this section.

5.4 Verifying and Proving Trigonometric Identities

In Section 5.3, we verified an identity numerically and graphically. To prove an identity for all values of the variable, we must use algebra.

Example 1

Consider the identity $1 - \cos^2 x = \cos^2 x \tan^2 x$.

a) Verify the identity numerically, when $x = \frac{\pi}{4}$ and when $x = 10$.

b) Verify the identity graphically.

c) Determine the values of x for which each side of the identity is defined.

d) Prove the identity algebraically.

Solution

a) When $x = \frac{\pi}{4}$,

$$1 - \cos^2 x = 1 - \cos^2\left(\frac{\pi}{4}\right)$$

$$= 1 - \left(\frac{1}{\sqrt{2}}\right)^2$$

$$= \frac{1}{2}$$

$$\cos^2 x \tan^2 x = \cos^2\left(\frac{\pi}{4}\right)\tan^2\left(\frac{\pi}{4}\right)$$

$$= \left(\frac{1}{\sqrt{2}}\right)^2 (1)^2$$

$$= \frac{1}{2}$$

Use a calculator to verify the identity when $x = 10$. The keystrokes required depend on the calculator. This screen shows the result for the TI-83. When $x = 10$, both sides of the equation have the same value.

```
1-cos(10)²
            .2959589691
cos(10)²tan(10)²
            .2959589691
```

b) Graph the functions $y = 1 - \cos^2 x$ and $y = \cos^2 x \tan^2 x$. The graphs appear identical.

```
Plot1  Plot2  Plot3
\Y1 ■ 1-cos(X)²
\Y2 ■ cos(X)²tan(X
)²
\Y3=
\Y4=
\Y5=
\Y6=
```

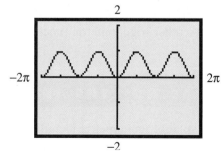

c) The expression on the left side, $1 - \cos^2 x$, is defined for all real values of x. Recall that the domain of $y = \tan x$ consists of all real numbers except $\dots, -\frac{3\pi}{2}, -\frac{\pi}{2}, \frac{\pi}{2}, \frac{3\pi}{2}, \dots$. Hence, the expression on the right side, $\cos^2 x \tan^2 x$, is defined for all real numbers except $\dots, -\frac{3\pi}{2}, -\frac{\pi}{2}, \frac{\pi}{2}, \frac{3\pi}{2}, \dots$

d) $1 - \cos^2 x = \cos^2 x \tan^2 x$

> ### Think…
>
> Use the Pythagorean identity on the left side. On the right side, use the definition of $\tan x$, restricting the solution to the values of x for which $\tan x$ is defined.

Left side	Reason
$1 - \cos^2 x = \sin^2 x$	Pythagorean identity

Right side	Reason
$\cos^2 x \tan^2 x = \cos^2 x \times \left(\dfrac{\sin x}{\cos x}\right)^2$ $= \cos^2 x \times \dfrac{\sin^2 x}{\cos^2 x}$ $= \sin^2 x$	Definition of $\tan x$, where $x \neq \ldots -\dfrac{3\pi}{2}, -\dfrac{\pi}{2}, \dfrac{\pi}{2}, \dfrac{3\pi}{2}, \ldots$

Since both sides simplify to the same expression, the identity is proved.

To prove identities algebraically, you must decide how to simplify algebraically. It is necessary to show that either both sides simplify to the same expression or that one side simplifies to the other side.

Example 2

Prove the identity $\sec x\, (1 + \cos x) = 1 + \sec x$.

a) Identify the values of x for which each side of the identity is defined.

b) Prove the identity algebraically.

Solution

a) Since $\sec x = \dfrac{1}{\cos x}$, the function $y = \sec x$ is defined for all real values of x except those for which $\cos x = 0$. Hence, the identity is defined for all real values of x except $\pm\dfrac{\pi}{2}, \pm\dfrac{3\pi}{2}, \pm\dfrac{5\pi}{2}, \ldots$.

b) Assume that $x \neq \pm\dfrac{\pi}{2}, \pm\dfrac{3\pi}{2}, \pm\dfrac{5\pi}{2}, \ldots$.

> ### Think…
>
> Since there are brackets on the left side but not on the right side, try expanding the left side.

Left side	Reason
$\sec x\,(1+\cos x) = \sec x + \sec x \cos x$ $= \sec x + 1$ $= \text{Right side}$	Expanding Since $\sec x$ and $\cos x$ are reciprocal functions

Since the left side simplifies to the right side, the identity is proved.

When working with identities, it is inconvenient to write the restrictions on the variable for every identity. When we write an identity stating that two expressions are equal, we mean that they are equal for all the values of the variable for which both expressions are defined.

Example 3

Prove the identity $\sec x = \tan x \csc x$.

Solution

Think...

Since the two sides involve reciprocal functions, use their definitions.

Left side	Reason
$\sec x = \dfrac{1}{\cos x}$	Definition of $\sec x$

Right side	Reason
$\tan x \csc x = \dfrac{\sin x}{\cos x} \times \dfrac{1}{\sin x}$ $= \dfrac{1}{\cos x}$	Definition of $\tan x$ and $\csc x$

Since both sides simplify to the same expression, the identity is proved.

Example 4

Consider the identity $\dfrac{\sin x}{1 + \cos x} = \dfrac{1 - \cos x}{\sin x}$.

a) Prove the identity algebraically.

b) Predict a similar identity for the expression $\dfrac{\cos x}{1 + \sin x}$.

Solution

a) $\dfrac{\sin x}{1 + \cos x} = \dfrac{1 - \cos x}{\sin x}$

> ### Think...
>
> When we multiply the numerator and denominator of the expression on the left side by $1 - \cos x$, the denominator will become $1 - \cos^2 x$, and we can use the Pythagorean identity.

Left side	Reason
$\dfrac{\sin x}{1 + \cos x} = \dfrac{\sin x}{1 + \cos x} \times \dfrac{1 - \cos x}{1 - \cos x}$	Multiplying by 1
$\quad = \dfrac{\sin x\,(1 - \cos x)}{1 - \cos^2 x}$	
$\quad = \dfrac{\sin x\,(1 - \cos x)}{\sin^2 x}$	Pythagorean identity
$\quad = \dfrac{1 - \cos x}{\sin x}$	Dividing numerator and denominator by $\sin x$
$\quad = $ Right side	

Since the left side simplifies to the right side, the identity is proved.

b) The pattern of the terms $\sin x$ and $\cos x$ in part a suggests that a similar identity might be $\dfrac{\cos x}{1 + \sin x} = \dfrac{1 - \sin x}{\cos x}$.

DISCUSSING THE IDEAS

1. Explain your answer to each question.

  a) When you verify an identity numerically, does it matter which number you substitute for x?

  b) Does it matter if you substitute an irrational number such as $\dfrac{\pi}{6}$ or a rational number such as 2.4?

  c) Do you need to substitute more than one number for x?

  d) Does verifying an identity numerically prove the identity?

2. a) Describe some advantages of using a graphing calculator to verify an identity.

  b) Does verifying an identity graphically prove the identity? Explain.

3. Suppose an equation involving the trigonometric functions is given.

  a) How could you tell if it might be an identity? Explain.

  b) How could you tell for certain if it is an identity? Explain.

5.4 EXERCISES

A **1.** Verify each identity numerically, when $x = \frac{\pi}{4}$ and when $x = 2$.

a) $\tan x \cos x = \sin x$

b) $\cos x = \frac{\sin x}{\tan x}$

2. Verify each identity graphically.

a) $\sin x \sec x \cot x = 1$

b) $\cos x (\sec x - 1) = 1 - \cos x$

3. Prove each identity algebraically.

a) $\sin x \cot x = \cos x$

b) $\tan x \csc x = \sec x$

c) $\sin x = \frac{\tan x}{\sec x}$

d) $\cos x = \frac{\cot x}{\csc x}$

4. Prove each identity algebraically.

a) $\csc x (1 + \sin x) = 1 + \csc x$

b) $\sin x (1 + \csc x) = 1 + \sin x$

c) $\frac{1 - \tan x}{1 - \cot x} = -\tan x$

d) $\frac{1 + \cot x}{1 + \tan x} = \cot x$

5. Prove the identity in *Example 4b*.

6. Choose one identity from each of exercises 1 to 4. For each identity, identify the values of the variable for which both sides of the identity are defined.

B **7.** Consider the identity $\frac{\sin x}{1 - \cos x} = \frac{1 + \cos x}{\sin x}$.

a) Verify the identity numerically, when $x = \frac{\pi}{6}$ and when $x = 1.5$.

b) Verify the identity graphically.

c) Prove the identity algebraically.

8. Consider the identity $\frac{1 + \sin x}{1 + \csc x} = \sin x$.

a) Verify the identity numerically, when $x = \frac{\pi}{3}$ and when $x = 2.5$.

b) Verify the identity graphically.

c) Prove the identity algebraically.

9. Look at the identity in exercise 8.

a) Predict a similar identity for $\frac{1 + \cos x}{1 + \sec x}$, and verify it graphically.

b) Predict a similar identity for $\frac{1 + \tan x}{1 + \cot x}$, and verify it graphically.

10. a) Prove the identity $\frac{1 + \cos x}{1 - \cos x} + \frac{1 + \sec x}{1 - \sec x} = 0$ algebraically.

b) Predict and verify two other identities suggested by the identity in part a.

11. a) Prove the identity $\frac{\cos x}{1 + \sin x} + \frac{\cos x}{1 - \sin x} = 2 \sec x$ algebraically.

b) Predict and verify two other identities suggested by the identity in part a.

12. Prove each identity.

a) $\sin x \tan x + \sec x = \frac{\sin^2 x + 1}{\cos x}$

b) $\frac{\sin x + \tan x}{\cos x + 1} = \tan x$

c) $\sin^2 x \cot^2 x = 1 - \sin^2 x$

d) $\csc^2 x - 1 = \csc^2 x \cos^2 x$

13. Determine which of these equations are identities. If an equation is an identity, prove the identity. If an equation is not an identity, explain why it is not.

a) $\tan x + \cot x = \sec x \csc x$

b) $\sec^2 x + \csc^2 x = \sec^2 x \csc^2 x$

c) $\sec^2 x - \csc^2 x = \frac{\sec^2 x}{\csc^2 x}$

d) $\sec^2 x + \csc^2 x = (\tan x + \cot x)^2$

e) $\cos^2 x = \sin x (\csc x + \sin x)$

f) $\sin^2 x = \cos x (\sec x - \cos x)$

14. a) Prove the identity $\frac{\sin x + \cos x}{\csc x + \sec x} = \sin x \cos x$.

b) Predict a similar identity for the expression $\frac{\sin x + \tan x}{\csc x + \cot x}$, then prove it algebraically.

c) Establish another identity like those in parts a and b.

d) How many identities in all do you think there are like these? Explain.

15. a) Prove the identity $\frac{1}{1 + \sin x} + \frac{1}{1 - \sin x} = 2 \sec^2 x$.

b) Establish a similar identity for $\frac{1}{1 + \cos x} + \frac{1}{1 - \cos x}$.

16. a) Prove the identity $\frac{\tan x}{\sec x + 1} = \frac{\sec x - 1}{\tan x}$.

b) Predict a similar identity for the expression $\frac{\cot x}{\csc x + 1}$, then prove it algebraically.

c) Establish another identity like those in parts a and b.

17. A graphing calculator produced these four graphs. The equations of the functions are $y = \sin^2 x$, $y = \cos^2 x$, $y = \sin x^2$, and $y = \cos x^2$. Without graphing the functions, identify the equation that corresponds to each graph. Explain.

a)

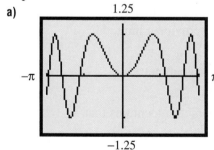

b)

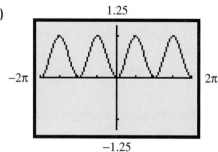

c)

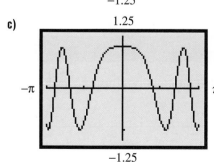

d)

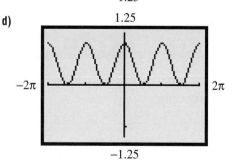

18. Refer to the graphing calculator screen in the solution of *Example 1b*. The graph appears to be that of a sinusoidal function.

　　a) What are the period and the amplitude of this function?

　　b) Predict an equation of this function.

　　c) Write two other identities that are suggested by this result. Use a graphing calculator to verify these identities.

C **19.** In exercise 15, identities were established for $\dfrac{1}{1+f(x)} + \dfrac{1}{1-f(x)}$, where $f(x) = \sin x$ and $f(x) = \cos x$. Establish similar identities where $f(x)$ represents each of the other four trigonometric functions.

20. Determine an identity that involves all six trigonometric functions.

　　a) Verify your identity numerically and graphically.

　　b) Prove your identity algebraically.

21. Write to explain how you found the identity in exercise 20.

22. a) Write an example of a trigonometric equation that is not an identity, but whose left side and right side produce graphs that appear to be identical when they are graphed on a graphing calculator.

　　b) Although the two graphs appear identical, find a way to use a calculator to demonstrate that the two graphs are different.

　　c) Modify your example in part a so that a calculator would not be able to demonstrate that the two graphs are different.

23. Recall that you encountered some identities in your study of logarithmic functions. With trigonometric functions, an extensive variety of identities can be formed. The identities in the preceding exercises provide a limited sample of this variety.

　　a) Explain why so many different identities occur with trigonometric functions.

　　b) Do you think there is a limit to the number of trigonometric identities that are possible? Explain.

COMMUNICATING THE IDEAS

Make a list of some strategies you could use to prove that an identity. Illustrate each strategy with an example.

In mathematics, we often combine the operation of evaluating a function with the operations of addition or subtraction. These operations may or may not give the same results if they are carried out in different orders. Consider the following examples.

Triple a sum.   The sum of the numbers tripled.
$3(x + y)$ $3x + 3y$

The square of a sum   The sum of the squares
$(x + y)^2$ $x^2 + y^2$

The square root of a sum   The sum of the square roots
$\sqrt{x + y}$ $\sqrt{x} + \sqrt{y}$

The reciprocal of a sum   The sum of the reciprocals
$\dfrac{1}{x + y}$ $\dfrac{1}{x} + \dfrac{1}{y}$

1. a) Which pairs of expressions above have the same values for all possible values of x and y for which the expressions are defined? Explain.

   b) Explain why the other pairs of expressions do not have the same values for all possible values of x and y for which the expressions are defined.

2. Determine whether $\sin(x + y)$ is equal to $\sin x + \sin y$ for all possible values of x and y. Explain.

3. Determine whether $\cos(x + y)$ is equal to $\cos x + \cos y$ for all possible values of x and y. Explain.

4. In all of the above examples, what happens if the addition sign is replaced with a subtraction sign? Explain.

In *Investigate*, you should have found that there was only one example in which the operations gave the same results when carried out in different orders; that is $3(x + y) = 3x + 3y$. We say that the operation of multiplication is *distributive* over addition. In general, function operations are not distributive over addition or subtraction.

To determine expressions for $\sin(x + y)$ and $\cos(x + y)$, we recall the definitions of the sine and cosine functions. Sin x and cos x are defined as the coordinates of a point P on a unit circle, where x represents the length of the arc from A(1, 0) to P.

We begin by using this definition to establish identities for $\cos(x - y)$ and $\sin(x - y)$.

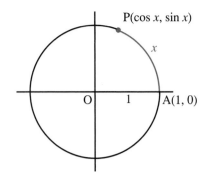

Deriving identities for $\cos(x - y)$ and $\sin(x - y)$

Step 1.
Let P($\cos x$, $\sin x$) and Q($\cos y$, $\sin y$) be points on a unit circle, where x and y represent the corresponding arc lengths from A(1, 0). Let R be on the circle such that OR is perpendicular to OQ. Since the slopes of OR and OQ are negative reciprocals, the coordinates of R may be represented by $(-\sin y, \cos y)$.

Step 2.
Rotate quadrilateral OQPR clockwise about the origin through ∠QOA. Then Q coincides with A, R coincides with B(0, 1), and P coincides with S. The arc length from A to S is $x - y$. Hence, the coordinates of S are $(\cos(x - y), \sin(x - y))$.

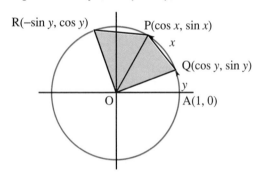

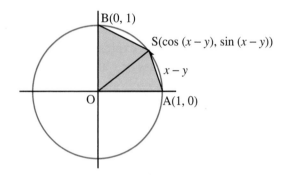

Step 3.
Since ∠POQ = ∠SOA, △POQ ≅ △SOA
Since the triangles are congruent,

$$PQ = SA$$

Use the distance formula.

$$\sqrt{(\cos x - \cos y)^2 + (\sin x - \sin y)^2} = \sqrt{(\cos(x - y) - 1)^2 + (\sin(x - y) - 0)^2}$$

Square each side to eliminate the radicals.

$$(\cos x - \cos y)^2 + (\sin x - \sin y)^2 = (\cos(x - y) - 1)^2 + \sin^2(x - y)$$

Expand the binomial squares.

$$\cos^2 x - 2\cos x \cos y + \cos^2 y + \sin^2 x - 2\sin x \sin y + \sin^2 y$$
$$= \cos^2(x - y) - 2\cos(x - y) + 1 + \sin^2(x - y)$$

$$(\cos^2 x + \sin^2 x) + (\cos^2 y + \sin^2 y) - 2(\cos x \cos y + \sin x \sin y)$$
$$= [\cos^2(x - y) + \sin^2(x - y)] + 1 - 2\cos(x - y)$$

Use the Pythagorean identity.

$$2 - 2(\cos x \cos y + \sin x \sin y) = 2 - 2\cos(x - y)$$

Hence,
$$\cos(x - y) = \cos x \cos y + \sin x \sin y \qquad \dots \text{①}$$

Similarly, using PR = SB, we can prove the following identity for $\sin(x - y)$:

$$\sin(x - y) = \sin x \cos y - \cos x \sin y \qquad \dots \text{②}$$

Equations ① and ② are the identities for $\cos(x - y)$ and $\sin(x - y)$ we have been seeking.

Example 1

Use a calculator to verify identities ① and ② for $x = 1.2$ and $y = 0.9$.

Solution

Substitute 1.2 for x and 0.9 for y into each identity.

Using identity ①:
$$\cos(1.2 - 0.9) = \cos 1.2 \cos 0.9 + \sin 1.2 \sin 0.9$$
or $\quad \cos 0.3 = \cos 1.2 \cos 0.9 + \sin 1.2 \sin 0.9$

Be sure the calculator is in radian mode.
Both sides of this equation are equal to approximately 0.955 336 489 1.

```
cos(1.2-.9)
           .9553364891
cos(1.2)cos(.9)+
sin(1.2)sin(.9)
           .9553364891
```

Using identity ②:
$$\sin(1.2 - 0.9) = \sin 1.2 \cos 0.9 - \cos 1.2 \sin 0.9$$
or $\quad \sin 0.3 = \sin 1.2 \cos 0.9 - \cos 1.2 \sin 0.9$

Both sides of this equation are equal to approximately 0.295 520 206 7.

```
sin(1.2-.9)
           .2955202067
sin(1.2)cos(.9)-
cos(1.2)sin(.9)
           .2955202067
```

Deriving identities for cos (x + y) and sin (x + y)

It is possible to obtain identities for $\cos(x + y)$ and $\sin(x + y)$ using diagrams similar to those on page 330. However, it is simpler to apply the odd-even identities to identities ① and ②. We can do this because the sum of two numbers can also be expressed as a difference; that is, $x + y = x - (-y)$.

Hence, we can write:

$$\cos(x + y) = \cos(x - (-y))$$
$$= \cos x \cos(-y) + \sin x \sin(-y)$$
$$= \cos x \cos y - \sin x \sin y$$

$$\sin(x + y) = \sin(x - (-y))$$
$$= \sin x \cos(-y) - \cos x \sin(-y)$$
$$= \sin x \cos y + \cos x \sin y$$

Sum and Difference Identities

$$\sin(x + y) = \sin x \cos y + \cos x \sin y$$
$$\sin(x - y) = \sin x \cos y - \cos x \sin y$$

$$\cos(x + y) = \cos x \cos y - \sin x \sin y$$
$$\cos(x - y) = \cos x \cos y + \sin x \sin y$$

Example 2

Consider the identity $\cos\left(\dfrac{\pi}{2} + x\right) = -\sin x$.

a) Verify the identity numerically, when $x = \dfrac{\pi}{4}$ and when $x = 2.5$.

b) Verify the identity graphically.

c) Prove the identity algebraically.

Solution

a) When $x = \dfrac{\pi}{4}$,

$$\cos\left(\frac{\pi}{2} + x\right) = \cos\left(\frac{\pi}{2} + \frac{\pi}{4}\right)$$
$$= \cos\frac{3\pi}{4}$$
$$= -\frac{1}{\sqrt{2}}$$
$$-\sin x = -\sin\frac{\pi}{4}$$
$$= -\frac{1}{\sqrt{2}}$$

Use a calculator to verify the identity when $x = 2.5$. When $x = 2.5$, both sides of the equation have the same value.

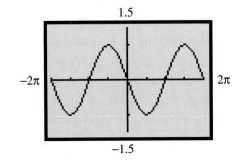

```
cos(π/2+2.5)
           -.5984721441
-sin(2.5)
           -.5984721441
```

b) Graph the functions $y = \cos\left(\dfrac{\pi}{2} + x\right)$ and $y = -\sin x$. The graphs appear identical.

```
Plot1 Plot2 Plot3
\Y1▊cos(π/2+X)
\Y2▊-sin(X)
\Y3=
\Y4=
\Y5=
\Y6=
\Y7=
```

c) $\cos\left(\frac{\pi}{2} + x\right) = -\sin x$

> **Think...**
>
> Use the identity for $\cos(x + y)$ to expand the left side.

Left side	Reason
$\cos\left(\frac{\pi}{2} + x\right) = \cos\frac{\pi}{2}\cos x - \sin\frac{\pi}{2}\sin x$ $= (0)\cos x - (1)\sin x$ $= -\sin x$ $= $ Right side	Identity for $\cos(x + y)$

Since the left side simplifies to the right side, the identity is proved.

DISCUSSING THE IDEAS

1. Recall the laws of exponents and the laws of logarithms. Do any of these laws illustrate an example of a function operation that is distributive over addition? Explain.

2. In the diagram in Step 1, page 330, why was P located farther from A along the unit circle than Q?

3. In *Example 2*, what other way is there to prove that $\cos\left(\frac{\pi}{2} + x\right) = -\sin x$? Explain.

5.5 EXERCISES

A 1. Use a calculator to verify that $\sin(x + y) = \sin x \cos y + \cos x \sin y$ for these values of x and y.

   a) $x = 2.2, y = 1.4$ **b)** $x = 1.8, y = -3.2$

2. Use a calculator to verify that $\cos(x + y) = \cos x \cos y - \sin x \sin y$ for these values of x and y.

   a) $x = 3.5, y = 4.8$ **b)** $x = 5.7, y = -2.4$

3. Verify that $\sin\frac{\pi}{2} = 1$ and $\cos\frac{\pi}{2} = 0$ by expanding and simplifying each expression.

   a) $\sin\left(\frac{\pi}{4} + \frac{\pi}{4}\right)$ **b)** $\cos\left(\frac{\pi}{4} + \frac{\pi}{4}\right)$

4. Determine the value of $\cos\frac{2\pi}{3}$ by expanding and simplifying each expression.

   a) $\cos\left(\frac{\pi}{3} + \frac{\pi}{3}\right)$ **b)** $\cos\left(\frac{\pi}{2} + \frac{\pi}{6}\right)$ **c)** $\cos\left(\pi - \frac{\pi}{3}\right)$

5. Determine the value of $\sin \frac{3\pi}{4}$ by expanding and simplifying each expression.

a) $\sin\left(\frac{\pi}{2} + \frac{\pi}{4}\right)$

b) $\sin\left(\pi - \frac{\pi}{4}\right)$

6. Choose exercise 4 or 5. Write to explain why more than one expression can be expanded and simplified to produce the desired result.

7. Visualize a point P rotating around a unit circle starting at A(1, 0). Visualize a point Q rotating at the same time such that the arc length from P to Q is always 1 unit measured counterclockwise. The points can move either clockwise or counterclockwise. Identify the graph below that shows:

a) the first coordinate of P

b) the second coordinate of P

c) the first coordinate of Q

d) the second coordinate of Q

Graph 1

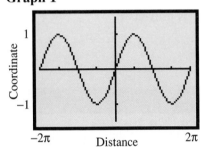

Graph 2

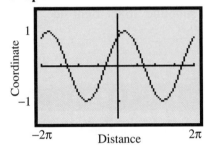

Graph 3

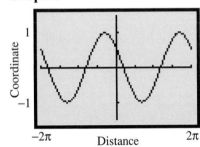

Graph 4

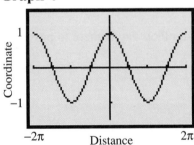

8. In exercise 7, write to explain how you identified the graphs.

9. In exercise 7, write two different equations for each graph that shows the coordinates of Q.

B **10.** Consider the identity $\sin\left(\frac{\pi}{2} + x\right) = \cos x$.

a) Verify the identity numerically, when $x = \frac{\pi}{4}$ and when $x = 2.7$.

b) Verify the identity graphically.   **c)** Prove the identity algebraically.

11. Consider the identity $\cos\left(\frac{\pi}{2} - x\right) = \sin x$.

a) Verify the identity numerically, when $x = \frac{\pi}{6}$ and when $x = 1.2$.

b) Verify the identity graphically.   **c)** Prove the identity algebraically.

12. Prove each identity algebraically.

a) $\sin\left(\frac{\pi}{2} - x\right) = \cos x$

b) $\cos\left(\pi + x\right) = -\cos x$

c) $\cos\left(\frac{3\pi}{2} + x\right) = \sin x$

d) $\sin\left(\frac{3\pi}{2} - x\right) = -\cos x$

13. Determine an identity for each expression.

a) $\cos\left(\pi - x\right)$ **b)** $\sin\left(\pi - x\right)$ **c)** $\sin\left(\pi + x\right)$ **d)** $\cos\left(x - \frac{\pi}{2}\right)$

14. These two expressions simplify to $\frac{\pi}{12}$: $\frac{\pi}{3} - \frac{\pi}{4}$ and $\frac{\pi}{4} - \frac{\pi}{6}$.

Determine the value of $\sin\frac{\pi}{12}$ by expanding each expression.

a) $\sin\left(\frac{\pi}{3} - \frac{\pi}{4}\right)$

b) $\sin\left(\frac{\pi}{4} - \frac{\pi}{6}\right)$

15. Determine the value of $\cos\frac{\pi}{12}$ by expanding each expression.

a) $\cos\left(\frac{\pi}{3} - \frac{\pi}{4}\right)$

b) $\cos\left(\frac{\pi}{4} - \frac{\pi}{6}\right)$

16. Find counterexamples to prove that $\cos\left(x - y\right) \neq \cos x - \cos y$ and $\sin\left(x - y\right) \neq \sin x - \sin y$.

17. a) Prove each identity.

$$\sin\left(x + \frac{0\pi}{6}\right) = 1\sin x + 0\cos x; \ \sin\left(x + \frac{\pi}{6}\right) = \frac{\sqrt{3}}{2}\sin x + \frac{1}{2}\cos x;$$

$$\sin\left(x + \frac{2\pi}{6}\right) = \frac{1}{2}\sin x + \frac{\sqrt{3}}{2}\cos x; \ \sin\left(x + \frac{3\pi}{6}\right) = 0\sin x + 1\cos x$$

b) Extend the list in part a to show similar identities for

$$\sin\left(x + \frac{4\pi}{6}\right), \ \sin\left(x + \frac{5\pi}{6}\right), \text{ and } \sin\left(x + \frac{6\pi}{6}\right).$$

18. Write identities for the expressions in each list. Describe any patterns.

a) $\sin\left(x + \frac{0\pi}{4}\right)$

$\sin\left(x + \frac{\pi}{4}\right)$

$\sin\left(x + \frac{2\pi}{4}\right)$

$\sin\left(x + \frac{3\pi}{4}\right)$

b) $\sin\left(x + \frac{0\pi}{2}\right)$

$\sin\left(x + \frac{\pi}{2}\right)$

$\sin\left(x + \frac{2\pi}{2}\right)$

$\sin\left(x + \frac{3\pi}{2}\right)$

c) $\sin\left(x + 0\pi\right)$

$\sin\left(x + \pi\right)$

$\sin\left(x + 2\pi\right)$

$\sin\left(x + 3\pi\right)$

19. $P(x, y)$ is a point on the unit circle. Q is a point on the unit circle whose distance from $A(1, 0)$ measured along the circumference is 1 greater than that of P.

a) Write the coordinates of Q in terms of x and y.

b) Is your answer to part a valid for all possible positions of P on the circle? Explain.

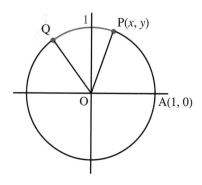

20. Visualize a regular polygon inscribed in a unit circle with centre O(0, 0). Let $P(x, y)$ represent the coordinates of one vertex of the polygon. Write the coordinates of the other vertices in terms of x and y, for each polygon.

a) an equilateral triangle

b) a square

c) a regular pentagon

d) a regular hexagon

21. Choose one regular polygon from exercise 20. Write to explain how you determined the expressions that represent the coordinates of one of the other vertices.

22. a) Prove that $\sin\left(\frac{\pi}{4} + x\right) + \sin\left(\frac{\pi}{4} - x\right) = \sqrt{2}\cos x$.

b) Find a similar expression for $\sin\left(\frac{\pi}{6} + x\right) + \sin\left(\frac{\pi}{6} - x\right)$.

c) Find a similar expression for $\sin\left(\frac{\pi}{3} + x\right) + \sin\left(\frac{\pi}{3} - x\right)$.

d) State a general result suggested by parts a, b, and c, then prove it.

23. Look at the expressions in exercise 22. What other expressions do they make you think of? Investigate similar results with these other expressions.

C 24. Determine whether there are any values of x and y such that:

a) $\sin(x + y) = \sin x + \sin y$

b) $\cos(x + y) = \cos x + \cos y$

c) $\tan(x + y) = \tan x + \tan y$

25. Choose one part of exercise 24. Write to explain how you determined the values of x and y.

26. In the equations below, k is a constant. Choose one equation. Determine whether there are any values of k for which the equation is an identity. If so, what are these values of k? If not, explain why. Support your answer with some diagrams or an explanation.

a) $\sin(x + k) = \sin x + \sin k$

b) $\cos(x + k) = \cos x + \cos k$

27. On the diagrams on page 330, use the fact that PR = SB to prove the identity for $\sin(x - y)$.

28. Use diagrams similar to those on page 330. Derive the identities for $\cos(x + y)$ and $\sin(x + y)$.

COMMUNICATING THE IDEAS

Your friend thinks that $\sin(x + y) = \sin x + \sin y$ and $\cos(x + y) = \cos x + \cos y$. How would you convince him that these are not correct? Write down at least two different ways you could do this.

Products Involving sin x and cos x

In this investigation, you will use a graphing calculator to explore the functions $f(x) = \sin x \cos x$, $g(x) = \cos x \cos x$, and $h(x) = \sin x \sin x$.

The function $f(x) = \sin x \cos x$

1. a) Graph the function $f(x) = \sin x \cos x$.

    b) Use the results of part a. Make conjectures about the period and the amplitude of the function.

    c) Use the results of parts a and b. Write $f(x)$ as a single trigonometric function.

2. a) Use the results of exercise 1. Write an expression for $\sin 2x$ in terms of $\sin x$ and $\cos x$.

    b) Use the identity $\sin (x + y) = \sin x \cos y + \cos x \sin y$. Prove that your expression is correct.

The function $g(x) = \cos x \cos x$

3. a) Graph the function $g(x) = \cos x \cos x$, or $g(x) = \cos^2 x$.

    b) Use the results of part a. Make conjectures about the period, the amplitude, and the vertical displacement of the function.

    c) Use the results of parts a and b. Write $g(x)$ as a single trigonometric function.

4. a) Use the results of exercise 3. Write an expression for $\cos 2x$ in terms of $\cos x$.

    b) Use the identity $\cos (x + y) = \cos x \cos y - \sin x \sin y$. Prove that your expression is correct.

The function $h(x) = \sin x \sin x$

5. a) Graph the function $h(x) = \sin x \sin x$, or $h(x) = \sin^2 x$.

    b) Use the results of part a. Make conjectures about the period, the amplitude, and the vertical displacement of the function.

    c) Use the results of parts a and b. Write $h(x)$ as a single trigonometric function.

6. a) Use the results of exercise 5. Write an expression for $\cos 2x$ in terms of $\sin x$.

    b) Use the identity $\cos (x + y) = \cos x \cos y - \sin x \sin y$. Prove that your expression is correct.

5.6 Identities for sin 2x and cos 2x

In Section 5.5, you learned identities for $\sin(x + y)$ and $\cos(x + y)$. Special cases of these identities occur when x and y are equal.

An identity for sin 2x

1. Write the identity for $\sin(x + y)$. Replace y with x in the identity and simplify. The result is an identity for $\sin 2x$.

2. Verify the identity you found in two ways:

   a) Substitute particular values for x in each side of the identity.

   b) Graph the function defined by the left side and the function defined by the right side. Use the trace feature to check that the two graphs are the same.

3. Which, if any, proves that the identity is correct? Explain.

   a) your algebraic calculations in exercise 1

   b) your numerical verifications in exercise 2a

   c) your graphical verification in exercise 2b

Identities for cos 2x

4. Write the identity for $\cos(x + y)$. Replace y with x in the identity and simplify. The result is an identity for $\cos 2x$.

5. Repeat exercise 2.

6. Which, if any, proves that the identity is correct? Explain.

   a) your algebraic calculations in exercise 4

   b) your numerical verifications in exercise 5a

   c) your graphical verification in exercise 5b

7. Use the Pythagorean identity $\sin^2 x + \cos^2 x = 1$. Write two other identities for $\cos 2x$.

8. Choose one identity from exercise 7.

   a) Verify the identity by substituting particular values for x in each side.

   b) Verify the identity by graphing the function defined by the left side and the function defined by the right side.

In Section 5.5, we developed identities for the sine and cosine of the sum of two numbers.

$$\sin(x + y) = \sin x \cos y + \cos x \sin y$$
$$\cos(x + y) = \cos x \cos y - \sin x \sin y$$

Special cases of these identities occur when the two numbers x and y are equal. These identities reduce to identities for $\sin 2x$ and $\cos 2x$.

$$\sin 2x = \sin(x + x) \qquad\qquad \cos 2x = \cos(x + x)$$
$$= \sin x \cos x + \cos x \sin x \qquad\qquad = \cos x \cos x - \sin x \sin x$$
$$= 2 \sin x \cos x \qquad\qquad = \cos^2 x - \sin^2 x \qquad \dots ①$$

We can use the Pythagorean identity $\sin^2 x + \cos^2 x = 1$ to express the identity for $\cos 2x$ in two other forms.

Since $\sin^2 x + \cos^2 x = 1$,   Since $\sin^2 x + \cos^2 x = 1$,
then $\sin^2 x = 1 - \cos^2 x$.   then $\cos^2 x = 1 - \sin^2 x$.
Substitute this expression in ①: Substitute this expression in ①:

$$\cos 2x = \cos^2 x - \sin^2 x \qquad\qquad \cos 2x = \cos^2 x - \sin^2 x$$
$$= \cos^2 x - (1 - \cos^2 x) \qquad\qquad = (1 - \sin^2 x) - \sin^2 x$$
$$= 2 \cos^2 x - 1 \qquad\qquad = 1 - 2 \sin^2 x$$

Identities for $\sin 2x$ and $\cos 2x$

$$\sin 2x = 2 \sin x \cos x \qquad\qquad\qquad \cos 2x = \cos^2 x - \sin^2 x$$
$$\cos 2x = 2 \cos^2 x - 1$$
$$\cos 2x = 1 - 2 \sin^2 x$$

Example 1

Use a calculator to verify that $\sin 1.2 = 2 \sin 0.6 \cos 0.6$.

Solution

Evaluate each side of the equation.
The results are both approximately $0.932\ 039\ 1$.

The patterns in the identities for $\sin 2x$ and $\cos 2x$ can be used to simplify certain trigonometric expressions.

Example 2

Consider the identity $\dfrac{1 + \cos 2x}{\sin 2x} = \cot x$.

a) Verify the identity numerically, when $x = \dfrac{\pi}{6}$ and when $x = 4$.

b) Verify the identity graphically.

c) Prove the identity algebraically.

Solution

a) When $x = \dfrac{\pi}{6}$,

$$\frac{1 + \cos 2x}{\sin 2x} = \frac{1 + \cos \frac{\pi}{3}}{\sin \frac{\pi}{3}}$$

$$= \frac{1 + \frac{1}{2}}{\frac{\sqrt{3}}{2}}$$

$$= \frac{3}{\sqrt{3}}, \text{ or } \sqrt{3}$$

$$\cot x = \frac{\cos \frac{\pi}{6}}{\sin \frac{\pi}{6}}$$

$$= \frac{\frac{\sqrt{3}}{2}}{\frac{1}{2}}$$

$$= \sqrt{3}$$

Use a calculator to verify the identity when $x = 4$. When $x = 4$, both sides of the equation have the same value.

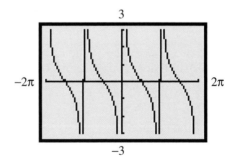

b) Graph the functions $y = \dfrac{1 + \cos 2x}{\sin 2x}$ and $y = \dfrac{1}{\tan x}$. The graphs appear identical.

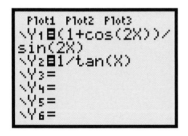

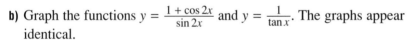

c) $\dfrac{1 + \cos 2x}{\sin 2x} = \cot x$

Think...

Since the left side involves double angles and the right side does not, use the identities for $\cos 2x$ and $\sin 2x$. There are three expressions we could substitute for $\cos 2x$. If we use $2\cos^2 x - 1$, the 1 in the numerator will be eliminated.

Left side	Reason
$\dfrac{1 + \cos 2x}{\sin 2x} = \dfrac{1 + (2\cos^2 x - 1)}{2\sin x \cos x}$	Identity for $\sin 2x$ and $\cos 2x$
$= \dfrac{2\cos^2 x}{2\sin x \cos x}$	
$= \dfrac{\cos x}{\sin x}$	Dividing numerator and denominator by $2\cos x$
$= \cot x$	Definition of $\cot x$
$= \text{Right side}$	

Since the left side simplifies to the right side, the identity is proved.

In *Example 2*, the expression $\dfrac{1 + \cos 2x}{\sin 2x}$ on the left side is not defined for values of x when $\sin 2x = 0$. Hence, the identity holds for all values of x except those for which $\sin 2x = 0$.

Example 3

Write each expression as a single trigonometric function.

a) $2\sin 0.45 \cos 0.45$ 　　　　　　**b)** $\cos^2 5 - \sin^2 5$

Solution

> **Think...**
>
> Compare the given expressions with the expressions on the right sides of the identities for $\sin 2x$ and $\cos 2x$.

a) The expression $2\sin 0.45 \cos 0.45$ can be obtained by substituting 0.45 for x in the right side of the identity $\sin 2x = 2\sin x \cos x$. Hence, we substitute 0.45 for x in the left side of the identity.
Therefore, $2\sin 0.45 \cos 0.45 = \sin 0.90$.

b) Similarly, by substituting 5 for x in the identity $\cos 2x = \cos^2 x - \sin^2 x$, we obtain $\cos^2 5 - \sin^2 5 = \cos 10$.

1. The identities for $\sin 2x$ and $\cos 2x$ are sometimes called *double-angle identities*. Explain why this name is appropriate.

2. Why are there three identities for $\cos 2x$ and only one for $\sin 2x$?

3. The second calculation in the graphing calculator screen in the solution of *Example 2a* shows that we calculated $\frac{1}{\tan 4}$. Explain why we did not calculate $\cot 4$.

4. a) In the first step in the solution of *Example 2c*, how did we know which of the three identities for $\cos 2x$ to use?

   b) Could we have used either of the other two identities for $\cos 2x$? Explain.

5.6 EXERCISES

1. Use a calculator to verify the identity $\sin 2x = 2 \sin x \cos x$ for each value of x.

    a) 0.45   **b)** 2   **c)** −0.68   **d)** −5   **e)** 100

2. Choose one identity for $\cos 2x$. Use a calculator to verify the identity you chose for each value of x.

    a) 0.69   **b)** 3   **c)** −0.42   **d)** −8   **e)** 200

3. a) Sketch the graphs of the functions $y = \sin 2x$ and $y = 2 \sin x$.

    b) Use the graphs to explain why $\sin 2x \neq 2 \sin x$.

4. a) Sketch the graphs of the functions $y = \cos 2x$ and $y = 2 \cos x$.

    b) Use the graphs to explain why $\cos 2x \neq 2 \cos x$.

5. Write each expression in each list as a single trigonometric function.

a)	b)	c)
$2 \sin 1 \cos 1$	$\cos^2 1 - \sin^2 1$	$\sin 3 \cos 3$
$2 \sin 2 \cos 2$	$\cos^2 2 - \sin^2 2$	$\sin 6 \cos 6$
$2 \sin 3 \cos 3$	$\cos^2 3 - \sin^2 3$	$\sin 12 \cos 12$
$2 \sin 4 \cos 4$	$\cos^2 4 - \sin^2 4$	$\cos 24 \sin 24$

6. Write each expression as a single trigonometric function.

    a) $2 \sin \frac{\pi}{6} \cos \frac{\pi}{6}$ **b)** $\cos^2 \frac{\pi}{10} - \sin^2 \frac{\pi}{10}$

    c) $2 \cos^2 0.5 - 1$ **d)** $1 - 2 \sin^2 3$

    e) $\cos^2 1 - \sin^2 1$ **f)** $1 - 2 \cos^2 4$

7. Visualize point P rotating around a unit circle, starting at A(1, 0). Visualize point Q rotating at the same time so that the distance travelled by Q is twice that of P. The points can move either clockwise or counterclockwise. Identify the graph below that shows:

a) the first coordinate of P

b) the second coordinate of P

c) the first coordinate of Q

d) the second coordinate of Q

Graph 1

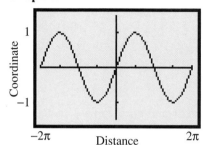

Graph 2

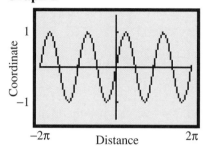

Graph 3

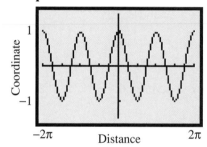

Graph 4

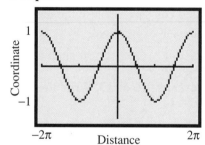

8. In exercise 7, write to explain how you identified the graphs.

9. In exercise 7, write two different equations for each graph that shows the coordinates of Q.

B **10.** Consider the identity $\sin^2 x = \frac{1 - \cos 2x}{2}$.

a) Verify the identity numerically, when $x = \frac{\pi}{6}$ and when $x = 2.3$.

b) Verify the identity graphically.

c) Prove the identity algebraically.

11. Look at the identity in exercise 10.

a) Predict a similar identity for $\cos^2 x$, then verify it graphically.

b) Prove the identity algebraically.

12. Write to explain how you found the identity in exercise 11a.

13. Prove each identity algebraically.

a) $1 + \sin 2x = (\sin x + \cos x)^2$

b) $\sin 2x = 2 \cot x \sin^2 x$

c) $\cos 2x = \dfrac{1 - \tan^2 x}{1 + \tan^2 x}$

d) $\sec^2 x = \dfrac{2}{1 + \cos 2x}$

14. Use the identity in exercise 10.

a) Substitute $\dfrac{\pi}{12}$ for x. Use the result to determine an expression for $\sin \dfrac{\pi}{12}$.

b) Substitute $\dfrac{\pi}{8}$ for x. Use the result to determine an expression for $\sin \dfrac{\pi}{8}$.

15. Use the identity in exercise 11.

a) Determine an expression for $\cos \dfrac{\pi}{12}$.

b) Determine an expression for $\cos \dfrac{\pi}{8}$.

16. Given $\sin \dfrac{\pi}{6} = \dfrac{1}{2}$ and $\cos \dfrac{\pi}{6} = \dfrac{\sqrt{3}}{2}$, use the identities for $\sin 2x$ and $\cos 2x$ to verify the value of each expression.

a) $\sin \dfrac{\pi}{3}$ and $\cos \dfrac{\pi}{3}$

b) $\sin \dfrac{2\pi}{3}$ and $\cos \dfrac{2\pi}{3}$

17. Given $\sin \dfrac{\pi}{4} = \dfrac{\sqrt{2}}{2}$ and $\cos \dfrac{\pi}{4} = \dfrac{\sqrt{2}}{2}$, use the identities for $\sin 2x$ and $\cos 2x$ to verify the value of each expression.

a) $\sin \dfrac{\pi}{2}$ and $\cos \dfrac{\pi}{2}$

b) $\sin \pi$ and $\cos \pi$

18. Choose one part of exercise 16 or 17. Write to explain how you verified the values.

19. a) Graph the function $f(x) = \dfrac{2 \tan x}{1 + \tan^2 x}$.

b) Use the result of part a to predict an identity for $\dfrac{2 \tan x}{1 + \tan^2 x}$.

c) Prove the identity algebraically.

20. Repeat exercise 19 using the function $g(x) = \dfrac{2 \cot x}{1 + \cot^2 x}$.

21. In exercises 19 and 20, both functions simplify to the same trigonometric function. That is, $\dfrac{\tan x}{1 + \tan^2 x} = \dfrac{\cot x}{1 + \cot^2 x}$ is an identity.

a) Predict two other identities similar to this.

b) Verify the identities graphically.

22. $P(x, y)$ is a point on the unit circle. Q is a point on the unit circle whose distance from $A(1, 0)$ measured along the circumference is double that of P.

a) Write the coordinates of Q in terms of x and y.

b) Is your answer to part a valid for all possible positions of P on the circle? Explain.

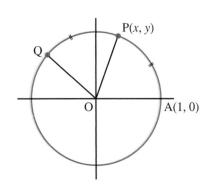

23. In *Example 2*, the identity $\frac{1 + \cos 2x}{\sin 2x} = \cot x$ was proved. Establish similar identities for each expression.

a) $\frac{1 - \cos 2x}{\sin 2x}$ **b)** $\frac{\sin 2x}{1 + \cos 2x}$ **c)** $\frac{\sin 2x}{1 - \cos 2x}$

24. a) Use the identity for $\sin 2x$ and an identity for $\cos 2x$. Verify that $\sin^2 2x + \cos^2 2x = 1$.

b) Write to explain your choice of the identity for $\cos 2x$ in part a.

25. When $\sin x + \cos x = 0.5$, determine the value of $\sin 2x$ in as many different ways as you can.

C 26. Solve each equation algebraically. Express the roots in exact form.

a) $2 \sin x \cos x - 1 = 0$ **b)** $4 \sin x \cos x - 1 = 0$

27. In exercise 17 on page 321, you solved the following equations algebraically. Solve these equations algebraically again, using a different method.

a) $\cos^2 x - \sin^2 x = 1$ **b)** $\sin^2 x - \cos^2 x = 1$

28. Consider the equation $\sin x = \sin 2x$, for $0 \le x < 2\pi$.

a) Solve the equation graphically.

b) Solve the equation algebraically.

29. Consider the equation $\cos x = \cos 2x$, for $0 \le x < 2\pi$.

a) Solve the equation graphically.

b) Solve the equation algebraically.

30. Consider the identities $\cos 2x = \cos^2 x - \sin^2 x$ and $\sin 2x = 2 \sin x \cos x$. Suppose we add corresponding sides of the two identities to obtain:
$\cos 2x + \sin 2x = \cos^2 x + 2 \sin x \cos x - \sin^2 x$
Observe the pattern on the right side.

a) Where in your previous study of mathematics have you encountered a pattern that is similar to this?

b) Explain why the pattern is not exactly the same as this one.

COMMUNICATING THE IDEAS

Write a short paragraph to explain how to obtain the identities for $\sin 2x$ and $\cos 2x$. Explain why there are three identities for $\cos 2x$ but not for $\sin 2x$, and how an identity for $\tan 2x$ can be determined.

Evaluating Trigonometric Functions

To keep pace with progress in navigation and astronomy in the 17th and 18th centuries, people required accurate values of trigonometric functions. Hence, mathematicians were faced with the problem of determining these values using basic arithmetic operations. The key to solving this problem was to connect the graphs of the trigonometric functions to the graphs of polynomial functions.

For example, the graphs of $y = \sin x$ and $y = \cos x$ are shown below. For values of x very close to 0:

- The graph of $y = \sin x$ appears to coincide with a straight line.
- The graph of $y = \cos x$ appears to coincide with a parabola.

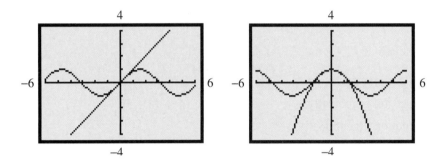

Use a scientific calculator, a graphing calculator, or a spreadsheet.

1. Recall that the equation of a straight line through the origin has the form $y = mx$, where m represents the slope of the line.

   a) Use estimation or systematic trial to determine the equation of the line in the first graph above.

   b) Determine the values of x so that the y-coordinates of points on the line and the sine graph differ by less than 0.01.

   c) Use the definition of $\sin x$ in Section 4.1. Explain why your equation in part a is correct.

2. Recall that the equation of a parabola with vertex (0, 1) and axis of symmetry the y-axis has the form $y = ax^2 + 1$.

   a) The parabola in the second graph above opens down. What does this tell you about the value of a?

   b) Use estimation or systematic trial to determine the equation of the parabola in the second graph above.

   c) Determine the values of x so that the y-coordinates of points on the parabola and the cosine graph differ by less than 0.01.

3. Use the results of exercises 1 and 2. Describe a simple way to approximate the values of $\sin x$ and $\cos x$ when x is very close to 0.

In exercises 1 and 2, you should have discovered the following results.

For values of x very close to 0,
$$\sin x \doteq x \qquad \cos x \doteq 1 - \tfrac{1}{2}x^2$$

For improved accuracy, more terms on the right sides of the equations are required. Recall that the graph of a polynomial function has hills and valleys, just like the graphs of the sine and cosine functions. When the coefficients are chosen carefully, the hills and valleys in the graphs of polynomial functions should coincide with the hills and valleys in the graphs of the sine and cosine functions. Determining these coefficients for trigonometric and other functions was an important mathematical problem in the 17th and 18th centuries.

Recall from *Problem Solving*, page 145, that a function $f(x)$ can sometimes be expressed as an infinite series of powers of x. The series for $\sin x$ and $\cos x$ are shown below. The denominators use the factorial notation that was introduced on page 145.

$$\sin x = x - \tfrac{1}{3!}x^3 + \tfrac{1}{5!}x^5 - \tfrac{1}{7!}x^7 + \dots$$
$$\cos x = 1 - \tfrac{1}{2!}x^2 + \tfrac{1}{4!}x^4 - \tfrac{1}{6!}x^6 + \dots$$

4. Write the first six terms of the series for $\sin x$ and for $\cos x$.

5. Use the result of exercise 4 to calculate each value. Check using the diagram on page 228, and the [SIN] or [COS] key on your calculator.

a) $\cos 1.2$ **b)** $\sin 1.2$ **c)** $\cos 2.7$ **d)** $\sin 2.7$

6. The equation of the line in the first graph on page 346 is $y = x$. The right side of this equation contains the first term in the series for $\sin x$.

a) Predict what the graphs will look like if you graph $y = \sin x$ and the following functions on the same grid.

i) $y = x - \tfrac{1}{3!}x^3$

ii) $y = x - \tfrac{1}{3!}x^3 + \tfrac{1}{5!}x^5$

iii) $y = x - \tfrac{1}{3!}x^3 + \tfrac{1}{5!}x^5 - \tfrac{1}{7!}x^7$

b) Graph the functions to check your predictions.

7. The equation of the parabola in the second graph on page 346 is $y = 1 - \frac{1}{2!}x^2$. The right side of this equation contains the first two terms in the series for $\cos x$.

a) Predict what the graphs will look like if you graph $y = \cos x$ and the following functions on the same grid.

i) $y = 1 - \frac{1}{2!}x^2 + \frac{1}{4!}x^4$

ii) $y = 1 - \frac{1}{2!}x^2 + \frac{1}{4!}x^4 - \frac{1}{6!}x^6$

iii) $y = 1 - \frac{1}{2!}x^2 + \frac{1}{4!}x^4 - \frac{1}{6!}x^6 + \frac{1}{8!}x^8$

b) Graph the functions to check your predictions.

8. Explain why the series for $\sin x$ contains only odd powers of x, while the series for $\cos x$ contains only even powers of x.

9. Determine how many terms of the series

$$\sin x = x - \frac{1}{3!}x^3 + \frac{1}{5!}x^5 - \frac{1}{7!}x^7 + \ldots$$

are needed to obtain values that are accurate to at least 5 decimal places over one complete cycle.

10. Determine how many terms of the series

$$\cos x = 1 - \frac{1}{2!}x^2 + \frac{1}{4!}x^4 - \frac{1}{6!}x^6 + \ldots$$

are needed to obtain values that are accurate to at least 5 decimal places over one complete cycle.

Recall from *Problem Solving*, page 145, the series for the function e^x.

$$e^x = 1 + x + \frac{1}{2!}x^2 + \frac{1}{3!}x^3 + \frac{1}{4!}x^4 + \frac{1}{5!}x^5 + \frac{1}{6!}x^6 + \frac{1}{7!}x^7 + \ldots$$

Except for some signs, the series for $\sin x$ and $\cos x$ contain the same terms as the series for e^x. When something like this happens in mathematics, it is usually not a coincidence. This suggests there might be a connection between exponential functions and trigonometric functions. Discover this connection in exercise 11.

11. The series for e^x is rewritten by grouping the terms with even exponents and the terms with odd exponents.

$$e^x = \left(1 + \frac{1}{2!}x^2 + \frac{1}{4!}x^4 + \frac{1}{6!}x^6 + \ldots\right) + \left(x + \frac{1}{3!}x^3 + \frac{1}{5!}x^5 + \frac{1}{7!}x^7 + \ldots\right)$$

Determine the value of k such that:

$$e^{kx} = \left(1 - \frac{1}{2!}x^2 + \frac{1}{4!}x^4 - \frac{1}{6!}x^6 + \ldots\right) + k\left(x - \frac{1}{3!}x^3 + \frac{1}{5!}x^5 - \frac{1}{7!}x^7 + \ldots\right)$$

$$= \cos x + k \sin x$$

1. Solve $\sin^2 x - 2 \sin x - 1 = 0$, then write the general solution.

2. Solve each equation algebraically for $0 \leq x \leq 2\pi$.

   a) $\sin^2 x - 1 = 0$ b) $\sin^2 x + 2 \sin x + 1 = 0$ c) $2 \cos^2 x - 3 \cos x - 2 = 0$

3. Write a trigonometric equation that has each given root over the domain $0 \leq x < 2\pi$.

   a) π b) $\frac{\pi}{2}$ c) $\frac{5\pi}{3}$ d) $\frac{2\pi}{3}$ e) 0

4. Prove each identity.

   a) $\sec^2 x - 1 = \sin^2 x \sec^2 x$ b) $\csc x - \cos x \tan x = \frac{\cos x}{\tan x}$

   c) $\frac{\cos x + \cot x}{1 + \sin x} = \cot x$ d) $\cos^2 x \tan^2 x = 1 - \cos^2 x$

5. Choose one identity from exercise 4. Write to explain how you proved it.

6. Write each expression as a single trigonometric function.

   a) $\sin \frac{\pi}{4} \cos \frac{\pi}{3} + \cos \frac{\pi}{4} \sin \frac{\pi}{3}$ b) $\cos 0.4 \cos 0.5 - \sin 0.4 \sin 0.5$

   c) $\cos \frac{\pi}{6} \cos \frac{\pi}{2} + \sin \frac{\pi}{6} \sin \frac{\pi}{2}$ d) $\sin 0.3 \cos (-0.5) - \cos 0.3 \sin (-0.5)$

7. Suppose $\sin x = \frac{3}{5}$ and $\cos y = \frac{5}{12}$, and x and y are in the first quadrant. Determine each value.

   a) $\cos (x + y)$ b) $\sin (x - y)$ c) $\sin 2x$ d) $\cos 2y$

8. Find an exact expression for each trigonometric function.

   a) $\sin \frac{\pi}{12}$ b) $\cos \frac{13\pi}{12}$ c) $\cos \frac{7\pi}{12}$ d) $\sin \frac{23\pi}{12}$

9. Choose one trigonometric function from exercise 8. Write to explain how you found the exact expression.

10. Solve each equation to 4 decimal places for $0 \leq x < 2\pi$.

    a) $8 \sin^2 x - 6 \sin x + 1 = 0$ b) $3 \cos^2 x = 4 \cos x + 4$

    c) $\sin 2x + \cos x = 0$ d) $3 \cos^2 x - 3 \sin^2 x + 2 = 0$

11. Choose one equation from exercise 10. Write to explain how you solved it.

12. a) Establish identities for $\sin 3x$ and $\cos 3x$ by expanding $\sin (2x + x)$ and $\cos (2x + x)$. Different results are possible, depending on which expression for $\cos 2x$ is used, and whether the Pythagorean identity is used. Try different possibilities, and decide which are the best expressions for $\sin 3x$ and $\cos 3x$.

    b) Write to explain the reason for your choice for the best expressions in part a.

1. a) Sketch the graph of the function $y = (x - 2)^2 + 4$ for $-5 \le x \le 5$.

   b) What are the domain and range of the function?

   c) What are the x- and y-intercepts of the graph?

2. Describe what happens to the equation of a function when you make each change to its graph.

   a) Translate the graph 4 units up.

   b) Translate the graph 7 units left.

   c) Translate the graph 5 units right and 3 units down.

   d) Translate the graph 6 units left and 12 units down.

3. Describe how the graph of the first function compares to the graph of the second function.

   a) $y = \frac{1}{2x}$ $y = \frac{1}{2x} + 4$

   b) $y = |x|$ $y = |x - 3|$

4. Sketch the graphs of these functions on the same grid.

   a) $y = x^3 - 3$ b) $y = (x + 2)^3 - 5$

   c) $y = (x - 2)^3$ d) $y = (x - 5)^3 + 6$

5. Choose one function from exercise 4. Write to explain how you sketched it.

6. The graph of a function $y = f(x)$ is transformed as described below. The equation of its image has the form $y = af(bx)$. Determine the values of a and b for each transformation.

   a) Compress by a factor of $\frac{1}{4}$ horizontally and expand by a factor of 3 vertically.

   b) Expand by a factor of 2 horizontally, by a factor of 4 vertically, and reflect in the y-axis.

   c) Compress by a factor of $\frac{1}{3}$ horizontally and a factor of $\frac{1}{5}$ vertically.

   d) Expand by a factor of 3 horizontally, compress by a factor of $\frac{1}{6}$ vertically, and reflect in the x-axis.

7. a) Sketch the graphs of $f(x) = x^3 + 8$ and $y = \frac{1}{f(x)}$.

   b) Identify the asymptotes of $y = \frac{1}{f(x)}$.

8. There are now 300 insects in a colony. The population of the colony doubles every 5 days.

   a) Express the population, P, of the colony as an exponential function of the elapsed time, d days; then graph the function.

   b) Calculate the number of days it would take for the population to reach 10 000 insects.

9. Determine each logarithm.

a) $\log_4 0.25$ b) $\log_7 343$ c) $\log_2 \sqrt{2}$

10. Solve for x.

a) $\log_x 64 = 2$ b) $\log_5 0.04 = x$ c) $\log_2 x = 5$

11. The half-life of a radioactive substance is 23 days. How long is it until each percent remains?

a) 10%   b) 3%

12. a) Graph the functions $y = 4^x$ and $y = 3(4^{x-1})$ on the same axes.

b) How was the graph of $y = 4^x$ transformed to obtain the graph of $y = 3(4^{x-1})$?

c) Identify the domain, range, asymptotes, and intercepts of each graph.

13. Convert each angle to degrees. Express each angle to 2 decimal places.

a) 5.2 radians   b) 1.3 radians   c) 3 radians

14. Determine the length of the arc that subtends each angle at the centre of a circle with radius 7 cm. Express each length to 1 decimal place.

a) 1.2 radians   b) 208°

15. Sketch each angle in standard position.

a) $\theta = -30°$ b) $\theta = 110°$

c) $\theta = -\frac{3\pi}{5}$ radians   d) $\theta = \frac{5\pi}{3}$ radians

16. In each part of exercise 15, determine two angles that are coterminal with θ.

17. Choose one angle from exercise 16. Write to explain how you determined two coterminal angles.

18. Graph each function.

a) $y = 2 \cos 3(x - \frac{\pi}{3}) - 5$ b) $g(x) = 3 \sin 2(x + \frac{\pi}{4}) + 1$

19. Write an equation for a sine function with these properties: amplitude: 2; period: 3; phase shift: 4; vertical displacement: -5

20. The graphing calculator screen shows the coordinates of one point on the graph of $y = \tan \pi x$, to the nearest thousandth. Use only the information on the screen. Write the x-coordinates of four other points on the graph that have the same y-coordinate as this point.

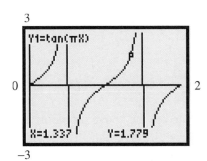

Analyzing Club Keno

CONSIDER THIS SITUATION

The British Columbia Lottery Corporation is endorsed by the Province of British Columbia. The revenue from ticket sales is returned to the province. Fifty percent is allocated to health care, and 50% is reserved for other community needs. One lottery is Club Keno.

To play in the Club Keno lottery in British Columbia, you choose numbers from 1 to 80. You can pick a minimum of 1 number up to a maximum of 10 numbers. The central computer system randomly selects 20 numbers from 1 to 80. Potential winnings, shown in the prizes table, depend on both how many numbers you picked and how many were selected by the central computer system.

Examine the prizes table.

- Explain why the amounts of the prizes tend to increase when more numbers are played.
- Find two different ways to win $5000. Compare the chances of winning. Does this seem fair?
- If you were playing Club Keno, how many numbers would you pick? Explain.

On pages 379 to 381, you will develop mathematical models for calculating the probability of winning any of the prizes, and your expected payoff if you play Club Keno.

FYI Visit www.awl.com/canada/school/connections

For information related to the above problem, click on MATHLINKS followed by AWMath. Then select a topic under Analyzing Club Keno.

Prizes

Pick	Match	Win for Each $1 Played ($)	Chances to win
1	1	2	1 : 4
2	2	10	1 : 17
3	3	20	1 : 73
	2	2	1 : 8
4	4	50	1 : 327
	3	5	1 : 24
	2	1	1 : 5
5	5	400	1 : 1551
	4	15	1 : 83
	3	2	1 : 12
6	6	1500	1 : 7753
	5	50	1 : 324
	4	5	1 : 36
	3	1	1 : 8
7	7	5000	1 : 40 980
	6	100	1 : 1366
	5	15	1 : 116
	4	2	1 : 20
	3	1	1 : 6
8	8	10 000	1 : 230 115
	7	400	1 : 6233
	6	50	1 : 423
	5	10	1 : 55
	4	2	1 : 13
9	9	25 000	1 : 1 380 688
	8	2500	1 : 30 682
	7	200	1 : 1691
	6	25	1 : 175
	5	4	1 : 31
	4	1	1 : 9
10	10	100 000	1 : 8 911 712
	9	5000	1 : 163 382
	8	500	1 : 7385
	7	50	1 : 621
	6	10	1 : 88
	5	3	1 : 20
	0	3	1 : 22

INVESTIGATE Counting without Counting

1. A cafe has a lunch special consisting of
   an egg or a ham sandwich (E or H);
   milk, juice, or coffee (M, J, or C);
   and yogurt or pie for dessert (Y or P).

   a) One item is chosen from each category.
      List all possible meals.

   b) How many possible meals are there?

   c) How can you determine the number of
      possible meals without listing all of them?

2. The cafe also features ice cream cones
   in 24 flavours. You can order regular,
   sugar, or waffle cones. Suppose you
   order a double cone with two
   scoops of ice cream.

   a) How many choices of type of cone
      do you have?

   b) How many choices of flavour for the
      first scoop of ice cream do you have?

   c) How many choices do you have for
      the second scoop?

   d) How many different double cones
      are possible?

3. In exercise 2, suppose you order a triple cone with 3 scoops of ice cream.
   How many different triple cones are possible?

4. Suppose one item can be selected in m ways, and for each way a second
   item can be selected in n ways. How many ways can the two items be selected?
   This result is known as the *fundamental counting principle*.

We are constantly confronted with making choices. When two or more choices
must be made together, we can determine the total number of possibilities
without counting all of them individually.

Example 1

A computer store sells 6 different computers, 4 different monitors, 5 different printers, and 3 different multimedia packages. How many different computer systems are available?

Solution

There are 6 computers.   For each computer, there are 4 monitors.   For each monitor, there are 5 printers.   For each printer, there are 3 multimedia packages

$$\boxed{6} \quad \times \quad \boxed{4} \quad \times \quad \boxed{5} \quad \times \quad \boxed{3}$$

The number of different systems is $6 \times 4 \times 5 \times 3$, or 360.

The counting in *Example 1* illustrates the following principle.

The Fundamental Counting Principle

If one item can be selected in m ways, and for each way a second item can be selected in n ways, then the two items can be selected in mn ways.

Example 2

How many even 2-digit whole numbers are there?

Solution

Since 0 cannot be the first digit, the first digit can be selected in 9 ways.
Since the second digit must be $0, 2, 4, 6,$ or 8, it can be selected in 5 ways.
Use the fundamental counting principle.

There are $9 \times 5 = 45$ ways to select the two digits.
Hence, there are 45 even 2-digit whole numbers.

Example 3

In each case, how many 2-digit whole numbers can be formed using the digits 0, 2, 4, 6, and 8?

a) Repetitions are allowed.   **b)** Repetitions are not allowed.

Solution

a) The tens digit can be 2, 4, 6, or 8. There are 4 ways to select the tens digit.
The units digit can be any of the 5 digits.
Use the fundamental counting principle.
There are $4 \times 5 = 20$ whole numbers.

b) The tens digit can be 2, 4, 6, or 8. There are 4 ways to select the tens digit.
The units digit can be any of the other 4 digits.
Hence, there are $4 \times 4 = 16$ whole numbers.

DISCUSSING THE IDEAS

1. Consider the fundamental counting principle. Explain why the number of ways the items can be selected are multiplied and not added.

2. Suppose a number such as 05 were considered a 2-digit number. How would the solutions to *Examples 2* and *3* change?

6.1 EXERCISES

 1. A student has 4 different shirts (S1, S2, S3, and S4), 2 different pairs of pants (P1 and P2), and 3 different pairs of shoes (H1, H2, and H3).

   a) List all possible outfits in such a way to ensure that all have been counted and none has been counted twice. How many possible outfits did you list?

   b) Use the fundamental counting principle. Determine the number of outfits there should be. Do your answers to parts a and b match?

2. A person has 5 shirts and 3 pairs of pants. In how many ways can he select an outfit?

3. A student has 5 blouses, 4 skirts, and 4 sweaters. In how many ways can she select an outfit comprising 3 items?

4. In exercises 2 and 3, what assumptions are you making?

5. How many odd 2-digit whole numbers are there?

6. In each case, how many even 2-digit whole numbers can be made using the digits $1, 2, 3, 4, 5, 6, 7,$ and 8.

   a) Repetitions are not allowed.   **b)** Repetitions are allowed.

7. Use the digits $1, 3, 5, 7,$ and 9, and no repetitions.

   a) How many 3-digit whole numbers can be formed?

   b) How many 4-digit whole numbers can be formed?

B **8.** A true-false test has 5 questions. The answer to each question is a guess.

   a) How many possible answers are there for each question?

   b) How many different ways are there to complete the test?

   c) What is the probability that all 5 questions will be answered correctly?

9. A multiple-choice test has 5 questions, with 4 possible answers for each question. Suppose the answer to each question is a guess.

   a) How many possible answers are there for each question?

   b) How many different ways are there to complete the test?

   c) What is the probability that all 5 questions will be answered correctly?

10. In the Pick 3 lottery, participants pick 3 digits from 0 to 9 and choose Straight Play or Box Play. The lottery corporation randomly selects 3 digits from 0 to 9. The three digits can be the same. Straight Play wins if the 3 digits picked match the 3 winning digits in the same order. Box Play wins if the 3 digits picked match the 3 winning digits in any order. According to the lottery corporation, the odds of winning are 1 in 1000 for Straight Play and 1 in 167 for Box Play. Write to explain how these chances of winning are determined for both games.

11. How many different ways are there to spell out each word vertically?

a)	**b)**	**c)**	**d)**
P	AAAA	B	C
OO	LLL	RR	OO
RRR	BB	I I I	LLL
TTTT	E	T T T T	UUUU
	RR	I I I	MMMMM
	NNNN	SS	B B B B B B
	I I I I I	H	I I I I I I I
			A A A A A A A A

12. Open the *Cars* database. Sort the data by Class.

   a) How many classes of cars are there?

   b) List the number of cars in each class.

   c) Use the fundamental counting principle. Determine the number of ways to select one car from each class.

13. Assume a car licence plate consists of 6 characters. Each character can be any of the letters from A to Z, or any numeral from 0 to 9.

   a) How many licence plates are possible?

   b) Write to explain why the answer to part a is different from the number of licence plates that would be produced.

14. In exercise 13, assume that a car licence plate can consist of up to 6 characters. Each character can be any letter from A to Z, or any numeral from 0 to 9. How many licence plates are possible?

15. In a hotel in Hong Kong, a room key is a card. The card has positions for holes that form a 5 by 10 array. Each position in the array is either punched with a hole or left blank.

   a) How many different keys are possible?

   b) Suppose all the possible keys were distributed equally among all the people on Earth. How many keys would there be for each person?

16. The book *Cent Mille Milliards de Poèmes* consists of 10 sonnets written in French. Each sonnet is sliced into 14 strips, one for each line. The strips can be flipped so that a sonnet can be created using any one of the 10 available strips for each line. It is said that so many sonnets are possible that you could probably read a sonnet that no one has ever read before, or will ever read again. Check the validity of this statement.

17. The final score in a hockey game is 5 − 2. How many different scores are possible at the end of the second period?

18. The dial on a 3-number combination lock contains markings to represent the numbers from 0 to 59. How many combinations are possible in each case?

   a) The first and second numbers must be different, and the second and third numbers must be different.

   b) The first and second numbers must differ by at least 3.

19. A "mission statement" is a brief description of a company's objective or purpose. A humorous newspaper column in 1998 contained a generic statement that could be used to create mission statements for departments of the federal government. The generic mission statement has this form:

"To (A) and (A) the (B) and (B) of the (C) department of Canada."

To create a mission statement, choose 2 words from column A, 2 from column B, and 1 from column C.

Column A	Column B	Column C
foster	development	Indian affairs
promote	competitiveness	health
support	growth	industry
assist	improvement	immigration
preserve	creation	employment
protect	potential	trade
enhance	prosperity	finance
stimulate	awareness	environment

For example, a possible mission statement is "To support and preserve the potential and growth of the immigration department of Canada."

a) Write 2 other examples of mission statements using this model.

b) How many different mission statements can be written?

20. A Bingo card has 5 columns, each with 5 spaces. The first column contains numbers from 1 to 15; the second column contains numbers from 16 to 30; the third column has its centre square shaded and contains numbers from 31 to 45; the fourth and fifth columns contain numbers from 46 to 60, and 61 to 75, respectively.

a) How many different Bingo cards can be printed?

b) Suppose every person on Earth counted these cards at the rate of 1 per second. How long would it take to count all of them?

c) Suppose the cards were printed on paper 0.1 mm thick and they were stacked one on top of another. How high would the stack be?

B	I	N	G	O
10	28	39	53	71
5	21	34	50	67
14	29	free	51	70
2	17	41	46	63
8	25	31	60	74

COMMUNICATING THE IDEAS

Write to explain what is meant by the fundamental counting principle, and include some examples of how it is used.

INVESTIGATE Permutations Involving Different Objects

1. Two letters, A and B, can be written in two different orders, AB and BA. These are *permutations* of A and B.

  a) List all the permutations of 3 letters A, B, and C. How can you be certain that you have listed all of them, and that you have not counted any permutation more than once? How many permutations are there?

  b) List all the permutations of 4 letters A, B, C, and D. How can you be certain that you have listed all of them, and that you have not counted any permutation more than once? How many permutations are there?

  c) Predict the number of permutations of 5 letters A, B, C, D, and E.

  d) Suppose you know the number of letters. How can you determine the number of permutations?

2. Instead of arranging letters in order, we can arrange objects if they are different. Explain your answer to each question.

  a) How many different ways can 5 people be arranged in a line?

  b) How many different ways can 5 different books be arranged on a shelf?

  c) How many permutations are there of the letters of the word PROVE?

3. Consider the letters A, B, C, D, and E. Instead of using all the letters to form permutations, we could use fewer letters. For example, DB is a 2-letter permutation of these 5 letters.

  a) List all the different 2-letter permutations of the 5 letters A, B, C, D, and E.

  b) How many 3-letter permutations are there?

4. From a committee of 5 people, a chair and a vice chair are to be chosen. How many different ways can this be done?

5. Your calculator may have a $_nP_r$ key or a menu item.

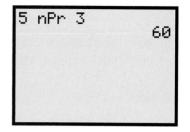

a) Find out how to use this feature. For example, to determine $_5P_3$ on the TI-83 graphing calculator, press 5 [MATH] [▶] [▶] [▶] 2 3 [ENTER].

b) Determine some values of $_nP_r$ for different values of n and r. Look at your solutions to exercises 1 to 4 for ideas of numbers to use.

c) Explain what $_nP_r$ means.

6. Use the results of exercise 5 to write an expression for $_nP_r$ in terms of n and r. What are the restrictions on n and r?

Many student lockers are secured with a 3-number combination lock. Knowing the 3 numbers is not sufficient to open the lock. The numbers must be used in the correct sequence. The order of the numbers is important.

An arrangement of a set of objects is a *permutation*. The order of the objects is important.

Example 1

How many permutations can be formed using all the letters of the word MUSIC?

Solution

There are five different letters. Visualize placing the letters in boxes.

☐ ☐ ☐ ☐ ☐

There are 5 choices for the first box.
For each of these 5 choices, there are 4 choices for the second box.
Hence, there are 5×4, or 20, choices for the first two boxes.
For each of these 20 choices, there are 3 choices for the third box.
Hence, there are $5 \times 4 \times 3$, or 60, choices for the first three boxes.
For each of these 60 choices, there are 2 and 1 choices for the remaining two boxes. Hence, the number of permutations is $5 \times 4 \times 3 \times 2 \times 1 = 120$.

Example 2

How many permutations can be formed using all the letters of the word CLARINET?

Solution

There are 8 different letters. The number of permutations is $8 \times 7 \times 6 \times 5 \times 4 \times 3 \times 2 \times 1 = 40\ 320$.

Recall the factorial notation introduced in the Problem Solving pages in Chapters 2 and 5.

Factorial notation

$1! = 1$
$2! = 2 \times 1,\text{or } 2$
$3! = 3 \times 2 \times 1,\text{or } 6$
$4! = 4 \times 3 \times 2 \times 1,\text{or } 24$
$$\vdots$$
$n! = n(n-1)(n-2)(n-3) \times \cdots \times 3 \times 2 \times 1,\text{where } n \geq 1.$

Your calculator should have a factorial key or menu item. For example, to determine 52! on the TI-83 graphing calculator, press 52 [MATH] [▶] [▶] [▶] 4 [ENTER]. The result is the number of ways a deck of 52 cards can be arranged, approximately 8.1×10^{67}.

Example 3

How many 3-letter permutations can be formed from the letters of the word CLARINET?

Solution

There are 8 different letters. Visualize placing the letters in three boxes.

There are 8 choices for the first box, 7 for the second, and 6 for the third. Hence, the number of 3-letter permutations is $8 \times 7 \times 6 = 336$.

In *Example 3*, we found the number of permutations of 8 objects taken 3 at a time. This is denoted by $_8P_3$. Hence,

$$_8P_3 = 8 \times 7 \times 6$$
$$= \frac{8 \times 7 \times 6 \times 5 \times 4 \times 3 \times 2 \times 1}{5 \times 4 \times 3 \times 2 \times 1}$$
$$= \frac{8!}{5!}$$

> The number of permutations of n different objects taken r at a time is:
>
> $$_nP_r = \frac{n!}{(n-r)!}$$

A special case of this formula occurs when $r = n$. The formula becomes $_nP_n = \frac{n!}{0!}$, but 0! has not been defined. An expression such as $_8P_8$ represents the number of permutations of 8 objects taken 8 at a time. In *Example 2*, we found that $_8P_8 = 8!$. Substituting 8 for n in the formula $_nP_n = \frac{n!}{0!}$, we obtain $_8P_8 = \frac{8!}{0!}$.

Comparing these two results, we see that 0! is equal to 1.

For the formula $_nP_r = \frac{n!}{(n-r)!}$ to have meaning when $r = n$, we define 0! = 1.

Use your calculator to verify this definition.

DISCUSSING THE IDEAS

1. In the formula $_nP_r = \frac{n!}{(n-r)!}$, explain why the n objects must be different. Use the examples in this section to illustrate your explanation.

2. For what values of n and r does the formula $_nP_r = \frac{n!}{(n-r)!}$ apply? Explain your answer in two different ways.

3. Explain why the expression for $_nP_r$ is $\frac{n!}{(n-r)!}$ and not $\frac{n!}{r!}$.

4. Explain the following definition of $n!$.
   $$0! = 1$$
   $$n! = n(n-1)(n-2)(n-3) \times \cdots \times 3 \times 2 \times 1, \text{where } n \geq 1.$$

6.2 **EXERCISES**

A **1.** Determine.

   a) 6!   **b)** 7!   **c)** 9!   **d)** $\frac{6!}{5!}$ **e)** $\frac{6!}{4!}$ **f)** $\frac{6!}{3!}$

2. Simplify.

   a) $_1P_1$

   b) $_2P_1$, $_2P_2$

   c) $_3P_1$, $_3P_2$, $_3P_3$

   d) $_4P_1$, $_4P_2$, $_4P_3$, $_4P_4$

   e) $_5P_1$, $_5P_2$, $_5P_3$, $_5P_4$, $_5P_5$

3. Write your answers to exercise 2 in a triangle of numbers similar to the shape at the right. The triangle can be continued indefinitely by adding rows. Find as many patterns in this triangle as you can. Describe each pattern.

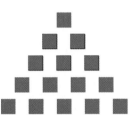

4. How many permutations are there of all the letters in each word?

   a) FRY   **b)** FISH   **c)** FIRST

5. Use the words in exercise 4.

   a) How many 2-letter permutations are there?

   b) How many 3-letter permutations are there?

6. These are the names of towns in western Canada. How many permutations are there of all the letters in each word?

   a) OLDS   **b)** OGEMA   **c)** OLIVER

   d) OAKBURN   **e)** RICHMOND   **f)** OSTERWICK

7. Use the words in exercise 6. How many ways can each number of letters of each word be arranged?

   a) 2   **b)** 3   **c)** 4

8. How many permutations are there of the words in this sentence?

   I DO NOT WANT LUNCH.

9. A bowl contains an apple, a peach, a pear, a banana, an apricot, a plum, and an orange. In how many ways can the fruit be distributed among 7 children?

10. A model train has an engine, a caboose, a tank car, a flat car, a boxcar, a refrigerator car, and a stock car. How many ways can all the cars be arranged between the engine and the caboose?

11. Which of the following expressions are not defined? For each expression you chose, write to explain why it is not defined.

a) $_9P_6$ b) $_6P_6$ c) $_6P_9$ d) $_{-6}P_3$ e) $_6P_{2.5}$ f) $_6P_0$

B 12. How many ways can four different books be arranged on a shelf? Why is it necessary that the books be different?

13. There are 10 different books. How many ways can 4 of these books be arranged on a shelf?

14. Calculate the number of ways that an executive consisting of 4 people (president, vice president, treasurer, and secretary) can be selected from a group of 20 people.

15. Open the *NBA* database. Find the records for the 96–97 season. Sort the data by Position.

a) How many permutations are there of 2 forwards (include forward/centre and guard/forward)?

b) How many permutations are there of 3 guards (include guard/forward)?

16. A map of the 4 western provinces is to be coloured, with a different colour for each province. Nine colours are available. How many different ways are possible?

17. Choose a value of n and r in $_nP_r$. Write a problem for which your value of $_nP_r$ is the solution. Give your problem to a classmate to solve.

18. Solve each equation for n.

a) $_nP_2 = 12$ b) $_nP_2 = 20$ c) $_nP_2 = 30$ d) $_nP_2 = 90$

e) $_nP_3 = 24$ f) $_nP_3 = 60$ g) $_nP_3 = 120$ h) $_nP_3 = 720$

19. Solve each equation for n.

a) $_6P_n = 30$ b) $_6P_n = 120$ c) $_6P_n = 360$ d) $_6P_n = 720$

20. Choose one equation from exercise 18 or 19. Write to explain how you solved the equation.

21. Write each expression without using the factorial symbol.

a) $\dfrac{(n+2)!}{n!}$ b) $\dfrac{(n-3)!}{n!}$ c) $\dfrac{(n+1)!}{(n-1)!}$ d) $\dfrac{(n+4)!}{(n+2)!}$

COMMUNICATING THE IDEAS

Write to explain what is meant by a permutation. Explain a strategy for listing all the permutations of the letters in a word that contains four or more different letters.

INVESTIGATE **Permutations Involving Identical Objects**

1. **a)** How many permutations are there of 4 different letters A, B, C, and D?

   b) Suppose some letters were identical. Would there still be the same number of permutations? Explain.

2. Consider the letters A, A, B, and C.

   a) List all the permutations of these 4 letters. Make sure you list all of them, and that you have not listed any permutation more than once. How many permutations are there?

   b) How does the number of permutations compare with the number of permutations of 4 different letters?

3. Repeat exercise 2 a and b for the letters A, A, A, and B.

4. Repeat exercise 2 a and b for the letters A, A, B, and B.

5. Compare your answers in part b of exercises 2, 3, and 4. Try to develop a general rule to determine the number of permutations of a number of objects when some of them are identical.

6. **a)** Test your rule in exercise 5. Use your rule to predict the number of permutations of each set of 5 letters.

   i) A, B, B, B, C   ii) A, B, B, B, B   iii) A, A, B, B, B

   b) Check your predictions by listing the permutations.

To find a pattern for calculating the number of permutations of objects when some are identical, we will compare the number of permutations of the 4 letters in the words FUEL, FULL, and LULL.

Two identical letters

Consider the permutations of the 4 letters in the word FUEL.

FUEL	FULE	FEUL	FELU	FLUE	FLEU
UFEL	UFLE	UEFL	UELF	ULFE	ULEF
EFUL	EFLU	EUFL	EULF	ELFU	ELUF
LFUE	LFEU	LUFE	LUEF	LEFU	LEUF

There are 4!, or 24 permutations.

If we change the E in FUEL to L, we get the word FULL. If we change each E to L in the list of permutations above, we obtain:

FULL	FULL	FLUL	FLLU	FLUL	FLLU
UFLL	UFLL	ULFL	ULLF	ULFL	ULLF
LFUL	LFLU	LUFL	LULF	LLFU	LLUF
LFUL	LFLU	LUFL	LULF	LLFU	LLUF

Only the permutations shown in colour are different. Visualize any permutation in the first list (such as FUEL) and the corresponding permutation in the second list (FULL, in this case). Interchanging E and L in the first permutation corresponds to interchanging L and L in the second permutation. This gives a different permutation in the first list (FULE, in this case) but the same permutation in the second list. Since there are 2! ways to interchange E and L, the number of different permutations in the second list is $\frac{4!}{2!}$, or 12.

Three identical letters

Consider the permutations of the 4 letters in the word FUEL again.

FUEL	FULE	FEUL	FELU	FLUE	FLEU
UFEL	UFLE	UEFL	UELF	ULFE	ULEF
EFUL	EFLU	EUFL	EULF	ELFU	ELUF
LFUE	LFEU	LUFE	LUEF	LEFU	LEUF

If we change the F and E in FUEL to L, we get the word LULL. If we change the Fs and Es to Ls in the first list of permutations, we obtain:

LULL	LULL	LLUL	LLLU	LLUL	LLLU
ULLL	ULLL	ULLL	ULLL	ULLL	ULLL
LLUL	LLLU	LULL	LULL	LLLU	LLUL
LLUL	LLLU	LULL	LULL	LLLU	LLUL

Only the permutations shown in colour are different. Visualize any permutation in the first list (such as FUEL) and the corresponding permutation in the second list (LULL, in this case). Rearranging F, E, and L in the first permutation corresponds to rearranging L, L, and L in the second permutation. This gives different permutations in the first list (FULE, EUFL, EULF, LUEF, and LUFE in this case) but the same permutation in the second list. Since there are 3! ways to rearrange F, E, and L, the number of different permutations in the second list is $\frac{4!}{3!}$, or 4.

Notice the pattern of these examples. When a letter occurs twice, the number of permutations is $\frac{1}{2!}$ of what it would be if the letters were all different. When a letter occurs 3 times, the number of permutations is $\frac{1}{3!}$ of what it would be if the letters were all different.

The number of permutations of *n* objects taken *n* at a time, if there are *a* alike of one kind, *b* alike of another kind, *c* alike of another kind, and so on, is:

$$\frac{n!}{a!b!c!\dots}$$

Example

Determine the number of permutations of all the letters in each word.
a) PARALLEL   **b)** PARALLELOGRAM

Solution

a) PARALLEL

If the 8 letters were all different, there would be 8! permutations. Since there are 2 As and 3 Ls, the number of permutations is $\frac{8!}{2!3!}$, or 3360.

b) PARALLELOGRAM

There are 13 letters. Since there are 3 As, 2 Rs, and 3 Ls, the number of permutations is $\frac{13!}{3!3!2!}$, or 86 486 400.

DISCUSSING THE IDEAS

1. Explain why there are fewer permutations of a number of objects if some of them are identical than there are if all of them are different.

2. In the expression $\frac{n!}{a!b!c!\dots}$ in the box above, explain why *a*!, *b*!, and *c*! in the denominator are multiplied.

6.3 EXERCISES

A 1. Determine.

    a) $\frac{5!}{2!2!}$ **b)** $\frac{8!}{3!2!}$ **c)** $\frac{12!}{3!3!2!}$ **d)** $\frac{10!}{2!2!4!}$

2. These are names of towns in western Canada. How many permutations are there of all the letters in each name?

    a) OROK   **b)** OXBOW   **c)** OTTHON

    d) OLALLA   **e)** OKOTOKS   **f)** OSOYOOS

B 3. There are two 1s, three 2s, and four 3s. How many 9-digit numbers can be formed?

4. There are 3 blue flags, 3 white flags, and 2 red flags. How many different signals can be constructed by making a vertical display of 8 flags?

5. How many ways can 4 red, 2 blue, and 3 green marbles be distributed among 9 children, if each child is to receive 1 marble?

6. A true-false test has 5 questions. How many answer keys are possible in each situation?

   a) All 5 answers are T.

   b) Four answers are T and 1 answer is F.

   c) Three answers are T and 2 answers are F.

   d) Two answers are T and 3 answers are F.

   e) One answer is T and 4 answers are F.

   f) All 5 answers are F.

7. Add your answers to the six parts of exercise 6. Write to explain why the sum is the same as the answer to exercise 8b on page 357.

8. On a 5-question true-false test, two answers are T and three answers are F. How many different answer keys are possible?

9. On a 10-question multiple-choice test, 3 answers are A, 2 answers are B, 4 answers are C, and 1 answer is D. How many different answer keys are possible? Write to explain how you calculated the number of answer keys.

10. An airline pilot reported her itinerary for 7 days. She spent 1 day in Winnipeg, 1 day in Regina, 2 days in Edmonton, and 3 days in Yellowknife.

    a) How many different itineraries are possible?

    b) What difference would it make if the first day and the last day had been spent in Yellowknife?

    c) Write to explain how you completed parts a and b.

11. Student A wants to visit student B. Roads are shown as lines on a grid. Only south and east travel directions can be used. The trip shown is described by the direction of each part of the trip: E S S E S E

    How many different paths can A take to get to B? Explain.

12. On each grid, how many different paths can A take to get to B? Explain.

a) A

B

b) A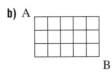

B

c) How many different paths would there be on each size of grid?
  i) 10 by 10   **ii)** x by x **iii)** 8 by 12   **iv)** x by y

13. On each grid, how many different paths are there from A to B?

a) A

B

b) A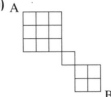

B

 14. Visualize grids made of cubes in 3 dimensions.

a) How many paths are there from A to B if each path must be as short as possible and follow the edges of the grid? Explain.

i) A

B

ii) A

B

iii) A

B

b) How many different paths would there be on each size of grid?
  i) 10 by 10 by 10   **ii)** x by x by x
  iii) 8 by 10 by 12   **iv)** x by y by z

15. How many 7-letter permutations are there of the letters in the word OKANAGAN? Explain.

COMMUNICATING THE IDEAS

Your friend missed today's mathematics lesson and telephoned you about it. How would you explain over the telephone?

a) the method of calculating the number of permutations of objects when some of them are identical

b) the reason why the method works

6.4 Combinations

To play Lotto 649, you must choose numbers from 1 to 49. To count the number of choices, visualize all 49 numbers in a line as shown.

1 2 3 4 5 6 7 8 9 10 11 12 13 14 15 16 17 18 19 20 21 22 23 24 25 26 27 28 29 30 31 32 33 34 35 36 37 38 39 40 41 42 43 44 45 46 47 48 49

N N N Y N N N N N N Y N N N N N N Y N N N N N N N Y N N N N N N N N N N N N N N Y N N Y N N N N N

Denote each number selected by Y (yes) and each number not selected by N (no). For example, the second line above represents the selection 4, 11, 18, 26, 41, and 44. For each selection, there must be 6 Ys and 43 Ns. Hence, the number of possible selections is the number of ways that 6 Ys and 43 Ns can be arranged. When we use the result of Section 6.3, the number of ways is $\frac{49!}{6!43!}$. This expression is represented by the symbol $_{49}C_6$.

Evaluating $_{49}C_6$ using the $_nC_r$ feature

Your calculator may have a $_nC_r$ key or menu item. For example, to determine $_{49}C_6$ on the TI-83 graphing calculator, press:

49 [MATH] [▶] [▶] [▶] 3 6 [ENTER]

Hence, $_{49}C_6 = 13\ 983\ 816$

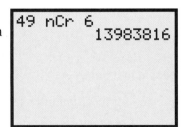

Evaluating $_{49}C_6$ using factorials

$_{49}C_6 = \frac{49!}{6!43!}$

If your calculator does not have a $_nC_r$ key or menu item, it should have a factorial key. Key in: 49 [x!] [÷] 6 [x!] [÷] 43 [x!] [=] to display 13983816.

Evaluating $_{49}C_6$ using arithmetic

$$_{49}C_6 = \frac{49!}{6!43!}$$
$$= \frac{49 \times 48 \times 47 \times 46 \times 45 \times 44}{6 \times 5 \times 4 \times 3 \times 2 \times 1}$$
$$= 13\ 983\ 816$$

There are 13 983 816 ways to select the 6 numbers in Lotto 649.

VISUALIZING	**Number of Selections in Lotto 649**

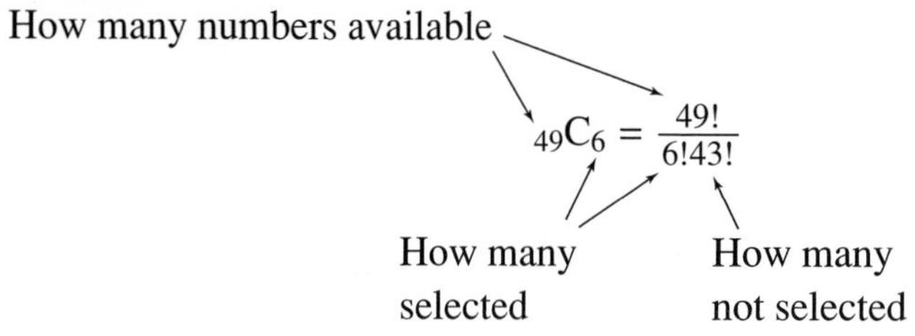

How many numbers available

$$_{49}C_6 = \frac{49!}{6!43!}$$

How many
selected

How many
not selected

The numbers in Lotto 649 are determined by the lottery corporation by jumbling 49 numbered balls in a cage and allowing 6 balls to drop out. The order in which the six numbers appear is not important.

Example 1

How many different committees of 3 people can be selected from 8 people?

Solution

Think...

Visualize the 8 people in a line: A   B   C   D   E   F   G   H

Denote people selected for the committee by Y, those not selected by N.

A typical committee is: N   Y   N   N   N   Y   Y   N

The number of committees is the same as the number of ways of arranging 3 Ys and 5 Ns.

There are 8 people available, where 3 are to be selected and 5 are not selected. The number of committees is $_8C_3$.

$$_8C_3 = \frac{8!}{3!5!}$$
$$= \frac{8 \times 7 \times 6}{3 \times 2 \times 1}$$
$$= 56$$

Hence, 56 different 3-person committees can be selected.

A selection from a group of objects without regard to order is a *combination*.

In the Lotto 649 example, $_{49}C_6$ represents the number of combinations of 49 different objects taken 6 at a time. In *Example 1*, $_8C_3$ represents the number of combinations of 8 different objects taken 3 at a time.

> The number of combinations of n different objects taken r at a time is:
>
> $$_nC_r = \frac{n!}{r!(n-r)!}$$

Example 2

A standard deck of 52 playing cards consists of 4 suits (spades, hearts, diamonds, and clubs) of 13 cards each.

a) How many different 5-card hands can be formed?

b) How many different 5-card red hands can be formed?

c) How many different 5-card hands can be formed containing at least 3 black cards?

Solution

a) The number of combinations of 5 cards chosen from 52 cards is:

$$_{52}C_5 = \frac{52!}{5!47!}$$
$$= 2\ 598\ 960$$

b) There are two red suits (hearts and diamonds) and a total of 26 red cards. The number of combinations of 5 cards chosen from 26 cards is:

$$_{26}C_5 = \frac{26!}{5!21!}$$
$$= 65\ 780$$

c)

> *Think ...*
>
> There could be 3, 4, or 5 black cards. Consider each case separately.

Case 1.   3 black cards and 2 red cards
The black cards can be chosen in $_{26}C_3$ ways, and for each of these ways the red cards can be chosen in $_{26}C_2$ ways.
The total number of combinations is:

$$_{26}C_3 \times _{26}C_2 = \frac{26!}{3!23!} \times \frac{26!}{2!24!}$$
$$= 845\ 000$$

Case 2.   4 black cards and 1 red card

The black cards can be chosen in $_{26}C_4$ ways, and for each of these ways the red cards can be chosen in $_{26}C_1$ ways.

The total number of combinations is:

$$_{26}C_4 \times {}_{26}C_1 = \frac{26!}{4!22!} \times \frac{26!}{1!25!}$$
$$= 388\ 700$$

Case 3.   5 black cards

The black cards can be chosen in $_{26}C_5$ ways, which is $\frac{26!}{5!\ 21!} = 65\ 780$.

The total number of combinations is:

$$845\ 000 + 388\ 700 + 65\ 780 = 1\ 299\ 480$$

The symbol $\binom{n}{r}$ is often used instead of $_nC_r$. The symbol $_nC_r$ is used in this book because it appears on calculator keys or menu items.

DISCUSSING THE IDEAS

1. How many different committees of 5 people can be selected from 8 people? Explain why the answer to this question is the same as the answer to *Example 1*.

2. How did we know that *Example 2c* had to be considered as three cases?

3. For what values of n and r does the formula $_nC_r = \frac{n!}{r!(n-r)!}$ apply? Use examples to illustrate your explanation.

4. Explain how $_nC_r$ is related to $_nP_r$.

6.4 EXERCISES

A 1. Simplify.

   a) $_0C_0$

   b) $_1C_0,\ _1C_1$

   c) $_2C_0,\ _2C_1,\ _2C_2$

   d) $_3C_0,\ _3C_1,\ _3C_2,\ _3C_3$

   e) $_4C_0,\ _4C_1,\ _4C_2,\ _4C_3,\ _4C_4$

   f) $_5C_0,\ _5C_1,\ _5C_2,\ _5C_3,\ _5C_4,\ _5C_5$

2. Write your answers to exercise 1 in a triangle of numbers similar to the shape at the right.

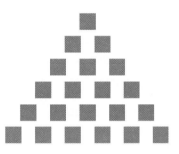

a) Find as many patterns in this triangle as you can. Describe each pattern.

b) Use patterns to write two more rows of the triangle.

3. How many 2-member committees can be chosen from 6 people?

4. a) How many ways can a committee of 6 students be chosen from 10 students?

b) How many ways can a committee of 4 students be chosen from 10 students?

c) Explain why the answers to parts a and b are the same.

5. These are names of towns in western Canada. How many 3-letter combinations can be formed from the letters in each word?

a) OYEN   b) OSAGE   c) OUTRAM   d) OAKBURN

6. Choose one word from exercise 5. List all possible 3-letter combinations that can be formed from the letters in the word.

7. How many different 10-question examinations can be formed from a test bank containing 25 questions?

8. In a Scratch & Win promotion, participants scratch any 3 spots on a card containing 9 spots. The person who has 3 matching spots wins the prize shown under the spots. How many different ways are there to scratch 3 spots?

9. A model train has an engine, a caboose, a tank car, a flat car, a boxcar, a refrigerator car, and a stock car. Different orders of the same cars do not make different trains. How many ways can a train be made with each number of cars between the engine and the caboose?

a) 1   b) 2   c) 3   d) 4   e) 5

10. Simplify, without using the triangle in exercise 2 or a calculator.

a) $_{10}C_0$ b) $_{10}C_1$ c) $_{10}C_2$ d) $_{11}C_2$ e) $_{12}C_2$

f) $_{10}C_3$ g) $_{11}C_3$ h) $_{12}C_3$ i) $_{10}C_4$ j) $_{11}C_4$

11. Write an expression for each number of combinations.

a) $_nC_0$ b) $_nC_1$ c) $_nC_2$ d) $_nC_3$ e) $_nC_4$

B **12.** From a group of 5 student representatives, 3 will be chosen to work on the dance committee.

   a) How many committees are possible?

   b) List all possible committees.

13. Refer to exercise 12. Each committee must have a chairperson. How many committees are possible?

14. Five boys and 5 girls were nominated for a homecoming celebration at a local school. How many ways can a king, a queen, and a court of 2 students be selected from those nominated?

15. A committee consists of 10 people.

   a) How many ways can a subcommittee of 3 people be selected from the committee?

   b) How many ways can an executive subcommittee consisting of 3 people (chairperson, treasurer, and secretary) be selected from the committee?

   c) Write to explain why the answers to parts a and b are different.

16. Open the *Cars* database.

   a) How many different ways can 9 cars be selected from these records?

   b) How many different ways can 3 compact cars be selected from these records?

   c) How many different ways can 6 cars be selected if no more than 1 car is a luxury car?

17. From a deck of 52 cards, how many 5-card hands can be formed in each case?

   a) There are only aces or face cards.

   b) There are only cards numbered 2, 3, 4, 5, 6, 7, 8, 9, and 10.

   c) There are only clubs.

   d) There are only red cards.

18. From a deck of 52 cards, the 12 face cards are removed. From these face cards, 4 are chosen. How many combinations that have at least two red cards are possible?

19. From a deck of 52 cards, how many different 5-card hands can be formed in each case?

   a) with exactly 3 red cards

   b) with at least 3 red cards

   c) with at most 3 red cards

20. To play in the Super 7 lottery, you must choose 7 numbers from 1 to 47. To play in the Lotto 649 lottery, you must choose 6 numbers from 1 to 49. To win each jackpot, the numbers chosen must match the numbers drawn by the lottery corporation.

   a) Without doing any calculations, which do you think is more likely, winning the Super 7 jackpot or winning the Lotto 649 jackpot?

   b) Determine how many times one is more likely than the other.

21. On May 17, 1998, the Powerball Lottery in Oregon had a main jackpot of $195 million U.S. In this lottery, participants choose 5 numbers from 1 to 49 and 1 number from 1 to 42.

   a) How many different ways are there to choose the numbers?

   b) What is the probability that all 6 numbers chosen are drawn in the lottery?

22. Refer to exercises 20 and 21. You might think that there are fewer ways to choose the numbers in the Powerball Lottery than in the Lotto 649 Lottery because the sixth number is chosen from the numbers from 1 to 42 instead of from 1 to 49. Explain why this is not so.

23. A deck of 52 cards is shuffled, and a hand containing x cards is dealt. The graphing calculator screen shows the number of possible hands, y, as a function of x.

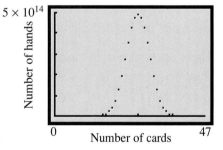

   a) For what value of x does the maximum value of y occur? Explain.

   b) Determine the coordinates of the maximum point. What does the y-coordinate of this point represent?

   c) Explain why the graph is symmetrical about the line $x = 26$.

   d) What is the domain of the function? Explain.

   e) Write the equation of the function.

24. Solve each equation for n.

   a) $_nC_2 = 10$ b) $_nC_3 = 10$ c) $_nC_4 = 35$ d) $_nC_4 = 70$

   e) $_5C_n = 10$ f) $_6C_n = 15$ g) $_6C_n = 20$ h) $_{10}C_n = 120$

25. Eight points are marked on a circle.

   a) How many triangles can be formed using any 3 of the 8 points?

   b) How many line segments can be formed using any 2 of the 8 points?

   c) Suppose the points are joined in order to form an octagon. How many diagonals does the octagon have?

26. Repeat exercise 25 for n points marked on a circle.

C **27.** Although Lotto 649 involves choosing 6 out of 49 numbers, participants can select 7, 8, or 9 numbers according to the following rule printed on the selection form.

> When you select 7, 8, or 9 numbers, you are playing the 7, 28, or 84 six-number combinations of the numbers selected.

Explain how the numbers 7, 28, and 84 are determined.

28. There are 8 boys and 12 girls in a drama club. How many ways can a committee of 5 be selected in each case?

a) There must be at least 2 boys.

b) There must be at least 2 girls.

c) There must be at least 2 boys and 2 girls.

29. In a student council election, there are 3 candidates for president, 3 for secretary, and 2 for treasurer. Each student may vote for at least one position. How many ways can a ballot be marked?

30. How many ways can 3 numbers be chosen from 1, 2, 3, 4, 5, 6, 7, 8, 9, and 10 so that no 2 of the 3 numbers are consecutive?

31. These are names of towns in western Canada. How many 4-letter combinations are there of the letters in each word?

a) ONOWAY   **b)** OSBORNE   **c)** OUTLOOK

32. Recall that a factor of a natural number n is any number that divides n, including 1 and n. How many factors of each number are there?

a) 36   **b)** 360   **c)** 3600

33. In exercise 8, the contest rules state that there are 12 718 300 contest cards. The largest prize is $100 000 cash, and there are two $100 000-prizes.

a) What is the probability of winning one of the two $100 000-prizes?

b) What is the probability that the contest organizers will have to award a $100 000-prize?

COMMUNICATING THE IDEAS

Write to explain the difference between a permutation and a combination. Include some examples to illustrate your explanation.

Analyzing Club Keno

On page 352, you considered these questions about the Club Keno lottery.

- Are the two ways of winning $5000 equally likely, or is one more likely than the other?
- How many numbers would you pick when playing Club Keno?

You can use combinations to develop a mathematical model to calculate the probability of winning any of the prizes in the Club Keno lottery.

 DEVELOP A MODEL

Use the prizes table on page 353. Assume all wagers are $1.

1. One way to win $5000 is to pick 10 numbers and have exactly 9 of them included among the 20 chosen by the central computer system.

   a) Suppose you pick 10 numbers and 9 of them are among those chosen by the computer.
      i) How many ways can 9 of your 10 numbers be chosen?
      ii) There are 70 numbers you did not pick, and the computer chose 11 of them. How many ways can 11 of the 70 numbers you did not pick be chosen?

   b) How many ways can 9 of the 10 numbers you picked and 11 of the 70 numbers you did not pick be among those chosen by the computer?

   c) There are 80 numbers, and the computer chooses 20 of them. How many ways can this be done?

   d) Use the results of parts b and c. Calculate the probability that exactly 9 of your 10 numbers are chosen by the computer. Represent this probability by P(pick 10, match 9).

2. Another way to win $5000 is to pick 7 numbers and have all 7 included among the 20 chosen by the computer.

   a) Suppose you pick 7 numbers and all 7 are among those chosen by the computer.
      i) How many ways can all your 7 numbers be chosen?
      ii) There are 73 numbers you did not pick, and the computer chose 13 of them. How many ways can 13 of the 73 numbers you did not pick be chosen?

   b) How many ways can all the 7 numbers you picked and 13 of the 73 numbers you did not pick be among those chosen by the computer?

c) Calculate the probability that all your 7 numbers are chosen by the computer. Represent this probability by P(pick 7, match 7).

3. Compare your answers to exercises 1d and 2c. Which way of winning $5000 is more likely?

4. Write a formula using combination symbols for P(pick *x*, match *y*).

Your formula in exercise 4 is a mathematical model for calculating the probability of winning any Club Keno prize.

5. Use the model to calculate each probability.

a) P(pick 10, match 8)	**b)** P(pick 8, match 5)
c) P(pick 4, match 2)	**d)** P(pick 2, match 2)

 LOOK AT THE IMPLICATIONS

Recall that if there is a payoff for each event in an experiment, the *expectation* is the average amount you would expect to win if you perform the experiment many times.

6. a) Use your result from exercise 5d. Determine the expectation for a person who always picks 2 numbers when playing Club Keno.

   b) Determine the expectation for a person who picks these numbers.

     i) 1 number   **ii)** 3 numbers   **iii)** 4 numbers

7. a) Start a new spreadsheet document. Enter the headings in rows 1 and 2. Enter the numbers from the prizes table on page 353 in columns A, B, and D. The spreadsheet has been started below.

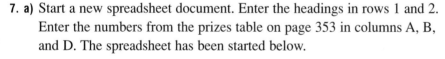

	A	B	C	D	E
1	Pick	Match	Probability	Win $	Expectation $
2	1	1		2	
3	2	2		10	
4	3	3		20	
5	3	2		2	
6	4	4		50	
7	4	3		5	
⋮					
32	10	10		100000	
33	10	9		5000	

b) The answer to exercise 1d, page 379, should have been
$\frac{{}_{10}C_9 \times {}_{70}C_{11}}{{}_{80}C_{20}}$. Since cell C33 corresponds to this situation, enter this
formula in cell C33:

• for Microsoft Excel:
=COMBIN(A33,B33)*COMBIN(80–A33, 20–B33)/COMBIN(80,20)
Copy this formula to cell C2, then copy it down to cell C38. Check the
results against your answers to some of the previous exercises.

c) In cell E2, enter the formula =C2*D2. Copy this formula down to cell E3.
In cell E5, enter the formula =C4*D4+C5*D5.
In cell E8, enter the formula =C6*D6+C7*D7+C8*D8. Enter similar
formulas in cells E11, E15, E20, E25, E31, and E38. Do not enter any
formulas in the other cells in column E.

8. Use your results from exercise 7.

a) To have the greatest expectation, how many numbers should you pick
when playing Club Keno?

b) What reasons might you give for picking a different number of numbers?
Do you think this would be a good idea? Explain.

REVISIT THE SITUATION

9. In some provinces, 70 numbers are used in Keno instead of 80.

a) How do you think the results in column E of the spreadsheet would differ
from those you calculated in exercise 7 (assuming the same prize structure)?

b) Check the answer to part a by modifying the formulas in column C to
account for 70 numbers instead of 80 numbers.

10. In 1994, a person won $620 000 in Montreal by correctly picking 19 out of
20 numbers in an electronic Keno game three times in a row. Calculate the
probability of this win, assuming that the numbers are chosen randomly.
Assume the numbers are chosen from 80 numbers.

11. a) In what ways is Club Keno similar to, yet different from, Lotto 649?

b) In Lotto 649, calculate the probability that each number of numbers you
picked is drawn by the lottery corporation.

i) 5   **ii)** 4   **iii)** 3

6.5 Pascal's Triangle

In exercise 2 on page 375, you wrote some values of $_nC_r$ in a triangular pattern. This triangular array of numbers has intrigued mathematicians for centuries. It was called the "Precious Mirror of the Four Elements" by Shih-Chieh, a Chinese mathematician of the thirteenth century. The triangle, *Pascal's triangle*, is named after the great French mathematician Blaise Pascal (1623–1662) because of his work with the properties of the triangle.

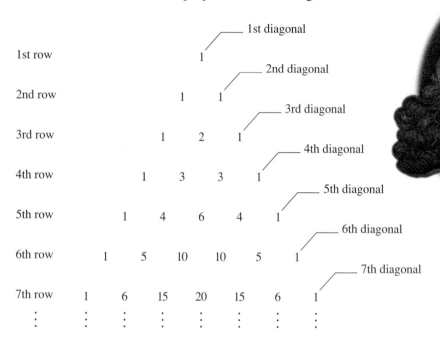

1st row

1st diagonal

2nd row

2nd diagonal

3rd row

3rd diagonal

4th row

4th diagonal

5th row

5th diagonal

6th row

6th diagonal

7th row

7th diagonal

```
1st row                     1
2nd row                  1     1
3rd row               1     2     1
4th row            1     3     3     1
5th row         1     4     6     4     1
6th row      1     5    10    10     5     1
7th row   1     6    15    20    15     6     1
```

In the $(n + 1)$th row, the $(r + 1)$th number is the number of combinations of n objects taken r at a time, $_nC_r = \dfrac{n!}{r!(n-r)!}$. For example, in the 7th row, the 4th number is the number of combinations of 6 objects taken 3 at a time, $_6C_3 = \dfrac{6!}{3!3!}$, or 20.

In exercise 2 on page 375, you found some patterns in Pascal's triangle. Here are two important patterns you probably discovered.

The Symmetrical Pattern

The numbers in each row read the same from left to right or from right to left. For example, in the 7th row, the first 15 is $_6C_2$ and the second 15 is $_6C_4$. We know that $_6C_2 = _6C_4$ because each expression equals 15. We can use the meaning of combinations to explain why $_6C_2 = _6C_4$.

$_6C_2$ represents the number of committees of 2 people that can be selected from 6 people. Visualize the 6 people in a line: A B C D E F. Denote each person selected for the committee by Y and each person not selected by N. For example: N Y N Y N N. The number of committees is the number of ways of arranging 2 Ys and 4 Ns. $_6C_4$ represents the number of committees of 4 people that can be selected from 6 people. For example: Y N Y N Y Y. The number of committees is the number of ways of arranging 2 Ns and 4 Ys. Since the number of ways of arranging 2 Ys and 4 Ns is the same as the number of ways of arranging 2 Ns and 4 Ys, $_6C_2 = {_6C_4}$.

In general, $_nC_r = {_nC_{n-r}}$; the proof is left to the exercises.

The Recursive Pattern

In each row, each number except the first and last is the sum of the two numbers immediately above it. For example, in the 7th row, the second 15 is the sum of the numbers 5 and 10 in the 6th row: $_6C_4 = {_5C_3} + {_5C_4}$. We can use combinations to explain this relationship. $_6C_4$ represents the number of committees of 4 people that can be selected from 6 people: A B C D E F

Either person A is on the committee or person A is not on the committee. If A is on the committee, the other 3 committee members must be chosen from the other 5 people; there are $_5C_3$ ways to do this. If A is not on the committee, all 4 committee members must be chosen from the other 5 people; there are $_5C_4$ ways to do this. Since there are no other possibilities, the number of ways to choose the committee is $_5C_3 + {_5C_4}$. Therefore, $_6C_4 = {_5C_3} + {_5C_4}$.

In general, $_nC_r = {_{n-1}C_{r-1}} + {_{n-1}C_r}$; the proof is left to the exercises.

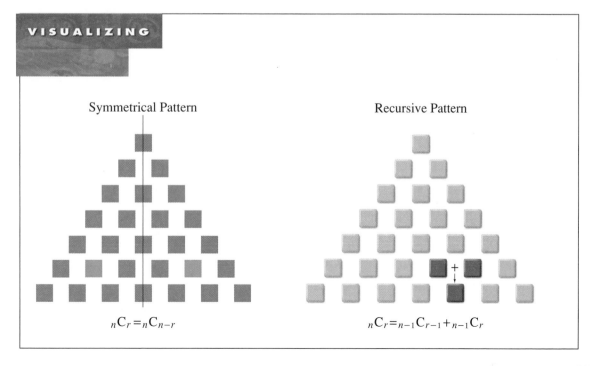

| VISUALIZING |

Symmetrical Pattern

Recursive Pattern

$$_nC_r = {_nC_{n-r}}$$

$$_nC_r = {_{n-1}C_{r-1}} + {_{n-1}C_r}$$

1. Some rows in Pascal's triangle contain an odd number of numbers. Other rows contain an even number of numbers. How can you tell if the number of numbers in a given row is odd or even?

2. In any given row, explain why the numbers increase toward the middle, then decrease toward the end.

3. On page 382, the $(r + 1)$th number in the $(n + 1)$th row is $_nC_r = \dfrac{n!}{r!(n - r)!}$. It might seem awkward to use the $(r + 1)$th number and the $(n + 1)$th row. What changes would be needed to refer to the rth number in the nth row?

4. In exercise 3 on page 364, you wrote some values of $_nP_r$ in a triangular pattern. This triangle is not as useful as Pascal's triangle. Explain why.

6.5 EXERCISES

Refer to Pascal's triangle on page 382 when you complete these exercises.

B **1.** Prove that $_7C_5 = {}_7C_2$ in these two ways.

   a) numerically, by evaluating each expression

   b) by reasoning, using the meaning of combinations

2. Prove that $_nC_r = {}_nC_{n-r}$ in these two ways.

   a) algebraically, by showing that the two expressions are equal

   b) by reasoning, using the meaning of combinations

3. Prove that $_7C_5 = {}_6C_4 + {}_6C_5$ in these two ways.

   a) numerically

   b) by reasoning, using the meaning of combinations

4. Prove that $_nC_r = {}_{n-1}C_{r-1} + {}_{n-1}C_r$ in these two ways.

   a) algebraically

   b) by reasoning, using the meaning of combinations

5. a) What is the second number in the 50th row?

   b) How can you determine the second number in any row? Explain.

6. a) What is the third number in the 50th row?

   b) How can you determine the third number in any row? Explain.

7. Explain in these two ways why the numbers in the second diagonal are the natural numbers.

   a) using the formula for $_nC_r$

   b) using the meaning of combinations

8. a) Add some consecutive numbers in the second diagonal, starting with the first number. Find this sum in the triangle. Explain.

   b) Use Pascal's triangle to determine a formula for the sum of the first n natural numbers.

9. Recall that these natural numbers are the *triangular numbers*.

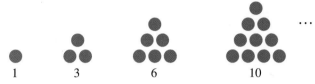

1 3 6 10

   a) Explain why the triangular numbers appear in the third diagonal of Pascal's triangle.

   b) Add some consecutive numbers in the third diagonal, starting with the first number. Find this sum in the triangle. Explain.

   c) Use Pascal's triangle to determine a formula for the sum of the first n triangular numbers.

10. In each pinball situation below, visualize numbering the exits at the bottom from left to right: 1, 2, 3, …

   a) Answer these questions for each situation.

    i) How many different paths are there that reach each exit?

    ii) What is the total number of paths from the top to the bottom?

    iii) What is the probability that the ball reaches each exit?

   b) What assumptions are you making?

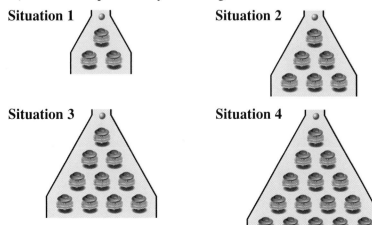

Situation 1   **Situation 2**

Situation 3   **Situation 4**

11. In any pinball situations similar to those in exercise 10, explain why the total number of paths from the top to the bottom is a power of 2.

12. In each pinball situation below, determine the number of different paths a ball could take when it falls from the top to the bottom. Explain.

a)

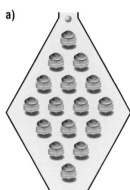

b)

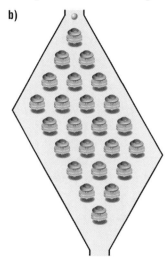

13. On a coordinate grid, visualize starting at $(0, 0)$ and using a pencil to move to any point in the first quadrant according to these rules:

- You must always move along the grid lines, without taking your pencil off the paper.
- You must always move either to the right or up.

a) How many different ways can you move to each point?

i) (1, 0), (0, 1)

ii) (2, 0), (1, 1), (0, 2)

iii) (3, 0), (2, 1), (1, 2), (0, 3)

iv) (4, 0), (3, 1), (2, 2), (1, 3), (0, 4)

b) Explain why the results of part a are the numbers in Pascal's triangle.

14. Refer to exercise 11, page 369. Use combinations to explain the results.

15. This diagram shows a regular dodecagon with its diagonals.

a) How many line segments are on the diagram? Explain.

b) How many of these line segments are diagonals?

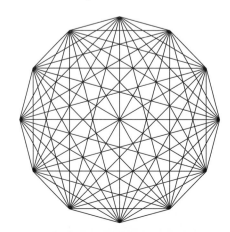

16. How many diagonals are there in a regular polygon with 20 sides? Explain.

17. What is the general formula for the number of diagonals in a regular polygon with n sides? Explain.

C 18. **a)** Add the numbers in each of several rows of Pascal's triangle. What do you notice?

   b) What does the sum of the numbers in the 6th row represent, in terms of choosing items from a set of 5 items?

   c) Use combinations to explain why the sum of the numbers in part b is 2^5.

   d) Explain why the sum of the numbers in the $(n + 1)$th row is 2^n.

19. Visualize a set of four elements A $= \{a, b, c, d\}$. A set such as B $= \{a, c\}$ is called a *subset* of A because every element that is in set B is also in set A. The *empty set*, denoted by ϕ, contains no elements and is considered as a subset of every set.

   a) List all the subsets of set A that contain each number of elements.

      i) 0   **ii)** 1   **iii)** 2   **iv)** 3   **v)** 4

   b) How many subsets are there with each number of elements? Where are these numbers found in Pascal's triangle? Use combinations to explain why these numbers in the triangle represent the number of subsets of set A with each number of elements.

   c) What is the total number of subsets of set A? Explain why this number is correct.

20. What is the total number of subsets in a set with n elements? Explain.

21. Find a formula for the greatest number in the nth row of Pascal's triangle.

COMMUNICATING THE IDEAS

Write to explain what Pascal's triangle is and why it is useful.

The Pigeonhole Principle

A certain problem-solving strategy is based on a very simple observation.

Suppose there are 4 pigeons and only 3 pigeonholes. If all the pigeons are put in the pigeonholes, then one pigeonhole will contain at least 2 pigeons.

Pigeonhole Principle

If $(n + 1)$ pigeons are put in n pigeonholes, then one pigeonhole will contain at least 2 pigeons.

Suppose there are 10 pigeons and only 3 pigeonholes. If all the pigeons are put in the pigeonholes, then one pigeonhole will contain at least 4 pigeons.

Extended Version of the Pigeonhole Principle

If $(kn + 1)$ pigeons are put in n pigeonholes, then one pigeonhole will contain at least $(k + 1)$ pigeons.

Although it is simple, the pigeonhole principle can be used to solve many problems. These problems can vary in difficulty from easy to hard.

In each problem, identify what represents the pigeons and what represents the pigeonholes. Then solve the problem.

1. Determine how many students there must be in your class to be certain that:

    a) at least 2 students have birthdays in the same month

    b) at least 3 students have birthdays in the same month

    c) at least 4 students have birthdays in the same month

2. How many students must there be in your school to be certain that at least two students have birthdays on the same day?

3. In 1998, the population of Canada reached 30 million. Prove that there are at least 81 968 Canadians who have birthdays on the same day.

4. Determine how many cards you must draw from a deck of 52 cards to be certain that you will have:

    a) at least 2 red cards or 2 black cards

    b) at least 2 cards from the same suit

    c) at least 2 cards with the same face value

5. A drawer contains 10 brown socks and 10 blue socks. Suppose you close your eyes and remove socks from the drawer. How many socks must you remove to be certain you will have 2 socks of the same colour?

6. Given any 5 different natural numbers, prove that at least two of them leave the same remainder when divided by 4.

7. Five points are randomly located in an equilateral triangle with sides 2 units long. Prove that at least two of the points are no more than 1 unit apart.

8. Five points are randomly located in a square with sides 2 units long. Prove that at least two of the points are no more than $\sqrt{2}$ units apart.

9. Suppose 6 numbers are chosen from the numbers 1, 2, 3, 4, 5, 6, 7, 8, 9, and 10. Prove that no matter which 6 numbers are chosen, there will always be two so that:

    a) Their difference is 1.　　　　　**b)** One of them is a factor of the other.

10. There are 6 people at a party. Prove that either at least 3 of them are mutual acquaintances or at least 3 of them are mutual strangers.

11. Here are two examples of 6 natural numbers, in increasing order:

    　　4 **5 11 15 17** 29　　　　　　7 10 **22 31 37** 50

    In each case, the numbers in colour are adjacent and have a sum that is divisible by 6. Prove that this always occurs. That is, prove that for any 6 natural numbers written in order, there will always be one number that is divisible by 6 or two or more adjacent numbers that have a sum that is divisible by 6.

Graphing the Functions $y = {}_nC_x$ and $y = {}_xC_r$

In the equation $y = {}_nC_r$ there are two variables on the right side. This equation can be considered as a function by allowing one variable to vary while the other remains fixed. It is convenient to represent the one that varies by x.

Graphing $y = {}_nC_x$

1. a) Predict what the graphs of the functions in the graphing calculator screen would look like. Note that the display Y₁=1 nCr X represents $Y_1 = {}_1C_x$.

```
Plot1  Plot2  Plot3
\Y1=1 nCr X
\Y2=2 nCr X
\Y3=3 nCr X
\Y4=4 nCr X
\Y5=5 nCr X
\Y6=6 nCr X
\Y7=
```

b) Use a graphing calculator to verify your predictions in part a. (Be sure to use an appropriate viewing window.)

c) What is the domain of each function? Explain.

2. Write to explain what the function $y = {}_nC_x$ represents, how its graph changes for different values of n, and how it is related to Pascal's triangle. Include some graphs to illustrate your explanation.

Graphing $y = {}_xC_r$

3. a) Predict what the graphs of the functions in the graphing calculator screen would look like.

```
Plot1  Plot2  Plot3
\Y1=X nCr 1
\Y2=X nCr 2
\Y3=X nCr 3
\Y4=X nCr 4
\Y5=X nCr 5
\Y6=X nCr 6
\Y7=
```

b) Use a graphing calculator to verify your predictions in part a.

c) What is the domain of each function?

4. Write to explain what the function $y = {}_xC_r$ represents, how its graph changes for different values of n, and how it is related to Pascal's triangle. Include some graphs to illustrate your explanation.

Comparing the Functions

5. a) All the functions in exercise 3 are examples of a certain type of function you have studied previously. What type of function is this? Explain.

b) Write the equations of the first four functions in the graphing calculator screen in exercise 3, without using the combination symbol.

6. Are the functions in exercise 1 examples of a certain type of function you have studied previously? Explain.

Patterns in Binomial Powers

You know how to expand the square of a binomial:
$$(x + 1)^2 = x^2 + 2x + 1$$

Since it is tedious to do the calculations by hand, computer software such as *Derive, Theorist*, and *Maple* can be used to expand powers such as $(x + 1)^3$, $(x + 1)^4$, $(x + 1)^5$, and so on. This screen shows a few results.

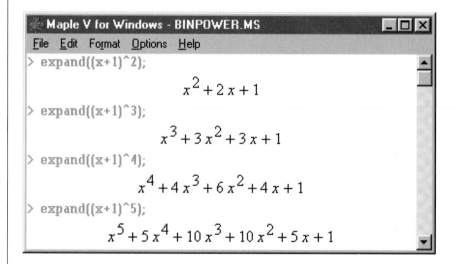

1. Find as many patterns as you can in the results. Describe each pattern.

2. Use the patterns to predict each expansion.

   a) $(x + 1)^6$ **b)** $(x + 1)^7$ **c)** $(x + 1)^8$

3. Use the patterns to write the expansion for each power.

   a) $(a + b)^1$ **b)** $(a + b)^2$ **c)** $(a + b)^3$

   d) $(a + b)^4$ **e)** $(a + b)^5$ **f)** $(a + b)^6$

4. Use the patterns to expand.

   a) $(x + 2)^3$ **b)** $(x + 2)^4$ **c)** $(x + 2)^5$

   d) $(x - 1)^3$ **e)** $(x - 1)^4$ **f)** $(x - 1)^5$

   g) $(x - 3)^3$ **h)** $(x - 3)^4$ **i)** $(x - 3)^5$

5. Use combinations to explain why Pascal's triangle appears in the coefficients of the expansions.

In *Investigate*, you learned that the coefficients of the terms in the expansion of a binomial power such as $(a + b)^4$ are numbers in Pascal's triangle. We can use combinations to explain why this happens.

Recall that $(a + b)^2 = (a + b)(a + b)$ ①

$$= a^2 + ab + ab + b^2$$ ②

$$= a^2 + 2ab + b^2$$

Observe that:

- Each term in ② is the product of 2 factors. In each term, an a or a b is taken from each binomial factor in ①.
- The first term is a^2, which is formed by choosing the a from both binomial factors.
- The second term contains ab, and is formed by choosing the a from one factor and the b from the other factor. Since there are 2 ways to do this, the second term is $2ab$.
- The third term is b^2, which is formed by choosing the b from both binomial factors.

Hence, $(a + b)^2 = a^2 + 2ab + b^2$

The coefficients are the numbers in the 3rd row of Pascal's triangle.

We can use similar reasoning to expand a binomial power such as $(a + b)^4$.

$$(a + b)^4 = (a + b)(a + b)(a + b)(a + b)$$

- Each term is the product of 4 factors. In each term, an a or a b is taken from each binomial factor.
- The first term is a^4. It is formed by choosing the a from each of the 4 binomial factors. There is only one way to do this.
- The second term contains a^3b. It is formed by choosing the b from any binomial factor and the three as from the other 3 binomial factors. The b can be chosen in $_4C_1$ ways, and for each way the three as can be chosen in only one way. Hence, the coefficient of a^3b is 4, and the second term is $4a^3b$.
- The third term contains a^2b^2. There are $_4C_2$ ways to choose 2 bs from the 4 binomial factors and only one way to choose the a from the remaining factors. Hence, the coefficient of a^2b^2 is 6, and the third term is $6a^2b^2$.
- Similarly, the fourth term is $4ab^3$ and the fifth term is b^3.

Therefore, $(a + b)^4 = a^4 + 4a^2b + 6a^2b^2 + 4ab^3 + b^4$

The coefficients are the numbers in the 5th row of Pascal's triangle.

The Binomial Expansion of $(a + b)^4$

$$(a + b)^4 = {}_4C_0a^4 + {}_4C_1a^3b + {}_4C_2a^2b^2 + {}_4C_3ab^3 + {}_4C_4b^4$$

$\uparrow$ $\uparrow$ $\uparrow$ $\uparrow$ $\uparrow$

0bs 1b 2bs 3bs 4bs

Number of ways to choose this many bs from 4 factors $(a + b)$

This reasoning applies to the expansion of any binomial power of the form $(a + b)^n$, where n is a natural number.

Example 1

Expand, then simplify $(2x - 3)^3$.

Solution

Think ...

Use the expansion $(a + b)^3 = {}_3C_0a^3 + {}_3C_1a^2b + {}_3C_2ab^2 + {}_3C_3b^3$, replacing a with $2x$ and b with -3.

$$(2x - 3)^3 = (1)(2x)^3 + (3)(2x)^2(-3) + (3)(2x)(-3)^2 + (1)(-3)^3$$
$$= 8x^3 - 36x^2 + 54x - 27$$

In *Example 1*, we obtained the coefficients of the expansion of $(2x - 3)^3$ using the numbers in the 4th row of Pascal's triangle. In general, we can obtain the coefficients of the expansion of $(a + b)^n$ from the numbers in the $(n + 1)$th row of Pascal's triangle:

$${}_nC_0 \quad {}_nC_1 \quad {}_nC_2 \quad {}_nC_3 \quad \cdots \quad {}_nC_k \quad \cdots \quad {}_nC_{n-2} \quad {}_nC_{n-1} \quad {}_nC_n$$

Observe that ${}_nC_0$ is the 1st number, ${}_nC_1$ the 2nd number, ${}_nC_2$ the 3rd number, and so on. Hence, ${}_nC_k$ is the $(k + 1)$th number.

The expansion of $(a + b)^n$ is the *binomial theorem*.

> ## The Binomial Theorem (using combinations)
> $(a + b)^n = {_nC_0}a^n + {_nC_1}a^{n-1}b + {_nC_2}a^{n-2}b^2 + {_nC_3}a^{n-3}b^3 + \cdots$
>
> $$+ {_nC_k}a^{n-k}b^k + \cdots + {_nC_{n-1}}ab^{n-1} + {_nC_n}b^n$$
>
> n is a whole number.
>
> The general term is the $(k + 1)$th term: $t_{k+1} = {_nC_k}a^{n-k}b^k$

We can use the general term to determine particular terms in a binomial expansion without writing all the terms.

Example 2

Determine the 7th term in the expansion of $(x - 2)^{10}$.

Solution

The general term is $t_{k+1} = {_{10}C_k}(x)^{10-k}(-2)^k$.
To determine t_7, substitute 6 for k.

$$t_7 = {_{10}C_6}(x)^4(-2)^6$$
$$= \frac{10!}{6!4!}x^4(64)$$
$$= \frac{10 \times 9 \times 8 \times 7}{4 \times 3 \times 2 \times 1} \times 64x^6$$
$$= 13\ 440x^6$$

Sometimes only the first few terms of a binomial expansion are required. In this situation, it is useful to express the first few binomial coefficients using algebraic expressions instead of combinations.

Recall from the definition of $_nC_r$, page 373, or exercise 11, page 375, that:

$_nC_0 = 1$

$_nC_1 = n$

$_nC_2 = \frac{n(n-1)}{2!}$

$_nC_3 = \frac{n(n-1)(n-2)}{3!}$

$\vdots$

$_nC_n = 1$

Using these expressions, we can write an alternative form of the binomial theorem.

> **The Binomial Theorem (using algebraic expressions)**
>
> $$(a + b)^n = a^n + na^{n-1}b + \frac{n(n-1)}{2!}a^{n-2}b^2 + \frac{n(n-1)(n-2)}{3!}a^{n-3}b^3 + \cdots + b^n$$
>
> n is a whole number.

Example 3

Write the first four terms of the binomial expansion of $(1 - x)^{20}$.

Solution

Use the binomial theorem above. Substitute $a = 1$, $b = -x$, and $n = 20$.

$$(1 - x)^{20} = (1)^{20} + 20(1)^{19}(-x) + \frac{20 \times 19}{2}(1)^{18}(-x)^2 + \frac{20 \times 19 \times 18}{6}(1)^{17}(-x)^3 + \cdots$$

$$= 1 - 20x + 190x^2 - 1140x^3 + \cdots$$

The binomial theorem is important because it has applications to probability and calculus. Applications to probability will be encountered in Chapter 7.

DISCUSSING THE IDEAS

1. Explain why expressions of the form $(a + b)^n$ are called binomial powers.

2. Explain why we obtain the coefficients in the expansion of $(a + b)^n$ from the numbers in the $(n + 1)$th row of Pascal's triangle and not the nth row.

3. *Visualizing*, page 393, shows how to visualize the coefficients in the expansion of $(a + b)^4$ in terms of counting the number of ways to choose different numbers of bs in the terms of the expansion. Could this have been written in terms of the as instead of the bs? Explain.

4. In *Example 2*, why did we substitute 6 for k in the general term expression to determine the 7th term?

5. Why is the statement "n is a whole number" included as part of the binomial theorem?

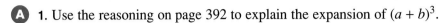

Ⓐ **1.** Use the reasoning on page 392 to explain the expansion of $(a + b)^3$.

2. Expand, using Pascal's triangle.

a) $(x + 1)^5$ b) $(x + 1)^6$ c) $(x - 1)^7$ d) $(x - 1)^8$

e) $(x + y)^5$ f) $(x + y)^6$ g) $(x - y)^7$ h) $(x - y)^8$

3. a) How many terms are there in the expansions of $(x + y)^9$ and $(x + y)^{10}$?

b) Which of the expansions in part a has a middle term?

c) Under what condition does the expansion of $(x + y)^n$ have a middle term?

Ⓑ **4.** Expand, using the binomial theorem.

a) $(x + 2)^3$ b) $(x - 3)^4$ c) $(1 + x)^6$ d) $(2 - x)^5$

e) $(a - 2b)^4$ f) $(2a + 3b)^3$ g) $(3a - 1)^5$ h) $(3a - 2b)^6$

5. Write the first four terms in each expansion.

a) $(c + d)^{10}$ b) $(x + 2)^{12}$ c) $(2 - x)^8$ d) $(1 - 2x)^9$

6. Determine the indicated term in each expansion.

a) the 8th term in the expansion of $(x - 2)^{10}$

b) the 4th term in the expansion of $(x - 5)^{13}$

c) the 10th term in the expansion of $(1 - 2a)^{12}$

d) the 11th term in the expansion of $(x - 2)^{13}$

e) the middle term in the expansion of $(x - 3)^7$

7. Expand and simplify.

a) $(x + y)^6 + (x - y)^6$ **b)** $(x + y)^6 - (x - y)^6$ **c)** $(x^2 - 1)^5$ **d)** $\left(1 - \frac{1}{x}\right)^7$

Ⓒ **8.** Write the binomial theorem using sigma notation.

9. Although the statement "n is a whole number" was included in the display of the binomial theorem using algebraic expressions (page 395), the terms of the expansion can be written for other values of n.

a) Use this form of the binomial theorem. Write expansions for other kinds of binomial powers such as $(1 - x)^{-1}$ and $(1 - x)^{0.5}$.

b) Use particular values of x to determine whether your expansions are meaningful.

COMMUNICATING **THE IDEAS**

Write to explain why combinations are involved in binomial expansions.

1. There are 8 horses in a race. How many possibilities are there for the win, place, and show results?

2. A sports club with 30 members wishes to pick a president, a vice-president, a secretary, and a treasurer. Assume that no person can hold two offices. How many ways can the selections be made?

3. A postal code consists of a letter, a digit, a letter, a digit, a letter, and a digit. The letters D, F, I, O, Q, and U are never used. In addition, W and Z are not used as the first letters of postal codes.

   a) How many different postal codes are possible?

   b) Suppose the post office removed the restrictions on the letters. How many new postal codes would be possible?

4. There are 7 empty seats on a bus and 4 people come on board. How many ways can they be seated?

5. A football team has 6 basic plays. How many arrangements of 3 different plays could be called?

6. How many ways can a Winter Carnival committee of 6 people be selected from 8 boys and 10 girls in each case?

   a) There are no restrictions.

   b) There are exactly 4 boys on the committee.

   c) There are exactly 4 girls on the committee.

7. Visualize joining different numbers of points with line segments in all possible ways. These diagrams show the line segments for 2, 3, 4, 5, and 6 points.

    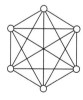

   a) Count the line segments on each diagram and record the results.

   b) Compare the results of part a with Pascal's triangle. Where are these numbers found in the triangle? Use combinations to explain why these numbers in the triangle represent the numbers of line segments on these diagrams.

   c) What is the general formula for the number of line segments when there are *n* points? Prove that your formula is correct.

1. Describe what happens to the graph of a function when you make each change to its equation.

   a) Replace x with $x - 5$.
   b) Replace x with $x + 7$.
   c) Replace y with $y + 4$.
   d) Replace y with $y - 2$.
   e) Replace x with $x + 2$ and y with $y - 1$.
   f) Replace x with $x - 3$ and y with $y + 6$.

2. a) Sketch the function $f(x) = 3(x^2 - 1)$.
   b) Write the equations of the functions $f(2x)$ and $f\left(\frac{1}{3}x\right)$.
   c) On the same axes as in part a, sketch the graphs of the functions $f(2x)$ and $f\left(\frac{1}{3}x\right)$.

3. Write in terms of $\log x$ and $\log y$.

   a) $\log xy^2$
   b) $\log x\sqrt{y}$
   c) $\log 10x^3y^2$
   d) $\log \sqrt[3]{xy^2}$
   e) $\log \frac{x}{\sqrt{y}}$
   f) $\log \frac{x^2}{\sqrt[3]{y}}$

4. When the pH value of the water in a lake falls below 4.7, nearly all species of fish in the lake are deformed or killed. How many times as acidic as clean rainwater, which has a pH of 5.6, is this lake?

5. Determine the sum of the first 10 terms of this geometric series.
   $1 + 3 + 9 + 27 + \cdots$.

6. In exercise 5, which term is 19 683?

7. Write an expression to represent any angle coterminal with each angle θ.

   a) $\theta = 25°$
   b) $\theta = -48°$
   c) $\theta = -\frac{\pi}{4}$ radians
   d) $\theta = \frac{7\pi}{2}$ radians

8. State each exact value.

   a) $\sin \frac{4\pi}{3}$
   b) $\sin \left(-\frac{5\pi}{2}\right)$
   c) $\sin \frac{2\pi}{3}$
   d) $\sin \frac{7\pi}{3}$

9. Repeat exercise 8, replacing each sine with a cosine.

10. State the amplitude, period, phase shift, and vertical displacement for each function.

    a) $y = 3 \sin 2\pi\frac{(t + 4)}{7} - 3$
    b) $y = 4 \cos 2\pi\frac{(t - 1)}{6} + 2$

11. Graph each function over two cycles. For each function, determine the domain, range, period, and the equations of the asymptotes.

    a) $y = \csc 3x$
    b) $y = \csc 3\pi x$

12. On a typical day at an ocean port, the water has a minimum depth of 10.0 m at 1:30 P.M. The maximum depth of 18.0 m occurs 5.9 h later. Assume the relation between the depth of the water and time is a sinusoidal function.

a) What is the period of the function?

b) Write an equation for the depth of the water at any time, t hours.

c) Estimate the depth of the water at 4:00 P.M.

d) Estimate the times when the water is 12.0 m deep.

13. Prove each identity.

a) $\sin x \tan x + \sec x = \dfrac{\sin^2 x + 1}{\cos x}$

b) $\dfrac{1 + \cos x}{1 - \cos x} = \dfrac{1 + \sec x}{\sec x - 1}$

14. A "count" in baseball is an ordered pair indicating the number of balls and strikes on the batter. For example, "2 and 1" means 2 balls and 1 strike. How many different counts are possible?

15. a) How many 3-digit whole numbers have digits that are all different?

b) How many 4-digit whole numbers have digits that are all different?

c) How many 5-digit whole numbers have digits that are all different?

16. Calculators such as the TI-92 and computer programs such as *Maple* can evaluate numerical expressions to hundreds of digits. This screen shows the results when the TI-92 determines 10!, 20!, and 30!.

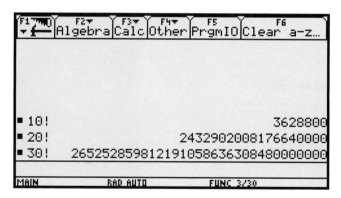

a) Explain why there are 2 zeros at the end of 10!, 4 zeros at the end of 20!, and 7 zeros at the end of 30!.

b) The TI-92 can determine large factorials such as 100!. This is a number with 158 digits. Determine the number of zeros at the end of 100!. Explain.

17. On each diagram in exercise 7, page 397, the points were drawn on a circle and then the circle was removed. Would the results be the same if the points had not been drawn on a circle? Explain.

Interpreting Medical Test Results

CONSIDER THIS SITUATION

Tests for some diseases are often conducted as part of a physical examination. The results from the medical laboratory could take a few days.

- Do you think these tests are always 100% accurate? Explain.

- Suppose you were tested for a serious disease. What would be your reaction in each case?
    a) The test was positive.

    b) The test was negative.

- Is either situation possible? Explain.
    a) You could test positive and yet not have the disease.

    b) You could test negative and yet have the disease.

- Are the preceding two situations equally likely?

On pages 434-435, you will use a mathematical model to explore questions such as these. Your model will involve theoretical probabilities.

 FYI Visit **www.awl.com/canada/school/connections**

For information related to the above problem, click on <u>MATHLINKS</u>, followed by <u>AWMath</u>. Then select a topic under Interpreting Medical Test Results.

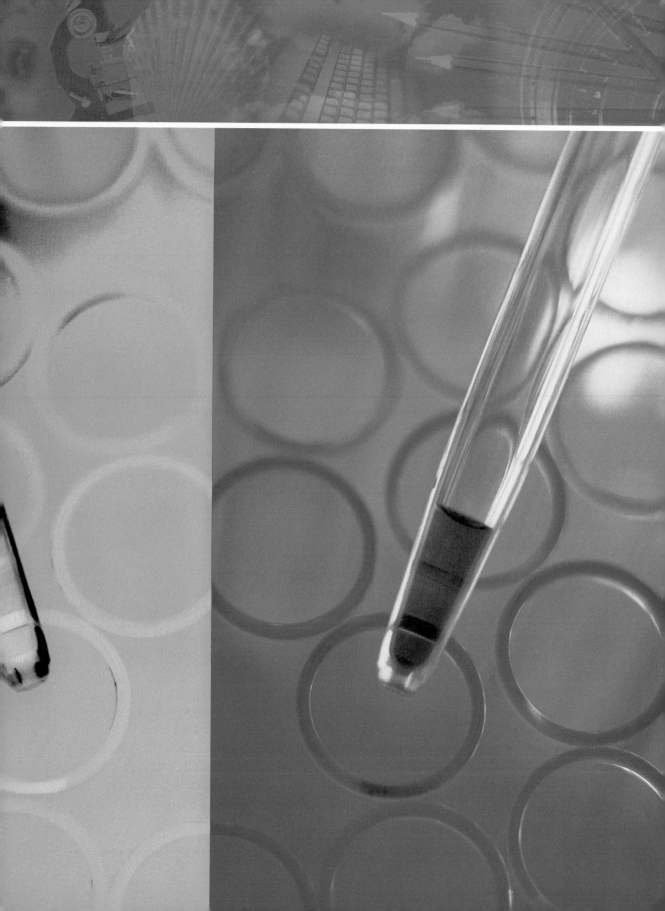

7.1 Experimental and Theoretical Probability

Recall that you have used two different ways to solve equations. To solve an equation such as $x^2 - 7x + 12 = 0$, you used certain rules or algorithms. This is the theoretical method, but not all equations can be solved this way. A different method uses systematic trial or graphing technology. This is the only way to solve an equation such as $2 \sin x = x$.

Probabilities can also be determined in two different ways.

INVESTIGATE

What is the probability that in a family of 3 children there are exactly 2 girls? Make a guess. Record your guess. Now, complete this investigation.

Estimating the probability

1. It is not possible to consider every family with 3 children to determine how many families have 2 girls. However, you can obtain a sample of a few families with 3 children.

   Find out how many students in your class have exactly 2 siblings. How many of these families have exactly 2 girls? Determine the probability of 2 girls out of 3 children by dividing the number of 3-children families with exactly 2 girls by the total number of families with 3 children. How does this answer compare to your guess?

2. Assume that the probability of a boy and the probability of a girl are equal; that is, P(boy) = P(girl) = $\frac{1}{2}$. Use 3 coins to represent the 3 children in a family. Simulate a family of 3 children by tossing the 3 coins. Let heads represent a girl; that is, P(girl) = P(heads) = $\frac{1}{2}$.

   Toss the 3 coins many times. Determine the relative frequency of 2 girls in a family of 3 children. How does this answer compare to your guess?

Calculating the probability

3. Draw a tree diagram to represent the possible arrangements of boys and girls in a family of 3 children. Determine the number of branches that contain 2 girls. Divide that number by the total number of branches to determine the probability that there are 2 girls in the family. How does this answer compare to your guess, and to your estimates in exercises 1 and 2? What assumption are you making about the possible combinations of boys and girls in a family?

In *Investigate*, you considered a probability problem in two different ways. You estimated the probability, using sampling and a simulation. Probabilities determined this way are *experimental probabilities*. You also calculated the probability by dividing the number of combinations of 2 girls and 1 boy by the total number of combinations of boys and girls in a family. Probabilities determined this way are *theoretical probabilities*.

Simple examples of theoretical probability involve tossing coins or rolling dice. For a coin, the theoretical probability of heads is P(heads) = $\frac{1}{2}$. This does not mean that if we toss a coin, say, 200 times there will be exactly 100 heads (we will see later in this chapter that the probability that this occurs is about 0.06). P(heads) = $\frac{1}{2}$ means that if we toss a coin many times, the fraction of times heads occurs gets closer and closer to $\frac{1}{2}$.

You can simulate tossing a coin many times with a TI-83 graphing calculator. To simulate 200 tosses, press [MATH] [►] [►] [►] 7 1 [,] .5 [,] 200 [ )] [ENTER]. A screen similar to that (below left) appears.

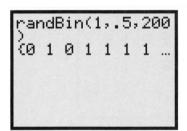
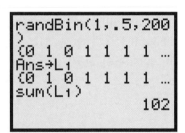

The instruction randBin(1, 0.5, 200) simulates 200 trials, where the probability of success is 0.5. When there is a success, the result is 1; otherwise, the result is 0. Let 1 represent getting heads when tossing a coin. In the screen (above left), heads appeared on the 2nd, 4th, 5th, 6th, and 7th trials. To determine the total number of heads, we store the numbers in memory, then add them. Press [STO►] [L₁] [ENTER] to store the numbers in list L1. To add the numbers, press [LIST] [►] [►] 5 [L₁] [ )] [ENTER]. A screen similar to that (above right) appears. In this simulation, 1 occurred 102 times in 200 trials, so the experimental probability of heads is $\frac{102}{200}$, or 0.51.

Example 1

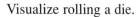

Visualize rolling a die.

a) What is the theoretical probability of getting a 5?

b) Simulate rolling a die 300 times and recording the number of times a 5 appears. From your simulation, what is the experimental probability of getting a 5? Compare this probability with the theoretical probability.

Solution

a) There are 6 equally likely outcomes: 1 2 3 4 5 6
$P(5) = \frac{1}{6}$, or approximately 0.167.

b) Let 1 represent getting a 5 when rolling a die.
To generate the data:
Press [MATH] [▶] [▶] [▶] 7 1 [,] 1 [÷] 6 [,] 300 [)] [ENTER].
To store the data: Press [STO▶] [L₁] [ENTER]
To add the data: Press [LIST] [▶] [▶] 5 [L₁] [)] [ENTER].

For this simulation, 1 appeared 46 times. The experimental probability of getting a 5 is $\frac{46}{300}$, or about 0.153. This is close to the theoretical probability, 0.167.

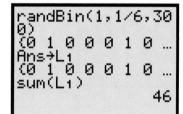

Recall that an *event* is any outcome, or set of outcomes, of an experiment. Rolling a die and getting 5 is an event. Drawing a face card from a deck of 52 cards is another event. Since 12 of the 52 cards in a deck are face cards, $P(\text{face card}) = \frac{12}{52}$, or $\frac{3}{13}$. This is the theoretical probability of drawing a face card. When we repeat the experiment of drawing a card many times (replacing the card each time), the fraction of times we draw a face card should get closer and closer to $\frac{3}{13}$.

When you work with probability, it is inconvenient to use the adjective "theoretical" every time. Unless otherwise stated, the word probability means theoretical probability.

Probability

If an experiment has n equally likely outcomes of which r outcomes are favourable to event A, then the probability of event A is: $P(A) = \frac{r}{n}$.

Example 2

Euchre is played with the 9, 10, jack, queen, king, and ace of each suit. Suppose one card is drawn from those cards. What is the probability that it will be each card?

a) a face card **b)** a red ace

Solution

There are 6 cards in each of 4 suits, or 24 cards.

a) There are 3 face cards in each suit, or 12 face cards.

$$P(\text{face card}) = \frac{12}{24}$$

$$= \frac{1}{2}$$

b) There are 2 red aces.

$$P(\text{red ace}) = \frac{2}{24}$$

$$= \frac{1}{12}$$

DISCUSSING THE IDEAS

1. We frequently see or hear references to probability. Are these references more likely to be theoretical probability or experimental probability? Explain.

2. Your friend thinks that when a coin is tossed 200 times, the probability that there will be 100 heads is $\frac{1}{2}$. How would you convince him that this is not correct?

7.1 EXERCISES

 1. At a traffic light, the red light is on for 30 s, amber for 5 s, and green for 45 s. What is the probability of arriving when the light is red?

2. A roulette wheel consists of 38 numbers: 1 to 36, 0, and 00. What is the probability that a roulette ball will come to rest on an even number other than 0 or 00?

3. A coin showed heads 40 times in 60 tosses.

   a) What is the experimental probability of a head?

   b) What is the theoretical probability of a head?

B **4.** Two coins are tossed. Consider the event "getting 2 heads."

　　a) Determine the experimental probability with 20 tosses of the 2 coins.

　　b) Determine the theoretical probability.

　　c) Compare the results in parts a and b. What do you notice?

5. A coin showed heads 4 times in a row.

　　a) What is the probability of this event?

　　b) Is it more likely, less likely, or just as likely for a tail to appear on the next toss as it was on the first toss? Explain.

MODELLING the Probability of Heads or Tails

When we write P(heads) = $\frac{1}{2}$, we are modelling the experiment of tossing a coin. The model assumes that when a coin is tossed, only two equally likely outcomes can occur.

- Although cylindrical in shape, a coin is not a perfect cylinder. Researchers once analyzed the distribution of mass in the design on the two sides of a coin and concluded that one side was slightly heavier than the other. Which side is slightly more likely to appear when a coin is tossed? Explain.

- Do you think the assumption that there are only two equally likely outcomes is reasonable? Explain.

---

6. A biased die was tossed 600 times. The number 1 occurred 150 times.

　　a) What is the experimental probability of a 1 occurring?

　　b) Estimate the probability of the other outcomes, assuming that they are equally likely.

7. Write to explain how you could use a die to simulate an experiment with each number of equally likely outcomes.

　　a) 2　　　　　　**b)** 3　　　　　　**c)** 4　　　　　　**d)** 5

8. A tetrahedral die with outcomes 1, 2, 3, and 4 is rolled 3 times.

　　a) What is the probability that each roll results in a 1?

　　b) Suppose you want to determine the experimental probability, but you do not have a tetrahedral die. Could you simulate the situation with 2 coins? If you can, explain; if you cannot, suggest an alternative method.

9. A card is drawn from a shuffled deck of 52 cards. What is the probability that it is each card?

　　a) a red card　　**b)** a face card　　**c)** a spade　　**d)** an ace

10. Use the randBin instruction to determine the experimental probability for 300 trials of each event in exercise 9.

11. In the game "In Between," three cards from a 52-card deck are dealt. To win, the value of the third card must be in between the first two.

   a) Determine the probability of winning given the first two cards already dealt.

     i) a 2 and a 6   **ii)** a 5 and a queen

     iii) a 7 and an 8   **iv)** a jack and a king

   b) Identify any assumptions you made in your answers to part a.

12. Open the *Cars* database. For a random selection, calculate the probability of each selection.

   a) a luxury car   **b)** a four-door car   **c)** a noise level below 70 dB

13. Probability theory was formalized by mathematicians 350 years ago to answer a gambler's questions about dice. Antoine Gombauld, also known as the Chevalier de Méré, asked Blaise Pascal this question, "What is more likely: obtaining at least one 6 on four rolls of a die, or obtaining at least one pair of 6s on 24 rolls of two dice?" To answer this question, Pascal and Pierre de Fermat began a systematic study of games of chance, and the theory of probability was founded. You can use a graphing calculator to simulate the two situations and answer the question yourself.

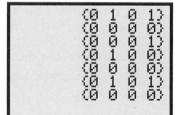

   a) Use a randBin instruction to simulate rolling a die 4 times, and record when at least one 6 appears. Repeat many times to determine the experimental probability of obtaining at least one 6 on four rolls of a die. Your screen will look similar to this.

   b) Simulate rolling 2 dice 24 times, and record when at least one pair of 6s appears. Repeat many times to determine the experimental probability of obtaining one pair of 6s on 24 rolls of 2 dice.

   c) What is your answer to the Chevalier de Méré's question?

14. Simulate tossing two coins 200 times and recording the number of times 2 heads appear. Compare the result with your answers to exercise 4.

COMMUNICATING THE IDEAS

Some people have argued that probability is an *experimental* science rather than a *theoretical* science. What do you think? Write to explain your thinking. Give reasons for your decision.

Simulating Exponential Decay

Visualize this experiment: toss 40 pennies and remove all those that show heads. Toss the remaining pennies and remove all those that show heads. Repeat several times until few pennies are left.

Instead of conducting the experiment, we can simulate it with a graphing calculator. The screen (below left) shows the results of one simulation. The plotted points show the percent of pennies remaining after each step. In the screen (below right), the third column shows the number of pennies remaining after each step. The second column shows these numbers as percents. These percents were used to plot the points on the graph.

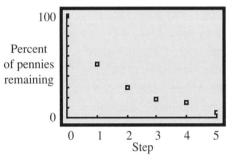

1. Use visual estimation. Check that the plotted points correspond to the data in the first two columns.

2. Visualize repeating the simulation many times, for the same number of pennies. Suppose you record the data for each simulation.

   a) What patterns would you expect to find in these data? Explain.

   b) How would these patterns affect the graph?

These screens show the exponential curve of best fit drawn for the data above, and the data for its equation.

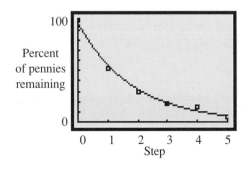

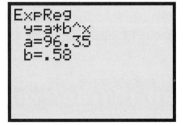

3. a) Write the equation of this curve.

   b) Explain why the curve does not pass through (0, 100).

4. Visualize repeating the simulation many times, for the same number of pennies. Suppose you record the values of *a* and *b* for the equation of the exponential curve of best fit.

a) What patterns would you expect to find in these values?

b) How would these patterns affect the graph?

5. Suppose the simulation was repeated with 100 pennies, 1000 pennies, 10 000 pennies, or even more pennies. What pattern would you expect to find in the results? Explain.

6. The results described above are experimental results because they are based on an experiment, or a simulation of an experiment.

a) What would a graph of theoretical results look like?

b) What is the theoretical equation of the exponential curve of best fit?

7. Recall that each exponential decay situation has a particular half-life.

a) Calculate the half-life for your equation in exercise 6b.

b) Explain the meaning of half-life in this situation.

Visualize conducting a similar experiment with dice. Suppose all dice that show 6 are removed at each step. These screens show the results of one simulation for 40 dice.

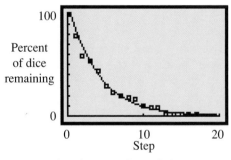
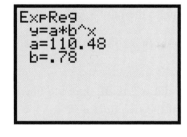

8. a) Write the equation of the exponential curve of best fit above.

b) What is the theoretical equation of the exponential curve of best fit?

9. a) Calculate the half-life for the equation in exercise 8b.

b) Explain the meaning of half-life in this situation.

10. The exponential decay situations above can be used as models for radioactive decay. In what ways are these situations similar to radioactive decay? In what ways are they different from radioactive decay?

To carry out the above simulations, ask your teacher for the programs SIMCOIN and SIMDICE from the Teacher's Resource Book.

Probability and Exponential Functions

7.2 Related Events

INVESTIGATE

Visualize rolling two different dice and recording the results. For example, when the black die shows 5 and the white die shows 2, we can record the result as an ordered pair (5, 2). All possible results can be recorded in a table similar to this.

	White die					
	1	**2**	**3**	**4**	**5**	**6**
1						
2						
3						
4						
5		(5, 2)				
6						

Black die (label at rows 3–4 on the left)

1. Copy and complete the table.

2. Graph the ordered pairs on a grid.

3. Let S, T, and V be these events.
   S: The white die is 2 more than the black die.
   T: The sum of the two dice is 7.
   V: The sum of the two dice is 6.
   For each event below, list the outcomes and state how many there are.
   a) S   **b)** T   **c)** V

   d) S or T   **e)** T or V   **f)** S or V

   g) S and T   **h)** T and V   **i)** S and V

4. Write the probability of each event in exercise 3.

5. Which events in exercise 3 can occur at the same time? Which cannot occur at the same time?

6. Look for patterns or relationships in the results of exercises 3 and 4. Explain any patterns or relationships you find.

In *Investigate*, you considered the experiment of rolling two different dice. You started by recording all the possible outcomes of this experiment. The completed table showing all possible outcomes that can occur when two different dice are rolled is the sample space for rolling two dice.

The list of all possible outcomes of an experiment is the *sample space*.

Example 1

List the sample space for each experiment.
a) rolling a die and tossing a coin

b) recording genders of children in families with 3 children

Solution

a) The sample space is:

    1H   2H   3H   4H   5H   6H   1T   2T   3T   4T   5T   6T

b) The sample space is:

    GGG   GGB   GBG   BGG   GBB   BGB   BBG   BBB

Since the sample space in *Investigate* comprises ordered pairs, we can plot them on a graph.

In *Investigate*, you considered events similar to those below. We can illustrate these events on the graph.

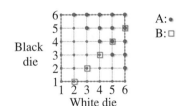

A: The sum of the two dice is greater than 7.
B: The black die is 1 less than the white die.

There are 15 outcomes favourable to A and 5 outcomes favourable to B.

$$P(A) = \frac{15}{36} \qquad P(B) = \frac{5}{36}$$

We can define other events that are related to these events.

The event that A does not occur

The event that A does not occur is the *complement* of event A, and is denoted by $\overline{A}$.

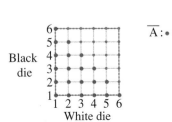

$\overline{A}$: The sum of the two dice is not greater than 7.

There are 21 favourable outcomes.

$$P(\overline{A}) = \frac{21}{36}$$

The event that both A and B occur

The event that both A and B occur is denoted by A and B.

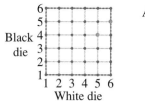

A and B: The sum of the two dice is greater than 7 and the black die is 1 less than the white die.

There are 2 favourable outcomes.
P(A and B) = $\frac{2}{36}$

The event that A occurs or B occurs (or both)

The event that either A or B occurs (or both) is denoted by A or B.

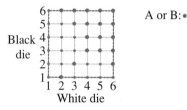

A or B: The sum of the two dice is greater than 7 or the black die is 1 less than the white die.

There are 18 favourable outcomes.
P(A or B) = $\frac{18}{36}$

In the above situations, the probabilities were determined by counting the favourable outcomes and dividing by the total number of outcomes. In this chapter, we will develop methods to calculate the probabilities P($\overline{A}$), P(A and B), and P(A or B) directly when P(A) and P(B) are known.

In any experiment, the outcomes of any event A plus the outcomes of its complementary event $\overline{A}$ form the sample space. When the experiment is conducted, it is certain that either event A or its complement $\overline{A}$ occurs.

For any event A, P(A) + P($\overline{A}$) = 1

We can use this equation to calculate the probability of an event when we know the probability of its complement. In some situations, it may be easier to determine the probability of the complementary event.

Example 2

One card is drawn from a well-shuffled deck of 52 cards. What is the probability of each event?

a) a red face card   **b)** not a red face card

Solution

The sample space comprises the 52 cards. There are 2 red suits, with 3 face cards in each suit; that is, 6 red face cards.

a) $P(\text{red face card}) = \frac{6}{52}$

b) $P(\text{not a red face card}) = 1 - \frac{6}{52}$

$$= \frac{46}{52}$$

In *Example 2*, we could have listed all the cards in the sample space. However, in many situations, it is impractical or impossible to list all the elements in a sample space. Instead, we use a graphical display. The sample space is represented by a rectangle, and the events are represented by ovals or other figures inside the rectangle. Verify that the situations in the diagrams below correspond to those in the four graphs on pages 411 and 412.

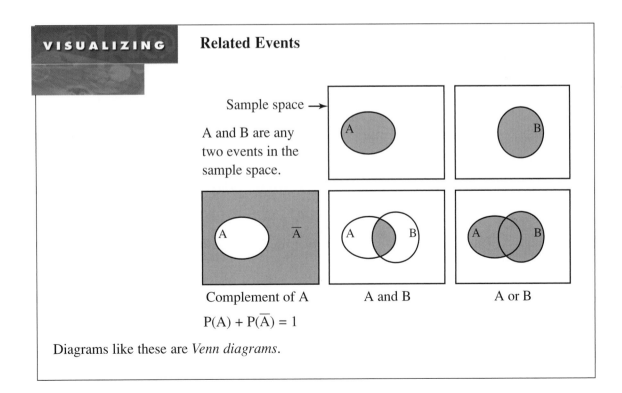

VISUALIZING   **Related Events**

Sample space →

A and B are any two events in the sample space.

Complement of A

$P(A) + P(\overline{A}) = 1$

A and B   A or B

Diagrams like these are *Venn diagrams*.

1. In *Investigate* and the dice examples on pages 411 and 412, would the results change if the two dice had the same colour? Explain.

2. In each part of *Example 1*, how can we be certain that we have listed all the outcomes?

3. How could you use a Venn diagram to explain why $0 \le P(A) \le 1$ for any event A?

4. A certain game cannot end in a draw. Suppose you know the probability of winning the game. How can you determine the probability of losing the game? Explain.

5. **a)** Does P(A and B) = P(B and A)? Explain.

   b) Does P(A or B) = P(B or A)? Explain.

7.2 EXERCISES

A 1. List the sample space for the roll of a cubical die and a tetrahedral die.

2. List the sample space for the toss of a coin and the choice of a random digit from 0 to 9.

3. Suppose the probability of precipitation is 0.4. What is its complement? What does it mean?

4. Suppose the probability of passing the driving test on the first try is 0.1. What is the probability of failing on the first try?

5. An inspector at an egg farm routinely inspects a dozen eggs from each consignment. The inspector records the number of broken eggs as part of the quality control process.

   a) Describe the outcomes in the event "at least 3 eggs are broken."

   b) Describe the outcomes in the event "no more than 2 eggs are broken."

   c) Describe the complement of the event "at most 1 egg is broken."

B 6. Two dice are rolled. Use a sample space to determine the probability of each event.

   a) The difference of the two dice is at least 4.

   b) The sum of the two dice is less than 4.

   c) The difference is at least 4 or the sum is less than 4.

   d) The difference is at least 4 and the sum is less than 4.

7. Two dice are rolled. Determine the probability of each event.

   a) The sum is an even number.   **b)** The sum is either 6 or 12.

   c) A double is rolled.   **d)** The sum is at least 9.

8. a) List the sample space for each experiment.

   i) tossing 1 coin   **ii)** tossing 2 coins

   iii) tossing 3 coins   **iv)** tossing 4 coins

b) Write to explain your method for completing part a.

9. A student says that in 3 tosses of a coin she can get at least 2 heads.

a) List the sample space for this experiment.

b) What is the complement of the event "at least 2 heads"?

c) What are the probabilities of the event and its complement?

10. Repeat exercise 9, but replace "at least 2 heads" with "at most 2 heads."

11. One card is drawn from a well-shuffled deck of 52 cards.

a) Describe the sample space.

b) Use the sample space to determine the probability of each event.

   i) A: getting an ace

   ii) B: getting a black card

   iii) H: getting a heart

c) Determine each probability.

   i) $P(\overline{H})$ **ii)** P(A or B)   **iii)** P(A and H)   **iv)** P(B or H)

12. About 350 years ago, the Chevalier de Méré asked Blaise Pascal this question, "What is the probability of throwing two dice and *not* getting a 1 or a 6?" Use a sample space to answer this question.

13. At the Juice Bar, a "custom triple" is a blend of any 3 fruit juices from apple (A), banana (B), cherry (C), orange (O), or papaya (P). Use the letter associated with each fruit.

a) List or describe the sample space.

b) List the outcomes for the event X: the blend contains no apple juice.

c) List the outcomes for the event $\overline{X}$.

d) Determine P(X) and $P(\overline{X})$.

14. A single letter is randomly selected from the word SASKATCHEWAN.

a) List the sample space.

b) Determine the probability of each event.

   i) X: The letter chosen is a consonant.

   ii) Y: The letter chosen is S.

   iii) Z: The letter chosen is A.

c) Determine each probability.

   i) $P(\overline{X})$ **ii)** P(X and Y)   **iii)** P(X and Z)   **vi)** P(X or Z)

15. Open the *Population and Food Production* database. Use the data for 1992.

    a) List the sample space for the random selection of a region.

    b) What is the probability of choosing Oceania?

    c) Suppose you choose Oceania. List the sample space for the random selection of a country.

    d) For part c, what is the probability of choosing Vanuatu?

16. A certain experiment has only 4 outcomes: $O_1, O_2, O_3,$ and O_4. The probability of each outcome is twice the probability of the preceding outcome. Determine the probability of each outcome.

C **17.** Anita leaves work and arrives at the subway station at a random time between 5:00 P.M. and 6:00 P.M. inclusive. Her mother lives west and her grandmother lives east of this subway station. Anita takes whichever train arrives first to have dinner with either her mother or her grandmother. The eastbound train arrives every 10 min starting on the hour. The westbound train arrives every 10 min starting at one minute past the hour. Anita's arrival time in minutes can be expressed as an ordered pair. The first number is the time until the arrival of the eastbound train. The second number is the time until the arrival of the westbound train. For example, if Anita arrives at 5:00, the ordered pair is (0, 1); if she arrives at 5:01, the ordered pair is (9, 0); and so on.

    a) List the sample space for Anita's arrival time.

    b) How many points are in the sample space?

    c) Which is more likely: Anita will eat with her mother or Anita will eat with her grandmother? Explain.

18. Human blood is classified according to two main antigen systems—the ABO system and the Rh system. When the A or B antigens are present the blood type contains the letter(s). When neither A nor B is present the letter O is used. Possible blood types are A, B, AB, and O. A positive or negative sign indicates the Rh antigen.

For example: A+ means antigen A and the Rh antigen are present;
AB− means antigen A and B antigens are present but Rh antigen is absent;
O+ means A and B antigens are absent but the Rh antigen is present;
O− means that all three antigens are absent.
Determine the sample space for the different blood types.

COMMUNICATING THE IDEAS

Write to explain the advantages and disadvantages of listing a sample space for determining probabilities. Include some examples to illustrate your explanation.

Non-Transitive Dice

Most dice games are games of chance, and winning depends more on luck than on strategy. However, this game is different. Determine if a winning strategy exists.

You and a friend each roll one of four cubical dice; whoever rolls the higher number wins. The dice have these numbers on their faces:

A: 1 1 1 5 5 5   B: 2 2 2 2 6 6   C: 3 3 3 3 3 3   D: 4 4 4 4 0 0

1. Suppose your friend chooses die A and you choose die B. Both of you roll the dice at the same time.

   a) Copy and complete this table. For each possible outcome, write W if you win and L if you lose.

   b) Which die is more likely to win? What is the probability of winning for this die?

   Die B

Die A	2	2	2	2	6	6
1						
1						
1						
5						
5						
5						

2. Now, your friend uses die B and you use die C.

   a) Repeat exercise 1 for die B and die C. Write the numbers on your die along the top. Write the numbers on your friend's die down the side.

   b) Which die is more likely to win? What is the probability of winning for this die?

3. Now, your friend uses die C and you use die D. Repeat exercise 2 for die C and die D.

4. Finally, your friend uses die D and you use die A. Repeat exercise 2 for die D and die A.

In arithmetic, it is self-evident that if $x = y$ and $y = z$, then $x = z$. We say that the equality relation is *transitive*.

5. **a)** Give another example of a transitive relation in arithmetic.

   b) Give an example of a non-transitive relation in arithmetic.

   c) Give an example from geometry of each relation.
   i) a transitive relation   **ii)** a non-transitive relation

6. Explain why the title of this page is appropriate.

A Three-Game Series

Suk-Yee is a promising chess player.
As encouragement, Suk-Yee's mother offers
a prize. Suk-Yee must win at least 2 consecutive
games in a series of 3 games to be played
alternately against her mother and her teacher.
Suk-Yee can choose either the series "m–t–m."
(mother–teacher–mother) or the series "t–m–t."
(teacher–mother–teacher). Suk-Yee's teacher
is a better player than her mother. Which series
should Suk-Yee choose?

Here are some approaches to solving this problem.

Using intuition

1. Just by guessing, which series do you think Suk-Yee is more likely
to win? Explain.

Simulating the series with dice

2. a) Use dice to simulate the outcomes of the chess games in the two
series. Repeat several times to determine the experimental
probability that Suk-Yee wins each series.

b) Which series should Suk-Yee choose? Does your answer agree
with your guess in exercise 1?

Simulating the series with a graphing calculator

Ask your teacher for the programs MTM and TMT from
the Teacher's Resource Book. Enter the programs in your
TI-83 calculator. Each program simulates 100 three-game
series, for a total of 300 games. You are prompted to enter
the probabilities of winning against the mother and against
the teacher. Enter two decimal numbers between 0 and 1,
ensuring that the probability of winning against the mother
is the greater of the two. The program generates numbers
at random between 0 and 1 and compares them with the
probabilities entered. The screens at the right show a
typical result for the series "t–m–t." The probabilities in
this example were P(win against mother) = P(m) = 0.8
and P(win against teacher) = P(t) = 0.45.

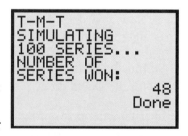

Press [STAT] [ENTER] to see the simulated results for each series. Each row in the first three columns corresponds to one series of three games, with a win indicated by 1. In any row, if there is a 1 in both L1 and L2, or in both L2 and L3, the program enters a 1 in list L4. The sum of the numbers in list L4 is the number of series won.

3. a) Choose a value for P(m) and a smaller value for P(t). Run the MTM program several times using these probabilities, and record the number of series won. Determine the mean of the results. This represents the experimental probability that Suk-Yee will win the "m–t–m" series.

  b) Repeat part a, using the program TMT with the same probabilities. Determine the probability that Suk-Yee will win the "t–m–t" series.

  c) Which series should Suk-Yee choose? Does your answer agree with your guess in exercise 1?

Analyzing the series numerically

4. a) List the ways Suk-Yee could win the prize in either series.

  b) Choose a value for P(m) and a smaller value for P(t). Use these probabilities to calculate the probability that Suk-Yee wins each series.

  c) Which series should Suk-Yee choose? Does your answer agree with your guess in exercise 1?

5. By now, you should have discovered the surprising result that Suk-Yee should choose the seemingly tougher series "t–m–t." Look at your calculations in exercise 4. Explain why Suk-Yee is more likely to win this series than the "m–t–m" series.

Analyzing the series algebraically

6. a) Use $P(m) = m$ and $P(t) = t$, where $m > t$, to determine the probability that Suk-Yee wins each series.

  b) Use the results of part a to explain why Suk-Yee is more likely to win the "t–m–t" series.

7. Your results in exercise 6a represent the theoretical probabilities that Suk-Yee will win each series. How do the probabilities you determined in exercises 2, 3, and 4 compare with these theoretical probabilities?

Looking back

8. Chess games can end in a draw. Did you allow for draws in your solutions to the above exercises? If so, explain how. If not, how could you modify your results?

9. Explain how the above results would be affected if Suk-Yee played all three games against the same person.

7.3 The Event A or B

In Section 7.2, we determined the probabilities P(A and B) and P(A or B) by counting outcomes in the sample space. Since this method can be inefficient, other methods to calculate these probabilities are needed. In this section, we will develop a method to calculate P(A or B) — the probability that event A occurs or event B occurs, or both.

Consider two situations involving drawing a card from a shuffled deck of 52 cards.

Situation 1 Drawing a spade or a red card

Let S and R represent these events.
S: The card is a spade.
R: The card is red.

Consider the event S or R: the card is a spade or is red.
Since all the 13 spades are different from the 26 red cards, there are 13 + 26, or 39 cards that are spades or red. Hence, P(S or R) = $\frac{39}{52}$. This is the sum of the probabilities of S and R. That is, P(S or R) = P(S) + P(R)

We say the events S and R are mutually exclusive because they cannot occur at the same time. Since they cannot occur at the same time, P(S and R) = 0.

Two events A and B that cannot occur at the same time are *mutually exclusive* events. They have no common outcomes.

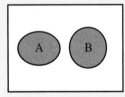
P(A or B) = P(A) + P(B)

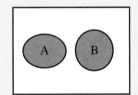
P(A and B) = 0

Situation 2 Drawing a spade or a face card

Let S and F represent these events.
S: The card is a spade.
F: The card is a face card.

Consider the event S or F: the card is a spade or a face card.
There are 13 spades and 12 face cards, but 3 face cards are spades. When we add 13 and 12, we count these 3 cards twice. So, there are 13 + 12 − 3, or 22 cards that are spades or face cards. Hence, P(S or F) = $\frac{22}{52}$.

This is the sum of the probabilities of S and F, less the probability that the card is both a spade and a face card.

$$P(S \text{ or } F) = P(S) + P(F) - P(S \text{ and } F)$$
$$= \frac{13}{52} + \frac{12}{52} - \frac{3}{52}$$
$$= \frac{22}{52}$$

The events S and F are not mutually exclusive because they have some common outcomes.

Two events A and B that are not mutually exclusive have some common outcomes.

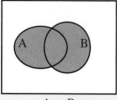

A or B   A and B

$$P(A \text{ or } B) = P(A) + P(B) - P(A \text{ and } B)$$

Example 1

There are 30 students in a class. Sixteen students surf the Internet and 10 students use e-mail. Of these students, 6 students do both. What is the probability that a randomly selected student in the class surfs the Internet or uses e-mail?

Solution

Suppose a student is randomly selected. Let S and E represent these events.
S: The student surfs the Internet.
E: The student uses e-mail.

The events S and E are not mutually exclusive.
$$P(S \text{ or } E) = P(S) + P(E) - P(S \text{ and } E)$$
$$= \frac{16}{30} + \frac{10}{30} - \frac{6}{30}$$
$$= \frac{20}{30}, \text{ or } \frac{2}{3}$$

The probability that a randomly selected student surfs the Internet or uses e-mail is $\frac{2}{3}$.

Example 2

Two dice are rolled. Determine the probability that the sum of the numbers on the dice is an even number or a prime number.

Solution

Construct a sample space to show the possible sums.

Second die

	1	2	3	4	5	6
1	2	3	4	5	6	7
2	3	4	5	6	7	8
3	4	5	6	7	8	9
4	5	6	7	8	9	10
5	6	7	8	9	10	11
6	7	8	9	10	11	12

First die (labels for rows)

Of the 36 outcomes, 18 are even and 15 are prime numbers. However, one prime is even. Therefore,

P(sum is even or sum is prime)

$$= P(\text{sum is even}) + P(\text{sum is prime}) - P(\text{sum is even and prime})$$

$$= \frac{18}{36} + \frac{15}{36} - \frac{1}{36}$$

$$= \frac{32}{36}$$

The probability of an even number or a prime number is $\frac{32}{36}$.

DISCUSSING THE IDEAS

1. Suppose A and B are two mutually exclusive events. Explain how we know that P(A or B) = P(A) + P(B) and P(A and B) = 0.

2. Is the converse of each statement true? Explain.

   a) If A and B are mutually exclusive, then P(A or B) = P(A) + P(B).

   b) If A and B are mutually exclusive, then P(A and B) = 0.

3. Suppose you obtained a number greater than 1 when you calculated a certain probability. What should you suspect? Explain.

4. Explain how you could use a Venn diagram to solve each example.

   a) *Example 1*   b) *Example 2*

A 1. Use the sample space in *Example 2*. When two dice are rolled, determine the probability of each event.

   a) The sum of the numbers is less than or equal to 7.

   b) The sum of the numbers is less than 5 or greater than 9.

   c) The sum of the numbers is less than 9 or greater than 5.

   d) The sum of the numbers is a perfect square or an even number.

   e) The sum of the numbers is a prime number or a 2-digit number.

2. In a survey, 42% of households contacted owned a home computer and 13% owned a home entertainment system. Eight percent of all households contacted owned both. What is the probability that a randomly selected household will own neither?

3. A market study found that 50% of a neighbourhood like Japanese food while 60% like Italian food. Thirty percent like both. What is the probability that a randomly selected resident will like Japanese food but not Italian food?

4. A TV station determined that 30% of its teenage viewers watch sports programs and 60% watch the soaps. Twenty percent watch neither. What is the probability that a randomly selected teenage viewer watches both?

5. A newspaper publisher determines, from its marketing survey, that 64.3% of its readers are interested in the business section and 37.5% of its readers are interested in the sports section. Fifteen percent are interested in neither section. What is the probability that a randomly selected reader would be interested in both sections?

6. Let A represent any event in an experiment. Determine each probability and explain what it represents.

   a) $P(A \text{ or } \overline{A})$ b) $P(A \text{ and } \overline{A})$

B 7. A card is drawn from a shuffled deck of 52 cards. Determine the probability of each event.

   a) The card is a heart or a spade.   b) The card is an ace or a face card.

   c) The card is a heart or a face card.   d) The card is an ace or a spade.

8. A computer store advertised its annual half-price sale in the newspaper and on TV. A survey of 200 customers indicated that 60 learned about the sale from the newspaper, 50 from TV, and 30 from both sources. What is the probability of each event?

   a) A randomly selected customer saw the advertisement in the newspaper and on TV.

b) A randomly selected customer saw the advertisement in at least one of the two media.

c) A randomly selected customer saw the advertisement in exactly one form.

9. Open the *NBA* database. Find the records for the 1994–1995 season.

a) How many players are listed?

b) How many players play more than one position?

c) Suppose a player is randomly selected. What is the probability of choosing a guard (include guard/forward)?

d) Suppose a player is randomly selected. What is the probability of choosing a centre or a forward (include forward/centre and guard/forward)?

10. A study of hand-eye coordination tested people on how quickly they could respond to a moving object on a screen. Thirty percent of the people responded in less than 0.3 s; 60% in 0.5 s or less; and 5% took more than 0.8 s. What is the probability of each event?

a) A randomly selected person from this group will take 0.8 s or less.

b) A randomly selected person from this group will take longer than 0.5 s.

c) A randomly selected person from this group will take between 0.3 s and 0.5 s inclusive.

C 11. In the Venn diagram, what regions are included in each section?

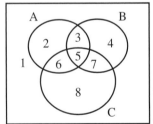

a) A and B

b) A and C

c) B and C

d) A or B

e) B or C

f) A or C

g) A and B and C

h) A or B or C

12. Develop a formula for P(A or B or C) similar to this formula.
P(A or B) = P(A) + P(B) − P(A and B)

7.4 The Event A and B

In this section we will develop a method to calculate P(A and B) — the probability that event A occurs on a first trial and event B occurs on a second trial. When we determine the probability of event B, we must always consider that event A has occurred. We use the symbol P(B | A). This is the *conditional probability* of B given A.

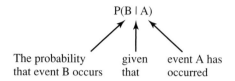

Note that P(B | A) may or may not be the same as P(B), depending on whether the occurrence of event A affects the probability of event B.

Consider two situations involving a deck of cards.

Situation 1 Drawing two cards from a deck, without replacement

A card is drawn from a shuffled deck of 52 cards, and not replaced. Then a second card is drawn. Let A and B represent these events.

A: The first card is a spade.
B: The second card is a spade.

P(A and B) represents the probability that both cards are spades.

There are 52 possibilities for the first card but, since it is not replaced, there are only 51 possibilities for the second card. The sample space comprises $52 \times 51 = 2652$ outcomes. If the first card is a spade, only 12 spades are left when the second card is drawn. The number of outcomes that are both spades is $13 \times 12 = 156$. Hence, there are 2652 outcomes, and 156 are favourable.

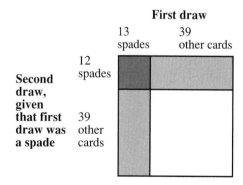

Therefore, P(A and B) = $\frac{156}{2652}$

Observe that P(A) = $\frac{13}{52}$ and P(B | A) = $\frac{12}{51}$, and $\frac{13}{52} \times \frac{12}{51} = \frac{156}{2652}$.

This example suggests that P(A and B) = P(A) × P(B | A). The events A and B are *dependent events* because the probability that event B occurs depends on the occurrence of event A.

Situation 2 Drawing two cards from a deck, with replacement

A card is drawn from a shuffled deck of 52 cards, then replaced.

Then a second card is drawn. Let A and B represent these events.

A: The first card is a spade.

B: The second card is a spade.

P(A and B) represents the probability that both cards are spades.

There are 52 possibilities for the first card and 52 for the second. The sample space comprises $52 \times 52 = 2704$ outcomes. The number that are both spades is $13 \times 13 = 169$. There are 2704 outcomes, and 169 are favourable.

Therefore, $P(A \text{ and } B) = \frac{169}{2704}$

Observe that $P(A) = \frac{13}{52}$ and $P(B \mid A) = \frac{13}{52}$,

and $\frac{13}{52} \times \frac{13}{52} = \frac{169}{2704}$.

First draw

13 spades   39 other cards

Second draw

13 spades

39 other cards

This example also suggests that $P(A \text{ and } B) = P(A) \times P(B \mid A)$. However, in this case, $P(B \mid A)$ is the same as $P(B)$. The events A and B are *independent events* because the probability that event B occurs does not depend on the occurrence of event A.

> For any events A and B, the calculation of P(A and B) must consider whether event A has occurred.
>
> $$P(A \text{ and } B) = P(A) \times P(B \mid A)$$
>
> - If the occurrence of event A affects the probability that event B occurs, then the events A and B are dependent.
> - If the occurrence of event A does not affect the probability that event B occurs, then the events A and B are independent. For independent events, $P(B \mid A) = P(B)$, and $P(A \text{ and } B) = P(A) \times P(B)$.

Example 1

Two cards are drawn without replacement from a shuffled deck of 52 cards. Let A and B represent these events.

A: The first card is a face card.

B: The second card is a face card.

Determine the probability of each event.

a) P(A and B)   **b)** $P(\overline{A} \text{ and } B)$

Solution

a) $P(A \text{ and } B) = P(A) \times P(B \mid A)$

$P(A)$ is the probability of drawing a face card from 52 cards, so $P(A) = \frac{12}{52}$.

$P(B \mid A)$ is the probability of drawing a face card from 51 cards, given that a face card has already been drawn. Hence, $P(B \mid A) = \frac{11}{51}$

$$
\begin{aligned}
P(A \text{ and } B) &= P(A) \times P(B \mid A) \\
&= \frac{12}{52} \times \frac{11}{51} \\
&= \frac{132}{2652}
\end{aligned}
$$

b) $P(\overline{A} \text{ and } B) = P(\overline{A}) \times P(B \mid \overline{A})$

$P(\overline{A})$ is the probability of not drawing a face card from 52 cards.

$$
\begin{aligned}
P(\overline{A}) &= 1 - P(A) \\
&= \frac{40}{52}
\end{aligned}
$$

$P(B \mid \overline{A})$ is the probability of drawing a face card from 51 cards, given that a face card has not already been drawn. Hence, $P(B \mid \overline{A}) = \frac{12}{51}$

$$
\begin{aligned}
P(\overline{A} \text{ and } B) &= P(\overline{A}) \times P(B \mid \overline{A}) \\
&= \frac{40}{52} \times \frac{12}{51} \\
&= \frac{480}{2652}
\end{aligned}
$$

Example 2

In one bag there are 2 white balls and 3 yellow balls. In a second bag there are 2 green balls and 1 orange ball. One ball is drawn from each bag. What is the probability of drawing 1 white ball and 1 green ball?

Solution

Let event A be the drawing of 1 white ball from the first bag.
Let event B be the drawing of 1 green ball from the second bag.
The probability of drawing 1 green ball does not depend on whether 1 white ball has been drawn. Hence, the events A and B are independent.

$$
\begin{aligned}
P(A \text{ and } B) &= P(A) \times P(B \mid A) \\
&= P(A) \times P(B) \\
&= \frac{2}{5} \times \frac{2}{3} \\
&= \frac{4}{15}
\end{aligned}
$$

The probability of drawing 1 white ball and 1 green ball is $\frac{4}{15}$.

Example 3

Determine the probability of obtaining at least one 6 on 4 rolls of a die.

Solution

> **Think ...**
>
> The successive rolls of a die are independent. The outcome of obtaining at least one 6 includes obtaining two, three, or four 6s; that is, all outcomes except "obtaining no 6s." So, we can use the complement.

Let A represent the event that a 6 is rolled, so $\overline{A}$ is the event that a 6 is not rolled. Then, $P(A) = \frac{1}{6}$ and $P(\overline{A}) = 1 - \frac{1}{6}$, or $\frac{5}{6}$

$P(\text{no 6 on 4 rolls}) = \frac{5}{6} \times \frac{5}{6} \times \frac{5}{6} \times \frac{5}{6}$

$$= \left(\frac{5}{6}\right)^4$$

$P(\text{at least one 6 on 4 rolls}) = 1 - \left(\frac{5}{6}\right)^4$

$$\doteq 0.518$$

The probability of obtaining at least one 6 on 4 rolls of a die is approximately 0.518.

Recall from Section 7.3 that mutually exclusive events cannot occur at the same time. When events occur at the same time, they can be either independent or dependent.

DISCUSSING THE IDEAS

1. In *Example 1*, are events A and B dependent or independent? Explain.

2. Consider *Example 2*.
    a) Explain why it is important that the white balls are only in the first bag and the green balls are only in the second bag.
    b) Would it make a difference if there were also some white balls in the second bag? Explain.

3. In *Example 3*, how would the probability of obtaining at least one 6 change if there were more than 4 rolls or fewer than 4 rolls of the die? Explain.

4. Is it possible for mutually exclusive events to be independent? Explain.

7.4 EXERCISES

A 1. Determine if the events in each part are dependent or independent. Explain.

   a) drawing a queen from a deck of cards then drawing another queen, when the experiment is carried out without replacement

   b) drawing a king for the first card and a jack for the second card, when the experiment is carried out with replacement

   c) tossing a coin and getting a head, then rolling a die and getting 3

   d) rolling a die and getting 6, then rolling it again and getting 1

   e) tossing a coin 10 times and getting a head every time

   f) winning at a solitaire game twice in a row

2. Thirty percent of seniors catch the flu every year. Fifty percent of seniors have yearly flu shots. Ten percent of seniors who have had flu shots get the flu. Are getting a flu shot and getting the flu independent events?

3. Forty percent of grade 12 students enrol in a tutorial program. Fifty percent of grade 12 students find the academic demands tougher than in grade 11. Twenty percent of grade 12 students are in the tutorial program and feel that way. Are these events independent: being enrolled in a tutorial program; and finding the academic demands tougher?

4. A survey of smokers found that of 1000 women, 200 were heavy smokers and 20 had emphysema. Of those who had emphysema, 13 were heavy smokers. Are being heavy smokers and having emphysema independent?

5. A bag contains 3 white balls and 2 black balls. The first draw produced a white ball. The first ball is not replaced. What is the probability that the second draw will produce another white ball?

6. A golf bag contains 6 white balls and 8 yellow balls. What is the probability of each event?

   a) drawing 3 white balls

   b) drawing 1 yellow ball and 1 white ball

7. In one bag, there are 2 white balls and 3 red balls. In a second bag, there are 4 white balls and 5 green balls. One ball is drawn from each bag.

   a) What is the probability of drawing each pair?

   i) 1 red ball and 1 green ball   **ii)** 1 red ball and 1 white ball

   iii) 1 white ball and 1 green ball   **iv)** 2 white balls

   b) What is the sum of the probabilities in part a? Explain.

8. A card is dealt from a deck of 52 cards. It is a spade. A second card is dealt. What is the probability that the second card is a spade?

9. Three cards are drawn without replacement from a shuffled deck of 52 cards. What is the probability of each event?

a) drawing 3 hearts **b)** drawing 3 jacks

c) drawing 3 red cards **d)** drawing 3 face cards

10. The probability that you are late for your carpool ride is $\frac{7}{10}$. The probability that the carpool ride arrives late is $\frac{1}{10}$. What is the probability that you are late and so is the carpool ride?

11. The probability that a student completes her mathematics assignment is $\frac{3}{5}$. The probability that she completes her French homework is $\frac{2}{5}$. Assume these events are independent. What is the probability of each event?

a) The student has both assignments completed.

b) The student has only one assignment completed.

c) The student has neither assignment completed.

B **12.** A biased coin with P(heads) = 0.7 was tossed 3 times. The coin came up heads first, then tails twice. What is the probability of this outcome?

13. In 5 tosses of a coin, the first two tosses resulted in 2 heads. What is the probability that the 5 tosses will produce 3 heads?

14. What is the probability that a randomly selected person was born on the same day of the week as you and is of the same gender?

15. Refer to *Example 1*.

a) Determine the probability of each event.

   i) P(A and $\overline{B}$) **ii)** P($\overline{A}$ and $\overline{B}$)

b) What is the sum of the probabilities P(A and B), P($\overline{A}$ and B), P(A and $\overline{B}$), and P($\overline{A}$ and $\overline{B}$)? Explain.

16. Two cards are drawn from a shuffled deck of 52 cards without replacement. Determine the probability of each event.

a) The cards are both spades.

b) Neither card is a spade.

c) Exactly one of the two cards is a spade.

17. Repeat exercise 16, replacing spade with each card.

a) a black card **b)** a face card **c)** an ace

18. Visualize rolling a die several times. These screens show how the probabilities that there are no 6s and that there is at least one 6 change as the number of rolls increases.

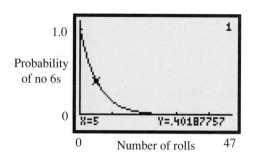

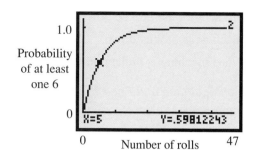

Probability of no 6s
0   1.0
Number of rolls   0   47
X=5   Y=.40187757

Probability of at least one 6
0   1.0
Number of rolls   0   47
X=5   Y=.59812243

a) Check that the results on the screens are the same as the results of *Example 3*.

b) Calculate to determine the probabilities of no 6s and at least one 6 on 6 rolls of a die.

19. The graphs in exercise 18 are graphs of two functions.

a) Write the equation of each function, where n is the number of rolls of the die.

b) For each function, state the domain and the equation of any asymptote.

c) Describe the range of each function.

20. Construct the graphs in exercise 18, then use them to complete this exercise.

a) Determine the probability that there will be no 6s on 10 rolls of a die.

b) Determine the least number of rolls so the probability that there will be at least one 6 is greater than 0.99.

21. To "cut" a deck of cards means to randomly divide it in two parts. Visualize cutting a deck of cards n times. Define the function $f(n)$ to represent the probability that there are no aces as the top card after n cuts.

a) Write the equation of the function.

b) Sketch the graph of the function. Describe how the graph is similar to, and different from, the first graph in exercise 18.

22. The probability that your new calculator battery will last one year is 0.7 and that it will last two years is 0.2. At the end of the first year, what is the probability that the battery will last until the end of the second year?

23. Teachers who use the office copiers feel that the probability that machine X is broken down is $\frac{1}{3}$, machine Y is $\frac{1}{2}$, and machine Z is $\frac{1}{6}$. What is the probability of each event?

a) All 3 machines are broken down.

b) No machine is broken down.

c) Only one machine is broken down.

24. Open the *Population and Food Production* database. Find the records for 1992. Suppose you select a country at random.

a) What is the probability of choosing a European country?

b) Suppose the first country selected is from North and Central America. You randomly select a second country from the remaining countries. What is the probability that you select the second country from Asia?

c) Suppose the first country selected is from Africa. You randomly select a second country from the remaining countries. What is the probability that you select the second country from Africa?

25. Open the *Population and Food Production* database. Write questions similar to those in exercise 24. Answer your questions. Give your questions to a classmate to answer. Check your answers against your classmate's answers.

26. Suppose there are 5 white golf balls, 3 yellow golf balls, and 4 orange golf balls in a pocket of a golf bag. Three balls are randomly drawn, and not replaced. What is the probability of each event?

a) drawing 3 white golf balls

b) drawing 3 yellow golf balls

c) drawing 3 orange golf balls

27. Use the pocket of golf balls in exercise 26. What is the probability of each event?

a) drawing 2 white golf balls and 1 yellow golf ball

b) drawing 2 yellow golf balls and 1 orange golf ball

c) drawing 2 orange golf balls and 1 white golf ball

d) drawing 2 yellow golf balls and 1 white golf ball

e) drawing one golf ball of each colour

28. Two soccer teams, A and B, have a kickoff to see which team wins the game. The teams take turns to attempt to score a goal. The first team to score wins. The probability of team A winning with 1 kick is 0.20. The probability of team B winning with 1 kick is 0.25.

a) Suppose team A kicks first. What is the probability that team B wins on its first kick?

b) Suppose team A kicks first. What is the probability that team A wins on its third kick?

c) Suppose team A kicks first. What is the probability that team A wins?

d) Suppose team B kicks first. What is the probability that team B wins?

C 29. Recall exercise 13, page 407, concerning the question the Chevalier de Méré asked Pascal 350 years ago. *Example 3* answers the first part of the question.

a) Calculate the probability of obtaining at least one pair of 6s on 24 rolls of two dice.

b) What is the answer to the question posed by the Chevalier de Méré?

30. In one bag, there are 2 white balls and 3 red balls. In a second bag, there are 4 white balls and 5 red balls. One ball is drawn from each bag. What is the probability of each event?

a) drawing 2 white balls

b) drawing 2 red balls

c) drawing 1 white ball and 1 red ball

COMMUNICATING THE IDEAS

Write to explain the differences between independent and dependent events. Include an explanation of how these are different from mutually exclusive events. Use examples from commonly available data such as sports scores and statistics.

Interpreting Medical Test Results

On page 400, we considered medical test results reported as positive or negative. We will develop a mathematical model to solve this problem:

Suppose a test for cancer is known to be 98% accurate. This means that the outcome of the test is correct 98% of the time. Suppose 0.5% of the population have cancer. What is the probability that a person who tests positive for cancer has cancer?

DEVELOP A MODEL

Suppose 1 000 000 randomly selected people are tested. There are four possibilities.

- A person with cancer tests positive.
- A person with cancer tests negative.
- A person without cancer tests positive.
- A person without cancer tests negative.

You can solve the problem by using the given information to determine the number of people for each possibility.

1. a) How many of the people tested have cancer?

   b) How many do not have cancer?

2. Assume the test is 98% accurate when the result is positive.

   a) How many people with cancer will test positive?

   b) How many people with cancer will test negative?

3. Assume the test is 98% accurate when the result is negative.

   a) How many people without cancer will test negative?

   b) How many people without cancer will test positive?

LOOK AT THE IMPLICATIONS

You can use the above results to solve the problem above, and other similar problems.

4. a) How many people tested positive for cancer?

   b) How many of these people have cancer?

   c) What is the probability that a person who tests positive for cancer has cancer?

REVISIT THE SITUATION

Use a spreadsheet to model this situation. Input the data and results from exercises 1 to 4.

5. a) Start a new spreadsheet document. Enter the text and data shown.

	A	B	C	D	E
1	P(cancer)	0.005	Accuracy:	0.98	
2	People with cancer				
3	People without cancer				
4					
5	People with cancer who test positive				
6	People with cancer who test negative				
7	People without cancer who test negative				
8	People without cancer who test positive				
9					
10	Total of people who test positive				
11	Total of people who test negative				
12					
13	P(person testing postive has cancer)				
14	P(person testing postive has no cancer)				
15	P(person testing negative has cancer)				
16	P(person testing negative has no cancer)				

b) Assume there are 1 000 000 people. Enter appropriate formulas in the shaded cells. Check the results against your answers to exercises 1 to 4.

c) Investigate how the probabilities change in each case.
   i) The test is less than 98% accurate.
   ii) The test is more than 98% accurate, including 100% accurate.
   iii) More than 0.5% of the population have cancer.
   iv) Less than 0.5% of the population have cancer.

6. Determine the probability that a person who tests negative for cancer has cancer.

7. In exercises 2 and 3, we assumed that 98% of the positive results were correct and that 98% of the negative results were correct. Make up some numbers to show that the 98% of the overall results can be correct although the percents of correct positive and negative results are not equal.

7.5 | Problems Involving Conditional Probability

This section presents some additional problems involving dependent events and conditional probability.

Example 1

A card is drawn from a shuffled deck of 52 cards, and not replaced. Then a second card is drawn. What is the probability that the second card is a face card?

Solution

> **Think ...**
>
> The first card may or may not be a face card. There will be a different probability calculation for the second card in each case.

Let A be the event that the first card is a face card.

Then $P(A) = \frac{12}{52}$

$$P(\overline{A}) = 1 - P(A)$$
$$= 1 - \frac{12}{52}$$
$$= \frac{40}{52}$$

Let B be the event that the second card is a face card.

If the first card was a face card, then $P(B \mid A) = \frac{11}{51}$.

If the first card was not a face card, then $P(B \mid \overline{A}) = \frac{12}{51}$.

For event B to occur, then either of these two events must occur.

A and B: The first card and the second card are face cards.

$\overline{A}$ and B: The first card is not a face card, and the second is a face card.

These events are mutually exclusive. So,

$$P(B) = P(A \text{ and } B) + P(\overline{A} \text{ and } B)$$
$$= P(A) \times P(B \mid A) + P(\overline{A}) \times P(B \mid \overline{A})$$
$$= \frac{12}{52} \times \frac{11}{51} + \frac{40}{52} \times \frac{12}{51}$$
$$= \frac{612}{2652}$$
$$\doteq 0.231$$

The relation in *Example 1* applies in many situations.

Relating Event B to Events A and $\overline{A}$

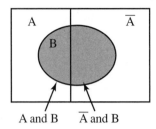

For any two events A and B: $P(B) = P(A \text{ and } B) + P(\overline{A} \text{ and } B)$

or: $P(B) = P(A) \times P(B \mid A) + P(\overline{A}) \times P(B \mid \overline{A})$

Example 2

A company has two factories that make computer chips. Suppose 70% of the chips come from factory 1 and 30% come from factory 2. In factory 1, 25% of the chips are defective; in factory 2, 10% of the chips are defective.

a) Suppose it is not known from which factory a chip came. What is the probability that the chip is defective?

b) Suppose a defective chip is discovered. What is the probability that the chip came from factory 1?

Solution

Represent these events.

A: A chip came from factory 1.
$\overline{A}$:   A chip came from factory 2.
B: A chip is defective.

a)   Determine P(B).
Relate event B to events A and $\overline{A}$.

$$P(B) = P(A \text{ and } B) + P(\overline{A} \text{ and } B)$$
$$= P(A) \times P(B \mid A) + P(\overline{A}) \times P(B \mid \overline{A})$$
$$= (0.70)(0.25) + (0.30)(0.10)$$
$$= 0.205$$

The probability that the chip is defective is about 20%.

b)

> ***Think ...***
>
> We need to determine the probability that the chip came from factory 1, given that it is defective; that is, $P(A \mid B)$. Compared to the conditional probabilities in part a, the positions of A and B are reversed. There is only one way to proceed — go back to the point where the conditional probability symbol was introduced.

Determine $P(A \mid B)$ from the conditional probabilities in part a.

Use the expression for $P(A$ and $B)$ on page 426 twice, once for $P(A$ and $B)$ and again for $P(B$ and $A)$.

$P(A$ and $B) = P(A) \times P(B \mid A)$

$P(B$ and $A) = P(B) \times P(A \mid B)$

The probabilities on the left sides are equal. So,

$$P(B) \times P(A \mid B) = P(A) \times P(B \mid A)$$

$$P(A \mid B) = \frac{P(A) \times P(B \mid A)}{P(B)}$$

$$= \frac{(0.70)(0.25)}{0.205}$$

$$\doteq 0.854$$

The probability that the defective chip came from factory 1 is about 85%.

In the solution to *Example 2b*, the formula $P(A \mid B) = \frac{P(A) \times P(B \mid A)}{P(B)}$ appeared. This formula was developed by Thomas Bayes in the 18th century, and is known as *Bayes' Law.*

DISCUSSING THE IDEAS

1. In the solution to *Example 1*, why are the events "A and B" and "$\overline{A}$ and B" mutually exclusive? Why is it important that they are mutually exclusive?

2. Compare the answer to *Example 1* with the answers to the two parts of *Example 1* in Section 7.4. What do you notice? Explain.

3. In the solution to *Example 2b*, how do we know that $P(A$ and $B) = P(B$ and $A)$?

4. In *Example 2*, describe two different ways to determine the probability that a defective chip came from factory 2.

5. What strategy was used to develop Bayes' Law? Would it be easier to remember the formula or the strategy that was used in the solution to *Example 2b*? Explain.

B 1. Two cards are drawn without replacement from a shuffled deck of 52 cards. What is the probability that the second card is each card?

   a) an ace   b) a club   c) a red card

   d) a red face card   e) a black jack   f) the queen of hearts

2. Two cards are drawn without replacement from a shuffled deck of 52 cards. Determine the probability of each event.

   a) The first card is a heart and the second card is the queen of hearts.

   b) The first card is the queen of hearts and the second card is a heart.

3. Two cards are drawn without replacement from a shuffled deck of 52 cards.

   a) Determine P(A and B), where A and B are these events.

      A: The first card is a spade.

      B: The second card is a face card.

   b) Determine P(A and B), where A and B are these events.

      A: The first card is a face card.

      B: The second card is a spade.

   c) Explain why the answers to parts a and b are the same.

4. How would the answer to *Example 2* change in each case? Explain.

   a) Fewer chips come from factory 1 and more chips come from factory 2.

   b) The percent of defective chips from factory 1 increases.

   c) The percent of defective chips from factory 2 increases.

5. There are 100 boys and 120 girls in the grade 12 year. Twenty boys and 30 girls have no siblings. A student is randomly selected.

   a) What is the probability that the student has no siblings?

   b) A student is chosen who has no siblings. What is the probability that the student is a girl?

6. This table shows the distribution of the Canadian population in four regions of Canada. The percent of people in each region who live in urban areas is also shown. Calculate the probability that a randomly selected Canadian lives in an urban area.

Region	Percent of population	Percent urban
Western Canada	30.1	74.8
Ontario	37.4	80.7
Quebec	24.5	77.0
Maritime Provinces	8.0	51.7

7. One bag contains 4 white balls and 6 black balls. Another bag contains 8 white balls and 2 black balls. A coin is tossed to select a bag, then a ball is randomly selected from that bag.

a) What is the probability that a white ball will be drawn?

b) Suppose a white ball was drawn. What is the probability that it came from the first bag?

8. In exercise 7, suppose there is a 70% chance that the first bag is selected. How would your answers to parts a and b change?

9. Five dimes and five nickels are arranged in 3 boxes as shown. A box is randomly selected, then a coin is randomly selected from that box. What is the probability that this coin will be a nickel?

10. In exercise 9, there are 10 coins altogether, and one-half of these are nickels.

a) Explain why the probability that the randomly selected coin is a nickel is not $\frac{1}{2}$.

b) Find a way to arrange the coins in the boxes so that the probability is $\frac{1}{2}$.

c) Assume there must be at least one coin in each box. Find a way to arrange the coins in each case.

  i) The probability is as large as possible.

  ii) The probability is as small as possible.

C **11.** Use the method of *Example 2* to solve the problem on page 434.

12. A new medical test for glaucoma is 95% accurate. Suppose 0.8% of the population have glaucoma. What is the probability of each event?

a) A randomly selected person will test negative.

b) A person who tests negative has glaucoma.

c) A person who tests positive does not have glaucoma.

COMMUNICATING THE IDEAS

Write to explain how the probabilities P(A | B) and P(B | A) are related. Include an example of a situation in which P(A | B) ≠ P(B | A) and another example of a situation in which P(A | B) = P(B | A).

7.6 Using Permutations and Combinations to Calculate Probabilities

Recall that problems involving probability arose in the study of permutations and combinations in Chapter 6. Permutations and combinations are useful for solving certain kinds of probability problems. For example, suppose two cards are drawn without replacement from a shuffled deck of 52 cards. We can calculate the probability that they are both aces in two different ways.

Using conditional probability

Visualize drawing the cards one at a time.

P(both cards are aces) = P(1st card is ace) $\times$ P(2nd card is ace | 1st card is ace)

$$= \frac{4}{52} \times \frac{3}{51}$$
$$\doteq 0.004\ 525$$

Using combinations

The sample space is the list of all possible ways to select 2 cards from the 52 cards. There are $_{52}C_2$ ways. There are $_4C_2$ ways to select the 2 aces. Hence, there are $_{52}C_2$ ways to select 2 cards, and $_4C_2$ of these are 2 aces.

$$\text{P(both cards are aces)} = \frac{_4C_2}{_{52}C_2}$$
$$= \frac{6}{1326}$$
$$\doteq 0.004\ 525$$

For some problems, you may find that it is more convenient to use combinations instead of conditional probability.

Example 1

A committee has 7 women and 5 men. A subcommittee of 4 is to be randomly selected. What is the probability of each event?

a) There are 4 women on the subcommittee.

b) There are exactly 3 women on the subcommittee.

c) There are exactly 2 women on the subcommittee.

Solution

a) The number of ways to select 4 women from 7 women is $_7C_4$.

The number of ways to select 4 people from 12 people is $_{12}C_4$.

The probability that there are 4 women on the subcommittee is:

$$P(4 \text{ women}) = \frac{_7C_4}{_{12}C_4}$$
$$= \frac{35}{495}$$
$$\doteq 0.070\ 707$$

b) The number of ways to select 3 women from 7 women is $_7C_3$.

The number of ways to select 1 man from 5 men is $_5C_1$.

Hence, the number of ways to select a subcommittee with exactly 3 women is $_7C_3 \times _5C_1$.

$$P(3 \text{ women}) = \frac{_7C_3 \times _5C_1}{_{12}C_4}$$
$$= \frac{175}{495}$$
$$\doteq 0.353\ 535$$

c) Similarly, there are $_7C_2$ ways to choose 2 women and $_5C_2$ ways to choose 2 men.

$$P(2 \text{ women}) = \frac{_7C_2 \times _5C_2}{_{12}C_4}$$
$$= \frac{210}{495}$$
$$\doteq 0.424\ 242$$

Look at the patterns in the solution to *Example 1*. In each part, the denominators are 495. This is the number of items in the sample space — the list of all possible committees. If we calculate the probabilities that other numbers of women are on the subcommittee, we will get similar results. We can write an expression for the probability that exactly x women are on the subcommittee.

$$P(x \text{ women}) = \frac{_7C_x \times _5C_{4-x}}{_{12}C_4}$$

This expression is a function of x, and we can write it using function notation as $f(x) = \frac{_7C_x \times _5C_{4-x}}{_{12}C_4}$. The domain of the function consists only of the numbers 0, 1, 2, 3, and 4, because these are the possible numbers of women on the subcommittee.

We can use the TI-83 graphing calculator to produce a table of values and the graph of the function. Enter this expression in the Y= list:

Y1= (7 nCr X) (5 nCr (4 − X)) / 12 nCr 4

Press TBLSET, and make sure that TblStart = 0, $\triangle$Tbl = 1, and that Auto is highlighted in the last two rows. Press TABLE to obtain the table of values in the first screen on the next page. You can see the probabilities for any number of women on the subcommittee.

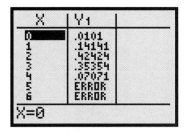

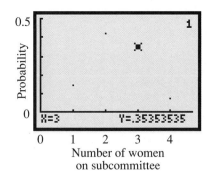

Since the function is defined only for a few integral values of x, be careful when you graph to ensure that points will appear on the screen. On the TI-83 graphing calculator, the display is 95 pixels wide. Hence, when we adjust the window so that x runs from 0 to 94, each pixel in the horizontal direction corresponds to an integral value of x. However, the maximum value of x is 4. Set the viewing window as $0 \leq x \leq 4.7, -0.05 \leq y \leq 0.5$, then press TRACE. Instead of using the arrow keys, it is more efficient to enter the value of x desired. Press 3 ENTER, and the cursor jumps to the point where $x = 3$. The screen (above right) shows that the probability there are exactly 3 women on the subcommittee is approximately 0.353 535.

In *Example 1*, combinations were used because the order in which the people are listed on the subcommittee was not important. When order is important, permutations are used.

Example 2

Three prizes are awarded in a raffle. Seven hundred fifty people hold one ticket each.

a) What is the probability that Alice, Ben, and Concetta win first, second, and third prizes, respectively?

b) What is the probability that Alice, Ben, and Concetta are the 3 prize winners (though not necessarily in that order)?

Solution

a) There is only one arrangement of the winners A, B, and C.

The number of ways to arrange first, second, and third prize winners from 750 ticket holders is $_{750}P_3$.

The probability that Alice, Ben, and Concetta win the first, second, and third prizes, respectively, is:

$$P(A, B, C) = \frac{1}{_{750}P_3}$$

$$= \frac{747!}{750!}$$

$$= \frac{1}{420\ 189\ 000}$$

b) There is only one selection of the winners A, B, and C.

The number of ways to select first-, second-, and third-prize winners from 750 ticket holders, without regard to order, is $_{750}C_3$.

The probability that Alice, Ben, and Concetta are the 3 prize winners is:

$$P(A, B, C \text{ in any order}) = \frac{1}{_{750}C_3}$$

$$= \frac{(747!)(3!)}{750!}$$

$$= \frac{1}{70\ 031\ 500}$$

DISCUSSING THE IDEAS

1. Look at the two ways on page 441 to calculate the probability that two cards randomly selected from a shuffled deck of 52 cards are both aces. Without calculating the final answers, explain why $\frac{_4C_2}{_{52}C_2}$ and $\frac{4}{52} \times \frac{3}{51}$ are equal.

2. In the solution to *Example 1b* and *c*, each numerator is the product of two combinations. Explain how the numerator in part a could be expressed as the product of two combinations.

3. In the first screen on page 443, explain why "ERROR" appears when $x > 4$.

4. Look at the screens on page 443. Visualize repeating similar calculations for the probabilities of different numbers of men on the committee. Would the results be the same as those on these screens? Explain.

7.6 EXERCISES

A 1. Refer to *Example 1*. What is the probability of each event?

a) There are 4 men on the subcommittee.

b) There are exactly 3 men on the subcommittee.

c) There are exactly 2 men on the subcommittee.

2. Refer to *Example 1*. Make a table and a graph similar to those on page 443. Show the probabilities for the different numbers of men on the subcommittee.

3. Three balls are randomly drawn from a bag containing 3 white balls and 5 black balls.

   a) How many ways are there to choose the 3 balls?

   b) How many ways are there to choose exactly 2 white balls?

   c) What is the probability that there are exactly 2 white balls?

4. A pizza store offers 15 different toppings on its pizzas.

   a) Suppose a pizza has 4 toppings. How many different pizzas can be made?

   b) What is the probability that 4 randomly selected toppings will include salami and green pepper?

5. Four people are to be randomly selected from a group of 8 boys and 6 girls. What is the probability of each event?

   a) Exactly 3 people will be girls.

   b) All 4 people will be boys.

   c) The four people chosen would alternate in gender.

6. Suppose both sexes are equally probable. What is the probability that a family of 4 children has exactly 3 boys?

B 7. Four balls are randomly drawn from a bag containing 3 white balls and 5 black balls. What is the probability that there are exactly 2 white balls?

8. Five cards are dealt from a shuffled deck of 52 cards. What is the probability of each event?

   a) A hand contains one pair of aces and 3 kings.

   b) A hand contains all hearts.

   c) A hand contains all hearts in sequence from an ace to 10.

   d) A hand contains 4 aces.

   e) A hand contains 4 queens and an ace.

9. What is the probability of being dealt all 4 kings and an ace in a 5-card hand?

10. What is the probability of a 5-card hand containing any "four-of-a-kind," one of which is the outcome in exercise 9?

11. A hand of 13 cards is dealt from a shuffled deck of 52 cards.

   a) What is the probability that it will contain each set of cards?
      i) no aces   ii) exactly 1 ace   iii) exactly 2 aces
      iv) exactly 3 aces   v) 4 aces

   b) What is the sum of the five probabilities in part a? Explain.

12. In exercise 11, explain how the probabilities will change in each case.

a) The hand contains fewer than 13 cards.

b) The hand contains more than 13 cards.

13. A hand of 13 cards is dealt from a shuffled deck of 52 cards. This screen shows the probability that the hand contains different numbers of spades.

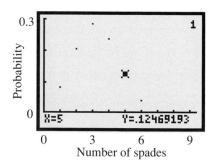

a) Calculate to confirm the result shown; that is, P(5 spades) ≐ 0.125.

b) Write to explain how you calculated in part a.

14. The graph in exercise 13 is the graph of a function.

a) Write the equation of the function, where n is the number of spades dealt.

b) State the domain of the function.

c) Describe the range of the function.

15. Construct the graph in exercise 13. Use the graph to determine the probability that the hand contains each number of spades.

a) 3 spades   **b)** 8 spades   **c)** 13 spades

16. A supermarket has 2 brands of potato chips on sale. The brands were randomly selected from 5 major brands: A, B, C, D, and E. What is the probability of each event?

a) The two choices include A.   **b)** The two choices include A but not B.

c) The two choices exclude D and E.

17. Open the *International Manufacturing* database. Find the data for the Mid 1970s. Assume countries are randomly selected.

a) What is the probability of selecting 3 countries that pay hourly?

b) What is the probability of selecting 3 countries so that the first pays hourly, the second pays daily, and the third pays monthly?

c) What is the probability of selecting 5 countries so that exactly 3 are from Europe?

18. In the Lotto 649 draw, 6 numbered balls from 49 balls are randomly selected without replacement. Before the draw, a person selects 6 numbers.

a) What is the probability of each event?
  i) The balls match all 6 selected numbers.
  ii) The balls match each of these selected numbers: 5, 4, 3, 2, 1, 0

b) What is the sum of the probabilities of the seven events in part a? Explain.

19. In a recent international test, 67% of the students answered Item 1 correctly, 53% answered Item 2 correctly, and 30% answered both items correctly. What is the probability that on a randomly selected test paper there is the correct answer to at least one of the two items?

20. Eight people enter a doubles tennis tournament, in which partners are randomly selected. What is the probability that two brothers, Sean and Kevin, are selected as partners?

21. The school computer club comprises 6 girls and 4 boys. Suppose the two co-captains are randomly selected. What is the probability of each event?

    a) The co-captains are both girls.

    b) The co-captains are both boys.

    c) The co-captains are 1 girl and 1 boy.

22. Of the 20 students on this year's student council, 14 are girls. Five students from the council are to be randomly selected to participate in a student exchange to Quebec. What is the probability of each event?

    a) All 5 students selected are girls.

    b) Only 1 student selected is a girl.

    c) At least 2 students selected are girls.

23. A Motor Vehicles Branch has 5 different written driving tests to administer randomly to prospective drivers. Two women and 3 men take the test. What is the probability of each event?

    a) The 3 men take the same test.

    b) Exactly 3 people take the same test.

    c) All 5 people take a different test.

C 24. In a bin of used golf balls, 30% of the balls are yellow and the rest are white. Three balls are randomly selected. Each one is replaced before the next one is drawn. What is the probability that at least 2 balls of the 3 balls selected are yellow?

25. A probability experiment has 3 independent outcomes: X, Y, and Z. Let XY denote the event in which outcome Y follows outcome X on two successive trials of the experiment. Is it true that $P(XY) = P(X) \times P(Y)$? Explain.

COMMUNICATING THE IDEAS

Write to explain how permutations and combinations are used to determine probabilities. Include comments on how combinations are used for events that have very large numbers of outcomes, such as a lottery.

What Are Your Chances of Guessing Correct Answers on Tests?

On true-false and multiple-choice tests, you can obtain correct answers by
random guessing. You can use a graphing calculator to explore the
probabilities of guessing the correct answers to questions on these tests.

True-false tests

A true-false test has 10 questions. Suppose you answer each question
simply by guessing. To simulate scoring the completed test, you could
toss a coin 10 times and record the number of heads. If there were 6
heads, your simulated score for the test is 6 out of 10, or 60%.

The TI-83 calculator can do this simulation
more efficiently. Press MATH ▶ ▶
▶ 7 10 , 0.5) ENTER. Press ENTER
a few more times. A screen similar to
this one will appear. Your screen may
have different random numbers.

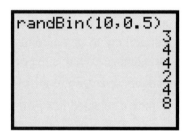

The command randBin(10,0.5) tells the calculator to do 10 trials. Each
trial results in success or failure, and the probability of success is 0.5. The
calculator returns the number of successes, which is a number from 0 to
10. Hence, the command randBin(10,0.5) simulates answering questions
on a true-false test with 10 questions. The screen above shows the scores
for six simulations: 3, 4, 4, 2, 4, and 8.

To display the results for more simulations,
use the command randBin(10,0.5,5). The
calculator will display the results of 5
simulations in a horizontal list. This screen
shows the results of repeating this command
to represent 25 simulations. Your screen
may be different.

randBin(10,.5,5)

{8 5 3 4 5}
{4 7 4 7 6}
{4 5 5 3 4}
{7 7 6 6 6}
{5 5 2 4 6}

1. a) Examine the numbers in the display. Which numbers occur most
   frequently? Which numbers occur less frequently? Explain.

   b) Suppose the simulations were repeated. Would you get the same
   results? Explain.

2. a) Use the randBin command to carry out 100 simulations. Record the
   results before they scroll off the screen.

b) Use your results to estimate the probabilities of obtaining scores of 0, 1, 2, 3, …, 9, and 10 on a true-false test containing 10 questions, by random guessing. Compare your estimates with those of other students.

3. Use your results from exercise 2. Estimate the probability of guessing the correct answers to 5 or more questions on a true-false test with 10 questions.

In exercise 2, you tallied the results by hand. Your calculator can perform this task and graph the data. Ask your teacher for the program called TFTEST from the Teacher's Resource Book. Enter the program in your calculator. Before running the program, clear any equations in the Y= list and set the graphing window to appropriate settings. The program will automatically set up the graph using Plot1 in the STAT PLOT menu.

The screens below show the results of 25 simulations for a 10-question test. The list of numbers at the bottom of the first screen is the number of times scores of 0, 1, 2, 3, 4, 5, and 6 occurred, but there is no room to display further results. The program pauses at this point. To continue running the program, press ENTER. The graph will be displayed. To see all the calculated data, press STAT ENTER. Lists L1 and L2 contain the scores and their frequencies. These are the ordered pairs that are plotted on the graph.

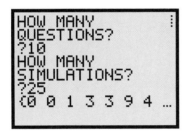
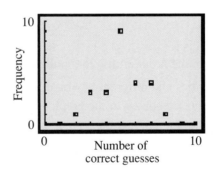

4. Suppose you run the program for 25 simulations.

a) What similarities and differences would you expect to find in the resulting graph, compared with the graph above?

b) Run the program a few times to check your predictions.

5. Suppose you run the program for 100 simulations.

a) What similarities and differences would you expect to find in the resulting graph, compared with the graphs in exercise 4?

b) Run the program to check your predictions.

c) Press [STAT] [ENTER] to view the data. Use list L2 to determine the experimental probabilities of obtaining scores of 0, 1, 2, 3, ..., 9, and 10 on a true-false test containing 10 questions, by guessing randomly. Compare the probabilities with those in exercise 2b.

Multiple-choice tests

A multiple-choice test has 10 questions, and there are 4 possible answers for each question. Suppose you answer each question by guessing randomly. Use the randBin function to simulate scoring the test. Just change the probability entered.

Ask your teacher for the program called MCTEST from the Teacher's Resource Book. Enter the program in your calculator. These screens show the results of 25 simulations for a 10-question test with 4 possible answers per question.

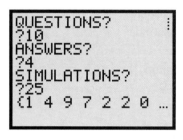
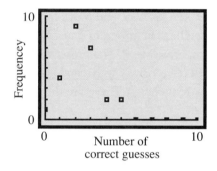

6. Suppose you run the program for 25 simulations.

a) What similarities and differences would you expect to find in the resulting graph, compared to the graph above?

b) Run the program a few times to check your predictions.

7. Suppose you run the program for 100 simulations.

a) What similarities and differences would you expect to find in the resulting graph, compared with the graphs in exercise 6?

b) Run the program to check your predictions.

c) Press [STAT] [ENTER] to view the data. Use list L2 to determine the experimental probabilities of obtaining scores of 0, 1, 2, 3, ..., 9, and 10 on a multiple-choice test containing 10 questions with 4 possible answers per question, by guessing randomly.

8. Use your results from exercise 7. Estimate the probability of guessing the correct answers to 5 or more questions on a multiple-choice test with 10 questions and 4 answers per question.

7.7 Using the Binomial Theorem to Calculate Probabilities

In Section 7.6, we saw how to use permutations and combinations to solve certain probability problems. The binomial theorem can also be used in certain situations. In *Investigate*, you will determine a probability experimentally. Later, you will use the binomial theorem to calculate the same probability.

INVESTIGATE

What is the probability of correctly guessing the outcome in exactly 1 out of 6 rolls of a die? Make a guess, then conduct the following experiment.

Step 1. Predict the outcome for one roll of a die, then roll the die. Repeat 6 times.

Step 2. Record whether you correctly guessed *exactly once* in the 6 rolls.

Step 3. Repeat Steps 1 and 2 nine times.

Combine your results with those of your classmates. Calculate the number of times the outcome was correctly guessed exactly once in 6 rolls.

Determine the total number of 6-roll experiments. Calculate the experimental probability of correctly guessing the outcome in exactly 1 out of 6 rolls of a die.

In *Investigate*, you determined an experimental probability. We can calculate the theoretical probability for the same outcome.

The probability of correctly guessing the outcome in 1 roll of a die is $\frac{1}{6}$.

The probability of incorrectly guessing the outcome is $\frac{5}{6}$.

The probability of 1 correct guess followed by 5 incorrect guesses is $\frac{1}{6} \times \left(\frac{5}{6}\right)^5$.

But the 1 correct guess could occur on any one of the 6 rolls.
There are $_6C_1$ ways to arrange 1 correct guess and 5 incorrect guesses.

So, P(exactly 1 correct guess in 6 rolls) $= {_6}C_1 \times \frac{1}{6} \times \left(\frac{5}{6}\right)^5$

$$\doteq 0.402$$

The above dice experiment has two outcomes: guessing correctly and guessing incorrectly. Any experiment that has two outcomes is a *binomial experiment*.

Recall that the binomial theorem is used to expand an expression such as $(p + q)^6$. If p and q represent the probabilities of the outcomes in a binomial experiment, then $p + q = 1$ and $(p + q)^6 = 1$. There are 7 terms in the expansion of $(p + q)^6$, and these terms have meaning in terms of probability.

For example, for the above experiment, let p represent the probability of correctly guessing an outcome, and let q represent the probability of incorrectly guessing an outcome. Then, the expansion

$$(p + q)^6 = {}_6C_6p^6 + {}_6C_5p^5q + {}_6C_4p^4q^2 + {}_6C_3p^3q^3 + {}_6C_2p^2q^4 + {}_6C_1pq^5 + {}_6C_0q^6$$

becomes

$$\left(\tfrac{1}{6} + \tfrac{5}{6}\right)^6 = {}_6C_6\left(\tfrac{1}{6}\right)^6 + {}_6C_5\left(\tfrac{1}{6}\right)^5\left(\tfrac{5}{6}\right) + {}_6C_4\left(\tfrac{1}{6}\right)^4\left(\tfrac{5}{6}\right)^2 + {}_6C_3\left(\tfrac{1}{6}\right)^3\left(\tfrac{5}{6}\right)^3 +$$

$$ {}_6C_2\left(\tfrac{1}{6}\right)^2\left(\tfrac{5}{6}\right)^4 + {}_6C_1\left(\tfrac{1}{6}\right)\left(\tfrac{5}{6}\right)^5 + {}_6C_0\left(\tfrac{5}{6}\right)^6$$

The terms on the right side represent these probabilities:

$$P(\text{guessing 6 correctly}) = {}_6C_6\left(\tfrac{1}{6}\right)^6$$

$$P(\text{guessing 5 correctly}) = {}_6C_5\left(\tfrac{1}{6}\right)^5\left(\tfrac{5}{6}\right)$$

$$P(\text{guessing 4 correctly}) = {}_6C_4\left(\tfrac{1}{6}\right)^4\left(\tfrac{5}{6}\right)^2$$

$$P(\text{guessing 3 correctly}) = {}_6C_3\left(\tfrac{1}{6}\right)^3\left(\tfrac{5}{6}\right)^3$$

$$P(\text{guessing 2 correctly}) = {}_6C_2\left(\tfrac{1}{6}\right)^2\left(\tfrac{5}{6}\right)^4$$

$$P(\text{guessing 1 correctly}) = {}_6C_1\left(\tfrac{1}{6}\right)\left(\tfrac{5}{6}\right)^5$$

$$P(\text{guessing 0 correctly}) = {}_6C_0\left(\tfrac{5}{6}\right)^6$$

It is left as an exercise in 7.7 Exercises to calculate each probability.
Since $\left(\tfrac{1}{6} + \tfrac{5}{6}\right)^6 = 1^6 = 1$, we know that the sum of the probabilities is 1.

Example 1

A true-false test has 12 questions. Suppose all questions are answered by random guessing. Determine the probability of each event, to 3 decimal places.

a) obtaining exactly 6 correct answers

b) obtaining 7 or 8 correct answers

Solution

Think ...

Use the binomial theorem for $(p + q)^{12}$, where p and q represent the probabilities of correctly and incorrectly guessing each answer.

The probability of obtaining the correct answer to one question is $\frac{1}{2}$.

The probability of obtaining an incorrect answer to one question is $\frac{1}{2}$.

Since there are 12 questions, use the expansion of $\left(\frac{1}{2} + \frac{1}{2}\right)^{12}$.

a) The probability of obtaining exactly 6 correct answers is:

$$P(6 \text{ correct}) = {}_{12}C_6\left(\frac{1}{2}\right)^6\left(\frac{1}{2}\right)^6$$

$$= \frac{12!}{6!6!}\left(\frac{1}{2^{12}}\right)$$

$$\doteq 0.225\ 586$$

The probability of obtaining exactly 6 correct answers is 0.226.

b) The events "obtaining 7 correct answers" and "obtaining 8 correct answers" are independent. Hence,

$$P(7 \text{ or } 8 \text{ correct}) = P(7 \text{ correct}) + P(8 \text{ correct})$$

$$= {}_{12}C_7\left(\frac{1}{2}\right)^7\left(\frac{1}{2}\right)^5 + {}_{12}C_8\left(\frac{1}{2}\right)^8\left(\frac{1}{2}\right)^4$$

$$= \frac{12!}{7!5!}\left(\frac{1}{2^{12}}\right) + \frac{12!}{8!4!}\left(\frac{1}{2^{12}}\right)$$

$$\doteq 0.193\ 359 + 0.120\ 850$$

$$\doteq 0.314\ 209$$

The probability of obtaining 7 or 8 correct answers is 0.314.

In *Example 1*, we can write an expression for the probability of getting x correct answers on the true-false test with 12 questions.

$$P(x \text{ correct}) = {}_{12}C_x\left(\frac{1}{2}\right)^x\left(\frac{1}{2}\right)^{12-x}$$

$$= {}_{12}C_x\left(\frac{1}{2}\right)^{12}$$

This expression is a function of x. We can write it using function notation as

$P(x) = {}_{12}C_x\left(\frac{1}{2}\right)^{12}$, where x represents the number of correct answers. The

domain of the function is the whole numbers from 0 to 12, inclusive.

We can use the TI-83 graphing calculator to produce the table of values and the graph of the function. In the Y= list, enter Y1 = 12 nCr X/2^12. Press TABLE to obtain the table of values in the screens below. You can see the probability of getting any number of correct answers on the test, including the results of *Example 1*.

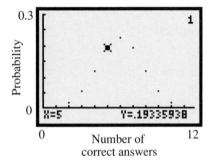

To graph the function, set the viewing window as $0 \le x \le 94 \div 8, -0.03 \le y \le 0.3$, then press TRACE 5 ENTER. The cursor jumps to the point where $x = 5$. The screen below shows that the probability of obtaining 5 correct answers by randomly guessing is about 0.193.

Example 2

A multiple-choice test has 12 questions. Each question has 5 choices, only one of which is correct. All questions are answered by randomly guessing. Determine the probability of each event, to 3 decimal places.

a) obtaining exactly 3 correct answers

b) obtaining at least 3 correct answers

Solution

The probability of obtaining the correct answer to one question is $\frac{1}{5}$.

The probability of obtaining an incorrect answer to one question is $\frac{4}{5}$.

Since there are 12 questions, consider the expansion of $\left(\frac{1}{5} + \frac{4}{5}\right)^{12}$.

Using a scientific calculator

a) $P(3 \text{ correct}) = {}_{12}C_3 \left(\frac{1}{5}\right)^3 \left(\frac{4}{5}\right)^9$

$$= \frac{12!}{3!9!} \left(\frac{4^9}{5^{12}}\right)$$

$$\doteq 0.236\ 223$$

The probability of obtaining exactly 3 correct answers is 0.236.

b) $P(0 \text{ correct}) = {}_{12}C_0 \left(\frac{1}{5}\right)^0 \left(\frac{4}{5}\right)^{12}$

$$\doteq 0.068\ 72$$

$P(1 \text{ correct}) = {}_{12}C_1 \left(\frac{1}{5}\right)^1 \left(\frac{4}{5}\right)^{11}$

$$\doteq 0.206\ 16$$

$P(2 \text{ correct}) = {}_{12}C_2 \left(\frac{1}{5}\right)^2 \left(\frac{4}{5}\right)^{10}$

$$\doteq 0.283\ 47$$

Add $P(0 \text{ correct}) + P(1 \text{ correct}) + P(2 \text{ correct})$ to obtain 0.558 35.

$P(\text{at least 3 correct}) \doteq 1 - 0.558\ 35$

$$\doteq 0.441\ 65$$

The probability of obtaining at least 3 correct answers is 0.441 65.

Using a graphing calculator

$P(x \text{ correct}) = {}_{12}C_x \left(\frac{1}{5}\right)^x \left(\frac{4}{5}\right)^{12-x}$

Consider the function $P(x) = {}_{12}C_x \left(\frac{1}{5}\right)^x \left(\frac{4}{5}\right)^{12-x}$.

Enter this equation, then press $\boxed{\text{TABLE}}$.
Y1 = 12 nCr X*(1/5)^X*(4/5)^(12–X)
Refer to the screen.

X	Y1
0	.06872
1	.20616
2	.28347
3	
4	.13288
5	.05315
6	.0155

Y1=.23622320128

a) $P(3 \text{ correct}) \doteq 0.236\ 223$

The probability of obtaining exactly 3 correct answers is 0.236.

b) $P(0 \text{ correct}) \doteq 0.068\ 72$

$P(1 \text{ correct}) \doteq 0.206\ 16$

$P(2 \text{ correct}) \doteq 0.283\ 47$

Add $P(0 \text{ correct}) + P(1 \text{ correct}) + P(2 \text{ correct})$ to obtain 0.558 35.

$P(\text{at least 3 correct}) \doteq 1 - 0.558\ 35$

$$\doteq 0.441\ 65$$

The probability of obtaining at least 3 correct answers is 0.441 65.

Binomial Probabilities

When: two outcomes, success or failure, are possible in an experiment
the probability of success is p
the experiment is performed n times

The probability of x successes is:

$$P(x) = {}_nC_x\, p^x(1-p)^{n-x}$$

In this section, we related binomial probabilities to functions. The connection between binomial probabilities and functions will be extended in Chapter 8.

DISCUSSING THE IDEAS

1. When p and q represent the probabilities of the outcomes in a binomial experiment, explain why $p + q = 1$.

2. Look at the graph on page 454. Visualize the line defined by $x = 6$. Explain, in these two ways, why this line is the axis of symmetry of the graph.

   a) using the equation of the function

   b) using the meaning of the graph in terms of the true-false test

3. Explain how the graph on page 454 would change in each case.

   a) There were more than 12 questions on the test.

   b) There were fewer than 12 questions on the test.

4. In *Example 2b*, explain why the probabilities of getting 0, 1, or 2 correct were added and the sum subtracted from 1.

5. Visualize using the TI-83 to graph the function in *Example 2*. Explain how this graph would be similar to and different from the graph on page 450.

7.7 EXERCISES

A **1.** Indicate the possible binomial outcomes in a survey involving each population.

a) students in grade 12

b) students attending your school

c) quality testing

d) educational testing

e) medical products

f) business and financial investments

2. Evaluate each term in the expansion of $\left(\frac{1}{6} + \frac{5}{6}\right)^6$ on page 452. Add the terms. What do you notice? Explain.

3. Assume that the births of a boy and a girl are equally likely. In a family of 3 children, determine the probability of each event.

a) 3 girls

b) 2 girls and 1 boy

c) 1 girl and 2 boys

d) 3 boys

4. Repeat exercise 3, assuming that P(girl) = 0.55.

5. A coin is tossed 5 times. Determine the probability of each event.

a) 5 heads

b) 4 heads, 1 tail

c) 3 heads, 2 tails

d) 2 heads, 3 tails

e) 1 head, 4 tails

f) 5 tails

6. Repeat exercise 5 for a biased coin, where P(heads) = 0.6.

B **7.** Suppose an archer could hit a target 80% of the time. What is the probability that the archer would hit the target 9 times out of 10?

8. The Bears and the Dinos are rival baseball teams. In the past few years, the Bears have won 60% of the games played between the two teams. What is the probability that the Bears will win exactly 5 of the next 7 games against the Dinos?

9. The Flames and the Avalanche are in the finals. Each team hopes to win the best-of-seven games. The probability of a Flames' win in any game is 80%. What is the probability that the Flames will win the series in exactly 6 games?

10. For exercise 9, what is the probability that the series will be over in exactly 6 games?

11. Paulene has scored a goal in 70% of the games she played. What is the probability that she would score a goal in at least 7 of her next 8 games?

12. Open the *NBA* database. Find the record for Charles Barkley in the 1996–97 season.

   a) How many free throws did he attempt?

   b) How many free throws did he score?

   c) What is the probability he will score a free throw?

   d) Assume the probability in part c. What is the probability he will score in his next 5 free throws?

   e) Assume the probability in part c. What is the probability he will score at least twice in the next 5 free throws?

13. Open the *NBA* database. Choose a player and a season. Make up questions similar to those in exercise 12. Answer your questions, then give them to a classmate to answer. Check your answers against your classmate's answers.

14. According to the weather statistics, the probability of rain for every day in March is 70%. What is the probability that it will rain at most 5 days out of 7 days in a week in March?

15. a) Calculate the probability of each event.
    i) Tossing a fair coin 2 times and getting exactly 1 head.
    ii) Tossing the coin 20 times and getting exactly 10 heads.
    iii) Tossing the coin 200 times and getting exactly 100 heads.

   b) Write to explain why the probability that exactly one-half the tosses are heads decreases when the number of tosses increases.

16. Visualize tossing a coin several times. The screen shows how the probability that exactly one-half the tosses are heads decreases as the number of tosses increases. Calculate to confirm the result shown; that is, for 30 tosses the probability that exactly one-half are heads is about 0.144 464.

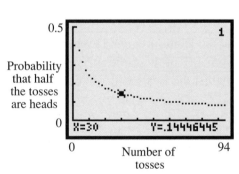

17. Construct the graph in exercise 16. Use the graph.

   a) Determine the probability that when a coin is tossed 40 times, there will be exactly 20 heads.

   b) Determine the least number of tosses required so the probability that exactly one-half are heads is less than $\frac{1}{10}$.

18. In exercises 15 to 17, you assumed that heads and tails were equally likely. Consider a biased coin, such that P(heads) = 0.6. Write to explain how the results of exercises 15 to 17 would change for this coin. Support your answer with some examples.

19. A multiple-choice test comprises 10 questions. There are 4 possible answers to each question. A student guesses the answer to each question. To pass the test, a student must get 9 or 10 correct answers. What is the probability that the student passes the test?

MODELLING Answering Multiple-Choice Tests

In *Example 2* and exercise 19, the binomial theorem was used as a model to predict results on a multiple-choice test.

- What assumptions does the model make about how the test is answered?
- Are these assumptions reasonable? Explain.
- Suppose you were answering a multiple-choice test with 12 questions. How would the probabilities that you obtain exactly 3 correct answers or at least 3 correct answers compare with those calculated in *Example 2*? Explain.

C **20.** The Bulldogs are in the best-of-seven playoffs against the Cougars. In past meetings, the Bulldogs have won 68% of their games against the Cougars. Assume that the probability of the Bulldogs winning any of the playoff games is 68%. What is the probability that the Bulldogs will win the playoffs in 5 games?

21. In exercise 20, what is the probability that the Bulldogs will lose the playoffs?

22. An examination comprises *n* true-false questions. To pass, you must answer at least one-half the questions correctly. Suppose you guess each answer. Explain your answer to each question.

a) Are you more likely to pass a test with few questions than you are to pass a test with many questions?

b) Does it matter if the number of questions is odd or even?

COMMUNICATING THE IDEAS

Write to explain why the binomial theorem is used to calculate probabilities in certain situations. Include an explanation of how you can tell whether the binomial theorem applies in a particular situation. Give examples to illustrate your explanations.

1. Suppose the probability that an experimental motion sensor malfunctions is approximately $\frac{1}{6}$. Use the roll of a die to simulate this sensor. Roll the die 60 times. Determine the experimental probability of a sensor malfunction.

2. The company that manufactures the sensors in exercise 1 uses two components in tandem. This reduces the probability of a malfunction, which now occurs only when both sensors malfunction. Simulate this tandem device with 60 rolls of 2 dice. Determine the experimental probability of a sensor malfunction.

3. In exercises 1 and 2, a motion sensor company used components with a probability of malfunction of $\frac{1}{6}$. Calculate the probability of a malfunction when the components were used in tandem, triggering a malfunction only when both components failed. Compare your answer to those of exercises 1 and 2.

4. Open the *International Manufacturing* database. Find the data for the Early 1990s. For a random selection, calculate the probability of each selection.

   a) a European country

   b) a country with a population greater than 10 million

   c) a currency that begins with the letter C

   d) a currency for which one unit of currency is greater than $1.00 U.S.

5. In vase A, there are 3 red balls and 5 white balls. In vase B, there are 6 red balls and 4 white balls. Two balls are randomly selected, without replacement, from each vase. What is the probability of each event?

   a) drawing 4 red balls

   b) drawing 4 white balls

   c) drawing 2 red balls from one vase and 2 white balls from the other

6. A viewer preference survey conducted by a cable-television network revealed that 46% of viewers watch sports, 31% watch comedy, and 33% watch drama. Of these viewers, 13% watch sports and comedy, 9% watch comedy and drama, and 11% watch sports and drama. Suppose 20% of viewers watch none of these 3 types of shows. What is the probability that a randomly selected person watches all three?

7. In a card game, a hand of 5 cards contains at least 2 spades. What is the probability that there are exactly 4 spades in that hand?

8. In a certain country, the probability that a child is a boy is 0.490. A family with 6 children has at least two girls. What is the probability that this family has exactly 4 girls?

1. Describe how the graph of $y = x^3 - 2$ compares to the graph of each function.

   a) $y = -x^3 - 2$ b) $y = 2 - x^3$ c) $x = y^3 - 2$

2. Describe what happens to the equation of a function when you make each change to its graph.

   a) Reflect the graph in the y-axis.   b) Reflect the graph in the line $y = x$.

   c) Reflect the graph in the x-axis.   d) Reflect the graph in both axes.

3. Simplify.

   a) $\log_2 64 - \log_2 32$ b) $\log_6 12 + \log_6 3$ c) $\log 2 + \log 50$

   d) $\log_{12} 18 + \log_{12} 8$ e) $\log_6 432 - \log_6 2$ f) $\log_3 108 - \log_3 4$

4. Refer to the decibel scale on page 100. How many times as loud as:

   a) a rock group is a jet engine?

   b) the threshold of hearing is the threshold of pain?

   c) an air conditioner is a heavy truck?

5. For each angle θ, calculate $\sin \theta$, $\cos \theta$, then $\tan \theta$.

   a) $\theta = 120°$ b) $\theta = -\frac{\pi}{6}$ c) $\theta = \frac{3\pi}{4}$

6. Determine each pair of values. Draw a diagram to explain the results.

   a) $\sin 0.4$ and $\cos 0.4$ b) $\sin 2$ and $\cos 2$

   c) $\sin (-4)$ and $\cos (-4)$ d) $\sin (-5.5)$ and $\cos (-5.5)$

7. a) Solve the equation $\tan x = 5$ to 4 decimal places for the domain $0 \le x < 2\pi$.

   b) Use a diagram to explain the results in part a.

   c) Suppose the domain is the set of all real numbers. Write an expression to represent all the roots of the equation $\tan x = 5$.

8. Suppose $\sin x = \frac{4}{5}$ and $\cos y = -\frac{12}{13}$, and x and y are in the second quadrant. Determine each expression.

   a) $\sin (x + y)$ b) $\cos (x - y)$ c) $\cos 2x$ d) $\sin 2y$

9. An area code consists of 3 numbers.

   a) How many different area codes are possible if there are no restrictions on the numbers?

   b) Suppose the first digit cannot be 0 or 1. How many fewer area codes would be possible with this restriction?

Converting Scores

CONSIDER THIS SITUATION

In British Columbia, some students who wrote the provincial mathematics examination in 1997 were concerned because they were unable to answer some questions. These students predicted their scores out of 100 would be low. To their surprise, their provincial score, which was a converted score out of 800, was much better than expected.

- Why were the scores converted?

- Do you agree that the scores should have been converted? Explain.

- Would it make a difference if the original scores had been too high and the Ministry of Education converted them to lower scores?

On pages 504–506, you will develop a mathematical model to convert scores so they have a predetermined mean and a predetermined percent of scores above a passing grade.

 FYI Visit www.awl.com/canada/school/connections

For information related to the above problem, click on MATHLINKS, followed by AWMath. Then select a topic under Converting Scores.

Simulating Tossing a Coin 12 Times

In this investigation, you will simulate tossing a coin 12 times to determine the number of heads that appear. You repeat this experiment many times, and convert the results to experimental probabilities. The purpose is to investigate what happens to the experimental probabilities as the number of repetitions of the experiment increases.

Gathering the data

- Clear all equations from the Y= list.
- Set the window: Xmin = 0, Xmax = 13, Xscl = 1, Ymin = –3, Ymax = 12, Yscl = 1.
- Press [STAT PLOT] **1**. On the first line, select On. On the second line, select the third graph type (histogram). Make sure L1 is beside Xlist, and 1 is beside Freq. Press [QUIT].

1. Press [MATH] [◄] **7** to select randBin(. Press 12 [,] 0.5 [,] 25 [ )] [STO►] [L₁] [ENTER]. The calculator will generate the results for 25 repetitions of the experiment and store them in list L1. Press [GRAPH] to obtain a graph similar to the one below.

2. Press [TRACE] and move the cursor to the top of each bar. Observe where the number of heads and the frequency appear. For example, this screen shows that 4 heads occurred 7 times. Record the results in the first row of a table similar to the one below. The results in the first row correspond to this screen. The results in the next two rows were obtained when the experiment was repeated.

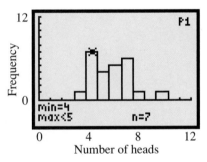

Frequencies of Different Numbers of Heads

Number of heads	0	1	2	3	4	5	6	7	8	9	10	11	12
In 25 repetitions	0	0	0	1	7	4	5	6	1	0	1	0	0
In next 25 repetitions	0	0	0	0	4	6	5	7	3	0	0	0	0
In next 25 repetitions	0	0	0	0	3	7	5	5	2	3	0	0	0
:		:	:	:	:	:	:	:	:	:	:	:	:

3. Repeat exercises 1 and 2 several times to obtain data for four or more sets of 25 repetitions.

4. Visualize adding the results in the first 2 rows to obtain results for 50 repetitions, then adding the results in the first 3 rows to obtain results for 75 repetitions, and so on. Make another table to show the frequencies for 25 repetitions, 50 repetitions, 75 repetitions, and so on.

Frequencies of Different Numbers of Heads

Number of heads	0	1	2	3	4	5	6	7	8	9	10	11	12
In 25 repetitions	0	0	0	1	7	4	5	6	1	0	1	0	0
In 50 repetitions	0	0	0	1	11	10	10	13	4	0	1	0	0
In 75 repetitions	0	0	0	1	14	17	15	18	6	3	1	0	0
⋮	⋮	⋮	⋮	⋮	⋮	⋮	⋮	⋮	⋮	⋮	⋮	⋮	⋮

5. Repeat exercise 4 to obtain frequencies for 100 or more repetitions of the experiment. Save these data for 8.3 Exercises.

Graphing the experimental probabilities

- Set the window: Xmin = 0, Xmax = 13, Xscl = 1, Ymin = 0, Ymax = 0.3, Yscl = 0.1.
- Press [STAT PLOT] **1**. On the first line, select On. On the second line, select the histogram. Make sure L1 is beside Xlist and L2 is beside Freq.
- Press [STAT] **1**, and clear list L1. Enter 0, 1, 2, 3, …, 12 in list L1.

Visualize converting the frequencies to experimental probabilities. Take the results for 25 repetitions and divide by 25. For 50 repetitions, divide by 50. For 75 repetitions, divide by 75, and so on. Your calculator can perform all the divisions as you enter the data.

6. Move the cursor to the first position in list L2. Enter the frequency for 0 heads in 25 repetitions, then press [÷] 25 [ENTER]. Enter all the frequencies for 25 repetitions in this way, dividing by 25 each time. To graph the data, press [GRAPH]. A graph similar to the one below appears. Sketch the graph or use the trace key to record the results.

7. Repeat exercise 6, using the frequencies for 50 repetitions (dividing by 50). Then repeat for 75 repetitions, and so on.

8. Describe how the graph of the experimental probabilities changes as the number of repetitions increases.

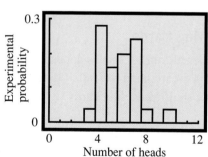

9. Combine your results with those of other students. Obtain graphs of experimental probabilities for as many repetitions of the experiment as possible.

8.1 | What Is a Distribution?

In statistics, the word "distribution" is used frequently, and much of this chapter deals with certain kinds of distributions. To understand the concept of a distribution, consider these examples.

Compact Discs Ownership

Number of CDs owned by students in a grade 12 class

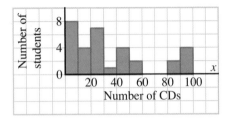

Probability that a randomly selected student in the class owns x CDs

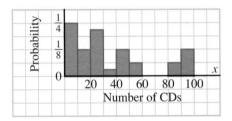

Rolling a Die

Number of times x appeared when a die was rolled 100 times

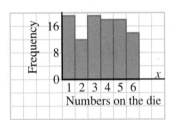

Theoretical probability that x will appear when a die is rolled

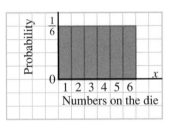

Drawing Cards

Number of times x spades appeared when a 5-card hand is dealt 100 times from a shuffled deck of cards

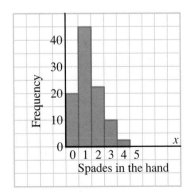

Theoretical probability that x spades will appear when a 5-card hand is dealt from a shuffled deck of cards

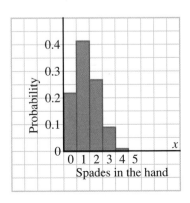

All the examples on page 466 are graphs of functions. The graphs on the left side represent frequency distributions because the quantities plotted along the vertical axis represent frequencies. The graphs on the right side represent probability distributions because the quantities plotted along the vertical axis represent probabilities. Since these graphs represent probabilities, they have certain properties that correspond to properties of probabilities.

A *probability distribution* is a function that provides the probability for every outcome of an experiment. It can be expressed in different ways. The examples below refer to the probability distribution on page 466 about the probability that x spades will appear when a 5-card hand is dealt.

- As a table of values
  The table of values for the third example on page 466 is shown. All the numbers in the second column lie between 0 and 1, and their sum is 1.

- As a graph of individual points

Number of spades	Probability
0	0.2215
1	0.4114
2	0.2743
3	0.0815
4	0.0107
5	0.0005

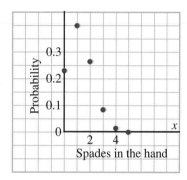

- As a histogram
  The total area of the bars in the histogram on page 466 is 1.

- As an equation (if possible)

$$P(x) = \frac{(_{13}C_x)(_{39}C_{5-x})}{_{52}C_5}$$

Two kinds of probability distributions are shown on page 466.

Uniform Distributions

A uniform distribution is a probability distribution where the probabilities of all the outcomes are equal. An example of a uniform distribution is the distribution on page 466 for the probability that x will appear when a die is rolled.

Binomial Distributions

In *Exploring with a Graphing Calculator*, pages 464–465, you simulated the experiment of tossing a coin 12 times. You displayed graphs of the experimental probabilities of obtaining different numbers of heads for many repetitions of the experiment. These graphs have the following properties:

- The heights of the bars represent probabilities.
- There are probabilities for 13 outcomes, from 0 heads to 12 heads.
- The sum of the probabilities is 1.
- Since the width of each bar is 1, the total area of all the bars is 1.
- As the number of repetitions increases, the graphs come closer and closer to a symmetrical graph representing the theoretical probabilities.

The theoretical probabilities are calculated using the binomial theorem. Recall from page 453 that the probability of getting x correct answers by random guessing on a true-false test with 12 questions is $P(x) = {}_{12}C_x\left(\frac{1}{2}\right)^{12}$. This is the same as the probability of getting x heads when 12 coins are tossed. A graph of this function is shown below. The histogram at the bottom right is also a graph of this function. The domain of the function comprises the whole numbers from 0 to 12, inclusive.

Experimental Probabilities for Tossing a Coin 12 Times

25 repetitions

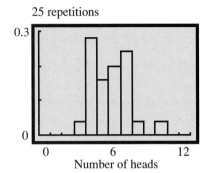

100 repetitions

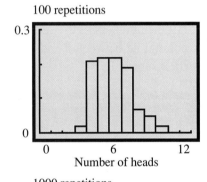

1000 repetitions

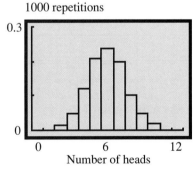

Binomial Distribution for x Successes in 12 Trials, where P(Success) = 0.5

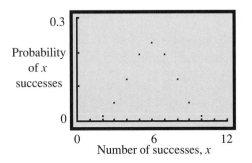

Theoretical Probabilities for Tossing a Coin 12 Times

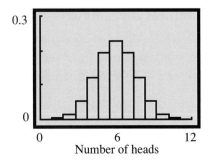

You can use a TI-83 graphing calculator to produce the graph on page 468 (bottom left). In Chapter 7, you constructed similar graphs by entering the corresponding equation in the Y= list. However, binomial distributions are used so frequently that a command is available that bypasses the need to enter the equation.

- Press [DISTR] **0** to select binompdf(. Press 12 [,] 0.5 [ )] [STO▶] [L₂] [ENTER].
- Press [STAT] **1**, and enter 0, 1, 2, 3, …, 12 in list L1. Make sure there are no other entries in this list.
- Press [STAT PLOT] **1** to select Plot 1. On the first line, select On. On the second line, select the first graph type. Make sure L1 is beside Xlist and L2 is beside Ylist. Set the plotting symbol to a dot.
- Press [GRAPH] to produce the graph.

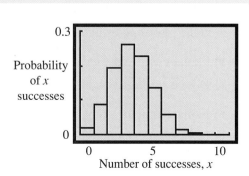

The command binompdf(12,0.5) calculates the probabilities of x successes for 12 trials, where the probability of success on each trial is 0.5. The calculator determines the probabilities for all possible outcomes, which are the whole numbers from 0 to 12 inclusive.

Example

a) Graph the binomial distribution corresponding to the experiment of rolling a die 10 times and recording the number of times a 1 or a 6 appears.

b) Verify that the probabilities lie between 0 and 1.

c) Verify that the sum of the probabilities is 1.

Solution

> ### Think …
> Change list L1 to show numbers from 0 to 10, and enter the data in list L2. Also change the window. The theoretical probability of getting a 1 or a 6 is $\frac{1}{3}$.

a) Change the window so that Xmax = 11.

Press [STAT PLOT] **1**. Select On and the histogram.

Press [STAT] **1**, and clear list L1. Enter 0, 1, 2, 3, …, 10 in list L1. Press [QUIT].

Press [DISTR] **0** 10 [,] 1 [÷] 3 [ )]

[STO▶] [L₂] [ENTER] [GRAPH].

b) Press [STAT] **1**, and verify that each number in list L2 is between 0 and 1.

c) Use the calculator to add the numbers in list L2. Press [LIST] [▶] [▶] **5** [L₂] [ )] to display sum(L2). Press [ENTER], and the calculator displays 1.

The properties in the *Example* part b and c are true for all binomial distributions. Since all the function values represent probabilities, they lie between 0 and 1; that is, $0 \le P(x) \le 1$. Since the probabilities for all outcomes are included, the sum of the function values is 1. We write this using sigma notation as $\Sigma P(x) = 1$. The expression $\Sigma P(x)$ is the sum of all possible values of $P(x)$.

Properties of a Binomial Distribution

A binomial distribution is a function $P(x)$. It is used when two outcomes, success or failure, are possible in an experiment.

$P(x)$ depends on two quantities:

the probability of success for one outcome: p
the number of trials in the experiment: n

There is a different binomial distribution for every pair of values of p and n. The domain of $P(x)$ comprises the whole numbers from 0 to n, inclusive. $P(x)$ represents the probability that there will be exactly x successes. For every value of x, $0 \le P(x) \le 1$ The sum of all the values of $P(x)$ is 1; that is, $\Sigma P(x) = 1$.

DISCUSSING   THE IDEAS

1. Explain why each graph on page 466 represents a function.

2. Think about some experiments in Chapter 7.

   a) Which experiments have a binomial distribution? Explain.

   b) Which experiments do not have a binomial distribution? Explain.

3. Add the probabilities in the table on page 467. Why is the sum not 1?

4. **a)** Explain why a histogram can represent the graph of a function.

   b) Does every function have a graph that is a histogram? Explain.

5. **a)** Explain why the graph of the binomial distribution on page 468 is symmetrical.

   b) Does every binomial distribution have a symmetrical graph? Explain.

A 1. Explain your answer to each question for the graphs below.

    a) Which distribution is a uniform distribution?

    b) Which distributions are binomial distributions?

    c) Which distribution is neither a uniform distribution nor a binomial distribution?

Distribution 1

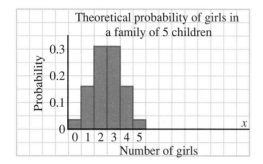

Distribution 2

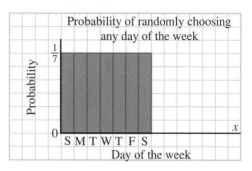

Distribution 3

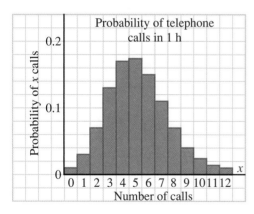

Distribution 4

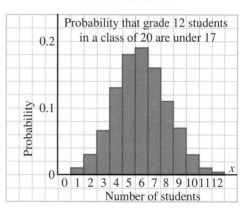

B 2. Use the sample space in *Example 2*, page 422.

    a) Construct a histogram to show the probability distribution of the sums of the numbers that appear when 2 dice are rolled.

    b) Is this a uniform distribution? Explain.

    c) Is this a binomial distribution? Explain.

3. **a)** List the sample space for the experiment of tossing a coin 3 times.

    b) Construct a histogram to show the probability distribution of heads when a coin is tossed 3 times.

4. Compare your histogram in exercise 3 with the one at the bottom of page 468. In what ways are the histograms similar? In what ways are they different?

5. a) Predict how a histogram that shows the probability of heads when a coin is tossed 30 times would differ from those on page 468.

b) Verify your prediction.

6. a) Graph the binomial distribution corresponding to the experiment of rolling a die 10 times and recording the number of times a 6 appears.

b) Predict how the graph in part a would change if the experiment was to record the number of times a 6 does not appear. Verify your prediction.

c) Predict how the graph in part a would change if the die was rolled more than 10 times. Verify your prediction.

d) Compare the graph in part a with the graph in the *Example*. In what ways are the graphs similar? In what ways are they different? Explain.

7. A game is such that the probability of winning, P(win) = 0.75. Describe an experiment to simulate the number of games won in 10 games. Conduct the experiment, then graph the frequency distribution and the probability distribution. Write to explain the results.

8. Simulate the toss of a coin 100 times by pressing MATH ◄ **5** 0 ⬚ , ⬚ 1 ⬚ , ⬚ 100 ⬚ ENTER. The screen lists 100 numbers, either 0 or 1 (to see the entire list, use ▶). Choose either 0 or 1 to represent heads and the other number to represent tails.

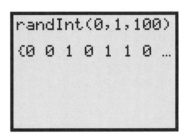

a) Count the number of runs of heads with each length. For example, if 1 represents heads, this screen shows one run with length 1 and one run with length 2.

b) Construct a frequency distribution. Explain the results.

C **9. a)** In exercise 8, a probability distribution could be constructed. Describe some difficulties involved in constructing this distribution.

b) In part a, is the distribution a binomial distribution? Explain.

10. Does $P(x) = \frac{x}{10}$ for x = 1, 2, 3, and 4 represent a probability distribution? Explain.

COMMUNICATING THE IDEAS

Write to explain how a probability distribution is different from, and similar to, a frequency distribution. Include as many differences and similarities as you can.

Engineering students from two colleges participated in a competition. They designed a new carton for eggs that would better withstand a drop from a predetermined height. The better carton would minimize the breakage of eggs. A sample of many cartons of each type was tested.

	Broken eggs	0	1	2	3	4	5	6
Carton A	Number of Cartons	4	12	30	36	28	10	6
Carton B	Number of Cartons	1	6	37	38	37	6	1

1. Which appears to be the better carton? Explain.

2. Draw a histogram for the number of broken eggs in each carton. Which appears to be the better carton? Explain.

3. Determine the mean, median, and mode for the number of broken eggs in each carton. Which appears to be the better carton? Explain.

4. As you learn more from the data, did your decision about which appeared to be the better carton change? Explain.

The histograms you drew in *Investigate* are important visual displays of the distributions of the number of broken eggs in the cartons.

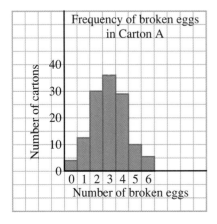

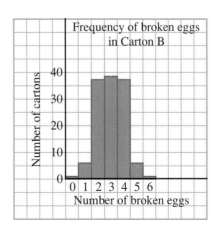

To describe data numerically, we use two numbers, which are the mean and the standard deviation.

- The *mean* is a measure of where the middle of the data occurs.
- The *standard deviation* is a measure of the extent to which data cluster around the mean.

Calculating the Mean of a Set of Data

Suppose there are n data values represented by $x_1, x_2, \ldots x_n$. The mean is the sum of the data values divided by the number of data values. In statistics, the mean is represented by the Greek letter μ (mu).

$$\mu = \frac{x_1 + x_2 + \ldots + x_n}{n}$$

When some data values are the same, we multiply to determine the sum. For example, in *Investigate*, to determine the total number of broken eggs in Carton A, you probably calculated $0(4) + 1(12) + 2(30) + 3(36) + 4(28) + 5(10) + 6(6)$.

> ### *Mean of a Data Set*
>
> Let $x_1, x_2, x_3, \ldots x_n$ represent any set of data values.
>
> Mean: $\mu = \dfrac{\sum x_i}{n}$

Calculating the Standard Deviation of a Set of Data

The mean is not an adequate description of a set of data. The mean cannot identify differences in variation among sets of data. Some indication is needed of how the data are clustered or spread out around the mean. There are different measures of this variation, but the most common is the standard deviation.

Since the standard deviation is a measure of how the data are spread out around the mean, we calculate the difference between each data value and the mean: $(x_1 - \mu), (x_2 - \mu), (x_3 - \mu), \ldots (x_n - \mu)$. The sum of these differences is not useful because it is always 0. Therefore, we square the differences before adding to obtain:

$$(x_1 - \mu)^2 + (x_2 - \mu)^2 + (x_3 - \mu)^2 + \ldots + (x_n - \mu)^2, \text{ or } \sum(x_i - \mu)^2$$

To determine the mean of the squared differences, we divide this sum by the number of differences, n, to obtain $\dfrac{\sum(x_i - \mu)^2}{n}$. To counterbalance the squaring,

we take the square root to obtain $\sqrt{\dfrac{\sum(x_i - \mu)^2}{n}}$. This number is the standard

deviation, and is represented by the Greek letter σ (sigma). Adding the squared differences, dividing by n, then taking the square root provides a number with a magnitude comparable to the original differences between the mean and the data values.

Standard Deviation of a Data Set

Let x_1, x_2, x_3, ... x_n represent any set of data values.

$$\text{Standard deviation: } \sigma = \sqrt{\frac{\sum(x_i - \mu)^2}{n}}$$

We shall calculate the standard deviation for each type of egg carton.

Carton A

The mean is 3.

The differences from the mean are:
$4(0 - 3), 12(1 - 3), 30(2 - 3), 36(3 - 3), 28(4 - 3), 10(5 - 3), 6(6 - 3)$

The sum of the squares of the differences is:
$4(9) + 12(4) + 30(1) + 0 + 28(1) + 10(4) + 6(9) = 236$

The number of data values is $4 + 12 + 30 + 36 + 28 + 10 + 6 = 126$

The mean of the squared differences is $\frac{236}{126}$.

The standard deviation is $\sqrt{\frac{236}{126}} \doteq 1.369$

Carton B

The mean is 3.

The differences from the mean are:
$1(0 - 3), 6(1 - 3), 37(2 - 3), 38(3 - 3), 37(4 - 3), 6(5 - 3), 1(6 - 3)$

The sum of the squares of the differences is:
$1(9) + 6(4) + 37(1) + 0 + 37(1) + 6(4) + 1(9) = 140$

The number of data values is $1 + 6 + 37 + 38 + 37 + 6 + 1 = 126$

The mean of the squared differences is $\frac{140}{126}$.

The standard deviation is $\sqrt{\frac{140}{126}} \doteq 1.054$

Just as the mean is a calculated number that represents the "typical" or "average" number in a set of data, the standard deviation is the "typical" or "standard" size of the difference (deviation) between the individual pieces of data and the mean.

Carton A has a standard deviation of 1.369; carton B has a standard deviation of 1.054. Since carton B has the lower standard deviation, the spread of the data for carton B is smaller; the data are closer to the mean. The smaller standard deviation means more consistent results for the number of broken eggs; that is, more data lie close to the mean. Carton B is probably the better carton.

Calculating the standard deviation as shown above is time consuming, and there is a good chance of making an error. For this reason, we use a graphing calculator. For the TI-83 calculator, first clear the lists.

To enter the data for carton A, press [STAT] **1**, and enter the whole numbers from 0 to 6 in list L1. Then, in list L2, enter the numbers of cartons from the table on page 473 (below left). Press [QUIT]. To calculate the standard deviation, press [STAT] [▶] **1** [L₁] [,] [L₂] [ENTER]. The screen (below right) appears.

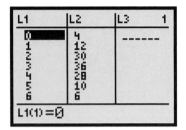

The TI-83 graphing calculator uses the symbols $\bar{x}$ for the mean and σx for the standard deviation. From the screen:

The mean is $\bar{x} = 3$.

The standard deviation is $\sigma x = 1.368\ 581\ 701$.

Example

A company needs various types of fasteners for general shop supplies. A buyer is examining two boxes, both of which contain 2000 fasteners with a mean length of 6.5 cm. According to its quality control records over the past year, the fasteners in crate A came from a population with a standard deviation of 2 cm, while those in crate B came from a population with a standard deviation of 0.01 cm. Which crate should the buyer choose?

Solution

The very small standard deviation of the fasteners in crate B suggests they have uniform size. If the buyer wants uniformity, she should choose crate B; if she wants a variety of sizes, crate A.

A number such as the mean or the standard deviation that measures or describes a distribution numerically, is a *statistic*.

In the explanation of standard deviation:

1. Why do we subtract the mean from each data value?

2. Why do we square the data values?

3. Why do we find the mean of the data values?

4. Why do we take the square root?

5. Explain why the sum of the differences $(x_1 - \mu) + (x_2 - \mu) + \ldots + (x_n - \mu)$ is 0.

8.2 EXERCISES

1. Calculate the mean and the standard deviation for each set of data. Give the answers to 2 decimal places, where necessary.

a) 20, 30, 30, 40 **b)** 18, 21, 24, 27, 30 **c)** 14, 19, 22, 22, 23 **d)** 17, 19, 19, 27, 50

2. Examine the frequency tables. Without calculation, determine which set of data is likely to have the greatest standard deviation and which the least standard deviation. Explain.

a)

Length	Frequency
14	2
15	5
16	12
17	5
18	1

b)

Length	Frequency
14	5
15	5
16	5
17	5
18	5

c)

Length	Frequency
14	1
15	3
16	18
17	3
18	0

3. Calculate the standard deviation for each set of data in exercise 2. How do the results compare with your answers to exercise 2?

4. Thirty-five randomly selected members of a high school basketball league were asked to make 10 free throws. The rate of success for each player was recorded. The standard deviation was computed for the group. The process was repeated using 35 randomly selected people at a shopping mall. Which group do you expect would have the greater standard deviation? Explain.

5. Open the *Cars* database. Calculate the mean and standard deviation for the acceleration of each type of car.

a) compact cars   **b)** small cars

B 6. Forty randomly selected people at a hockey game were asked to play a certain video game. Each score was recorded and the standard deviation was computed for the group. The process was repeated using 40 randomly selected people at a video-game arcade. Which group do you expect would have the greater standard deviation? Explain.

7. Examine the histograms. Without calculation, determine which set of data is likely to have the greatest standard deviation and which the least standard deviation. Explain.

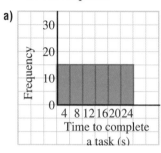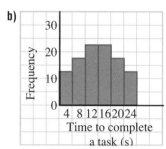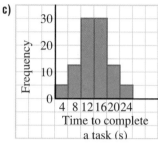

8. The lifetimes of 30 batteries from each of two brands of batteries are given.

Measured lifetimes of 30 brand X batteries (in years)

| 5.0 | 7.2 | 6.8 | 4.6 | 4.5 | 6.1 | 6.3 | 5.4 | 4.8 | 6.8 | 5.9 | 4.7 | 4.0 | 5.2 | 8.0 |
| 6.2 | 7.4 | 4.9 | 5.6 | 9.2 | 3.2 | 3.0 | 4.2 | 5.8 | 6.5 | 5.7 | 4.9 | 6.0 | 4.5 | 5.6 |

Measured lifetimes of 30 brand Y batteries (in years)

| 5.3 | 6.2 | 4.9 | 5.8 | 5.5 | 4.6 | 5.9 | 3.2 | 6.5 | 5.9 | 4.9 | 6.4 | 5.7 | 5.3 | 4.8 |
| 5.6 | 6.7 | 5.5 | 4.8 | 5.9 | 4.8 | 5.6 | 6.1 | 7.4 | 5.7 | 6.7 | 5.8 | 5.2 | 5.5 | 5.8 |

a) Calculate the mean lifetime for each brand.

b) Calculate the standard deviation of the lifetimes for each brand.

c) Which is the better battery? Explain.

9. Work with another student.

a) Construct one question that requires a numerical response. Give the question to a random sample of 60 people. Record their responses.

b) Calculate the mean and the standard deviation. Write to explain the results.

COMMUNICATING THE IDEAS

Write to explain how the standard deviation and the mean help to describe a set of data. Explain the difficulties that could arise if you know one value and not the other.

8.3 Mean and Standard Deviation for a Binomial Distribution

INVESTIGATE **Mean Number of Heads of a Coin**

In *Exploring with a Graphing Calculator*, pages 464–465, you simulated many repetitions of the experiment of tossing a coin 12 times. You can determine the mean number of heads for any number of repetitions.

Use your data from exercise 5 on page 465.

1. a) Determine the total number of heads that appeared when the experiment was repeated 25 times.

   b) Divide by 25 to determine the mean number of heads that appeared when the experiment was repeated 25 times.

2. Determine the mean number of heads when the experiment was repeated each number of times.

   a) 50 times   b) 75 times   c) 100 times

3. What seems to be happening to the mean number of heads for large numbers of repetitions of the experiment? Explain why this is reasonable.

4. The results for the experiment of tossing a coin 12 times and noting the number of heads that appear form a binomial distribution. For this distribution, $n = 12$ and $p = 0.5$. How is the answer to exercise 3 related to the values of n and p?

Recall that the mean and standard deviation of a set of data are determined using these formulas.

Mean: $\mu = \dfrac{\sum x_i}{n}$ Standard deviation: $\sigma = \sqrt{\dfrac{\sum (x_i - \mu)^2}{n}}$

The bars in the histogram of a binomial distribution have a middle. The bars are distributed about this middle in a certain way. Hence, we can define the mean and the standard deviation for a binomial distribution. Since a binomial distribution is not a set of data, the above formulas cannot be used to calculate the mean and standard deviation.

A binomial distribution is defined by two quantities: the number of trials, n, in an experiment, and the probability, p, of success in each trial. Hence, formulas for the mean and standard deviation will depend only on the values of n and p.

Mean for a Binomial Distribution

In *Investigate*, you determined the mean number of heads that appeared for many repetitions of the binomial experiment of tossing a coin 12 times. The theoretical results for this experiment form a binomial distribution with $n = 12$ and $p = 0.5$. The theoretical mean number of heads should depend only on these two values.

The histograms from page 468 are repeated at the right. This table shows the total number of heads and the mean number of heads corresponding to the first three histograms.

Number of repetitions	Total number of heads	Mean number of heads
25	141	5.64
100	586	5.86
1000	5958	5.958

The mean number of heads appears to be getting closer and closer to 6. This is the average number of heads you would expect when you perform the experiment many times. This average is determined by multiplying the number of coins, 12, by the probability of heads, 0.5. It also corresponds to the highest bar on the histogram of theoretical probabilities.

Although this is only one example, it does suggest that, for a binomial distribution with n trials and the probability of success on each trial is p, the mean is np. It can be shown that this is true for any binomial distribution (see *Mathematics File*, page 484).

Mean for a Binomial Distribution

For a binomial distribution with n trials, where the probability of success in each trial is p, the mean is:

$$\mu = np$$

Experimental Probabilities
for Tossing a Coin 12 Times

25 repetitions

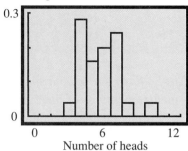

100 repetitions

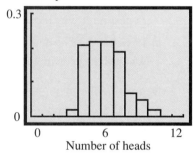

1000 repetitions

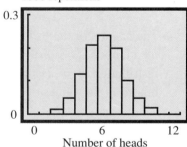

Theoretical Probabilities
for Tossing a Coin 12 Times

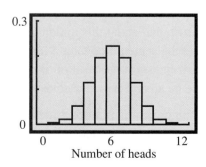

Standard Deviation for a Binomial Distribution

For the binomial distribution at the bottom of page 480, the standard deviation is a measure of how the probabilities for different numbers of heads are distributed about the mean, 6. We can use a TI-83 graphing calculator to calculate the standard deviation of the data for the first three histograms. The data for these histograms are given below.

Frequencies of Different Numbers of Heads

Number of heads	0	1	2	3	4	5	6	7	8	9	10	11	12
In 25 repetitions	0	0	0	1	7	4	5	6	1	0	1	0	0
In 100 repetitions	0	0	0	2	21	22	22	19	7	5	2	0	0
In 1000 repetitions	0	1	16	49	122	210	235	196	103	48	18	2	0

To enter the data for 25 repetitions, press [STAT] **1**, and clear lists L1 and L2. Enter the whole numbers from 0 to 12 in list L1. Enter the frequencies for 25 repetitions in list L2. Press [QUIT]. Press [STAT] [▶]**1** [L₁] [,] [L₂] [ENTER]. The screen (below right) appears.

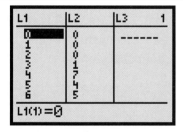 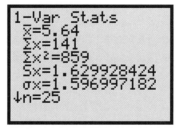

From the screen:
The mean is $\bar{x} = 5.64$, the same number in the table on page 480.
The standard deviation is $\sigma x = 1.596\ 997\ 182$, or approximately 1.597.

The standard deviations for 100 repetitions and for 1000 repetitions can be obtained the same way. The results for the data above are summarized below.

Number of repetitions	Total number of heads	Standard deviation
25	141	1.597
100	586	1.569
1000	5958	1.672

More repetitions are needed to determine a reasonable estimate of the standard deviation. However, we can estimate the standard deviation another way.

The heights of the bars in the bottom histogram on page 480 represent the theoretical probabilities of the different numbers of heads. We can enter these probabilities in list L2 and calculate the standard deviation.
Press [DISTR] **0** 12 [,] 0.5 [ )] [STO►] [L₂] to enter the probabilities in list L2.
To see the probabilities, press [STAT] **1**.
To calculate the standard deviation, press [STAT] [►] **1** [L₁] [,] [L₂] [ENTER].

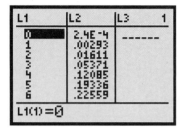

The standard deviation is approximately 1.732. This is the number we would have obtained if we had used many more repetitions of the experiment than we used on page 481.

Mathematicians have determined a formula for the standard deviation for a binomial distribution. See *Mathematics File*, page 484, for an explanation of the formula below.

Standard Deviation for a Binomial Distribution

For a binomial distribution with n trials, where the probability of success in each trial is p, the standard deviation is:

$$\sigma = \sqrt{np(1-p)}$$

We can use this formula to calculate the standard deviation for the binomial distribution on page 480. Substitute 12 for n and 0.5 for p in the formula.

$$\begin{aligned} \sigma &= \sqrt{12 \times 0.5 \times 0.5} \\ &= \sqrt{3} \\ &\doteq 1.732 \end{aligned}$$

DISCUSSING THE IDEAS

1. Use the histograms on page 480. Explain, for the binomial distribution with $n = 12$ and $p = 0.5$, why it is reasonable that the mean is 6.

8.3 EXERCISES

A 1. Use the table on page 481.

    a) Determine the mean number of heads for each number of repetitions.

      i) 25 repetitions   **ii)** 100 repetitions   **iii)** 1000 repetitions

    b) Check that the results in part a are the same as those on page 480.

    c) Repeat parts a and b using standard deviations.

2. Use your data from exercise 5, page 465.

    a) Determine the mean and the standard deviation for the greatest number of repetitions you have.

    b) Combine your data from exercise 5 with the data from other students to obtain data for the greatest number of repetitions. Also include the data in the bottom row of the table on page 481. Then determine the mean and standard deviation. Compare these with the results on pages 481 and 482.

3. Calculate the mean and standard deviation for each binomial distribution.

    a) $n = 100$, $p = 0.6$ **b)** $p = 0.7$, $n = 320$ **c)** $p = 0.05$, $n = 2000$

B 4. Calculate the mean and standard deviation for the binomial distribution defined by each experiment.

    a) rolling a die 10 times and recording how often 6 appears

    b) rolling a die 10 times and recording how often 6 does not appear

5. Choose one part of exercise 4.

    a) Predict what will happen to the mean and standard deviation when the die is rolled more than 10 times.

    b) Use the formulas to check your prediction for the die rolled:

      i) 20 times   **ii)** 60 times   **iii)** 300 times

    c) Graph the mean and standard deviation as functions of the number of times, x, the experiment is repeated. Describe each function.

COMMUNICATING THE IDEAS

Explain what is meant by the mean and standard deviation for a binomial distribution. Include examples to indicate how these measures are determined.

Where Do the Formulas $\mu = np$ and $\sigma = \sqrt{np(1-p)}$ Come From?

In Section 8.3, the formulas $\mu = np$ and $\sigma = \sqrt{np(1-p)}$ for the mean and standard deviation for a binomial distribution were presented. To explain the formulas, consider the first graphing calculator screen on page 482. The numbers in the first two columns were used to make this table.

The letter x is a *random variable* because its value depends on chance. Each number in the second column is the probability that x takes on the corresponding value in the first column.

Recall how you determined the mean and standard deviation for the number of broken eggs per carton in *Investigate*, page 473. The same methods are used to determine the mean and standard deviation of the random variable, x, in this table or in any other similar situation. To obtain the mean, we multiply each value of x by the corresponding probability, add the results, then divide by the total of the probabilities. The total of the probabilities is $\sum P(x) = 1$, where the sum is assumed to be over all possible values of x. We follow a similar method to determine the standard deviation. We can represent the calculations symbolically as follows.

x	$P(x)$
0	0.000 02
1	0.002 93
2	0.016 11
3	0.053 71
4	0.120 85
5	0.193 36
6	0.225 59
7	0.193 36
8	0.120 85
9	0.053 71
10	0.016 11
11	0.002 93
12	0.000 02

$$\mu = \frac{\sum xP(x)}{\sum P(x)} \qquad \sigma = \sqrt{\frac{\sum (x-\mu)^2 P(x)}{\sum P(x)}}$$

Substitute $\sum P(x) = 1$.

$$\mu = \sum xP(x) \quad \dots \text{①} \qquad \sigma = \sqrt{\sum (x-\mu)^2 P(x)} \quad \dots \text{②}$$

Recall from Section 8.1 that there are different kinds of probability distributions, and the binomial distribution is just one of them. Formulas ① and ② apply to any probability distribution. Therefore, to determine the formulas for a binomial distribution, we use the fact that the values of $P(x)$ are binomial probabilities. According to the binomial probability formula on page 456, $P(x) = {}_nC_x\, p^x(1-p)^{n-x}$. Mathematicians have substituted this formula into equations ① and ②, and simplified the results to obtain $\mu = np$ and $\sigma = \sqrt{np(1-p)}$. The algebra involved is difficult, and beyond the scope of this course.

8.4 Normal Distributions

The Crispy Chips Company produces bags of potato chips. The masses of 500 bags were measured to the nearest gram. These data are displayed in the two graphs below. The frequency graph shows the number of bags with each mass. The probability graph shows the probability that a randomly selected bag from the 500 bags has each mass. Data for this graph were obtained by dividing each frequency in the frequency graph by 500.

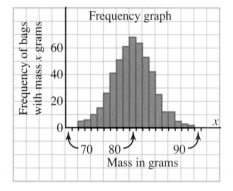

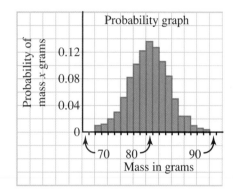

The Crispy Chips Company makes thousands of bags of potato chips. Visualize measuring the mass of each bag to the nearest tenth of a gram and creating a probability graph similar to the one above. The midpoints of the tops of the bars would form a curve similar to the one in the graph at the right.

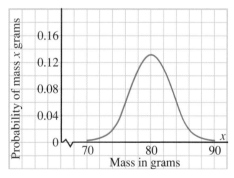

This famous bell-shaped curve is the graph of a *normal distribution.* Many characteristics of manufactured products have frequency graphs that closely approximate a normal distribution. Human characteristics such as height, mass, intelligence, and various kinds of abilities have frequency graphs that closely approximate a normal distribution. For example, the diagram below shows three different normal distributions of test scores.

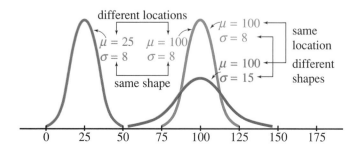

The differences are caused by changes in the mean and standard deviation.

- A change in the mean of the population, μ, causes the curve to be shifted right or left.
- A change in the standard deviation, σ, causes the curve to become broader or narrower.

The properties of normal distributions have been studied in great detail. Normal distributions are useful because certain percents of the data values fall within specific distances from the mean. Some examples of these relationships form the *68-95-99 Rule* below.

Properties of a Normal Distribution

Every normal distribution has a mean, μ, and a standard deviation, σ.
The graph is symmetrical about the mean.
Almost all the population lies within 3 standard deviations of the mean.
The horizontal axis is an asymptote.
The total area under the curve is 1.

68-95-99 Rule

About 68% of the population are within 1 standard deviation of the mean.
About 95% of the population are within 2 standard deviations of the mean.
About 99.7% of the population are within 3 standard deviations of the mean.

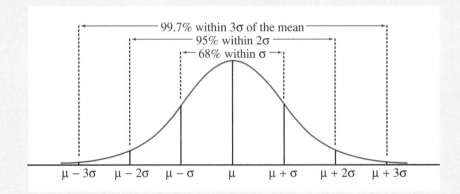

The percents 68%, 95%, and 99% are determined from the equation of the curve (see *Exploring with a Graphing Calculator*, page 496). It is not necessary to know the equation of the curve to use a normal distribution.

Populations that have a normal distribution are said to be *normally distributed*. The 68-95-99 rule can be used to make predictions about normally distributed populations.

Example

In a shipment of oranges, the oranges have diameters that are normally distributed with a mean of 8.0 cm and a standard deviation of 1.5 cm.

a) Sketch a graph to show the information about the oranges that is provided by the properties of a normal distribution on page 486.

b) Sketch graphs to show other information about the oranges that can be deduced from the results in part a.

Solution

a) The mean is 8.0 cm and the standard deviation is 1.5 cm. Since the oranges are normally distributed, about 68% of them have diameters that are within 1 standard deviation of the mean. This means that about 68% of the oranges have diameters between (8.0 − 1.5) cm and (8.0 + 1.5) cm; that is, between 6.5 cm and 9.5 cm (below left).

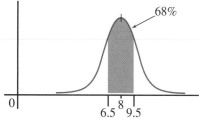

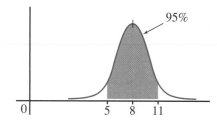

About 95% of the oranges have diameters that are within 2 standard deviations of the mean. This means that 95% of the oranges have diameters between (8.0 − 2 × 1.5) cm and (8.0 + 2 × 1.5) cm; that is, between 5.0 cm and 11.0 cm (above right).

b) Since the graph is symmetrical about the mean, several other related facts can be deduced. For example:

84% of the oranges have diameters less than 9.5 cm.

13.5% of the oranges have diameters between 9.5 cm and 11.0 cm.

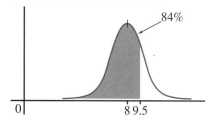

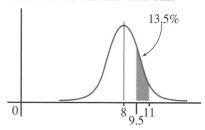

Recall that the range of a set of data is determined by subtracting the lowest value from the highest value. In the *Example*, the range cannot be determined using this definition because the diameters of the individual oranges are not known. However, we can estimate the range. We know that about 99.7% of the oranges have diameters within 3 standard deviations of the mean. This means that most of the oranges have diameters between (8.0 − 3 × 1.5) cm and (8.0 + 3 × 1.5) cm; that is, between 3.5 cm and 12.5 cm. Hence, a reasonable estimate of the range is 12.5 cm − 3.5 cm, or 9.0 cm.

DISCUSSING THE IDEAS

1. Look at the normal curves on page 485. Explain how the first curve would change in each case.

   a) The mean is the same and the standard deviation is 16.

   b) The mean is 75 and the standard deviation is the same.

   c) The mean is 50 and the standard deviation is 11.

2. a) Explain how the results in part b of the *Example* were determined.

   b) What other information about the oranges could have been determined? Explain.

3. Following the *Example*, an estimate of the range, 9.0 cm, was found by subtracting 3.5 cm from 12.5 cm.

   a) What is a faster way to obtain the 9.0 cm number? Explain.

   b) What are the advantages and disadvantages of estimating the range this way? Explain.

4. Describe the kinds of problems that can be solved using the information about a normal distribution that is presented above.

8.4 EXERCISES

A 1. Three sets of data are described. Match each data set with the graph most likely to represent its data distribution. Use each graph only once. Explain.

   a) the masses of the posts in a bundle of 200 wooden fence posts

   b) the ages of people in a municipal park on Sunday afternoon

   c) the marks received by students on a standardized achievement test

   i) 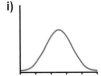   ii) 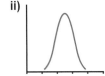   iii)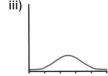

B 2. A barge contains a load of gravel. Assume the masses of the stones are normally distributed. The mean mass is 5.0 g and the standard deviation is 0.5 g.

   a) Sketch a normal curve to show the distribution of the masses of the stones. Mark the points that are 1, 2, and 3 standard deviations from the mean.

b) What percent of the shipment has the masses indicated?

    i) greater than 0.5 g   **ii)** less than 0.5 g   **iii)** between 4.5 g and 5.5 g

    iv) between 4.0 g and 6.0 g   **v)** between 5.0 g and 5.5 g   **vi)** between 5.5 g and 6.0 g

    vii) greater than 4.5 g   **viii)** less than 4.0 g

c) Estimate the range for the masses of the stones.

3. A shipment of logs has a mean diameter of 15 cm and a standard deviation of 2 cm. Assume the diameters of the logs are normally distributed.

    a) By how many standard deviations does a log with a diameter of 11 cm differ from the mean?

    b) What percent of the logs have a diameter less than 11 cm?

    c) Estimate the range for the diameters of the logs.

    d) Explain why the range cannot be determined exactly.

4. Open the *Population and Food Production* database. Sort the data by Year and Region. Assume the data are normally distributed.

    a) For African nations, determine the mean and standard deviation of calories per capita per day.

    b) How many African nations do you expect to be within 1 standard deviation of the mean?

    c) From the database, determine how many African nations do fall within 1 standard deviation of the mean. Compare this answer to the answer to part b.

5. A soft-drink manufacturer has a machine that fills 375-mL cans. The machine is designed to place 377.0 mL in the can with a standard deviation of 0.5 mL.

    a) What percent of the cans will have less than 376.0 mL?

    b) What percent of the cans will have less than 375.5 mL?

    c) Cans containing less than 375.5 mL are rejected. In a production run of 50 000, how many cans will be rejected?

    d) What is the probability that a randomly selected can will contain a volume greater than 376.0 mL?

COMMUNICATING THE IDEAS

Write to explain what is meant by a normal distribution and why normal distributions are useful. Include some examples to illustrate your explanation.

In Section 8.4, we used normal distributions to relate data values to the mean in terms of standard deviations. Every problem used a different normal distribution, depending on the mean and standard deviation of the population under consideration. Also, the problems were limited to data values that were exactly 1, 2, or 3 standard deviations from the mean. In this section, we introduce the standard normal distribution that can be used in every problem for any data values.

The *standard normal distribution* has a mean of 0 and a standard deviation of 1. When you use the standard normal distribution, it is customary to use the letter z instead of x to represent the numbers along the horizontal axis. These numbers represent the number of standard deviations from the origin. The horizontal axis is a *standard deviation scale*.

Standard Normal Distribution

The graph is symmetrical about the mean.   The mean is 0.
The horizontal axis is an asymptote.   The standard deviation is 1.
The total area under the curve is 1.

68-95-99 Rule

About 68% of the area is within 1 unit of the mean.
About 95% of the area is within 2 units of the mean.
About 99.7% of the area is within 3 units of the mean.

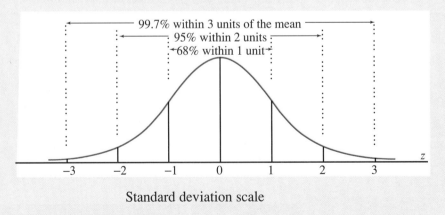

Standard deviation scale

To work with other areas, we define a new function $A(z)$.

$A(z)$ = area under the standard normal curve on the left side of a
vertical line corresponding to z

Example where $z < 0$

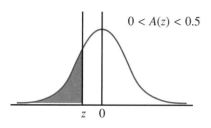

Example where $z > 0$

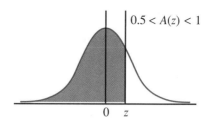

Values of the area function can be found with the TI-83 graphing calculator. You can also use the table on pages 494, 495.

Using a graphing calculator

For the TI-83 graphing calculator, clear all equations in the Y= list. Press [STAT PLOT] **1**, then select Off. Set the window to Xmin = –4.7, Xmax = 4.7, Xscl = 1, Ymin = –0.15, Ymax = 0.5, Yscl = 0. These settings can be used for all problems, unless otherwise stated.

Finding the area between two specified values
To find the area between –1.4 and 1.4, press [DISTR]
[▶]**1** to display ShadeNorm(. Then continue with
–1.4 [,] 1.4 [ )] [ENTER] to display this screen.
About 0.8385 of the area lies between –1.4 and 1.4.
This is 83.85% of the total area under the curve.
Since the z-axis is a standard deviation scale, this
means that about 83.8% of the area is within 1.4
standard deviations of the mean.

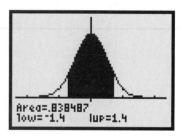

To complete another calculation, clear the drawing screen by pressing [DRAW] [ENTER].

Finding the area to the left of a specified value
You must always enter two values, one for the left side of
the area, the other for the right side. The area continues
indefinitely to the left, but the area beyond –5 is so small
that it can be ignored (see exercise 10). To find the area
to the left of 0.54, press [DISTR] [▶]**1** –5 [,] 0.54 [ )] [ENTER]
to display this screen. About 0.7054 of the area lies between
–5 and 0.54. This is 70.54% of the total area under the curve.

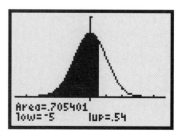

Using the table on pages 494 and 495

Finding the area to the left of a specified value
To find the area to the left of 0.54, move down the left column of the table to 0.5, then along the line to the entry in the column headed 0.04 to obtain 0.7054. About 0.7054 of the area lies to the left of 0.54. This is 70.54% of the total area under the curve.

Finding the area between two specified values

To find the area between −1.4 and 1.4, complete the preceding step twice and subtract the areas. The area to the left of 1.4 is 0.9192. The area to the left of −1.4 is 0.0808. The area between −1.4 and 1.4 is 0.9192 − 0.0808 = 0.8384. This is 83.84% of the total area under the curve.

Since the z-axis is a standard deviation scale, this means that about 83.8% of the area is within 1.4 standard deviations of the mean.

DISCUSSING THE IDEAS

Use the table on pages 494 and 495. Explain your answer to each question.

1. In the column headings on page 494, why do the numbers decrease from 0.09 to 0.00?

2. What area would you use in each case? Explain.

a) when $z > 3.49$ **b)** when $z < -3.49$

3. On page 494, the values of z are negative. Why are the areas not negative?

4. How are the numbers on page 494 related to the numbers on page 495? Explain.

8.5 EXERCISES

A **1.** Use the table on pages 494 and 495. Determine each area.

a) to the left of 2.31 **b)** to the left of −1.23 **c)** to the right of −1.88

d) between −1.41 and 1.41 **e)** between −2.3 and 2.3 **f)** between −0.66 and 0.66

2. Determine each area.

a) between 0.4 and 1.0 **b)** between 1.8 and 2.6 **c)** between −2.8 and −1.8

d) to the left of 1.6 **e)** to the left of −1.45 **f)** to the right of 2.4

3. Choose one part of exercise 1 and one part of exercise 2. Illustrate each percent as a shaded area under the standard normal curve.

B **4.** Use either the table on pages 494, 495, or a graphing calculator. Determine the area that satisfies the inequalities in each list. Explain any patterns you find.

a) $-1 < z < 1$
    $-0.1 < z < 0.1$
    $-0.01 < z < 0.01$

b) $-5 < z < 5$
    $-0.5 < z < 0.5$
    $-0.05 < z < 0.05$

c) $-2.5 < z < -1.5$
$-1.5 < z < -0.5$
$-0.5 < z < 0.5$
$0.5 < z < 1.5$
$1.5 < z < 2.5$

d) $z < 1.6$
$z > 1.6$
$-1.6 < z < 1.6$
$z > -1.6$
$z < -1.6$

5. Use the table on pages 494, 495, or a graphing calculator to verify the 68-95-99 rule.

6. Determine the area within 0.5 standard deviations of each z-value.

a) $z = 0$ **b)** $z = 1$ **c)** $z = -1$ **d)** $z = 2$ **e)** $z = -2$

7. Determine each area.

a) within 1 standard deviation of $z = 0.5$

b) within 0.1 standard deviations of $z = 0.5$

c) within 0.01 standard deviations of $z = 0.5$

8. When you use the shadeNorm command, the TI-83 calculator draws the appropriate graph before it displays the area. You can obtain the area without the graph by using the normalcdf command in the distribution menu.

a) To calculate the area in the first screen on page 491, press [DISTR] **2** to display normalcdf(. Then continue with -1.4 [,] 1.4 [ )] [ENTER] to display the result. Check that this is the area in the screen on page 491.

b) Choose one part of exercise 1 and one part of exercise 2. Use normalcdf to determine each area.

C **9.** Refer to the definition of the area function $A(z)$ on page 490.

a) Visualize how $A(z)$ changes as z increases from -3.5 to 3.5.

b) Sketch a graph of $y = A(z)$ for $-3.5 \le z \le 3.5$.

c) Describe the graph and list some of its properties.

10. On page 491, in the example about finding the area to the left of 0.54, we stated that -5 could be used on the left side of the region because the area to the left of -5 is so small that it can be ignored. Complete parts a and b to demonstrate this is true.

a) Find the area between -10 and 0.54, and compare it with the result on page 491.

b) Find the area between -10^{99} and -5 to determine the area that was ignored.

COMMUNICATING THE IDEAS

Write to explain how the standard normal distribution is different from, and similar to, a binomial distribution. Include as many differences and similarities as you can.

Areas under a Standard Normal Curve to the left of z									
z	0.09	0.08	0.07	0.06	0.05	0.04	0.03	0.02	0.01	0.00
−3.4	0.0002	0.0003	0.0003	0.0003	0.0003	0.0003	0.0003	0.0003	0.0003	0.0003
−3.3	0.0003	0.0004	0.0004	0.0004	0.0004	0.0004	0.0004	0.0005	0.0005	0.0005
−3.2	0.0005	0.0005	0.0005	0.0006	0.0006	0.0006	0.0006	0.0006	0.0007	0.0007
−3.1	0.0007	0.0007	0.0008	0.0008	0.0008	0.0008	0.0009	0.0009	0.0009	0.0010
−3.0	0.0010	0.0010	0.0011	0.0011	0.0011	0.0012	0.0012	0.0013	0.0013	0.0013
−2.9	0.0014	0.0014	0.0015	0.0015	0.0016	0.0016	0.0017	0.0018	0.0018	0.0019
−2.8	0.0019	0.0020	0.0021	0.0021	0.0022	0.0023	0.0023	0.0024	0.0025	0.0026
−2.7	0.0026	0.0027	0.0028	0.0029	0.0030	0.0031	0.0032	0.0033	0.0034	0.0035
−2.6	0.0036	0.0037	0.0038	0.0039	0.0040	0.0041	0.0043	0.0044	0.0045	0.0047
−2.5	0.0048	0.0049	0.0051	0.0052	0.0054	0.0055	0.0057	0.0059	0.0060	0.0062
−2.4	0.0064	0.0066	0.0068	0.0069	0.0071	0.0073	0.0075	0.0078	0.0080	0.0082
−2.3	0.0084	0.0087	0.0089	0.0091	0.0094	0.0096	0.0099	0.0102	0.0104	0.0107
−2.2	0.0110	0.0113	0.0116	0.0119	0.0122	0.0125	0.0129	0.0132	0.0136	0.0139
−2.1	0.0143	0.0146	0.0150	0.0154	0.0158	0.0162	0.0166	0.0170	0.0174	0.0179
−2.0	0.0183	0.0188	0.0192	0.0197	0.0202	0.0207	0.0212	0.0217	0.0222	0.0228
−1.9	0.0233	0.0239	0.0244	0.0250	0.0256	0.0262	0.0268	0.0274	0.0281	0.0287
−1.8	0.0294	0.0301	0.0307	0.0314	0.0322	0.0329	0.0336	0.0344	0.0351	0.0359
−1.7	0.0367	0.0375	0.0384	0.0392	0.0401	0.0409	0.0418	0.0427	0.0436	0.0446
−1.6	0.0455	0.0465	0.0475	0.0485	0.0495	0.0505	0.0516	0.0526	0.0537	0.0548
−1.5	0.0559	0.0571	0.0582	0.0594	0.0606	0.0618	0.0630	0.0643	0.0655	0.0668
−1.4	0.0681	0.0694	0.0708	0.0721	0.0735	0.0749	0.0764	0.0778	0.0793	0.0808
−1.3	0.0823	0.0838	0.0853	0.0869	0.0885	0.0901	0.0918	0.0934	0.0951	0.0968
−1.2	0.0985	0.1003	0.1020	0.1038	0.1056	0.1075	0.1093	0.1112	0.1131	0.1151
−1.1	0.1170	0.1190	0.1210	0.1230	0.1251	0.1271	0.1292	0.1314	0.1335	0.1357
−1.0	0.1379	0.1401	0.1423	0.1446	0.1469	0.1492	0.1515	0.1539	0.1562	0.1587
−0.9	0.1611	0.1635	0.1660	0.1685	0.1711	0.1736	0.1762	0.1788	0.1814	0.1841
−0.8	0.1867	0.1894	0.1922	0.1949	0.1977	0.2005	0.2033	0.2061	0.2090	0.2119
−0.7	0.2148	0.2177	0.2206	0.2236	0.2266	0.2296	0.2327	0.2358	0.2389	0.2420
−0.6	0.2451	0.2483	0.2514	0.2546	0.2578	0.2611	0.2643	0.2676	0.2709	0.2743
−0.5	0.2776	0.2810	0.2843	0.2877	0.2912	0.2946	0.2981	0.3015	0.3050	0.3085
−0.4	0.3121	0.3156	0.3192	0.3228	0.3264	0.3300	0.3336	0.3372	0.3409	0.3446
−0.3	0.3483	0.3520	0.3557	0.3594	0.3632	0.3669	0.3707	0.3745	0.3783	0.3821
−0.2	0.3859	0.3897	0.3936	0.3974	0.4013	0.4052	0.4090	0.4129	0.4168	0.4207
−0.1	0.4247	0.4286	0.4325	0.4364	0.4404	0.4443	0.4483	0.4522	0.4562	0.4602
−0.0	0.4641	0.4681	0.4721	0.4761	0.4801	0.4840	0.4880	0.4920	0.4960	0.5000

				Areas under a Standard Normal Curve to the left of z						
z	0.00	0.01	0.02	0.03	0.04	0.05	0.06	0.07	0.08	0.09
0.0	0.5000	0.5040	0.5080	0.5120	0.5160	0.5199	0.5239	0.5279	0.5319	0.5359
0.1	0.5398	0.5438	0.5478	0.5517	0.5557	0.5596	0.5636	0.5675	0.5714	0.5753
0.2	0.5793	0.5832	0.5871	0.5910	0.5948	0.5987	0.6026	0.6064	0.6103	0.6141
0.3	0.6179	0.6217	0.6255	0.6293	0.6331	0.6368	0.6406	0.6443	0.6480	0.6517
0.4	0.6554	0.6591	0.6628	0.6664	0.6700	0.6736	0.6772	0.6808	0.6844	0.6879
0.5	0.6915	0.6950	0.6985	0.7019	0.7054	0.7088	0.7123	0.7157	0.7190	0.7224
0.6	0.7257	0.7291	0.7324	0.7357	0.7389	0.7422	0.7454	0.7486	0.7517	0.7549
0.7	0.7580	0.7611	0.7642	0.7673	0.7704	0.7734	0.7764	0.7794	0.7823	0.7852
0.8	0.7881	0.7910	0.7939	0.7967	0.7995	0.8023	0.8051	0.8078	0.8106	0.8133
0.9	0.8159	0.8186	0.8212	0.8238	0.8264	0.8289	0.8315	0.8340	0.8365	0.8389
1.0	0.8413	0.8438	0.8461	0.8485	0.8508	0.8531	0.8554	0.8577	0.8599	0.8621
1.1	0.8643	0.8665	0.8686	0.8708	0.8729	0.8749	0.8770	0.8790	0.8810	0.8830
1.2	0.8849	0.8869	0.8888	0.8907	0.8925	0.8944	0.8962	0.8980	0.8997	0.9015
1.3	0.9032	0.9049	0.9066	0.9082	0.9099	0.9115	0.9131	0.9147	0.9162	0.9177
1.4	0.9192	0.9207	0.9222	0.9236	0.9251	0.9265	0.9279	0.9292	0.9306	0.9319
1.5	0.9332	0.9345	0.9357	0.9370	0.9382	0.9394	0.9406	0.9418	0.9429	0.9441
1.6	0.9452	0.9463	0.9474	0.9484	0.9495	0.9505	0.9515	0.9525	0.9535	0.9545
1.7	0.9554	0.9564	0.9573	0.9582	0.9591	0.9599	0.9608	0.9616	0.9625	0.9633
1.8	0.9641	0.9649	0.9656	0.9664	0.9671	0.9678	0.9686	0.9693	0.9699	0.9706
1.9	0.9713	0.9719	0.9726	0.9732	0.9738	0.9744	0.9750	0.9756	0.9761	0.9767
2.0	0.9772	0.9778	0.9783	0.9788	0.9793	0.9798	0.9803	0.9808	0.9812	0.9817
2.1	0.9821	0.9826	0.9830	0.9834	0.9838	0.9842	0.9846	0.9850	0.9854	0.9857
2.2	0.9861	0.9864	0.9868	0.9871	0.9875	0.9878	0.9881	0.9884	0.9887	0.9890
2.3	0.9893	0.9896	0.9898	0.9901	0.9904	0.9906	0.9909	0.9911	0.9913	0.9916
2.4	0.9918	0.9920	0.9922	0.9925	0.9927	0.9929	0.9931	0.9932	0.9934	0.9936
2.5	0.9938	0.9940	0.9941	0.9943	0.9945	0.9946	0.9948	0.9949	0.9951	0.9952
2.6	0.9953	0.9955	0.9956	0.9957	0.9959	0.9960	0.9961	0.9962	0.9963	0.9964
2.7	0.9965	0.9966	0.9967	0.9968	0.9969	0.9970	0.9971	0.9972	0.9973	0.9974
2.8	0.9974	0.9975	0.9976	0.9977	0.9977	0.9978	0.9979	0.9979	0.9980	0.9981
2.9	0.9981	0.9982	0.9982	0.9983	0.9984	0.9984	0.9985	0.9985	0.9986	0.9986
3.0	0.9987	0.9987	0.9987	0.9988	0.9988	0.9989	0.9989	0.9989	0.9990	0.9990
3.1	0.9990	0.9991	0.9991	0.9991	0.9992	0.9992	0.9992	0.9992	0.9993	0.9993
3.2	0.9993	0.9993	0.9994	0.9994	0.9994	0.9994	0.9994	0.9995	0.9995	0.9995
3.3	0.9995	0.9995	0.9995	0.9996	0.9996	0.9996	0.9996	0.9996	0.9996	0.9997
3.4	0.9997	0.9997	0.9997	0.9997	0.9997	0.9997	0.9997	0.9997	0.9997	0.9998

Where Does the 68-95-99 Rule Come From?

The equation of the standard normal curve is $y = \dfrac{1}{\sqrt{2\pi}} e^{-\frac{x^2}{2}}$, where

$e \doteq 2.718\,28$ is the base of the natural logarithms (see *Mathematics File*, page 141). You can use the TI-83 graphing calculator to graph this function and verify the 68-95-99 rule, and any other area calculations with the standard normal curve.

1. Set the graphing window to Xmin = –3.5, Xmax = 3.5, Ymin = –0.15, Ymax = 0.5. Enter the equation Y1 = $(1/\sqrt{(2\pi)})e\wedge(-X^2/2)$, then press GRAPH to obtain the standard normal curve.

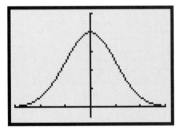
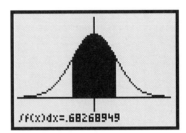

2. To calculate an area under the curve, press [CALC] **7.** You will be prompted for a lower limit. Press –1 [ENTER], and the cursor will jump to this point. You will be prompted for an upper limit. Press 1 [ENTER]. The cursor will draw the area under the curve between –1 and 1, and calculate the area as shown above. The symbol $\int f(x)dx$ is introduced in calculus, and represents an area under the graph of the function $y = f(x)$. This screen shows that the area under the curve from –1 to 1 is about 0.6827, or 68.27% of the total area.

3. Use the same method.

   a) i) Show that the area under the curve from –2 to 2 is about 95% of the total area.

   ii) Show that the area under the curve from –3 to 3 is about 99.7% of the total area.

   b) Verify the areas calculated on page 491.

4. The TI-83 graphing calculator has a function that graphs the normal curve. Press [Y=] [DISTR] **1** [X,T,θ,n] [ )] [ENTER] to enter the equation Y2 = normalpdf(X) in the Y= list. Then press GRAPH to graph this equation along with the equation above. Compare the graphs of the two equations. Use the trace key to show that they are the same.

8.6 Modelling Real Situations Using Normal Distributions

The heights of Canadian women between the ages of 18 and 24 approximate a normal distribution with mean 162.6 cm and standard deviation 6.4 cm. We can use this information to determine the percent of these women with heights in any given interval. For example, we can determine the percent of Canadian women between the ages of 18 and 24 who are over 170 cm tall.

Consider a graph of a normal distribution with mean $\mu = 162.6$ and standard deviation $\sigma = 6.4$. Visualize one standard deviation above the mean, which is $(162.6 + 6.4)$ cm, or 169.0 cm.

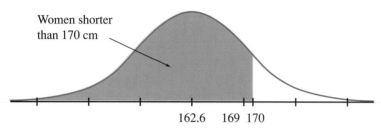

Normal curve with $\mu = 162.6$, $\sigma = 6.4$

Women shorter than 170 cm

162.6 169 170

Observe that 170 cm is $(170 - 162.6)$ cm, or 7.4 cm above the mean. This is slightly more than one standard deviation. Since one standard deviation is 6.4 cm, we can determine how many standard deviations 170 cm is above the mean by dividing the difference by 6.4.

$$\frac{170 - 162.6}{6.4} \doteq 1.16$$

Hence, 170 cm is 1.16 standard deviations above the mean. This is the same number of standard deviations above the mean as a z-score of 1.16 represents on the graph of the normal curve. Hence, the shaded area under the two curves represents the same fraction of the total area.

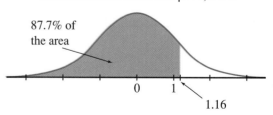

Standard normal curve with $\mu = 0$, $\sigma = 1$

87.7% of the area

0   1

1.16

We can use the table on pages 494 and 495 or a graphing calculator to determine the area to the left of 1.16. The result is 0.8770; so, the shaded area represents 87.70% of the total area. The same is true for the first graph. Hence, about 87.7% of the women are less than 170 cm tall. This means that 12.3% of Canadian women between the ages of 18 and 24 are over 170 cm tall.

In the above solution, the key step is to calculate the z-score, which is needed to determine the corresponding area under the standard normal curve. This was done by subtracting the mean from the given value, then dividing by the standard deviation. That is, the formula for calculating the z-score is $z = \frac{x - \mu}{\sigma}$.

These steps summarize the procedure.

Step 1. Sketch a normal curve with the given mean and standard deviation. Label the interval being considered, and shade the area above it.

Step 2. Convert the endpoints of the interval to z-scores, using the formula $z = \frac{x - \mu}{\sigma}$. Mark these endpoints on the graph.

Step 3. Use the table on pages 494 and 495 or a graphing calculator to determine the required area.

Step 4. Answer the problem.

Example 1

In a shipment of eggs, the eggs have a normal distribution with a mean height of 5.20 cm and a standard deviation of 0.12 cm. What is the probability that a randomly selected egg has a height between 5.00 cm and 5.30 cm?

Solution

Step 1.   Normal curve with $\mu = 5.20, \sigma = 0.12$

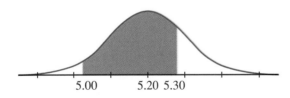

Step 2.   Convert each endpoint to a z-score.

$$z = \frac{x - \mu}{\sigma} \qquad\qquad z = \frac{x - \mu}{\sigma}$$

$$= \frac{5.00 - 5.20}{0.12} \qquad\qquad = \frac{5.30 - 5.20}{0.12}$$

$$\doteq -1.667 \qquad\qquad \doteq 0.8333$$

Step 3.   *Using the table*
The area for a z-score of 0.83 is 0.7967.
The area for a z-score of -1.67 is 0.0475.
Hence, the required area is $0.7967 - 0.0475 = 0.7492$

Using the TI-83 graphing calculator
Press DRAW ENTER ENTER to clear the
drawing screen.
Press DISTR ▶ 1 −1.667 , 0.833) ENTER.
The area is approximately 0.749 821.

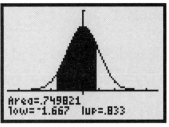

Area=.749821
low=-1.667 lup=.833

Step 4.   The probability that a randomly selected egg has a height
           between 5.00 cm and 5.30 cm is about 0.75, or 75%.

Since the total area under the standard normal curve is 1, the areas of regions
under the curve can be considered as probabilities. For example, since one-half
the area lies to the right of the mean, the probability that $z > 0$ is 0.5. We write
$P(z > 0) = 0.5$. Similarly, since 95% of the area lies between −2 and 2, we write
$P(-2 < z < 2) = 0.95$.

Example 2

A variable, z, is normally distributed. Determine each probability to 4 decimal
places and illustrate it as a shaded area under the standard normal curve.
a) $P(-0.64 < z < 0.64)$ **b)** $P(z > 0.64)$

Solution

Use a graphing calculator or the table on pages 494 and 495.
a) $P(-0.64 < z < 0.64) \doteq 0.7389 - 0.2611$
$\doteq 0.4778$

b) $P(z > 0.64) = 1 - P(z < 0.64)$
$\doteq 1 - 0.7389$
$\doteq 0.2611$

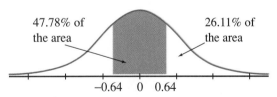

47.78% of
the area

26.11% of
the area

−0.64 0 0.64

Example 3

a) Explain the meaning of the equation $P(z < a) = 0.90$.

b) Determine the value of a to 2 decimal places.

Solution

a) The equation $P(z < a) = 0.90$ means that the area under the standard normal curve to the left of a is 0.90. That is, 90% of the area is to the left of a.

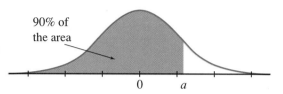

90% of the area

0   a

b)

> ### *Think ...*
>
> This is the inverse of the type of problems in *Example 2*.

Using the TI-83 graphing calculator

Press [DISTR] **3** 0.90 [ )] [ENTER] to display invNorm(.9) = 1.281551567.
Therefore, $P(z < 1.28) \doteq 0.9$, and $a \doteq 1.28$

Using the table on pages 494 and 495

Look in the table for the area closest to 0.90. This is 0.8997. Read the number at the far left of the line, 1.2, then the number at the top of the column (0.08) to obtain an area of 1.28.
Therefore, $P(z < 1.28) \doteq 0.9$, and $a \doteq 1.28$

DISCUSSING THE IDEAS

1. In the calculation of a z-score, does it matter whether we find $x - \mu$ or $\mu - x$? Explain.

2. Explain why the formula $z = \frac{x - \mu}{\sigma}$ expresses a value of x in terms of the number of standard deviations it is from the mean.

3. Explain why the area of a region under the standard normal curve represents a probability.

B **1.** The Stanford-Binet IQ scores of Canadians are normally distributed with a mean of 100 and a standard deviation of 16.

a) According to this test, what percent of Canadians lie in each interval?
   i) an IQ less than 100   **ii)** an IQ less than 116
   iii) an IQ greater than 116   **iv)** an IQ between 84 and 116

b) Canada has a population of about 30 000 000. How many of these people do you expect would have IQs above 140?

c) What is the probability that a randomly selected person has an IQ below 80?

2. The eggs in the shipment in *Example 1* are to be placed in containers that can accommodate heights between 4.92 cm and 5.48 cm.

a) Show this information on a graph.

b) What percent of the eggs must be stored in another container?

3. Bean plants in a field have heights that are normally distributed with a mean height of 20.0 cm and a standard deviation of 3.5 cm. Determine the percent of plants that would be expected to have each set of height characteristics.

a) between 20.0 cm and 24.9 cm tall

b) between 15.8 cm and 24.2 cm tall

c) taller than 29.1 cm

4. The results of a district-wide grade 12 mathematics achievement test were normally distributed with a mean of 220 and a standard deviation of 40. From the results, the district decided to do mathematics remediation work with all students who scored less than 116.

a) What percent of the grade 12 students were scheduled for remediation?

b) What is the probability that a randomly selected student is not scheduled for remediation?

5. A sample of 148 people produces a mean body temperature of 37.0°C, with a standard deviation of 0.35°C. Assume the temperatures are normally distributed.

a) Determine the expected number of people in the sample with body temperatures above 37.5°C.

b) Determine the expected number of people in the sample with body temperatures below 36.5°C.

In exercise 5, you assumed that the body temperatures are normally distributed.

- Do you think this assumption is reasonable for a sample this size? Explain.
- What would you expect to happen to the distribution of body temperatures if a much larger sample was involved? Explain.
- Is it possible for the body temperatures in a group of 148 people to be exactly normally distributed? Explain.

6. For entry into the Canadian Armed Forces, the standards for heights were different for men and women. The heights for men had to be between 158 cm and 194 cm, and for women between 152 cm and 184 cm. Assume the populations are normally distributed. The men's heights have a mean of 176 cm with a standard deviation of 8 cm. The women's heights have a mean of 163 cm with a standard deviation of 7 cm. Are the two height standards equivalent? Explain.

7. Conduct this activity with a partner. Have your partner suspend a 30-cm ruler as shown. Place your fingers at the bottom but do not allow them to touch the ruler. Your partner will drop the ruler at a random time. Catch the ruler as quickly as you can between your thumb and forefinger. Do not move your hand downward; brace it against a table if necessary. Conduct the activity 10 times. Each time, note and record the position where the ruler was caught.

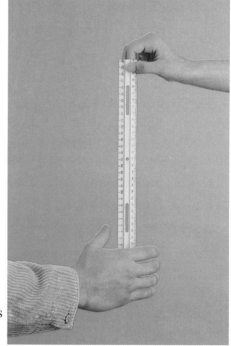

   a) Use your 10 results to calculate the mean position at which you caught the ruler.

   b) Determine the mean position for the whole class.

   c) Determine the standard deviation for the class.

   d) Compute your z-score based on the answers to parts a, b, and c.

   e) Interpret your result to part d. How do your reflexes compare to those of your classmates?

8. The variable z is normally distributed. Determine each probability.

   a) $P(-0.32 < z < 0.32)$

   b) $P(-1.44 < z < 1.44)$

   c) $P(z < 1.75)$

   d) $P(z < -1.75)$

   e) $P(z > -0.69)$

   f) $P(z > 2.69)$

9. Open the *NBA* database and sort the data by Season. Isolate the 1992–93 season.

   a) Calculate the mean and standard deviation for the number of games played.

   b) What is the z-score for a player who played in 75 games?

   c) What percent of players have played between 65 and 75 games?

10. The variable z is normally distributed.

   a) Determine the probabilities in each list and illustrate them as shaded areas under the standard normal curve.

     i) $P(0.40 < z < 0.60)$ **ii)** $P(1.40 < z < 1.60)$

         $P(0.45 < z < 0.55)$ $P(1.45 < z < 1.55)$

         $P(0.49 < z < 0.51)$ $P(1.49 < z < 1.51)$

   b) Visualize continuing the patterns in each list in part a. Determine each probability. Explain.

     i) $P(z = 0.50)$ **ii)** $P(z = 1.50)$

11. The variable z is normally distributed. Determine each value of a.

   a) $P(0 < z < a) = 0.2$ **b)** $P(0 < z < a) = 0.35$

   c) $P(z < a) = 0.85$ **d)** $P(z < a) = 0.65$

   e) $P(z > a) = 0.3$ **f)** $P(z > a) = 0.1$

12. a) This is a standard normal curve. What is the total area under this curve?

   b) Suppose $P(a < z < b) = 0.5$. What is the area under the curve for the interval $a < z < b$?

   c) Suppose $P(z < b) = 0.8$. What is the value of b?

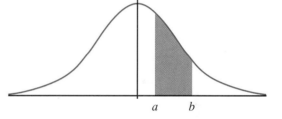

C **13.** In exercise 10b, you should have found the probability that z takes on any specific value is 0. In other words, for any real value of a, the probability that z equals a is 0. However, the sum of all the probabilities is 1. Explain this paradox.

COMMUNICATING THE IDEAS

Write to explain how the standard normal distribution is used to solve problems involving any normal distribution. Make up an example that is different from any of the examples or exercises in this section to illustrate your explanation.

Converting Scores

On page 462, you considered the problem of converting scores to have a predetermined mean and a predetermined percent of scores above a passing grade. We will assume that a passing grade is 50%. Different models for converting scores are possible. You will investigate some of these models below.

The actual scores obtained on an examination are the *raw scores*. We will make these assumptions.

- All scores are out of 100, unless stated otherwise.
- The raw scores are normally distributed with mean 46 and standard deviation 14. The distribution of the marks is shown.

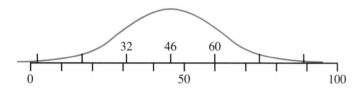

DEVELOP A MODEL

1. Use the standard normal distribution to determine the percent of the students who had scores of 50 or more.

One possible model is to add the same amount to every score.

2. Suppose every score is increased by 10. The distribution of the marks would be as shown.

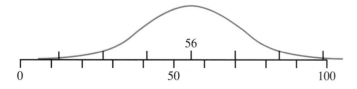

   a) What are the mean and the standard deviation for the converted scores?

   b) What percent of the converted scores are 50 or more?

3. By how much should every score be increased so that 90% of the scores are 50 or more?

4. Do you think this model is fair to all students? Explain.

An improved model uses the standard normal distribution. We can design the converted scores to fit any normal distribution by selecting an appropriate mean and standard deviation. The mean represents the desired average mark. The standard deviation represents how little or how much we want the scores to be spread around the mean.

5. Suppose we want a mean of 62 and a standard deviation of 10. The distribution of the marks would be as shown.

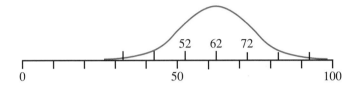

a) What percent of the converted scores are passes?

b) One raw score is 55.
   i) What percent of the raw scores are less than 55?
   ii) Determine the converted score.

c) Determine the converted score for each raw score.
   i) 35   ii) 75

d) In parts b and c, compare what happened to the raw scores of 35, 55, and 75. Does this seem fair? Explain.

6. In exercise 5, what percent of the converted scores are over 50?

 LOOK AT THE IMPLICATIONS

In exercises 5 and 6, we selected a standard deviation, then determined the percent of the scores that would be over 50. We usually select the percent of scores over 50, then determine the standard deviation.

7. Suppose we want a mean of 62 and 95% of the scores over 50. The distribution of the marks would be as shown.

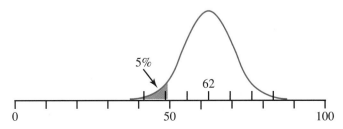

a) Determine the standard deviation for the converted scores.

b) A raw score is 55. Determine the converted score.

c) Determine the converted score for each raw score.
  i) 35   **ii)** 75

d) In parts b and c, compare what happened to the raw scores of 35, 55, and 75. Does this seem fair? Explain.

8. Write a set of instructions for another student to be able to change any raw score to a converted score.

 REVISIT THE SITUATION

Recall that the scores in the British Columbia 1997 mathematics examination were converted to a score out of 800.

9. One raw score is 55 out of 100. One model for converting this score to a score out of 800 is to multiply by 8 to obtain 440. Do you think it would be a good idea to convert all the scores this way? Explain.

10. Suppose we want a mean of 500 and a standard deviation of 200.

a) One raw score is 55. Determine the converted score.

b) Determine the converted score for each raw score.
  i) 35   **ii)** 75

c) What percent of the converted scores are over 400?

11. Suppose we want a mean of 500 and 90% of the converted scores over 400.

a) Determine the standard deviation for the converted scores.

b) One raw score is 55. Determine the converted score.

c) Determine the converted score for each raw score.
  i) 35   **ii)** 75

12. Write to describe how raw scores are converted to standardized scores. Explain how the conversions enable you to better analyze the data. Use examples of commonly available standardized scores, such as government examination scores, SAT scores, and IQ scores.

8.7 The Normal Approximation to a Binomial Distribution

In this chapter, you encountered two important probability distributions, binomial distributions and normal distributions. Recall that the total area of the bars in the graph of a binomial distribution is 1, and that the area under a normal curve is also 1. These properties, and the property that the two graphs have the same bell shape, suggest there may be some problems where the two kinds of distributions can be used together. Here is an example.

At the beginning of the school term, a mathematics teacher announces that she will give surprise quizzes, and the probability of a quiz on any given day is 0.2. What is the probability that the mathematics teacher gives at least 7 surprise quizzes in the next 30 days? We could solve this problem this way.

P(at least 7 quizzes) = P(7 quizzes) + P(8 quizzes) + P(9 quizzes) + … + P(30 quizzes)

Twenty-four different terms must be calculated, then added. We can determine the terms using the TI-83 graphing calculator. Press [DISTR] **0** 30 [,] 0.2 [ )] [STO▶] [L₂] [ENTER] to calculate and store the probabilities for any number of quizzes in list L2. Then press [STAT] **1** to see the probabilities. To add the probabilities, we must add the 8th to 31st terms in this list. Press [LIST] [▶] [▶] **5** [L₂] [,] 8 [,] 31 [ )] [ENTER]. The sum is shown in the screen below. Hence, the probability that there will be at least 7 surprise quizzes is about 0.39.

This method provides the correct answer, but it is inefficient because many calculations are involved. It also requires a graphing calculator.

A graphical approach suggests another way to solve the problem. The histogram on page 508 was drawn using the numbers in list L2 above. The required probability is the sum of the areas of the bars from 7 to 30. We can estimate this area using a normal curve. A normal curve with approximately the same shape as the histogram is shown on the diagram. There is a significant difference between a binomial distribution and a normal distribution. The graph of a binomial distribution consists of individual points or bars. The graph of a normal distribution is a smooth curve, which is not conveniently subdivided into bars. On the binomial distribution, the bar representing 7 quizzes has width

1 and it occupies the area starting at 6.5. Hence, the area under the curve from 6.5 to 30 is a reasonable estimate for the probability required. This means that we should use 6.5 instead of 7 as the lower endpoint of the interval. This adjustment is called a *continuity correction*.

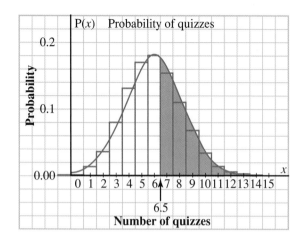

To work with a normal curve, we require its mean and standard deviation. Recall the formulas for the mean, μ, and standard deviation, σ, for a binomial distribution with n trials where the probability of success on each trial is p:

Mean: $\mu = np$ Standard deviation: $\sigma = \sqrt{np(1 - p)}$

We use these formulas to determine the mean and standard deviation of a normal curve that approximates the outline of the histogram above.

In this case, since there are 30 days: $n = 30$
The probability of a quiz on any day is: $p = 0.2$
Substitute in the formulas.

$$\mu = np$$
$$= 30 \times 0.2$$
$$= 6$$

$$\sigma = \sqrt{np(1 - p)}$$
$$= \sqrt{30 \times 0.2 \times 0.8}$$
$$\doteq 2.19$$

Sketch a normal curve with mean 6 and standard deviation 2.19. Mark the interval to the left of 6.5, and shade the area.

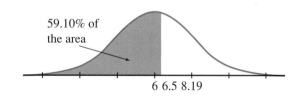

59.10% of the area

6 6.5 8.19

The endpoint of the interval is 6.5. Convert this to a z-score:

$$z = \frac{x - \mu}{\sigma}$$
$$= \frac{6.5 - 6}{2.19}$$
$$\doteq 0.23$$

Using the table on pages 494–495 or a graphing calculator, the area for a *z*-score of 0.23 is 0.5910. Assume the normal curve fits the histogram; the probability that there are fewer than 7 surprise quizzes is about 0.59. Therefore, the probability that there are 7 or more surprise quizzes is 1 – 0.59, or 0.41. This is only an estimate of the required probability because a normal curve cannot fit the histogram exactly.

To approximate binomial probabilities using the normal distribution, follow these steps.

Step 1. *Work with the binomial distribution.*

    a) Determine the values of *n* and *p*.

    b) Calculate the mean and standard deviation using
$$\mu = np \text{ and } \sigma = \sqrt{np(1 - p)}$$

Step 2. *Work with the normal distribution.*

    a) Sketch the normal curve, and shade the area of the appropriate interval.

    b) Apply the continuity correction by visualizing how the areas of the bars of the histogram are related to the area to be calculated.

    c) Convert the endpoints of the interval to *z*-scores.

    d) Use the table on pages 494–495 or a graphing calculator to determine the required area.

Step 3. *Work with the binomial distribution.*

    Answer the problem.

DISCUSSING THE IDEAS

1. Suppose you were the mathematics teacher. How could you make sure that the probability you give a surprise quiz each day is 0.2?

2. Refer to the histogram on page 508. The area under the normal curve from 6.5 to 30 does not equal the total area of the bars from 7 to 30.

    a) Will this cause the estimated probability to be greater than or less than the actual probability? Explain.

    b) Describe a situation in which the estimated probability should be closer to the actual probability than it is in the example.

3. In the second diagram on page 508, why did we mark 6.5 at the end of the interval?

4. Do you always subtract 0.5 when applying the continuity correction? Explain.

8.7 EXERCISES

(A) 1. Calculate the mean and standard deviation for each binomial distribution.
   a) $n = 600$, $p = 0.4$ b) $n = 400$, $p = 0.5$
   c) $p = 0.1$, $n = 120$ d) $p = 0.01$, $n = 1000$

(B) 2. Suppose 20 people are randomly selected to participate in a survey. Assume that $P(man) = P(woman) = 0.5$. Use a normal approximation to estimate the probability that at least 12 women will be chosen.

3. A true-false test has 40 questions. Assume all questions are answered by random guessing. Use a normal approximation to estimate the probability of guessing 25 or more answers correctly.

4. Use a normal approximation to estimate each probability.
   a) P(4, 5, or 6 heads when a coin is tossed 10 times)
   b) P(40 to 60 heads, inclusive, when a coin is tossed 100 times)
   c) P(400 to 600 heads, inclusive, when a coin is tossed 1000 times)
   d) Explain why the answers in parts a, b, and c are significantly different.

5. Use a normal approximation to estimate each probability.
   a) P(6 appears 9, 10, or 11 times when a die is rolled 60 times)
   b) P(6 appears 90 to 110 times, inclusive, when a die is rolled 600 times)
   c) P(6 appears 900 to 1100 times, inclusive, when a die is rolled 6000 times)
   d) Explain why the answers in parts a, b, and c are significantly different.

6. Choose exercise 4 or 5.
   a) Predict how the estimated probabilities would be affected if the continuity correction were not applied.
   b) Carry out some calculations to verify your prediction.
   c) How is the continuity correction affected if the number of repetitions of the experiment is increased? Explain.

7. Suppose the probability that a student in your mathematics class is wearing running shoes is 0.7. What is the probability that at most 25 out of 30 students in your mathematics class are wearing running shoes?

8. Use the normal distribution to approximate each probability.
   a) P(correctly guessing exactly 11 answers in a 20-item true-false test)
   b) P(exactly 200 boys in a group of 400 students) when P(boy) = 0.5
   c) P(exactly 80 defective in 1000 manufactured items) when P (defective) = 0.01
   d) P(exactly 15 sick people in 10 000 people) when P(sick) = 0.001

MODELLING a Binomial Distribution with a Normal Distribution

In the exercises above, we used a normal distribution to model a binomial distribution.

- What kinds of problems are appropriate for using a normal model?
- Give some reasons why a normal model gives only approximate results.
- The accuracy of the estimates depends on the values of n and p. It can be shown that the normal model should be used only when both np and $n(1-p)$ are greater than 5. Choose some exercises and check that $np > 5$ and $n(1-p) > 5$.
- Suggest some reasons why a normal model should not be used if either np or $n(1-p)$ is less than 5.

C 9. All the probabilities in exercise 8 involve a binomial distribution. However, not all these probabilities can be calculated efficiently using a binomial distribution. Explain your answer to each question.

a) Which probabilities can be calculated using a binomial distribution?

b) Which probabilities cannot be calculated using a binomial distribution?

10. Consider the problem about the surprise quizzes on pages 507–509. A function $P(x)$ can be defined as follows.

$P(x)$ = probability there will be x or fewer surprise quizzes in the next 30 classes
On page 507, we used the binomial distribution to show that $P(6) \doteq 0.61$.
On page 509, we used the normal distribution to estimate $P(6) \doteq 0.59$.

a) Visualize the graph of $y = P(x)$. Describe its properties.

b) Graph the function $y = P(x)$ as a bar graph. In list L1, input the numbers from 0 to 30. Press [DISTR] **A** 30 [,] 0.2 [ )] [STO►] [L₂] to calculate and store the results in list L2. Set up a bar graph and display the graph in an appropriate window. Does the result agree with your prediction in part a?

c) Graph the function $y = P(x)$ as a continuous curve. In the Y= list, clear any plots and equations. Move the cursor to Y1, and press [DISTR] 2 −5 [,] [X,T,θ,n] [,] 6 [,] 2.19 [ )] [GRAPH]. Use the trace function to determine some values of $P(x)$. Does the result agree with your prediction in part a?

COMMUNICATING THE IDEAS

Write to explain how a normal distribution can be used to approximate a binomial distribution. Describe some similarities and differences between the two kinds of distributions, and explain how they affect the approximation.

Using the Functions Toolkit

In this chapter, there are similarities between how the normal curves changed and the transformations of functions. For example, consider the graph of the standard normal curve and the graph of some other normal curve.

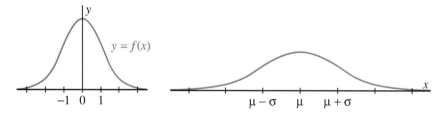

Visualize the second curve as the image of the standard normal curve under certain transformations. The area under both curves is 1.

1. a) By what factor has the standard normal curve been expanded or compressed:
　　i) horizontally?　　　　　　　　**ii)** vertically?

　b) By what amount has the curve been translated horizontally?

　c) Let $y = f(x)$ represent the equation of the standard normal curve. What is the equation of the image curve?

2. Consider the experiment of rolling a die 30 times. The screen (below left) shows the normal distribution approximating the probability that 6 appears x times. The screen (below right) shows the normal distribution approximating the probability that 6 does not appear x times.

　a) For what values of x do these graphs have meaning?

　b) Determine the mean and the standard deviation for each distribution.

　c) Explain why the means are different but the standard deviations are the same.

　d) Suppose the equation of the first curve is $y = f(x)$. What is the equation of the second curve? Explain.

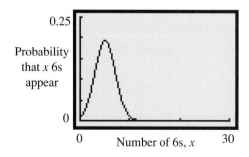 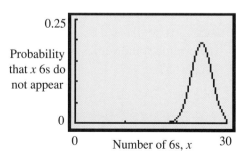

3. Turn to page 496 to find the equation of the standard normal curve. Write the equation of the image curve in exercise 1, and each normal curve in exercise 2.

1. A spinner is divided into 5 equal sectors labelled 10, 20, 30, 40, and 50. Construct a histogram to show the probability distribution of the outcomes for one spin of the spinner. Which type of probability distribution is this?

2. Devise an experiment to simulate the number of heads in 8 tosses of a coin when P(heads) = 0.3. Conduct the experiment, then graph the frequency distribution and probability distribution. Write to explain the results.

3. A sample of 20 bags of potato chips produced by the Crispy Chips Company was found to have these masses in grams.

80	85	88	92	76	85	94	76	78	80
84	88	75	78	77	68	70	72	80	74

A sample of 20 bags of potato chips produced by the Special Spuds Company was found to have these masses in grams.

85	96	90	69	78	77	79	82	80	81
78	81	77	80	75	79	76	81	80	76

  a) Calculate the mean and the standard deviation for each brand of chips.

  b) Which brand has the smaller standard deviation?

  c) What problems may be encountered if the standard deviation gets too high? Explain.

4. Work with a group of students. Choose a characteristic that can be measured, such as length, height, mass, or temperature. For example, take body temperatures using a thermometer or an infrared probe.

  a) Collect the data from each member of your class.

  b) Calculate the mean and standard deviation. Write to explain the results.

5. Open the *NBA* database and sort the data by Name. Use the records for Charles Barkley and Patrick Ewing.

  a) For each player:

   i) Calculate the average points per game for the 90-91 season.

   ii) Calculate the mean and standard deviation points per game for the player's career.

  b) Use the results of part a. Who is the better point scorer?

6. The mean score on a mathematics test is 115 and the standard deviation is 20. The teacher decides to give an A to all students who score 135 or more. Assume the scores are normally distributed. What percent of the students will receive an A?

1. Describe how each function compares to the graph of $y = x^2$.

   a) $y = 4x^2$ b) $y = \frac{1}{4}x^2$ c) $y = (-x)^2$ d) $2y = x^2$

2. Suppose you invest \$2000 at 5% interest compounded annually.

   a) Express the value of your investment, A dollars, after n years as an exponential function.

   b) Graph the function for $0 \le n \le 15$.

   c) Estimate how long it would take for your investment to be worth \$3000.

3. A product of a nuclear explosion is plutonium-239, which has a half-life of 24 000 years. What percent of plutonium-239 remains after 1000 years?

4. Determine the 6th and nth terms of this sequence: 2, 12, 72, 432, ...

5. In exercise 4, which term is 559 872?

6. Write the equation of the inverse of each logarithmic function.

   a) $y = \log_3 x$ b) $y = \log_{0.5} x$ c) $f(x) = \log_7 x$

7. Determine the length of the arc that subtends each angle at the centre of a circle with radius 9.0 cm.

   a) 3.1 radians   b) 333°

8. Write an expression to represent any angle coterminal with each angle θ.

   a) $\theta = -60°$ b) $\theta = 115°$ c) $\theta = \frac{-3\pi}{5}$ d) $\theta = \frac{2\pi}{3}$

9. Graph each function.

   a) $y = \cos 2\left(x - \frac{\pi}{2}\right)$ b) $g(x) = 3 \sin 3\left(x - \frac{2\pi}{3}\right) - 1$

10. Write each expression as a single trigonometric function.

   a) $\cos^2 \frac{\pi}{6} - \sin^2 \frac{\pi}{6}$ b) $2 \sin 0.8 \cos 0.8$

   c) $2 \cos^2 0.35 - 1$ d) $1 - 2 \sin^2 \frac{\pi}{4}$

11. Solve each equation to 4 decimal places for $0 \le x < 2\pi$.

   a) $3 \sin^2 x + 4 \sin x + 1 = 0$ b) $9 \cos^2 x + 6 \cos x + 1 = 0$

   c) $\sin^2 x - 2 \sin x - 3 = 0$ d) $4 \cos^2 x - 7 \cos x - 2 = 0$

12. Write the general solution for each equation in exercise 11.

13. Determine the indicated term in each expression.

   a) the 4th term in the expansion of $(x - 1)^7$

   b) the 6th term in the expansion of $(2 - x)^5$

c) the middle term in the expansion of $(x - 3)^8$

d) the 5th term in the expansion of $(2x + 1)^6$

e) the 7th term in the expansion of $(x + 2)^{13}$

14. There are 10 boys and 12 girls in a mathematics club. How many ways can 6 people be selected to participate in a contest in each case?

a) There are no restrictions.

b) There must be the same number of boys as girls.

15. From a standard deck of 52 playing cards:

a) How many different 6-card hands can be formed?

b) How many different 6-card hands can be formed if each hand contains only hearts?

c) How many different 6-card hands can be formed with no more than 4 black cards?

16. How many diagonals does a regular 25-sided polygon have?

17. Open the *Cars* database. Assume the data are normally distributed.

a) Calculate the mean and standard deviation for the top speeds of all cars.

b) Use the result of part a. What percent of cars have top speeds greater than 255.34 km/h?

c) Use the result of part a. What percent of cars have top speeds between 89.79 km/h and 222.23 km/h?

d) Use the database to repeat parts b and c. Explain any differences in the results.

18. Use the normal curve in the graph on page 508. Estimate each probability.

a) P(7 quizzes)   **b)** P(8 quizzes)   **c)** P(9 quizzes)

19. Participants in a swim team were timed as they swam two lengths of the pool. The times were normally distributed with a mean of 21.5 s and a standard deviation of 3.5 s. The coach chooses swimmers with the lowest times. She chooses 16% of the participants. What is the maximum time that a person in the group of chosen swimmers took?

20. You can graph the function $y = A(z)$ on a TI-83 graphing calculator.

a) Enter the equation Y1 = normalcdf(−5,X) and set an appropriate graphing window. Graph the function.

b) Trace along the graph. How do the results compare with the values in the table on pages 494 and 495?

9 CONIC SECTIONS

Designing a Dish Antenna

 CONSIDER THIS SITUATION

Satellite communications are an essential part of our modern world. On May 19, 1998, the *Galaxy IV* communications satellite suffered a disruption in operation that lasted a few hours. People found that their cellular phones and electronic pagers did not work, and some television broadcasts were disrupted.

The signals from a communications satellite reflect off the surface of a parabolic dish antenna and are concentrated at the point where the receiver is located.

- What are some advantages of satellite communications?

- What are some disadvantages?

- A dish antenna is aimed at a communications satellite, which is in orbit around Earth. The satellite travels at approximately 24 000 km/h. Explain why the dish antenna does not have to move to follow the satellite.

On pages 558 to 560, you will develop a mathematical model to calculate the distance from the centre of the dish antenna to the point where the receiver should be located. Your model will be the graph of a parabola.

 FYI Visit **www.awl.com/canada/school/connections**

For information related to the above problem, click on <u>MATHLINKS</u>, followed by <u>AWMath</u>. Then select a topic under Designing a Dish Antenna.

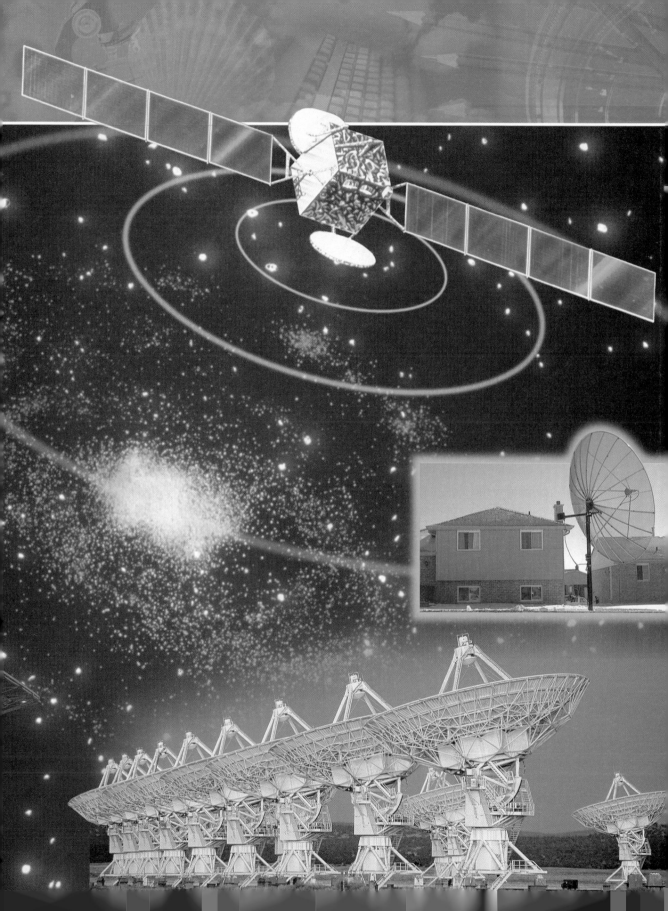

1. **a)** Hold a flashlight perpendicular to a wall. Describe the curve formed by the edge of the shadow.

   b) Predict how the curve will change when you move the flashlight farther from the wall, or closer to the wall, while still holding it perpendicular to the wall. Verify your predictions.

   c) Move the flashlight closer and closer to the wall until it touches the wall. Describe what happens to the curve.

2. **a)** Predict how the curve will change if you tilt the flashlight upward. Verify your prediction. The curve is an ellipse. Describe an ellipse.

   b) Predict how the ellipse will change if you increase the angle of elevation of the flashlight. Verify your prediction.

   c) Keep increasing the angle of elevation. How does the ellipse change?

3. **a)** Hold the flashlight vertical, close to the wall. The curve is a hyperbola. Describe a hyperbola, especially the parts that are farthest from the flashlight.

   b) Predict how the hyperbola will change if you decrease the angle of elevation. Verify your prediction.

   c) Keep decreasing the angle of elevation. How does the hyperbola change?

4. As the angle of elevation increases, the curve changes from an ellipse to a hyperbola.

   a) Try to determine the position where the curve changes from an ellipse to a hyperbola.

   b) At this position in part a, the curve is neither an ellipse nor a hyperbola. Look closely at the curve. What curve do you think this is?

   c) What is unusual about the position of the flashlight at that position?

5. **a)** Hold the flashlight to form an ellipse on the wall. Predict how the ellipse will change if you move the flashlight farther from, or closer to, the wall, without changing its angle of elevation. Verify your predictions.

   b) Move the flashlight closer and closer to the wall until it touches the wall. Describe what happens to the ellipse.

   c) Repeat parts a and b for the hyperbola.

6. Explain how a hyperbola differs from a parabola.

A small hole is drilled near the end of a pencil and a bent piece of wire is inserted. Visualize spinning the pencil rapidly between the palms of your hands. The wire traces a three-dimensional surface called a *cone*. Each position of the wire is a line on the cone. These lines are *generators* of the cone. The pencil forms the *axis* of the cone. All the generators intersect at a point called the *vertex*. The cone has two symmetrical parts, or *nappes*, on either side of the vertex.

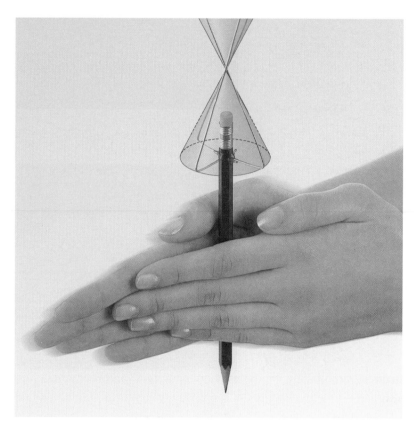

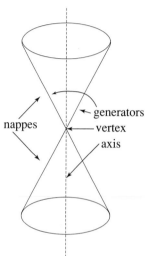

The ancient Greeks defined a cone as the surface generated when a line is rotated about a fixed point P on the line. The curves that result when a plane intersects a cone are *conic sections*. In *Investigate*, the curves produced on a wall when a flashlight is held at different angles are examples of conic sections.

The Greeks discovered many properties of conic sections, but they were not interested in practical applications. In the seventeenth century, Isaac Newton proved that the path of a body revolving around another in accordance with the law of gravitation is a conic section.

The Circle

In this drawing, the plane is parallel to the base of the cone. In this case, the curve of intersection is a circle. Hence, a circle is an example of a conic section.

Both the sun and the moon appearing to us as circular disks of about the same size cause a total solar eclipse.

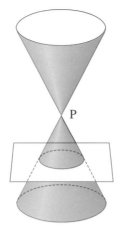

The Ellipse

When the intersecting plane is inclined to the base of the cone, the curve of intersection is an *ellipse*.

Satellites, planets, and some comets travel in elliptical orbits. Most asteroids circle the sun in orbits beyond that of Mars. In 1998, astronomers found an asteroid circling the sun inside Earth's orbit. The asteroid, named 1998 DK36, has a diameter of about 40 m. The closest DK36 has come to Earth is about 1.3 million kilometres.

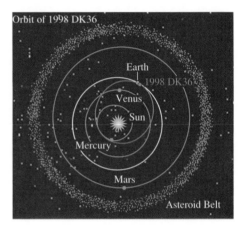

The Parabola

When the intersecting plane is parallel to a generator AB on the cone, the curve of intersection is a *parabola*.

Parabolas have many applications in astronomy. The mirrors in some telescopes have surfaces whose cross sections are parabolas.

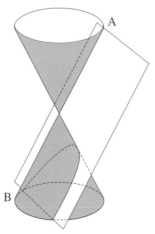

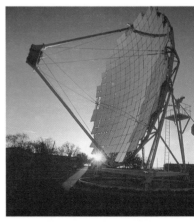

The Hyperbola

When the plane intersects both nappes of the cone, the curve of intersection is a *hyperbola*. A hyperbola has two distinct parts, called *branches*.

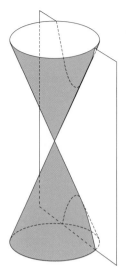

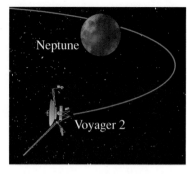

Neptune

Voyager 2

The *Voyager 2* space probe, launched in 1977, provides examples of hyperbolic paths. The arrangement of Jupiter, Saturn, Uranus, and Neptune during the 1980s allowed *Voyager 2* to pass by one planet on its way past the next. On each segment of this journey, *Voyager 2* followed a hyperbolic path. *Voyager 2* is now heading out of the solar system into interstellar space, at about 57 000 km/h. The diagram shows the hyperbolic path of *Voyager 2* as it passed by Neptune in August, 1989.

VISUALIZING

Refer to the diagram on page 520 that shows a plane intersecting a cone to form a circle. Look at the point on this circle that is farthest right. Visualize rotating the plane slowly about this point, until the plane takes the positions in the other diagrams on page 520 and above. What happens when the plane continues to rotate, beyond the position in the diagram above, until it has made a complete rotation?

DISCUSSING THE IDEAS

1. Explain why the curves caused by the shining of a flashlight on a wall are examples of conic sections.

2. Do the ellipse and the hyperbola appear to have any axes of symmetry? Explain, using either the flashlight model or the plane intersecting a cone.

3. Visualize a plane intersecting a cone. Explain how you can tell whether the points of intersection form each conic section.

   a) a circle   b) an ellipse   c) a parabola

   d) a hyperbola   e) none of the above

A **1.** A table lamp is near a wall. Describe the shadow formed on the wall when the lamp is turned on.

2. Refer to the diagram on page 520 that shows a plane intersecting a cone to form a circle. Assume the plane moves parallel to its original position. Describe what happens to the curve of intersection when the plane moves as described.

a) farther from the vertex of the cone

b) closer to the vertex of the cone

c) across the vertex of the cone

3. Repeat exercise 2, using the diagrams on pages 520 and 521 that show a plane intersecting a cone to form each conic section.

    a) an ellipse   **b)** a parabola   **c)** a hyperbola

4. Suppose a plane intersects a cone. Does every plane parallel to this plane intersect the cone? Explain.

5. Visualize a set of parallel planes intersecting a cone.

    a) Suppose the conic section on one of the planes is a circle. Describe the conic sections on the other planes. Explain how they are related to the first circle.

    b) Repeat part a, replacing the word circle with:
        i) ellipse   **ii)** parabola   **iii)** hyperbola

B **6.** Visualize shining a flashlight on a basketball on the floor. When the flashlight is directly above the centre of the basketball, the shadow of the basketball forms a circular region on the floor. Assume the light is always aimed toward the centre of the ball. Describe the position of the light relative to the ball for the shadow to be each conic section.

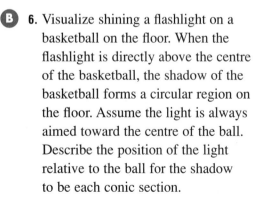

    a) an ellipse   **b)** a parabola   **c)** a hyperbola

7. A jet breaking the sound barrier creates a shock wave that has the shape of a cone. As the jet flies overhead, the shock wave affects points on the ground that form a curve. Describe the shape of this curve for each position of the jet.

a) flying parallel to the ground

b) gaining altitude

c) losing altitude

d) flying straight up

8. List as many different ways as possible in which a plane can intersect a cone and not form a curve. Describe how each is obtained.

9. Some comets follow elliptical orbits around the sun. Other comets originate from interstellar space, follow hyperbolic orbits around the sun, and return to interstellar space. Some comets have been observed on only one journey around the sun. Explain why it could be difficult to tell if the orbits of these comets are elliptical or hyperbolic.

10. The Edmonton Space and Science Centre features a large elliptical panel as a prominent part of its exterior design. Use the photograph to visualize how the ellipse is formed by the intersection of a plane with a cylinder.

a) Describe the different ways in which a plane can intersect a cylinder. Illustrate your description with sketches.

b) Is it possible for a plane to intersect a cylinder and form a parabola or a hyperbola? Explain.

c) Describe how a light source and a ball could be used to demonstrate that an ellipse is formed when a plane intersects a cylinder.

C 11. A flashlight is directed toward a wall so that some of the light falls on the wall. Suppose the angle formed at the flashlight by the rays of light is 60°.

a) Assume the angle of elevation of the flashlight increases from 0° to 90°. For what angles of elevation does each conic section appear on the wall?

  i) a circle   ii) an ellipse   iii) a parabola   iv) a hyperbola

b) For what other angles of elevation does one of the conic sections in part a appear on the wall? Explain. Include a statement of any assumptions you made to answer this question.

12. Refer to exercise 6. Suppose the basketball is resting on a glass tabletop (below left). Visualize the flashlight rotating in a circle around the basketball so that the angle formed by the rays of light is always 40°. Let θ represent the angle of rotation of the flashlight, $0° \le \theta \le 360°$. For what values of θ does the edge of the shadow form each conic section?

a) a circle   **b)** an ellipse   **c)** a parabola   **d)** a hyperbola

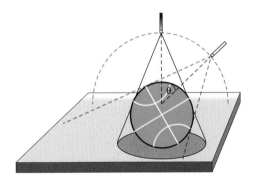

13. Suppose a plane intersects a cone with vertex V (above right). Let α represent the angle between the axis of the cone and one generator. Let β represent the angle between the axis of the cone and the perpendicular VN to the plane. Visualize the plane rotating around the vertex of the cone, so that $0° \le \beta \le 360°$. For what values of β is each curve of intersection produced?

a) a circle   **b)** an ellipse   **c)** a parabola   **d)** a hyperbola

14. For each conic section, describe the relationship among the plane, the central axis of the cone, and a generator.

15. An ellipse is a curve formed when a plane intersects a cone. Explain why an ellipse also occurs when a plane intersects a cylinder.

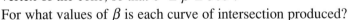

COMMUNICATING   THE IDEAS

Summarize, with diagrams, the different ways in which a plane can intersect a cone.

9.2 Graphing Circles and Rectangular Hyperbolas

Lines, parabolas, and circles are examples of conic sections. In earlier grades, you graphed lines, parabolas, and circles whose equations were written in certain forms. In the examples below, *standard form* refers to the form of the equation that is useful for graphing. All the numbers in these equations represent particular features of the graph, which can be used to sketch the graph. Each equation can be written in another form, called *general form*, with all the terms on the left side. This form illustrates the kinds of terms that occur in the equation, but the numbers in the equations do not usually represent features of the graph.

Line

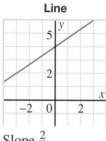

Slope $\frac{2}{3}$
y-intercept 4

Parabola

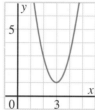

Vertex $(3, 1)$
Axis of symmetry
$x - 3 = 0$
Congruent to $y = 2x^2$

Circle

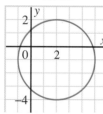

Centre $(2, -1)$
Radius $\sqrt{9} = 3$

Equations of above graphs in standard form:

$$y = \frac{2}{3}x + 4 \qquad\qquad y = 2(x - 3)^2 + 1 \qquad\qquad (x - 2)^2 + (y + 1)^2 = 9$$

Equations of above graphs in general form:

$$2x - 3y + 12 = 0 \qquad\qquad 2x^2 - 12x - y + 19 = 0 \qquad x^2 + y^2 - 4x + 2y - 4 = 0$$

An equation in general form containing only first-degree terms represents a straight line. For example, in the equation of the line above, every increase in x of 3 units corresponds to an increase in y of 2 units. The presence of second-degree terms, such as $2x^2$ in the equation of the parabola above, usually causes the graph to be curved. Parabolas and circles are examples of conic sections and have second-degree equations. This suggests that ellipses and hyperbolas might also have second-degree equations. In this and the following sections, you will investigate the graphs of second-degree equations.

Graphing Circles and Rectangular Hyperbolas

Some second-degree equations are not usually written in a form that is solved for y. Hence, graphing these equations is more involved than graphing functions whose equations have the form $y = f(x)$. Instructions are given below for *Graphmatica* and for the TI-83 calculator.

Using *Graphmatica*

You can enter second-degree equations directly into *Graphmatica*. Before you use this software, note:

• Graphs of circles should look like circles on the screen. This means that both axes must have the same scale. To obtain a viewing window with $-8 \le x \le 8$, the same scale on both axes, and the origin at the centre of the screen, follow these steps.

Step 1	*Step 2*	*Step 3*
Select Grid Range in the View menu. Enter $-8, 8$, and 0 in the Left, Right, and Bottom boxes. Leave the Top box blank. Check "Autoscale fourth coordinate", and click OK.	Select Grid Range again. Note that the computer calculated a value of 9.38 for the Top box. This is the vertical distance needed for equal scales on the axes.	Divide 9.38 by 2 to obtain 4.69. Enter 4.69 in the Top box and -4.69 in the Bottom box. Click OK.

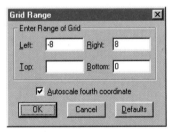 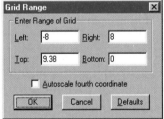 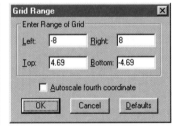

• *Graphmatica* can only graph equations in which either x or y occurs only once. For example, if you enter $2x^2 - 12x - y + 19 = 0$ or $(x - 2)^2 + (y + 1)^2 = 9$, the graphs will be displayed. However, if you enter $x^2 + y^2 - 4x + 2y - 4 = 0$, you will get an error message.

Using the TI-83

Ask your teacher for the program called CONICS from the Teacher's Resource Book. This program graphs any general second-degree equation, after it prompts for the coefficients. Before you use the program, note:

- The program prompts you to enter the coefficients of x^2, y^2, x, and y, and the constant term. Hence, the equations must be expressed in general form. For example, to graph the circle defined by $(x - 2)^2 + (y + 1)^2 = 9$, you must write the equation in the form $x^2 + y^2 - 4x + 2y - 4 = 0$.

- Graphs of circles should look like circles on the screen. This means that both axes should have the same scale. Use $-9.4 \le x \le 9.4$, $-6.2 \le y \le 6.2$, or select ZSquare in the zoom menu.

- The program uses the drawing features of the calculator, not the Y= list. The equation is not stored in the calculator's memory. This means that you cannot use the trace or tables features.

- To change the window, erase the screen and graph the equations again.

- To erase the screen, press $\boxed{\text{2nd}}$ $\boxed{\text{PRGM}}$ **1** to select ClrDraw.

- You can graph more than one equation by running the program again.

1. a) Use graphing technology to graph the relation $x^2 + y^2 = 16$. Describe the graph, and account for its shape.

b) In part a, change the constant 16 to other numbers, and repeat. Use both positive and negative constants, and zero. Are the results what you expected?

2. a) What information about the graph of $x^2 + y^2 = k$ does the value of k provide? Explain.

b) What happens to the graph of $x^2 + y^2 = k$ when k is changed? Consider both positive and negative values of k, and zero.

3. Without using graphing technology, sketch the graph of each relation.

a) $x^2 + y^2 = 49$ **b)** $x^2 + y^2 = 30$

An equation such as $x^2 + y^2 = 16$ represents a circle. What would the graph of an equation such as $x^2 - y^2 = 16$ look like?

4. **a)** Use graphing technology to graph the relation $x^2 - y^2 = 16$. Describe the graph.

   b) In part a, change the constant 16 to other numbers, and repeat. Use both positive and negative constants, and zero. Are the results what you expected?

5. **a)** What information about the graph of $x^2 - y^2 = k$ does the value of k provide? Explain.

   b) What happens to the graph of $x^2 - y^2 = k$ when k is changed? Consider both positive and negative values of k, and zero.

6. In what ways are the graphs of equations of the form $x^2 + y^2 = k$ different from the graphs of equations of the form $x^2 - y^2 = k$? List as many differences as you can.

7. Without using graphing technology, sketch the graph of each relation.

   a) $x^2 - y^2 = 49$ **b)** $x^2 - y^2 = 30$

   c) $x^2 - y^2 = -49$ **d)** $x^2 - y^2 = -30$

Each graph in exercise 7 is a hyperbola.

When a plane is perpendicular to the axis of a cone, the curve of intersection is a circle with centre on the axis of the cone. Recall that the equation of a circle with centre $(0, 0)$ and radius r is $x^2 + y^2 = r^2$.

Standard Equation of a Circle with Centre (0, 0)

The equation of a circle with centre $(0, 0)$ and radius r is:

$$x^2 + y^2 = r^2$$

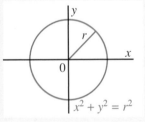

The equation of a circle contains the expression $x^2 + y^2$. Hence, an equation that contains the expression $x^2 - y^2$ cannot represent a circle. We can use graphing technology to determine what such an equation represents.

The graph of the relation $x^2 - y^2 = 9$ is shown (below left). A TI-83 calculator using the CONICS program, with the equation written in the form $x^2 - y^2 - 9 = 0$, produced this graph. As in *Investigate*, the graph is a hyperbola.

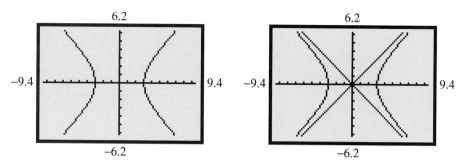

We can account for the shape of the graph. First, solve the equation for y.

$$x^2 - y^2 = 9$$
$$y^2 = x^2 - 9$$
$$y = \pm\sqrt{x^2 - 9}$$

Since $x^2 - 9$ occurs under the radical sign, $x^2 - 9 \geq 0$.
Hence, values of y are defined only when $x \geq 3$ or when $x \leq -3$.

When $x = \pm 3$, then $y = \pm\sqrt{(\pm 3)^2 - 9}$
$$= 0$$

Hence, $(3, 0)$ and $(-3, 0)$ are on the graph.

For each value of $x > 3$ or $x < -3$, there are two values of y, one positive and the other negative. When $|x|$ is large, x^2 is very large compared with 9, and so

$$y = \pm\sqrt{x^2 - 9}$$
$$\doteq \pm\sqrt{x^2}$$
$$\doteq x \text{ or } -x$$

Hence, the graph comes closer and closer to the lines defined by $y = x$ and $y = -x$. The graph is a hyperbola, and these lines are its asymptotes (see the second graph above. The point $(0, 0)$ is the *centre* of the hyperbola. The points $(3, 0)$ and $(-3, 0)$ are the *vertices*. The line segment joining the vertices is the *transverse axis*. Since the asymptotes are perpendicular, the hyperbola is a *rectangular hyperbola*.

The explanation above suggests that an equation of the form $x^2 - y^2 = a^2$ represents a rectangular hyperbola with centre $(0, 0)$ and vertices $(\pm a, 0)$.

Example 1

Sketch the rectangular hyperbola defined by $x^2 - y^2 = 16$.

Solution

The equation $x^2 - y^2 = 16$ represents a rectangular hyperbola with centre $(0,0)$ and vertices $(\pm 4, 0)$.

Draw lines with slopes ± 1 through $(0,0)$; these are the asymptotes.

Use the vertices and the asymptotes to sketch the hyperbola.

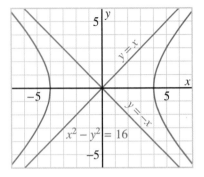

Example 2

a) Use graphing technology to graph the relation $x^2 - y^2 = -16$.

b) Determine the y-intercepts of the graph.

c) Describe the graph.

Solution

a) Use a TI-83 calculator and the CONICS program with the equation written in the form $x^2 - y^2 + 16 = 0$.

b) Substitute 0 for x to obtain $y^2 = 16$ or $y = \pm 4$. The y-intercepts are ± 4.

c) The graph appears to be a rectangular hyperbola with centre $(0,0)$ and vertices $(0, \pm 4)$ on the y-axis.

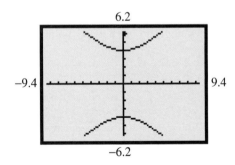

This screen shows the graph of $x^2 - y^2 = -16$ from *Example 2* superimposed on the graph of $x^2 - y^2 = 16$ above. The lines defined by $y = \pm x$ are also asymptotes of the graph of $x^2 - y^2 = -16$. We can explain this using an argument similar to that on page 529, or we can explain it by interchanging x and y in the equation $x^2 - y^2 = 16$.

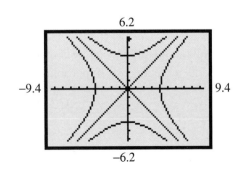

This gives $y^2 - x^2 = 16$ or $x^2 - y^2 = -16$. Hence, the graph of $x^2 - y^2 = -16$ can be obtained from the graph of $x^2 - y^2 = 16$ by reflection in the line $y = x$.

We can write the equation of any rectangular hyperbola with centre $(0,0)$ and vertices on the coordinate axes. The form of the equation depends on whether the vertices are on the x-axis or the y-axis.

Standard Equations of Rectangular Hyperbolas with Centre (0, 0)

The equation of a rectangular hyperbola with centre $(0,0)$ and vertices on the x-axis is $x^2 - y^2 = a^2$.

The equation of a rectangular hyperbola with centre $(0,0)$ and vertices on the y-axis is $x^2 - y^2 = -a^2$.

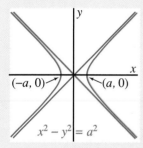

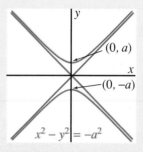

Vertices: $(a, 0)$ and $(-a, 0)$
Asymptotes: $y = x$ and $y = -x$

Vertices: $(0, a)$ and $(0, -a)$
Asymptotes: $y = x$ and $y = -x$

We can tell whether the vertices are on the x-axis or the y-axis from the standard equation.

DISCUSSING THE IDEAS

1. a) Explain why the equation $x^2 + y^2 = r^2$ represents a circle with centre $(0,0)$ and radius r.

   b) Explain why the equation $x^2 - y^2 = a^2$ represents a rectangular hyperbola with centre $(0,0)$ and vertices $(\pm a, 0)$.

   c) Explain why the equation $x^2 - y^2 = -a^2$ represents a rectangular hyperbola with centre $(0,0)$ and vertices $(0, \pm a)$.

2. Explain how you can tell from the standard equation of a rectangular hyperbola whether the vertices are on the x-axis or the y-axis.

3. Explain what each equation represents.

   a) $x^2 + y^2 = 0$ b) $x^2 - y^2 = 0$

1. Determine whether each point is on the rectangular hyperbola defined by $x^2 - y^2 = 15$.

a) $(-4, 1)$ b) $(7, 8)$ c) $(8, 7)$ d) $(0, \sqrt{15})$

2. State the coordinates of the vertices, the length of the transverse axis, and the equations of the asymptotes of the rectangular hyperbola defined by each equation.

a) $x^2 - y^2 = 25$ b) $x^2 - y^2 = 64$ c) $x^2 - y^2 = -81$

d) $x^2 - y^2 = 2$ e) $x^2 - y^2 = -5$ f) $x^2 - y^2 = -20$

3. The coordinates of one vertex of a rectangular hyperbola are given. The centre is $(0, 0)$. Write an equation of each rectangular hyperbola.

a) $(7, 0)$ b) $(0, 4)$ c) $(0, -6)$ d) $(-10, 0)$

4. Sketch the graphs of these relations on the same grid.

a) $x^2 - y^2 = 4$ b) $x^2 - y^2 = -4$ c) $x^2 - y^2 = 16$

d) $x^2 - y^2 = -16$ e) $x^2 - y^2 = 36$ f) $x^2 - y^2 = -36$

5. These graphs were produced by *Graphmatica*. For each graph, write a set of equations that could be used to make the pattern.

a)

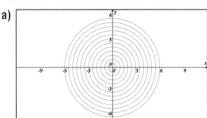

b)

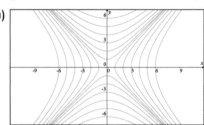

6. Make patterns similar to those in exercise 5.

7. a) Predict what the graphs of these equations would look like.

i) $x^2 + y^2 = 16$
$x^2 + y^2 = 4$
$x^2 + y^2 = 1$
$x^2 + y^2 = \frac{1}{4}$
$x^2 + y^2 = \frac{1}{16}$

ii) $x^2 - y^2 = 16$
$x^2 - y^2 = 4$
$x^2 - y^2 = 1$
$x^2 - y^2 = \frac{1}{4}$
$x^2 - y^2 = \frac{1}{16}$

b) Check your predictions.

c) Predict what happens if the constant term becomes 0. Check your predictions.

d) Predict what happens if the constant term becomes negative. Check your predictions.

e) Explain the above results in terms of the way in which a plane can intersect a cone.

8. For exercise 7, explain how the results would change if x and y were interchanged in the equations.

9. a) Sketch the graphs of these relations on the same grid, if possible.

    i) $x^2 + y^2 = 9$ **ii)** $x^2 + y^2 = 0$ **iii)** $x^2 + y^2 = -9$

b) Write to describe how you graphed the relations in part a.

10. a) Sketch the graphs of these relations on the same grid.

    i) $x^2 - y^2 = 9$ **ii)** $x^2 - y^2 = 0$ **iii)** $x^2 - y^2 = -9$

b) Visualize how the graph of $x^2 - y^2 = k$ changes as k decreases from 9 to −9. Write to describe the changes.

C **11.** Refer to the diagram on page 521 that shows a plane intersecting a cone to form a hyperbola. Assume the plane is parallel to the axis of the cone. When the hyperbola is a rectangular hyperbola, what property must the cone have? Explain.

12. Is it always possible for a plane to intersect a cone to form a rectangular hyperbola? Explain.

13. Consider the equation $Ax^2 + By^2 + C = 0$. What conditions must be satisfied by A, B, and C for this equation to represent each conic section?

a) a circle with centre $(0, 0)$

b) a rectangular hyperbola with centre $(0, 0)$ and vertices on:
    i) the x-axis   **ii)** the y-axis

COMMUNICATING THE IDEAS

Write to explain the differences between the graphs of $x^2 + y^2 = k$ and $x^2 - y^2 = k$. Include an explanation of the values of k that are possible in each equation, and why.

9.3 Stretching Graphs of Conic Sections

The graphs of conic sections are essentially three different types.

circular or elliptical (closed curve)	*hyperbolic* (two branches and two asymptotes)	*parabolic* (only one branch and no asymptotes)
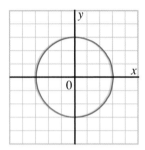	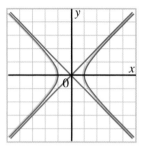	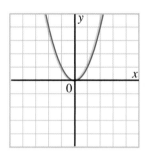

When the curves are drawn in a central position on a coordinate grid, as shown above, the equations have certain characteristic forms.

Circle with centre $(0,0)$

$$x^2 + y^2 = r^2$$

Squared terms separated by a + sign

Rectangular hyperbola with centre (0, 0) and
- vertices on x-axis
$$x^2 - y^2 = a^2$$
- vertices on y-axis
$$x^2 - y^2 = -a^2$$

Squared terms separated by a − sign

Parabola with vertex $(0,0)$ and axis of symmetry:
- the y-axis
$$y = x^2$$
- the x-axis
$$x = y^2$$

Only one squared term

To obtain the graphs of the other conic sections, we expand or compress these basic graphs. For example, we can obtain an ellipse graphically by stretching a circle horizontally and vertically. When we know the equation of the circle, we can use the tools from the functions toolkit in Chapter 1 to determine the equation of the ellipse.

Stretching Graphs of Conic Sections

The graphs of the circle $x^2 + y^2 = 1$ and the rectangular hyperbola $x^2 - y^2 = 1$ are shown below. In this investigation, you will graph other related equations on the same screen. You will use the tools from the functions toolkit in Chapter 1 to explain the relationship between the changes in the graphs and the changes in the equations.

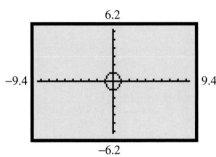

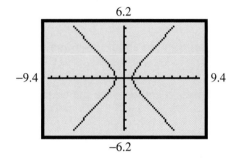

If you are using *Graphmatica*, see page 526.

If you are using the TI-83 calculator, use the CONICS program. You will have to write the equations in general form. For example, to write an equation such as $\left(\frac{x}{4}\right)^2 + \left(\frac{y}{2}\right)^2 = 1$ in general form, first write $\frac{x^2}{16} + \frac{y^2}{4} = 1$. In general form, the equation is $\frac{1}{16}x^2 + \frac{1}{4}y^2 - 1 = 0$, or $x^2 + 4y^2 - 16 = 0$. Either equation may be used.

Transforming the graph of the circle $x^2 + y^2 = 1$

1. a) Graph $\left(\frac{x}{3}\right)^2 + \left(\frac{y}{2}\right)^2 = 1$. Describe how the graph is related to the graph of $x^2 + y^2 = 1$. Explain this relationship.

  b) Repeat part a, using other numbers in the denominators of the squared terms.

2. What information about the graph of $\left(\frac{x}{a}\right)^2 + \left(\frac{y}{b}\right)^2 = 1$ do a and b provide? Explain.

3. In exercises 1 and 2, write the equation of each ellipse in general form.

4. Graph each relation. Describe its graph.

  a) $\left(\frac{x}{5}\right)^2 + y^2 = 1$ **b)** $x^2 + \left(\frac{y}{5}\right)^2 = 1$

  c) $\frac{x^2}{16} + \frac{y^2}{4} = 1$ **d)** $\frac{x^2}{4} + \frac{y^2}{16} = 1$

Transforming the graph of the rectangular hyperbola $x^2 - y^2 = 1$

5. a) Graph $\left(\frac{x}{3}\right)^2 - y^2 = 1$ and $x^2 - \left(\frac{y}{2}\right)^2 = 1$. Describe how each graph is related to the graph of $x^2 - y^2 = 1$. Explain each relationship.

b) Graph $\frac{x^2}{9} - \frac{y^2}{4} = 1$. Describe how the graph is related to the graph of $x^2 - y^2 = 1$. Explain.

c) Repeat part b, using other numbers in the denominators of the squared terms.

6. What information about the graph of $\frac{x^2}{a^2} - \frac{y^2}{b^2} = 1$ do a and b provide? Explain.

7. In exercises 5 and 6, write the equation of each hyperbola in general form.

8. Graph each relation. Describe each graph.

a) $\left(\frac{x}{5}\right)^2 - y^2 = 1$ **b)** $x^2 - \left(\frac{y}{5}\right)^2 = 1$

c) $\frac{x^2}{16} - \frac{y^2}{4} = 1$ **d)** $\frac{x^2}{4} - \frac{y^2}{16} = 1$

Transforming the graph of the rectangular hyperbola $x^2 - y^2 = -1$

9. a) Graph $\left(\frac{x}{3}\right)^2 - y^2 = -1$ and $x^2 - \left(\frac{y}{2}\right)^2 = -1$. Describe how each graph is related to the graph of $x^2 - y^2 = -1$. Explain each relationship.

b) Graph $\frac{x^2}{9} - \frac{y^2}{4} = -1$. Describe how the graph is related to the graph of $x^2 - y^2 = -1$. Explain.

c) Repeat part b, using other numbers in the denominators of the squared terms.

10. What information about the graph of $\frac{x^2}{a^2} - \frac{y^2}{b^2} = -1$ do a and b provide? Explain.

11. In exercises 9 and 10, write the equation of each hyperbola in general form.

12. Graph each relation. Describe each graph.

a) $\left(\frac{x}{5}\right)^2 - y^2 = -1$ **b)** $x^2 - \left(\frac{y}{5}\right)^2 = -1$

c) $\frac{x^2}{16} - \frac{y^2}{4} = -1$ **d)** $\frac{x^2}{4} - \frac{y^2}{16} = -1$

Transforming the graph of the parabola $y = x^2$

13. a) Graph $y = (2x)^2$. Describe how the graph is related to the graph of $y = x^2$. Explain this relationship.

b) Repeat part a, using other numbers in the equation.

c) What information about the graph of $y = (kx)^2$ does k provide? Explain.

14. a) Graph $x = (2y)^2$. Describe how the graph is related to the graph of $x = y^2$. Explain this relationship.

b) Repeat part a, using other numbers in the equation.

c) What information about the graph of $x = (ky)^2$ does k provide? Explain.

15. In exercises 13 and 14, write the equation of each parabola in general form.

16. Graph each relation. Describe each graph.

a) $y = \left(\dfrac{x}{2}\right)^2$ **b)** $x = \left(\dfrac{y}{2}\right)^2$

In *Investigate*, you made certain changes to the equations $x^2 + y^2 = 1$, $x^2 - y^2 = \pm 1$, $y = x^2$, and $x = y^2$, then used the tools from the functions toolkit in Chapter 1 to explain what happened to the graphs.

When a circle is expanded or compressed in perpendicular directions, the image graph is an ellipse. The longest chord in the ellipse is the *major axis*. The midpoint of the major axis is the *centre*. The chord through the centre perpendicular to the major axis is the *minor axis*.

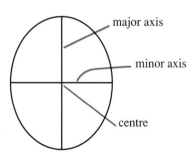

Sketching Ellipses

Expand the unit circle $x^2 + y^2 = 1$ horizontally by a factor of a, and vertically by a factor of b.

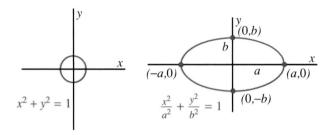

Example 1

Define the conic section defined by the relation $x^2 + 4y^2 = 36$, and sketch its graph.

Solution

$x^2 + 4y^2 = 36$

Think...

There are two second-degree terms separated by a + sign. The graph is an ellipse with centre $(0,0)$. The standard equation of an ellipse has a 1 on the right side.

Divide both sides of the equation by 36.

$$\frac{x^2}{36} + \frac{4y^2}{36} = \frac{36}{36}$$

$$\frac{x^2}{36} + \frac{y^2}{9} = 1$$

$$\left(\frac{x}{6}\right)^2 + \left(\frac{y}{3}\right)^2 = 1$$

The equation represents an ellipse with centre $(0,0)$, obtained by expanding the unit circle by a factor of 6 horizontally and a factor of 3 vertically. Sketch an ellipse with centre $(0,0)$, x-intercepts ±6, and y-intercepts ±3.

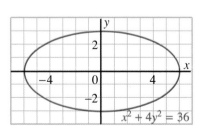

When a rectangular hyperbola is expanded or compressed in perpendicular directions, the image graph is another hyperbola. Unlike an ellipse, a hyperbola does not have both *x*- and *y*-intercepts. To sketch a hyperbola, we use its asymptotes. The asymptotes of the original hyperbola are expanded or compressed in the same way as the hyperbola.

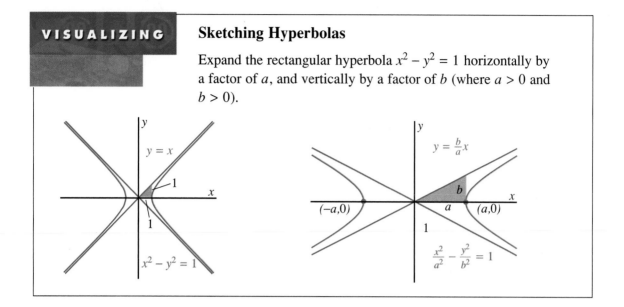

VISUALIZING

Sketching Hyperbolas

Expand the rectangular hyperbola $x^2 - y^2 = 1$ horizontally by a factor of a, and vertically by a factor of b (where $a > 0$ and $b > 0$).

Example 2

Describe the conic section defined by the relation $4x^2 - y^2 + 16 = 0$, and sketch its graph.

Solution

$4x^2 - y^2 + 16 = 0$

> **Think...**
>
> There are two second-degree terms separated by a − sign. The graph is a hyperbola. The standard equation of a hyperbola has 1 or −1 on the right side.

Write the equation as $4x^2 - y^2 = -16$.
Divide both sides by 16.

$$\frac{4x^2}{16} - \frac{y^2}{16} = -\frac{16}{16}$$

$$\frac{x^2}{4} - \frac{y^2}{16} = -1$$

$$\left(\frac{x}{2}\right)^2 - \left(\frac{y}{4}\right)^2 = -1$$

The equation represents a hyperbola with centre $(0, 0)$, obtained by expanding the rectangular hyperbola $x^2 - y^2 = -1$ by a factor of 2 horizontally and a factor of 4 vertically.

Since the right side has the same sign as the y^2 term, the vertices are on the y-axis. Their coordinates are $(0, \pm 4)$.

The equations of the asymptotes are $\frac{y}{4} = \pm\frac{x}{2}$, or $y = \pm 2x$. Graph the asymptotes and use them as a guide to sketch the hyperbola.

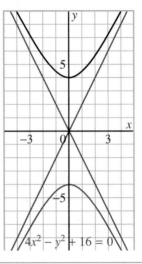

DISCUSSING THE IDEAS

1. When an equation is written in standard form, how can you tell if it represents an ellipse, a hyperbola, or a parabola? Explain.

2. When the equation of an ellipse is written in standard form, how can you tell if its major axis is horizontal or vertical? Explain.

3. When the equation of a hyperbola is written in standard form, how can you tell if its transverse axis is horizontal or vertical? Explain.

4. Is the general form of the equation of a conic section unique? Explain.

5. In *Example 1*, why did we begin by dividing both sides by 36?

6. In *Example 2*, explain how we know that the equations of the asymptotes are $\frac{y}{4} = \pm\frac{x}{2}$, or $y = \pm 2x$.

Standard Equations of Conic Sections Centred at (0, 0)

Ellipse with centre (0, 0)

Major axis on the x-axis

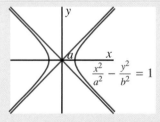

$$\frac{x^2}{a^2} + \frac{y^2}{b^2} = 1 \quad (a > b)$$

Major axis on the y-axis

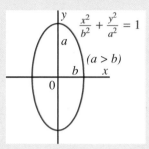

$$\frac{x^2}{b^2} + \frac{y^2}{a^2} = 1$$

$(a > b)$

The squared terms are separated by a + sign. The term in which the larger denominator occurs indicates which axis contains the major axis. If $a = b$, the equation represents a circle. Hence, a circle is a special case of an ellipse.

Hyperbola with centre (0, 0)

Transverse axis on the x-axis

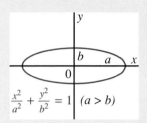

$$\frac{x^2}{a^2} - \frac{y^2}{b^2} = 1$$

Transverse axis on the y-axis

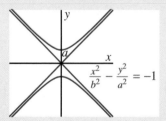

$$\frac{x^2}{b^2} - \frac{y^2}{a^2} = -1$$

The squared terms are separated by a − sign. The term that has the same sign as the constant term indicates which axis contains the transverse axis. When $a = b$, the equation represents a rectangular hyperbola.

Parabola with vertex (0, 0)

Axis of symmetry the x-axis
Opens right if $a > 0$
Opens left if $a < 0$

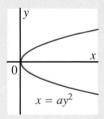

$x = ay^2$

Axis of symmetry the y-axis
Opens up if $a > 0$
Opens down if $a < 0$

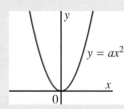

$y = ax^2$

A **1.** Convert each equation to general form.

a) $x^2 + y^2 = 9$

b) $x^2 - y^2 = -16$

c) $\left(\dfrac{x}{3}\right)^2 - y^2 = -1$

d) $x^2 + \left(\dfrac{y}{2}\right)^2 = 1$

e) $\left(\dfrac{x}{5}\right)^2 - \left(\dfrac{y}{2}\right)^2 = 1$

f) $\left(\dfrac{x}{4}\right)^2 + \left(\dfrac{y}{2}\right)^2 = 1$

2. Graph the equations in each list on the same grid.

a) $x^2 + y^2 = 1$

$\left(\dfrac{x}{2}\right)^2 + y^2 = 1$

$\left(\dfrac{x}{4}\right)^2 + y^2 = 1$

$\left(\dfrac{x}{8}\right)^2 + y^2 = 1$

b) $x^2 + y^2 = 1$

$x^2 + \left(\dfrac{y}{2}\right)^2 = 1$

$x^2 + \left(\dfrac{y}{4}\right)^2 = 1$

$x^2 + \left(\dfrac{y}{8}\right)^2 = 1$

3. Choose one list from exercise 2.

a) What happens to the graph when the number in the denominator becomes very large? Explain.

b) Explain the results of part a in terms of the way in which a plane can intersect a cone.

4. Graph the equations in each list on the same grid.

a) $x^2 + y^2 = 1$

$\left(\dfrac{x}{0.5}\right)^2 + y^2 = 1$

$\left(\dfrac{x}{0.25}\right)^2 + y^2 = 1$

$\left(\dfrac{x}{0.125}\right)^2 + y^2 = 1$

b) $x^2 + y^2 = 1$

$x^2 + \left(\dfrac{y}{0.5}\right)^2 = 1$

$x^2 + \left(\dfrac{y}{0.25}\right)^2 = 1$

$x^2 + \left(\dfrac{y}{0.125}\right)^2 = 1$

5. Choose one list from exercise 4.

a) What happens to the graph when the number in the denominator becomes very small? Explain.

b) Explain the results of part a in terms of the way in which a plane can intersect a cone.

B **6. a)** Sketch the relation $\dfrac{x^2}{4} + \dfrac{y^2}{9} = 1$.

b) Choose one denominator from the equation in part a. Graph two other equations of this type, by replacing this denominator with other positive numbers.

c) What conic section is represented by this type of graph?

d) What happens if one denominator is replaced with a negative number?

7. a) Sketch the relation $\dfrac{x^2}{4} - \dfrac{y^2}{9} = 1$.

b) Repeat exercise 6b to d.

8. For each part, predict what the graphs of these equations would look like if they were graphed on the same grid. Sketch the graphs of the equations.

a) $\dfrac{x^2}{1^2} + \dfrac{y^2}{5^2} = 1$

$\dfrac{x^2}{2^2} + \dfrac{y^2}{4^2} = 1$

$\dfrac{x^2}{3^2} + \dfrac{y^2}{3^2} = 1$

$\dfrac{x^2}{4^2} + \dfrac{y^2}{2^2} = 1$

$\dfrac{x^2}{5^2} + \dfrac{y^2}{1^2} = 1$

b) $\dfrac{x^2}{1^2} - \dfrac{y^2}{5^2} = 1$

$\dfrac{x^2}{2^2} - \dfrac{y^2}{4^2} = 1$

$\dfrac{x^2}{3^2} - \dfrac{y^2}{3^2} = 1$

$\dfrac{x^2}{4^2} - \dfrac{y^2}{2^2} = 1$

$\dfrac{x^2}{5^2} - \dfrac{y^2}{1^2} = 1$

9. Verify your predictions in exercise 8.

10. These graphs were produced using a graphing calculator. For each screen, write a set of equations that could be used to make the pattern.

a)
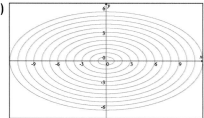

b)
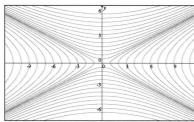

11. These graphs were produced by *Graphmatica*. For each screen, write a set of equations that could be used to make the pattern.

a)
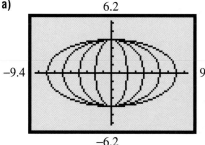

b)
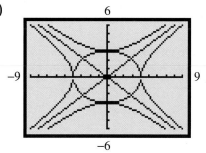

12. Make patterns similar to those in exercise 11.

13. State which equations represent each conic section.

i) a circle   **ii)** an ellipse   **iii)** a hyperbola   **iv)** a parabola

a) $\dfrac{x^2}{9} + \dfrac{y^2}{36} = 1$

b) $\dfrac{x^2}{5} - \dfrac{y^2}{10} = 1$

c) $y^2 = -4x$

d) $\dfrac{x^2}{4} - \dfrac{y^2}{4} = 1$

e) $\dfrac{x^2}{20} - \dfrac{y^2}{25} = -1$

f) $x^2 = \dfrac{1}{3}y$

14. Describe and sketch the graph of each relation.

a) $\dfrac{x^2}{25} + \dfrac{y^2}{16} = 1$ **b)** $\dfrac{x^2}{16} + \dfrac{y^2}{36} = 1$ **c)** $\dfrac{x^2}{16} - \dfrac{y^2}{9} = -1$

d) $\dfrac{x^2}{36} - \dfrac{y^2}{6} = 1$ **e)** $\dfrac{x^2}{9} + \dfrac{y^2}{6} = 1$ **f)** $\dfrac{x^2}{49} - \dfrac{y^2}{25} = -1$

g) $x = 3y^2$ **h)** $y = -2x^2$ **i)** $9x^2 + y^2 = 9$

15. Describe the conic section defined by each relation, and sketch its graph.

a) $x^2 + 9y^2 - 36 = 0$ **b)** $9x^2 + y^2 - 36 = 0$

c) $x^2 - 4y^2 = 36$ **d)** $x^2 - 4y^2 = -36$

e) $5x^2 + y^2 = 20$ **f)** $5x^2 - y^2 = 20$

g) $2x^2 + 3y^2 - 24 = 0$ **h)** $2x^2 - 3y^2 + 24 = 0$

16. a) Graph each relation.

    i) $x^2 - y^2 = 0$ **ii)** $(x - y)^2 = 0$ **iii)** $x^2 + y^2 = 0$

b) Explain how the graphs of the relations in part a could result when a plane intersects a cone.

C **17.** These designs were produced by *Graphmatica*. Write a set of instructions for someone to make patterns similar to these.

a) 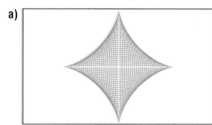   **b)**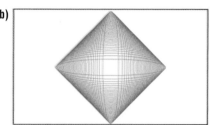

18. Investigate whether a square can be inscribed in each conic section. If it can, determine its area.

a) the ellipse $\dfrac{x^2}{a^2} + \dfrac{y^2}{b^2} = 1$ **b)** the hyperbola $\dfrac{x^2}{a^2} - \dfrac{y^2}{b^2} = 1$

19. Two parabolas have perpendicular axes of symmetry. Prove that their points of intersection lie on a circle.

COMMUNICATING THE IDEAS

Write to explain the difference between the standard form of an equation and the general form of an equation. Include in your explanation a summary of how the standard form of the equation of a conic section is useful for sketching its graph.

9.4 Translating Graphs of Conic Sections

To translate the graph of a conic section, we use the translation tool from the functions toolkit in Chapter 1. For example, suppose the ellipse defined by $\frac{x^2}{9} + \frac{y^2}{4} = 1$ is translated 5 units right and 1 unit down. According to the translation tool, the equation of the translated ellipse is $\frac{(x-5)^2}{9} + \frac{(y+1)^2}{4} = 1$.

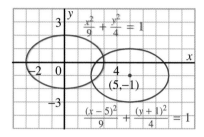

Suppose we write the equations of the two ellipses in general form. The equation of the original ellipse is $4x^2 + 9y^2 = 36$. The equation of the translated ellipse is $4(x-5)^2 + 9(y+1)^2 = 36$, or $4x^2 + 9y^2 - 40x + 18y - 73 = 0$. The two equations in general form contain the same quadratic terms. That is, the quadratic terms were not affected by the translation.

Example

Describe and sketch each relation.

a) $\frac{(x-3)^2}{25} - \frac{(y-4)^2}{4} = 1$

b) $\frac{(x-3)^2}{25} - \frac{(y-4)^2}{4} = -1$

Solution

a) $\frac{(x-3)^2}{25} - \frac{(y-4)^2}{4} = 1$

> *Think...*
>
> Since the squared terms are separated by a − sign, the equation represents a hyperbola. Its graph is obtained from the graph of $x^2 - y^2 = 1$ after expanding by a factor of 5 horizontally and a factor of 2 vertically, then translating 3 units right and 4 units up. The asymptotes of the hyperbola $x^2 - y^2 = 1$ are transformed the same way.

The centre of the hyperbola is $(3, 4)$. The vertices are 5 units to the left and right of this point. Hence, the coordinates of the vertices are $(8, 4)$ and $(-2, 4)$. The asymptotes have slopes $\pm\frac{2}{5}$. Draw lines through the centre with these slopes to represent the asymptotes. Use the asymptotes as a guide to sketch the hyperbola.

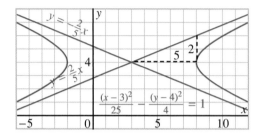

b) $\dfrac{(x - 3)^2}{25} - \dfrac{(y - 4)^2}{4} = -1$

> ### Think...
>
> The graph of this hyperbola is obtained from the graph of $x^2 - y^2 = -1$, using the same transformations as in part a.

The centre of the hyperbola is $(3, 4)$. The vertices are 2 units above and below this point. Hence, the coordinates of the vertices are $(3, 6)$ and $(3, 2)$. The asymptotes have slopes $\pm\frac{2}{5}$. Draw lines through the centre with these slopes to represent the asymptotes. Use the asymptotes as a guide to sketch the hyperbola.

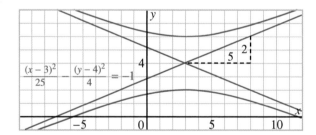

DISCUSSING THE IDEAS

1. In the *Example* parts *a* and *b*, how do we know the slopes of the asymptotes are $\pm\frac{2}{5}$?

2. In the *Example* parts *a* and *b*, how would the graphs differ if the 25 and 4 in the denominators were interchanged?

Standard Equations of Conic Sections Centred at (h, k)

Ellipse with centre (h, k)

Major axis horizontal

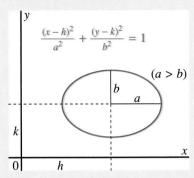

Major axis vertical

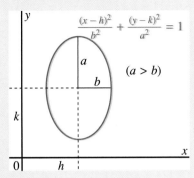

A circle is a special case of an ellipse, in which $a = b$.

Hyperbola with centre (h, k)

Transverse axis horizontal

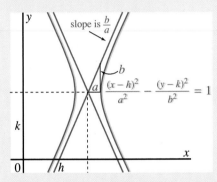

Transverse axis vertical

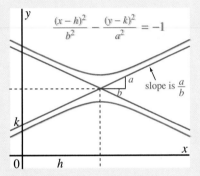

A rectangular hyperbola is a special case of a hyperbola, in which $a = b$.

Parabola with vertex (h, k)

Axis of symmetry horizontal

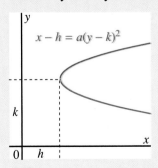

Axis of symmetry vertical

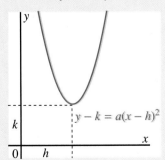

A 1. Write an equation to represent each conic section.

a)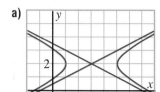

b)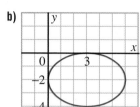

2. In each list, the graphs form a pattern. Describe each pattern.

a) $x^2 + (y + 4)^2 = 1$

$x^2 + (y + 2)^2 = 1$

$x^2 + (y + 0)^2 = 1$

$x^2 + (y - 2)^2 = 1$

$x^2 + (y - 4)^2 = 1$

b) $(x - 8)^2 + (y + 2)^2 = 4$

$(x - 4)^2 + (y + 2)^2 = 4$

$(x - 0)^2 + (y + 2)^2 = 4$

$(x - 4)^2 + (y + 2)^2 = 4$

$(x - 8)^2 + (y + 2)^2 = 4$

3. Describe and sketch each relation.

a) $\dfrac{(x - 4)^2}{9} + \dfrac{(y + 1)^2}{4} = 1$

b) $\dfrac{(x - 4)^2}{4} + \dfrac{(y + 1)^2}{9} = 1$

c) $\dfrac{(x - 4)^2}{9} - \dfrac{(y + 1)^2}{4} = 1$

d) $\dfrac{(x - 4)^2}{4} - \dfrac{(y + 1)^2}{9} = 1$

e) $\dfrac{(x - 4)^2}{9} - \dfrac{(y + 1)^2}{4} = -1$

f) $\dfrac{(x - 4)^2}{4} - \dfrac{(y + 1)^2}{9} = -1$

B 4. Sketch each relation.

a) $\dfrac{(x - 2)^2}{16} + \dfrac{(y + 3)^2}{49} = 1$

b) $\dfrac{(x + 5)^2}{16} + \dfrac{(y - 3)^2}{4} = 1$

c) $\dfrac{(x - 4)^2}{9} - \dfrac{(y + 6)^2}{4} = 1$

d) $\dfrac{(x - 2)^2}{9} - \dfrac{y^2}{25} = 1$

e) $\dfrac{(x + 3)^2}{18} + \dfrac{(y + 3)^2}{9} = 1$

f) $\dfrac{(x - 5)^2}{36} - \dfrac{(y - 1)^2}{9} = -1$

5. Sketch each relation.

a) $\dfrac{(x + 2)^2}{9} + (y + 3)^2 = 1$

b) $(x + 1)^2 + \dfrac{(y - 1)^2}{4} = 1$

c) $\dfrac{(x - 5)^2}{16} - (y + 2)^2 = 1$

d) $(x - 1)^2 - \dfrac{(y - 3)^2}{6} = -1$

e) $y + 2 = (x - 3)^2$

f) $y - 4 = \dfrac{(x + 1)^2}{4}$

6. These graphs were produced by *Graphmatica*. For each screen, write a set of equations that could be used to make the patterns.

a)
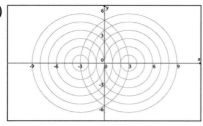

b)
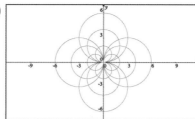

c)
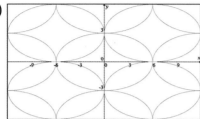

d)
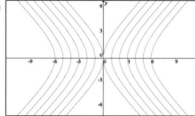

7. Make patterns similar to those in exercise 6.

8. Sketch each relation.

a) $4(x - 3)^2 + 9(y - 2)^2 = 36$

b) $4(x + 1)^2 - (y - 3)^2 = 16$

c) $(x + 5)^2 - 4(y + 4)^2 = -36$

d) $3(x - 2)^2 + 4(y + 2)^2 = 24$

e) $3x^2 - (y - 2)^2 = 27$

f) $(x + 5)^2 + 6y^2 = 36$

g) $4(y - 1) = (x - 2)^2$

h) $4(x + 2) = (y - 5)^2$

C **9.** These designs were produced by *Graphmatica*. For each screen, write a set of instructions that someone could use to make designs similar to these.

a)
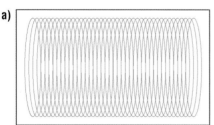

b)

COMMUNICATING THE IDEAS

Suppose an equation of a hyperbola is given in standard form. Write to explain how to determine the coordinates of its centre, the equations of its asymptotes, and how to tell if it opens to the left and right or if it opens up and down. Use examples to illustrate your explanations.

9.5 Graphing Second-Degree Equations in General Form

Recall that the graph of the equation $Ax + By + C = 0$ is a straight line. This is the general equation of the first degree in x and y. The general equation of the second degree in x and y is $Ax^2 + Bxy + Cy^2 + Dx + Ey + F = 0$.

I N V E S T I G A T E Graphing Second-Degree Equations

In these exercises, you will investigate the kinds of graphs the second-degree equation represents.

Graphing $Ax^2 + Cy^2 + Dx + Ey + F = 0$, where A and C have the same sign

1. a) Graph $Ax^2 + y^2 - 16 = 0$ for a few values of A, where:
   i) $A > 1$ ii) $0 < A < 1$ iii) $A = 0$

   b) Draw a conclusion based on the results of part a. Use the tools from the functions toolkit in Chapter 1 to explain your conclusion.

2. The graph of the equation $x^2 + y^2 - 16 = 0$ is transformed. Write the equation of each image in the form $Ax^2 + y^2 - 16 = 0$. Compare the value of A with the results of exercise 1.

   a) Compress horizontally by:
   i) a factor of $\frac{1}{2}$ ii) a factor of $\frac{1}{3}$

   b) Expand horizontally by:
   i) a factor of 2   ii) a factor of 3

3. Which conic sections are represented by graphs whose equations have the form $Ax^2 + y^2 + F = 0$, for these values of A?
   a) $A = 1$ b) $A \neq 0$ c) $A = 0$

4. Exercises 1, 2, and 3 involve the equation $Ax^2 + y^2 + F = 0$. Similar exercises could have involved $x^2 + Cy^2 + F = 0$. In what way would the results be different from those of exercises 1, 2, and 3? Explain.

5. a) Graph $x^2 + y^2 + Dx - 16 = 0$ for a few values of D, where:
   i) $D = 0$ ii) $D > 0$ iii) $D < 0$

   b) Graph $x^2 + y^2 + Ey - 16 = 0$ for a few values of E, where:
   i) $E = 0$ ii) $E > 0$ iii) $E < 0$

   c) Draw a conclusion from the results of parts a and b. Explain.

6. a) Graph $2x^2 + 3y^2 + Dx - 16 = 0$ for a few values of D, where:
   i) $D > 0$ ii) $D < 0$

b) Graph $2x^2 + 3y^2 + Ey - 16 = 0$ for a few values of E, where:
   i) $E > 0$ **ii)** $E < 0$

c) Draw a conclusion from the results of parts a and b. Explain.

7. Summarize as many properties as you can of the second-degree equation $Ax^2 + Cy^2 + Dx + Ey + F = 0$ in exercises 1 to 6.

Graphing $Ax^2 + Cy^2 + Dx + Ey + F = 0$, where A and C have opposite signs

8. a) Graph $Ax^2 - y^2 - 16 = 0$ for a few values of A, where:
   i) $A > 1$ **ii)** $0 < A < 1$ **iii)** $A = 0$

b) Repeat part a using the equation $Ax^2 - y^2 + 16 = 0$.

c) Draw a conclusion from the results of parts a and b. Explain.

9. The graphs of the equations $x^2 - y^2 - 16 = 0$ and $x^2 - y^2 + 16 = 0$ are transformed. Write the equation of each image in the form $Ax^2 - y^2 - 16 = 0$ or $Ax^2 - y^2 + 16 = 0$. Compare the values of A with the results of exercise 8. Explain.

a) Compress horizontally by:
   i) factor of $\frac{1}{2}$ **ii)** a factor of $\frac{1}{3}$

b) Expand horizontally by:
   i) factor of 2   **ii)** a factor of 3

10. Which conic sections are represented by graphs whose equations have the form $Ax^2 - y^2 + F = 0$ for these values of A?
   a) $A = 1$ **b)** $A \neq 0$ **c)** $A = 0$

11. Exercises 8, 9, and 10 involve the equation $Ax^2 - y^2 + F = 0$. Similar exercises could have involved $x^2 - Cy^2 + F = 0$. In what way would the results be different from those of exercises 8, 9, and 10? Explain.

12. a) Graph $x^2 - y^2 + Dx - 16 = 0$ for a few values of D, where:
   i) $D = 0$ **ii)** $D > 0$ **iii)** $D < 0$

b) Graph $x^2 - y^2 + Ey - 16 = 0$ for a few values of E, where:
   i) $E = 0$ **ii)** $E > 0$ **iii)** $E < 0$

c) Draw a conclusion from the results of parts a and b. Explain.

13. a) Graph $2x^2 - 3y^2 + Dx - 16 = 0$ for a few values of D, where:
   i) $D > 0$ **ii)** $D < 0$

b) Graph $2x^2 - 3y^2 + Ey - 16 = 0$ for a few values of E, where:
   i) $E > 0$ **ii)** $E < 0$

c) Draw a conclusion from the results of parts a and b. Explain.

14. Summarize as many properties as you can of the second-degree equation $Ax^2 + Cy^2 + Dx + Ey + F = 0$ in exercises 8 to 13.

Graphing $Ax^2 + Cy^2 + Dx + Ey + F = 0$, where either $A = 0$ or $C = 0$

15. a) Graph $y^2 + 4x + Ey + 4 = 0$ for a few values of E, where:

    i) $E = 0$ **ii)** $E > 0$ **iii)** $E < 0$

  b) Graph $x^2 + 4y + Dx + 4 = 0$ for a few values of D, where:

    i) $D = 0$ **ii)** $D > 0$ **iii)** $D < 0$

  c) Draw a conclusion from the results of parts a and b. Explain.

16. The general equation of the second degree in x and y is $Ax^2 + Bxy + Cy^2 + Dx + Ey + F = 0$. In all the above equations, $B = 0$.

  a) Graph some equations in which $B \neq 0$.

  b) For an equation in which $B \neq 0$, what property does the graph have?

We will consider only general second-degree equations in which there is no xy term. These equations have the form $Ax^2 + Cy^2 + Dx + Ey + F = 0$. In *Investigate*, you should have discovered the following properties of the graph of this equation.

Some Properties of the Graph of $Ax^2 + Cy^2 + Dx + Ey + F = 0$

We assume that A and C are not both 0.

- If the graph of the equation exists, it is a conic section.
- If the conic section is an ellipse, then $AC > 0$. The major axis is parallel to a coordinate axis.
- If the conic section is a hyperbola, then $AC < 0$. The transverse axis is parallel to a coordinate axis.
- If the conic section is a parabola, then $A = 0$ or $C = 0$. The axis of symmetry is parallel to a coordinate axis.

We can use graphing technology to graph second-degree equations in general form. We can use the tools from the functions toolkit in Chapter 1 to determine the equation of the image of any graph under expansions, compressions, and translations.

Example

a) Graph the conic section defined by $x^2 + 4y^2 - 9 = 0$.

b) Describe the graph and determine its intercepts.

c) Determine the equation of the image of the graph in part a after each transformation.

    i) a horizontal compression by a factor of $\frac{1}{2}$ and a vertical expansion by a factor of 3

    ii) a translation 5 units right and 2 units down

d) Sketch the graphs of the image equations in part c.

Solution

a) The graph of $x^2 + 4y^2 - 9 = 0$ is shown.

b) The graph is an ellipse with centre $(0,0)$. To determine the x-intercepts, substitute 0 for y in the equation.

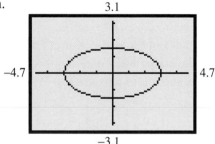

$$x^2 + 4(0)^2 - 9 = 0$$
$$x^2 = 9$$
$$x = \pm 3$$

The x-intercepts are ± 3.

To determine the y-intercepts, substitute 0 for x in the equation.

$$0^2 + 4y^2 - 9 = 0$$
$$y^2 = \frac{9}{4}$$
$$y = \pm\frac{3}{2}$$

The y-intercepts are ± 1.5.

c) Use the tools from the functions toolkit in Chapter 1. Start with the equation $x^2 + 4y^2 - 9 = 0$ in each case.

    i) For a horizontal compression of factor $\frac{1}{2}$, replace x with $2x$.
For a vertical expansion of factor $\frac{1}{3}$, replace y with $\frac{y}{3}$.
The equation of the image is:

$$(2x)^2 + 4\left(\frac{y}{3}\right)^2 - 9 = 0$$
$$4x^2 + \frac{4}{9}y^2 - 9 = 0$$
$$36x^2 + 4y^2 - 81 = 0$$

    ii) For a translation 5 units right, replace x with $x - 5$.
For a translation 2 units down, replace y with $y + 2$.
The equation of the image is:

$$(x - 5)^2 + 4(y + 2)^2 - 9 = 0$$
$$x^2 - 10x + 25 + 4y^2 + 16y + 16 - 9 = 0$$
$$x^2 + 4y^2 - 10x + 16y + 32 = 0$$

d) i) Start with the graph in part a. Compress the graph horizontally by a factor of $\frac{1}{2}$, and expand it vertically by a factor of 3.

The centre is $(0,0)$.

The coordinates of the points where the original ellipse intersects the x-axis are $(\pm 3, 0)$. The coordinates of the images of these points are $\left(\frac{\pm 3}{2}, 0\right)$, or $(\pm 1.5, 0)$.

The coordinates of the points where the original ellipse intersects the y-axis are $(0, \pm 1.5)$. The coordinates of the images of these points are $(0, \pm 1.5 \times 3)$, or $(0, \pm 4.5)$.

Sketch an ellipse with centre $(0,0)$ and passing through the points $(\pm 1.5, 0)$ and $(0, \pm 4.5)$.

ii) Start with the graph in part a. Translate the graph 5 units right and 2 units down.

The centre is $(5, -2)$.

The coordinates of the vertices are $(5 \pm 3, -2)$, or $(8, -2)$ and $(2, -2)$.

These are the endpoints of the major axis. The coordinates of the endpoints of the minor axis are $(5, -2 \pm 1.5)$, or $(5, -0.5)$ and $(5, -3.5)$.

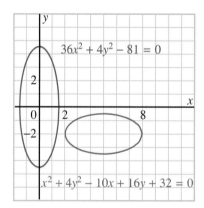

DISCUSSING THE IDEAS

1. In the box on page 552, explain why we can assume that A and C are not both 0.

2. In the *Example* part b, explain why we determine the x-intercept by letting $y = 0$, and why we determine the y-intercept by letting $x = 0$.

3. a) In the *Example* part d i, explain why the coordinates of the vertices are $(0, \pm 1.5 \times 3)$, and why the coordinates of the endpoints of the minor axis are $(\frac{\pm 3}{2}, 0)$.

b) In the *Example* part d ii, explain why the coordinates of the vertices are $(5 \pm 3, -2)$, and why the coordinates of the endpoints of the minor axis are $(5, -2 \pm 1.5)$.

1. a) Graph the equation $4x^2 + 25y^2 - 100 = 0$.

b) Choose either the coefficient of x^2 or the coefficient of y^2 in the equation in part a. Graph two other similar equations by replacing this coefficient with other numbers that have the same sign.

c) What conic section does this type of graph represent?

d) What happens if the coefficient you chose is replaced with other numbers that have the opposite sign? Explain.

2. Use the equation in exercise 1.

a) Graph two other similar equations by replacing the constant term with other numbers that have the same sign.

b) What happens if the constant term is replaced with other numbers that have the opposite sign? Explain.

3. a) Graph the equation $2x^2 - 3y^2 - 12 = 0$.

b) Repeat exercise 1b to d.

4. Use the equation in exercise 3. Repeat exercise 2a and b.

B **5. a)** Graph $x^2 + Cy^2 - 16 = 0$ for a few values of C, where:
   i) $C > 1$ **ii)** $0 < C < 1$ **iii)** $C = 0$

b) Draw a conclusion based on the results of part a. Explain.

6. The graph of the equation $x^2 + y^2 - 16 = 0$ is transformed as described below. Write the equation of each image in the form $x^2 + Cy^2 - 16 = 0$. Compare the value of C with the results of exercise 5.

a) Compress vertically by: **i)** a factor of $\frac{1}{2}$ **ii)** a factor of $\frac{1}{3}$
b) Expand vertically by: **i)** a factor of 2   **ii)** a factor of 3

7. a) Graph $x^2 + Cy^2 - 16 = 0$ for a few values of C, where:
   i) $C < -1$ **ii)** $-1 < C < 0$ **iii)** $C = 0$

b) Repeat part a using the equation $x^2 + Cy^2 + 16 = 0$.

c) Draw a conclusion based on the results of parts a and b. Explain.

8. The graphs of the equations $x^2 - y^2 - 16 = 0$ and $x^2 - y^2 + 16 = 0$ are transformed as described below. Write the equation of each image in the form $x^2 + Cy^2 - 16 = 0$ or $x^2 + Cy^2 + 16 = 0$. Compare the value of C with the results of exercise 7.

a) Compress vertically by: **i)** a factor of $\frac{1}{2}$ **ii)** a factor of $\frac{1}{3}$
b) Expand vertically by: **i)** a factor of 2   **ii)** a factor of 3

9. a) Graph the conic section defined by $x^2 + 9y^2 - 9 = 0$.

b) Describe the graph and determine its intercepts.

c) Determine the equation of the image of the graph in part a after each transformation.
   i) a horizontal compression by a factor of $\frac{1}{3}$ and a vertical expansion by a factor of 2
   ii) a translation 2 units left and 1 unit up

d) Sketch the graphs of the image equations in part c.

10. a) Predict what the graphs of these equations would look like.

i)
$$x^2 + y^2 = 25$$
$$10x^2 + y^2 = 25$$
$$100x^2 + y^2 = 25$$
$$1000x^2 + y^2 = 25$$

ii)
$$x^2 + y^2 = 25$$
$$0.1x^2 + y^2 = 25$$
$$0.01x^2 + y^2 = 25$$
$$0.001x^2 + y^2 = 25$$

b) Check your predictions.

c) Predict what happens when the coefficient of x^2 becomes 0. Check your prediction.

d) Predict what happens when the coefficient of x^2 becomes negative. Check your prediction.

e) Explain the above results in terms of the way in which a plane can intersect a cone.

11. In exercise 10, explain how the results would change if x and y were interchanged in the equations.

12. Repeat exercise 10 for these equations.

i)
$$x^2 - y^2 = 25$$
$$10x^2 - y^2 = 25$$
$$100x^2 - y^2 = 25$$
$$1000x^2 - y^2 = 25$$

ii)
$$x^2 - y^2 = 25$$
$$0.1x^2 - y^2 = 25$$
$$0.01x^2 - y^2 = 25$$
$$0.001x^2 - y^2 = 25$$

13. In exercise 12, explain how the results would change if x and y were interchanged in the equations.

14. Draw diagrams to illustrate what happens to the graph of each relation as C varies through all real numbers.

a) $x^2 + Cy^2 - 25 = 0$ **b)** $x^2 + Cy^2 + 25 = 0$

15. Draw diagrams to illustrate what happens to the graph of each relation as F varies through all real numbers.

a) $x^2 + 4y^2 + F = 0$ **b)** $x^2 - 4y^2 + F = 0$

16. Draw diagrams to illustrate what happens to the graph of each relation as D varies through all real numbers.

a) $x^2 + 4y^2 + Dx = 0$ **b)** $x^2 - 4y^2 + Dx = 0$

17. State which equations represent each conic section.

    i) a circle   **ii)** an ellipse   **iii)** a hyperbola   **iv)** a parabola

a) $x^2 - 6y - 3 = 0$ **b)** $4x^2 - 4y^2 - 9 = 0$ **c)** $x^2 + 4y^2 - 8 = 0$

d) $3x^2 + 3y^2 - 5 = 0$ **e)** $x = 6y^2$ **f)** $2x^2 - 3y^2 + 6 = 0$

18. Consider the equation $Ax^2 + Cy^2 + Dx + Ey + F = 0$. What conditions must be satisfied by the coefficients for this equation to represent each conic section?

a) a circle with centre $(0,0)$

b) a rectangular hyperbola with centre $(0,0)$ and vertices on the x-axis

c) a rectangular hyperbola with centre $(0,0)$ and vertices on the y-axis

d) a parabola with vertex $(0,0)$ and a horizontal axis of symmetry

e) an ellipse with centre $(0,0)$ and vertices on the y-axis

C **19.** Look at the general equation of the second degree in x and y.

a) What do you think is the general equation of the third degree in x and y?

b) Graph each example of a third-degree equation.

    i) $y^2(5 + x) = x^2(5 - x)$ **ii)** $y^2 = x^3 - 15x + 25$ **iii)** $(x + y)^2 + y^3 = 25$

c) Experiment with other third-degree equations.

20. a) Recall that the equation $Ax + By + C = 0$ represents a straight line. Special cases of this line occur when x or y is missing from the equation. Describe the graph of each equation. Illustrate your descriptions with examples.

    i) $By + C = 0$ **ii)** $Ax + C = 0$

b) Similarly, special cases of the equation $Ax^2 + Bxy + Cy^2 + Dx + Ey + F = 0$ occur when there are no terms involving one variable. Describe the graph of each equation. Illustrate your descriptions with examples.

    i) $Ey + F = 0$ **ii)** $Cy^2 + Ey + F = 0$

    iii) $Dx + F = 0$ **iv)** $Ax^2 + Dx + F = 0$

c) Explain how the results in part b can occur when a plane intersects a cone.

COMMUNICATING THE IDEAS

Recall that the general equation $Ax + By + C = 0$ always represents the same geometric figure, a straight line. We could say that the general equation $Ax^2 + Cy^2 + Dx + Ey + F = 0$ does not always represent the same type of geometric figure, and we could also say that it does. Explain this apparent inconsistency. Include examples to illustrate your explanation.

Designing a Dish Antenna

On page 516, you considered the problem of designing a dish antenna with a parabolic cross section to receive satellite signals. Parallel signals are reflected off the surface and concentrated at a point called the *focus*, where the receiver is located. Given the diameter and the depth of the antenna, how can we determine where to place the receiver?

 DEVELOP A MODEL

 You will use a parabola as a model of the cross section of a dish antenna. You can use computer graphing software to graph a parabola and some lines parallel to the axis of symmetry. On a printout of the graph, use a protractor to draw the reflected rays, which pass through the focus. The computer instructions below are for *Graphmatica*. You can modify these instructions for use with other software.

1. Start *Graphmatica*, and enter an equation of the form $y = ax^2$ for some value of a. You will need to experiment with this value to get a satisfactory parabola. Use the Grid Range option in the View menu to ensure the units of length are the same along both axes (see page 526). Your objective is to obtain a graph similar to the one below, where the parabola fills a large part of the graphing window.

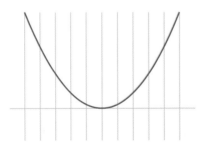

2. Draw several vertical lines. In *Graphmatica*, use an instruction similar to this:
   $x = a \{a: -10, 10, 2\}$
   This makes the computer graph the line $x = a$ for values of a from -10 to 10 in steps of 2, that is, the vertical lines $x = -10$, $x = -8$, $x = -6$, ..., $x = 10$.

3. When you are satisfied with the result, print one or more copies.

4. Visualize the vertical lines representing signals that are reflected by the parabola.

a) Look at the point where one line intersects the parabola. Use visual estimation to draw a tangent to the parabola at that point. Then use a protractor to draw a line representing the reflected signal. The angle between the tangent and the incoming signal is equal to the angle between the tangent and the reflected signal.

b) Repeat part a for several other points.

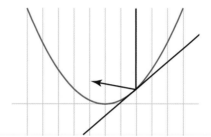

![MM] **LOOK AT THE IMPLICATIONS**

In exercise 4, you should have determined the position of the focus. The position is not exact, but it should be close enough for you to discover a pattern that determines the position of the focus of every parabola. Complete exercise 5 to discover the pattern.

5. a) Draw a line through the focus, F, which is parallel to the x-axis. At the point, G, where this line intersects the parabola, draw a perpendicular to the x-axis. You have now constructed the rectangle below.

b) Measure the length and width of this rectangle, and determine its length : width ratio. Round this value to the nearest whole number.

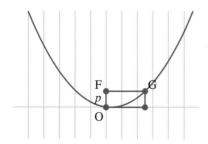

6. a) What simple relationship exists between the x-coordinate and the y-coordinate of G?

b) Let p represent the distance from the vertex to the focus. Write the coordinates of G in terms of p.

c) List some reasons why you cannot be certain that your coordinates are correct.

You should have found that the x-coordinate of G is double the y-coordinate. That is, the coordinates of G are $(2p, p)$.

 REVISIT THE SITUATION

In the dish antenna problem, you know the diameter and the depth of the antenna. This means that you can determine the equation of the cross-sectional parabola in the form $y = ax^2$. Then you can use the result of exercise 6 to determine the distance from the vertex to the focus.

7. A parabolic dish antenna has a diameter of 60 cm and a depth of 7 cm.

a) Let the equation of the cross-sectional parabola be $y = ax^2$. Use the given dimensions to determine the coordinates of H in the diagram below.

b) Determine the value of a for this parabola. Write the equation of the parabola.

c) Let G be the point on the parabola that has the same y-coordinate as the focus. You found in exercise 6 that the coordinates of G are $(2p, p)$. Use the result of part b to determine p for this parabola.

d) For this antenna, where should the receiver be located?

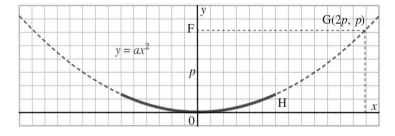

8. Repeat exercise 7 for an antenna with a diameter of d centimetres and a depth of h centimetres. Write a formula in terms of d and h for the distance from the vertex to the focus.

9.6 Converting Equations from General to Standard Form

When the equation of a quadratic relation is written in standard form, the numbers in the equation indicate certain properties of its graph. For example, we can tell that the equation $\dfrac{(x-2)^2}{9} + \dfrac{(y-1)^2}{16} = 1$ represents an ellipse with centre $(2, 1)$. If we write the equation in the form $\left(\dfrac{x-2}{3}\right)^2 + \left(\dfrac{y-1}{4}\right)^2 = 1$, we can tell that the major axis is vertical and that its vertices are 4 units above and below the centre.

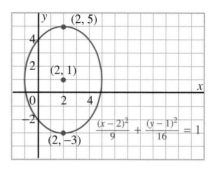

The above equation can also be written in general form.

$$\frac{(x-2)^2}{9} + \frac{(y-1)^2}{16} = 1$$

> Multiply each term by 9×16, or 144

$$16(x-2)^2 + 9(y-1)^2 = 144$$
$$16(x^2 - 4x + 4) + 9(y^2 - 2y + 1) = 144$$
$$16x^2 + 9y^2 - 64x - 18y - 71 = 0$$

When an equation is written in general form, we cannot determine the properties of the graph from the equation as readily as we can when it is in standard form. To write the equation in standard form, we must reverse the process that occurs when the binomial squares in an equation in standard form are expanded and the equation is rewritten in general form.

Example 1

Consider the conic section defined by $4x^2 - 9y^2 + 32x + 18y + 91 = 0$.

a) Write the equation in standard form. b) Identify the conic section.

c) Graph the conic section. Label its centre and vertices.

Solution

a) $4x^2 - 9y^2 + 32x + 18y + 91 = 0$

Collect the terms containing x, and the terms containing y.

$4x^2 + 32x - 9y^2 + 18y + 91 = 0$

Remove the coefficient of x^2 and the coefficient of y^2 as common factors.

$4(x^2 + 8x) - 9(y^2 - 2y) + 91 = 0$

Complete each square.

$4(x^2 + 8x + 16 - 16) - 9(y^2 - 2y + 1 - 1) + 91 = 0$

$4(x + 4)^2 - 64 - 9(y - 1)^2 + 9 + 91 = 0$

$4(x + 4)^2 - 9(y - 1)^2 = -36$

Divide each side by 36. $\dfrac{(x+4)^2}{9} - \dfrac{(y-1)^2}{4} = -1$

b) The equation represents a hyperbola with centre $(-4, 1)$ and vertices 2 units above and below the centre. The asymptotes have slopes $\pm\dfrac{2}{3}$.

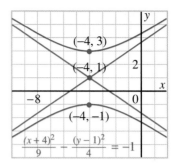

An equation that represents a parabola contains only one squared term. Hence, only one square can be completed.

Example 2

Consider the conic section defined by $x^2 + 10x + 4y + 13 = 0$.

a) Write the equation in standard form.

b) Identify the conic section. Graph it, then describe the properties of its graph.

Solution

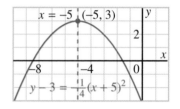

a)
$$x^2 + 10x + 4y + 13 = 0$$
$$x^2 + 10x + 25 - 25 + 4y + 13 = 0$$
$$(x+5)^2 + 4y - 12 = 0$$
$$4(y-3) = -(x+5)^2$$
$$y - 3 = -\tfrac{1}{4}(x+5)^2$$

b) The equation represents a parabola. Vertex: $(-5, 3)$; direction of opening: down; axis of symmetry: $x = -5$; congruent to $y = \tfrac{1}{4}x^2$

1. Refer to *Example 1*.

 a) Explain why "+ 16 − 16" and "+ 1 − 1" were inserted in the solution.

 b) Explain why both sides were divided by 36 in the last step.

 c) Explain how we could have predicted that the equation represents a hyperbola.

 d) Explain how you can tell from the equation that the asymptotes have slopes $\pm\dfrac{2}{3}$.

A 1. Write each equation in general form. What are the values of $A, C, D, E,$ and F?

    a) $3(x-1)^2 + (y+2)^2 = 9$ **b)** $(x+5)^2 - 2(y-1)^2 = 10$

    c) $y - 2 = 3(x-4)^2$ **d)** $\dfrac{(x+2)^2}{9} + \dfrac{(y-3)^2}{4} = 1$

    e) $\dfrac{(x+1)^2}{6} - \dfrac{(y+2)^2}{3} = -1$ **f)** $x + 4 = -2(y-3)^2$

2. Each equation represents one conic section in the list at the right. Identify the conic section that corresponds to each equation.

    a) $9x^2 + 4y^2 - 54x + 16y + 61 = 0$ circle

    b) $2x^2 - 3y^2 - 8x - 6y + 11 = 0$ ellipse

    c) $y^2 - 4x + 6y - 23 = 0$ hyperbola

    d) $x^2 + y^2 + 4x + 5y = 0$ parabola

B 3. Write each equation in standard form, then identify the conic section.

    a) $2x^2 + y^2 + 12x - 2y + 15 = 0$ **b)** $4x^2 + 9y^2 - 8x + 36y + 4 = 0$

    c) $x^2 - 9y^2 - 4x + 18y - 14 = 0$ **d)** $x^2 - 4y^2 - 2x - 3 = 0$

    e) $y^2 - 4x + 8y + 3 = 0$ **f)** $x^2 + 2x + 3y + 4 = 0$

4. The diagram shows three overlapping squares.

    a) Suppose the green region has the same area as the yellow square. What is the relation between x and y?

    b) Graph the relation between x and y.

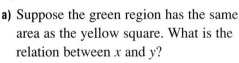

C 5. Consider any conic section with equation $Ax^2 + Cy^2 + Dx + Ey + F = 0$. Prove that the expression $Ax^2 + Cy^2$ does not change when the graph of the conic section is translated.

6. Suppose the graph of a circle were rotated. Would an xy term be introduced in its equation? Explain.

COMMUNICATING THE IDEAS

Write to explain the difference between the standard form and the general form of the equation of a conic section. Use examples to illustrate how to convert a given equation of a conic section from general form to standard form, and vice versa.

Why Is an Ellipse an Ellipse?

Recall that an ellipse has been described in two different ways earlier in this chapter. In Section 9.1, the closed curve of intersection of a plane and a cone is an ellipse. In Section 9.3, the curve obtained by expanding or compressing a circle in perpendicular directions is also an ellipse. Although both curves look the same, we cannot be certain that they are just by observation.

This raises the following question.

> How can we be sure that the curve that results when a circle is expanded or compressed in perpendicular directions is the same as the closed curve that results when a plane intersects a cone?

To solve this problem, we begin by defining an ellipse as follows.

When a plane intersects a cone to form a closed curve, this curve is called an *ellipse*.

Using this definition, you can prove that the curve obtained by expanding or compressing a circle in perpendicular directions is an ellipse.

To model a plane intersecting a cone, visualize a ball resting on a glass tabletop (see diagram on page 565). A cone of light shining on the ball forms an elliptical shadow on the glass. The light source represents the vertex of the cone. The edge of the shadow represents an ellipse. The lines passing through the ellipse and the light source are the generators of the cone. The glass tabletop represents a plane intersecting the cone. The ball represents a sphere that is tangent to the cone and also to the plane. Visualize a second sphere that is also tangent to the cone and the plane, but on the other side of the plane.

Represent points as follows:
  F_1 and F_2 are the points of contact of the spheres with the plane.
  P is any point on the ellipse.
  Q_1 and Q_2 are the points on the spheres that lie on the generator through P.

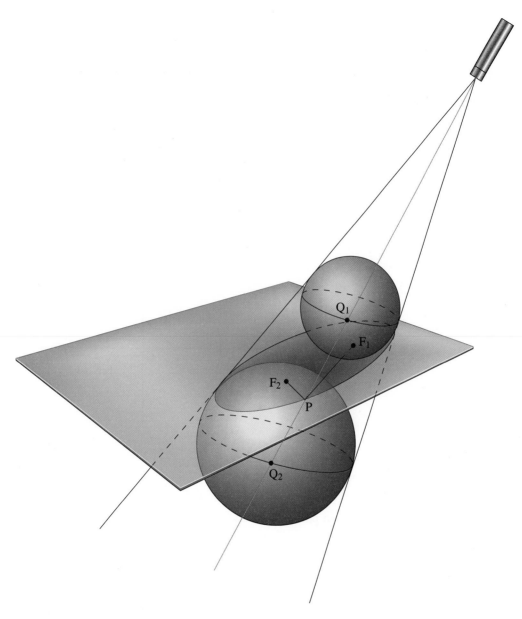

1. **a)** Explain why $PF_1 = PQ_1$ and $PF_2 = PQ_2$.

   b) Use the result of part a. Prove that $PF_1 + PF_2$ is constant for all positions of P on the ellipse.

2. For any position of the plane that forms an ellipse, explain why there will always be a tangent sphere on:

   a) the same side of the plane as the vertex of the cone

   b) the other side of the plane

Exercise 1 establishes this property of an ellipse.

Constant Sum Property

There are two points F_1 and F_2 inside an ellipse such that $PF_1 + PF_2$ is constant for all points P on the ellipse. The points F_1 and F_2 are the *foci* of the ellipse.

You can use the constant sum property to construct an ellipse by placing a loop of string around two thumbtacks on a piece of cardboard. Move a pencil around the tacks while keeping the string taut — the pencil draws an ellipse.

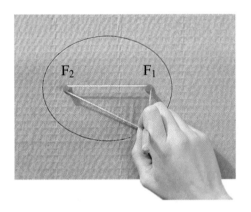

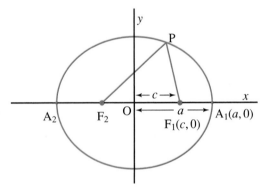

3. **a)** Explain what happens to the ellipse in each case.
   i) The length of the loop of string is changed.
   ii) The distance between the thumbtacks is changed.

   b) Explain how the changes in part a affect the position of the plane in the diagram on page 565.

Visualize the ellipse in the diagram on page 565 as seen from above. The line through F_1 and F_2 intersects the ellipse at two points A_1 and A_2, which are the *vertices*. The line segment A_1A_2 is the *major axis*. The midpoint, O, of the line segment F_1F_2 is the *centre*. Let a and c represent the distances OA_1 and OF_1, respectively. Draw coordinate axes on the figure, where O is the origin, and the coordinates of F_1 and A_1 are $(c, 0)$ and $(a, 0)$, respectively.

You can use the constant sum property to determine the equation of the ellipse in the above diagram in terms of a and c.

4. a) Prove that $PF_1 + PF_2 = 2a$ for all points P on the ellipse.

  b) Let the coordinates of any point P on the ellipse be (x, y). Use the distance formula to write the equation in part a in terms of x and y.

  c) Your equation should contain two radical expressions. Recall that the strategy for solving a radical equation is to eliminate a radical by isolating it on one side of the equation and squaring each side. Use this strategy, and simplify the equation as far as possible.

  d) Your result in part c should be the equation $\dfrac{x^2}{a^2} + \dfrac{y^2}{a^2 - c^2} = 1$.

  Explain why this equation can be written in the form $\dfrac{x^2}{a^2} + \dfrac{y^2}{b^2} = 1$.

5. Explain why your solution to exercise 4 proves that the closed curve resulting when a plane intersects a cone also occurs when a circle is expanded or compressed in perpendicular directions.

6. An ellipse and its major axis are given.

  a) Describe a geometric construction that could be used to determine the positions of the foci F_1 and F_2.

  b) Prove that your construction is correct.

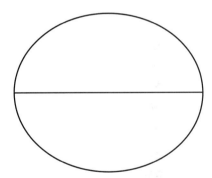

7. In exercises 1 to 5, you defined an ellipse in terms of a plane intersecting a cone, and proved that it also occurs when a circle is expanded or compressed in perpendicular directions. Conversely, you could have defined an ellipse as the curve that results when a circle is expanded or compressed in perpendicular directions. Explain how you could prove that this curve is formed when a plane intersects a cone.

1. These graphs were produced by *Graphmatica*. For each screen, write a set of equations that could be used to make the pattern.

   a)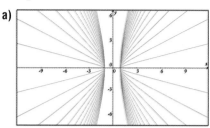

   b)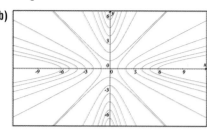

2. a) Graph the conic section defined by $3x^2 + y^2 - 6 = 0$.

   b) Describe the graph and determine its intercepts.

   c) Determine the equation of the image of the graph in part a after each transformation.
      i) a horizontal expansion by a factor of 2 and a vertical compression by a factor of $\frac{1}{2}$
      ii) a translation 3 units right and 2 units up

   d) Sketch the graphs of the image equations in part c.

3. Sketch each relation and identify the curve.

   a) $x^2 + y^2 = 9$ b) $4x^2 + y^2 = 16$ c) $4x^2 - y^2 = 16$

   d) $y = \frac{x^2}{8}$ e) $4x^2 + 25y^2 = 100$ f) $4x^2 - 25y^2 = 100$

4. Write to explain why the type of conic section defined by the general equation $Ax^2 + Cy^2 + Dx + Ey + F = 0$ depends only on the expression $Ax^2 + Cy^2$.

5. The equation $Ax^2 + Bxy + Cy^2 + Dx + Ey + F = 0$ is a more general equation than the one in exercise 4. This equation also represents conic sections. If $B \neq 0$, explain how the conic sections represented by this equation differ from those represented by the equation $Ax^2 + Cy^2 + Dx + Ey + F = 0$.

6. Write each equation in standard form, then identify the conic section.

   a) $x^2 + y^2 - 8x + 6y + 9 = 0$ b) $x^2 + 4y^2 - 2x + 16y + 13 = 0$

   c) $3x^2 + 4y^2 + 18x - 16y + 31 = 0$ d) $3x^2 - 2y^2 - 36x + 96 = 0$

   e) $y^2 - 8x - 8y = 0$ f) $6x^2 + 24x - y + 19 = 0$

7. Describe the graph of each relation.

   a) $3x^2 + 2y^2 - 6x + 16y + 35 = 0$ b) $2x^2 + y^2 - 12x - 8y + 36 = 0$

   c) $4x^2 - y^2 + 16x - 4y + 12 = 0$ d) $x^2 + 2xy + y^2 + 2x + 2y + 1 = 0$

1. Describe what happens to the equation of a function when you make each change to its graph.

   a) Expand vertically by a factor of 5.

   b) Compress horizontally by a factor of $\frac{1}{3}$, then reflect in the y-axis.

   c) Expand horizontally by a factor of 3.

   d) Compress vertically by a factor of $\frac{1}{2}$, then reflect in the x-axis.

2. Chemists use the symbols $[H^+]$ and $[OH^-]$ to represent the concentration of hydrogen ions and hydroxyl ions, respectively, in a solution. Concentration is measured in units called moles per litre. When $[H^+] > [OH^-]$, the solution is acidic. When $[H^+] < [OH^-]$, the solution is alkaline.

   a) Chemists have determined that $[H^+][OH^-] = 10^{-14}$. In pure water, $[H^+]$ and $[OH^-]$ are equal. What are the values of $[H^+]$ and $[OH^-]$ for pure water?

   b) pH is defined as $pH = -\log_{10}[H^+]$. Explain why this means that the pH value of pure water is 7.

   c) What is the hydrogen-ion concentration, $[H^+]$, of a weak vinegar solution with a pH value of 3.1?

3. The midpoints of the sides of a square with sides 1 m long are joined to form another square. Then the midpoints of the sides of the second square are joined to form a third square. Suppose this process were continued indefinitely to form an infinite set of smaller and smaller squares converging on the centre of the original square.

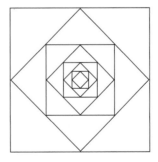

   a) Determine the total length of the segments that form the sides of all the squares.

   b) Visualize drawing the diagram on a large piece of paper, starting with a square with sides 1 m long. Estimate the total length of the segments you could draw, before the segments become so small that you cannot draw any more.

   c) Write to explain how you estimated the length in part b. Identify any assumptions you made to arrive at your estimate.

4. a) Establish an identity for $\tan 2x$.

   b) Establish identities for $\sin 4x$ and $\cos 4x$.

5. Write an equation for a cosine function with each set of properties.

  a) amplitude: 3.5   period: 2.5   phase shift: −6   vertical displacement: −2

  b) maximum: 3.5   minimum: 2.5   period: 6   phase shift: 2

6. Graph each function over two cycles. For each function, determine the domain, range, period, and the equations of the asymptotes.

  a) $y = 3\cot 2(x - \frac{\pi}{4})$

  b) $y = \frac{1}{2}\sec\frac{4\pi}{5}x$

7. Prove each identity.

  a) $\sin^4 x - \cos^4 x = 2\sin^2 x - 1$

  b) $\dfrac{\csc x}{\sec^2 x} = \csc x - \sin x$

  c) $\dfrac{\sin x + \tan x}{1 + \cos x} = \tan x$

  d) $\dfrac{\cos x}{1 - \sin x} + \dfrac{\cos x}{1 + \sin x} = 2\sec x$

8. Choose an identity from exercise 7. Write to explain how you proved it.

9. Determine an exact expression for each trigonometric function.

  a) $\cos\frac{\pi}{12}$

  b) $\sin\frac{13\pi}{12}$

  c) $\sin\frac{7\pi}{12}$

  d) $\cos\frac{23\pi}{12}$

10. Solve each equation algebraically for $0 \le x < 2\pi$. Express the roots in exact form.

  a) $\sin x = \cos^2 x\,(\sec^2 x - 1)$

  b) $\cos x = \sin^2 x\,(\csc^2 x - 1)$

11. Determine the number of permutations of all the letters in each word.

  a) COMMITTEE

  b) ADDRESS

12. There are one 1, three 5s, and four 8s. How many 8-digit numbers can be formed?

13. Write the first five terms of the binomial expansion of $(2 - x)^{50}$.

14. Determine the indicated term in each expression.

  a) the 4th term in the expansion of $(1 - 2x)^6$

  b) the last term in the expansion of $(2x + 1)^4$

  c) the 9th term in the expansion of $(x + y)^{12}$

15. People are either left-eye or right-eye dominant. Open both eyes and cover a far corner of a room with a finger. Alternately, close one eye, then the other. When the dominant eye is closed, the finger appears to move off the corner. (In most people, if the dominant eye is the right eye, then the dominant hand is the right hand. For a left-eye dominant person, the dominant hand is the left hand.) How would you determine the probability of left-eye dominant people in the population?

16. An examination comprises 10 true-false questions. A student guesses the answer to each question. What is the probability of each event?

    a) The student answers exactly 5 questions correctly.

    b) The student answers exactly 6 questions correctly.

    c) The student answers at least 7 questions correctly.

17. Open the *Cars* database. Assume the data are normally distributed.

    a) Calculate the mean and standard deviation for the prices of the 1994 cars.

    b) Use part a to determine the percent of cars that cost more than $40 000.00.

    c) Use the database to determine the percent of cars that cost more than $40 000.00

    d) Compare your answers to parts b and c. Explain what you notice.

18. The standard normal curve is a function. You can graph this function using the normalpdf command in the distribution menu.

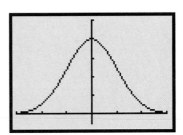

    a) Enter the equation Y1 = normalpdf(X) and set an appropriate graphing window. Graph the function.

    b) Trace to determine the vertical intercept.

    c) Visualize an isosceles $\triangle ABC$ with its base AB on the horizontal axis and vertex C at the point where the curve intersects the vertical axis, so that CA = CB. Suppose the area of the triangle is equal to the area under the curve. What are the coordinates of A and B?

19. These graphs were produced by Graphmatica. For each screen, write a set of equations in general form that could be used to make the pattern.

a)
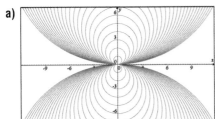

b)
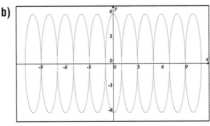

20. Make patterns similar to those in exercise 19.

21. Describe and sketch the graph of each relation.

    a) $16x^2 + 9y^2 = 144$ **b)** $4x^2 - 25y^2 = 100$ **c)** $9x^2 - 25y^2 = -400$

    d) $x^2 + 3y^2 = 12$ **e)** $4x^2 + y^2 = 20$ **f)** $3x^2 + 4y^2 = 36$

    g) $3x^2 - y^2 = 18$ **h)** $5x^2 - 4y^2 = 40$ **i)** $8x^2 - 6y^2 = 48$

Answers

Chapter 1

Mathematical Modelling: The Distance to the Horizon, page 8

$d \doteq 3.568\sqrt{h}$

1.1 Exercises, page 10

1. a) 8.51 m, 7.04 m, 1.16 m
   b) i) 0.49 m   ii) 1.96 m   iii) 7.84 m
   c) The distance quadruples.   d) 4.29 s

2. a) $5000; $2500; $1250   b) It is halved.

3. a) 113 km, 159.8 km, 226 km
   b) The distance is increased by a factor of $\sqrt{2}$.
   c) The height is increased by a factor of 4.

4. a) $y = \sqrt{3x + 10} - 2$ b) $y = \frac{5}{6 - x}$
   c) $y = |2x + 5| - 1$ d) $y = 0.2(x + 1)^2 - 3$
   e) $y = \sqrt{4 - (x - 2)^2}$

5. b) The graph of $y = \sqrt{k - x^2}$ is half a circle, centre (0, 0), and radius $\sqrt{k}$. As k changes, the radius of the half-circle changes.
   c) The graph is the point (0, 0).
   d) $k \geq 0$

6. c) $y = |x|$ d) All real values of k

7. a) 3, 4.5, −2.7
   b) Slope: 1, x-intercept : 0, y-intercept : 0, domain: all real numbers, range: all real numbers

8. a) All equal 5
   b) Slope : 0, x-intercept: none, y-intercept : 5, domain: all real numbers, range: $y = 5$
   c) Answers may vary.

9. b) Descriptions may vary. As a increases, the slope of the graph for $x > 1$ increases. The graph gets closer to the y-axis. As a decreases, the slope of the graph for $x > 1$ decreases.

12. Answers may vary.

Exploring with a Graphing Calculator: Lines and Curves of Best Fit, page 15

1. a) 24.3 years   b) 13.9 years
   c) 8.6 years   d) 4.9 years

2. Explanations may vary.

3. b) PwrReg   c) $y \doteq 2x^{\frac{1}{2}}$

Investigate, page 16

1. b) The graph of $y = (x - 2)^3$ is the image of $y = x^3$ after a translation of 2 units right.
   c) Yes
   d) For points with the same y-coordinate, the x-coordinate of the point on $y = (x - 2)^3$ is 2 greater than the x-coordinate of the point on $y = x^3$.

2. Yes

3. Yes; explanations may vary.

4. a) k provides information about the horizontal translation.
   b) When k is increased, the graph is translated right by the amount of the increase. When k is decreased, the graph is translated left by the amount of the decrease.
   c) Explanations may vary.

6. b) The graph of $y - 2 = x^3$ is the image of $y = x^3$ after a translation 2 units up.
   c) Yes
   d) For points with the same x-coordinate, the y-coordinate of the point on $y - 2 = x^3$ is 2 greater than the y-coordinate of the point on $y = x^3$.

7. Yes

8. Yes; explanations may vary.

9. a) k provides information about the vertical translation.
   b) When k is increased, the graph is translated up by the amount of the increase. When k is decreased, the graph is translated down by the amount of the decrease.
   c) Explanations may vary.

10. a) $y = \sqrt{x} + 1$, $y = \sqrt{x} - 1$
    b) $y = 2x + 5$, $y = 2x + 1$
    c) $y = \sqrt{25 - x^2} + 3$, $y = \sqrt{25 - x^2} - 3$

1.2 Exercises, page 20

1. a) There is a translation of 3 units left.
   b) There is a translation of 4 units right.
   c) There is a translation of 3 units down.
   d) There is a translation of 4 units up.
   e) There are translations of 2 units right and 2 units up.
   f) There are translations of 5 units left and 1 unit down.

2. a) $y = f(x)$ becomes $y = f(x - 5)$.
   b) $y = f(x)$ becomes $y = f(x + 8)$.
   c) $y = f(x)$ becomes $y - 7 = f(x - 2)$.
   d) $y = f(x)$ becomes $y + 5 = f(x + 4)$.

3. a) $c = 0, d = -4$ b) $c = 4, d = 0$
   c) $c = -3, d = 2$ d) $c = 8, d = -6$

4. a) The graph of $y = x^2 - 3$ is the image of $y = x^2$ after a translation of 3 units down.
   b) The graph of $y = (x - 3)^2$ is the image of $y = x^2$ after a translation of 3 units right.
   c) The graph of $y = (x + 3)^2$ is the image of $y = x^2$ after a translation of 3 units left.

5. a) The graph of $y = |x| + 4$ is the image of $y = |x|$ after a translation of 4 units up.
   b) The graph of $y = |x + 4|$ is the image of $y = |x|$ after a translation of 4 units left.
   c) The graph of $y = |x - 4|$ is the image of $y = |x|$ after a translation of 4 units right.

6. a) The graph of $y = \sqrt{x - 5}$ is the image of $y = \sqrt{x}$ after a translation of 5 units right.
   b) The graph of $y = \frac{1}{2}(x - 3) - 1$ is the image of $y = \frac{1}{2}x - 1$ after a translation of 3 units right.
   c) The graph of $y = \frac{1}{x + 1}$ is the image of $y = \frac{1}{x}$ after a translation of 1 unit left.

7. a) $y = x^4$ **b)** $y = (x + 1)^4 - 3$
c) $y = (x - 3)^4 + 1$

10. Explanations may vary.

12. a) The graph of $y = 2(x - 5) + 3$ is a the image of $y = 2x$, after a translation of 5 units right and 3 units up.
b) The graph of $y = (x - 5)^2 + 3$ is the image of $y = x^2$, after a translation of 5 units right and 3 units up.
c) The graph of $y = \sqrt{x - 5} + 3$ is the image of $y = \sqrt{x}$, after a translation of 5 units right and 3 units up.

14. a) Translate the graph of $f(x)$: 2 units right to get the image graph of $f(x - 2)$, 2 units down to get the image graph of $f(x) - 2$, and 2 units right and 2 units down to get the image graph of $f(x - 2) - 2$.
b) Translate the graph of $f(x)$: 3 units left and 1 unit down to get the image graph of $f(x + 3) - 1$, 3 units right and 1 unit up to get the image graph of $f(x - 3) + 1$, and 5 units left and 3 units up to get the image graph of $f(x + 5) + 3$.

15. a) The graph of $y = \log(x - 2) + 7$ is the image of $y = \log x$ after translations of 2 units right and 7 units up.
b) Translations describe how a graph moves on a grid, they are independent of the shape of the graph.

16. Predictions may vary.

17. a) Explanations may vary. **b)** Descriptions may vary.

18. a) The temperature of the gas when the volume is 0 mL
b) $y \doteq 0.56x + 151.76$ **c)** $-271.3°C$
d) $y = 0.56x$

19. a) $f(x) = \frac{2}{3}x$ **b)** An infinite number

20. a) $V = L(d - 4)^2$

Investigate, page 25

1. b) The graph of $y = \frac{1}{4}(-x) + 3$ is the image of $y = \frac{1}{4}x + 3$, after a reflection in the y-axis.
c) Yes
d) For points with the same y-coordinate, the x-coordinate of the point on $y = \frac{1}{4}(-x) + 3$ is the opposite of the x-coordinate of the point on $y = \frac{1}{4}x + 3$.

2. Yes; explanations may vary.

3. a) The graph is reflected in the y-axis.

5. b) The graph of $-y = \frac{1}{4}x + 3$ is the image of $y = \frac{1}{4}x + 3$, reflected in the x-axis.
c) Yes
d) For points with the same x-coordinate, the y-coordinate of the point on $-y = \frac{1}{4}x + 3$ is the opposite of the y-coordinate of the point on $y = \frac{1}{4}x + 3$.

6. Yes; explanations may vary.

7. a) The graph is reflected in the x-axis.

8. a) $y = -(2x + 3)$ **b)** $y = -x^3$ **c)** $y = -\sqrt{x}$

9. b) The graph of $x = \frac{1}{4}y + 3$ is the image of $y = \frac{1}{4}x + 3$ after a reflection in the line $y = x$.
c) Yes
d) The x- and y-coordinates of a point on $y = \frac{1}{4}x + 3$ are the y- and x-coordinates of a point on $x = \frac{1}{4}y + 3$.

10. Yes; explanations may vary.

11. a) The graph is reflected in the line $y = x$.

12. All the equations are functions because they pass the vertical line test.

1.3 Exercises, page 31

1. a) The graph is reflected in the y-axis.
b) The graph is reflected in the x-axis.
c) The graph is reflected in the y-axis and in the x-axis.
d) The graph is reflected in the line $y = x$.

2. a) $y = f(x)$ becomes $-y = f(x)$.
b) $y = f(x)$ becomes $y = f(-x)$.
c) $y = f(x)$ becomes $x = f(y)$.
d) $y = f(x)$ becomes $-y = f(-x)$.

3. Explanations may vary.

4. a) The graph of $y = 2(-x) + 1$ is the image of $y = 2x + 1$ after a reflection in the y-axis.
b) The graph of $-y = 2x + 1$ is the image of $y = 2x + 1$ after a reflection in the x-axis.
c) The graph of $x = 2y + 1$ is the image of $y = 2x + 1$ after a reflection in the line $y = x$.

5. a) $y = -x^4$ **b)** $y = -\frac{5}{x^2 + 1}$ **c)** $y = -f(x)$

6. a) $y = -x - 1$ **b)** $y = \sqrt{4 + x}$ **c)** $y = f(-x)$

7. a) $y = -\frac{2}{5}(x + 3)$ **b)** $y = \pm\sqrt{2 - x} + 2$ **d)** $y = \frac{1}{x + 1} - 2$

8. Explanations may vary.
a) $y = \frac{1}{1 - x}$ **b)** $y = -\frac{1}{x}$ **c)** $y = \frac{1}{1 + x}$

9. a) $x = 1, y = 0$ **b)** $x = 0, y = 0$ **c)** $x = -1, y = 0$

11. All except $x = |y - 2|$

12. Explanations may vary.

14. c) i) $y = \sqrt{25 - (-x - 5)^2}$
ii) $y = -\sqrt{25 - (x - 5)^2}$ **iii)** $x = \sqrt{25 - (y - 5)^2}$

15. a) The graph of $y = 2(-x) - 5$ is the image of $y = 2x - 5$ after a reflection in the y-axis.
b) The graph of $-y = 2x - 5$ is the image of $y = 2x - 5$ after a reflection in the x-axis.
c) The graph of $x = 2y - 5$ is the image of $y = 2x - 5$ after a reflection in the line $y = x$.

16. a) The graph of $y = f(-x)$ is the image of $y = f(x)$ after a reflection in the y-axis.
b) The graph of $y = -f(x)$ is the image of $y = f(x)$ after a reflection in the x-axis.
c) The graph of $y = -f(-x)$ is the image of $y = f(x)$ after a reflection in the x- and y-axes.
d) The graph of $x = f(y)$ is the image of $y = f(x)$ after a reflection in the line $y = x$.

17. a) Answers may vary. $y = x^2$, $y = 2x^2 + 3$

b) The graph is symmetrical about the y-axis. Explanations may vary. When the graph of an even function is reflected in the y-axis, it is the same graph.

18. a) Answers may vary. $y = x^3$, $y = x$

b) The graph is symmetrical about the origin. Explanations may vary. When the graph of an odd function is rotated $180°$ about the origin, it is the same graph.

19. a) n is even and $a_{n-1} = a_{n-3} = a_{n-5} = \ldots = 0$.

b) n is odd and $a_{n-1} = a_{n-3} = \ldots = 0$.

20. a) The graph of $y = \tan(-x)$ is the image of $y = \tan x$ after a reflection in the y-axis.

b) Reflections describe how a graph changes position on a grid. These changes are independent of the graph.

21. a) $y = -|x - k|$, $k = 0, \pm 1, \pm 2, \pm 3, \pm 4, \pm 5$

b) $y = \sqrt{k \pm x}$, $k = 0, \pm 1, \pm 2, \pm 3, \pm 4, \pm 5$

22. Patterns may vary.

23. Examples may vary. $y = x$

24. b) Explanations may vary.

c) The graph is reflected in the y-axis and translated k units right.

Investigate, page 35

1. b) The graph of $y = \dfrac{1}{(2x)^2 + 1}$ is the image of $y = \dfrac{1}{x^2 + 1}$ after a horizontal compression by a factor of $\frac{1}{2}$.

c) Yes

d) For points with the same y-coordinate, the x-coordinate of the point on $y = \dfrac{1}{(2x)^2 + 1}$ is half the x-coordinate of the point on $y = \dfrac{1}{x^2 + 1}$.

2. Answers may vary. Yes

3. Yes; explanations may vary.

4. a) k provides information about the horizontal expansion or compression of the graph.

b) Descriptions may vary. When $|k|$ increases, there is a horizontal compression. When $|k|$ decreases, there is a horizontal expansion. When k changes sign, there is also a reflection in the y-axis.

c) Explanations may vary.

6. b) The graph of $2y = \dfrac{1}{x^2 + 1}$ is the image of $y = \dfrac{1}{x^2 + 1}$ after a vertical compression by a factor of 2.

c) Yes

d) For points with the same x-coordinate, the y-coordinate of the point on $2y = \dfrac{1}{x^2 + 1}$ is $\frac{1}{2}$ of the y-coordinate of the point on $y = \dfrac{1}{x^2 + 1}$.

7. Yes

8. Yes; explanations may vary.

9. a) k provides information about the vertical expansion or compression of the graph.

b) Descriptions may vary. When $|k|$ increases, there is a vertical compression. When $|k|$ decreases, there is a vertical expansion. When k changes sign, there is also a reflection in the x-axis.

c) Explanations may vary.

1.4 Exercises, page 41

1. a) There is a horizontal compression by a factor of $\frac{1}{3}$.

b) There is a vertical compression by a factor of $\frac{1}{3}$.

c) There is a horizontal expansion by a factor of 2.

d) There is a vertical expansion by a factor of 2.

e) There is a horizontal compression by a factor of $\frac{1}{2}$ and reflection in the y-axis.

f) There is a vertical compression by a factor of $\frac{1}{2}$ and reflection in the x-axis.

g) There is a horizontal and vertical compressions by a factor of $\frac{1}{2}$.

h) There is a horizontal compression by a factor of $\frac{1}{3}$, reflection in the y-axis, and a vertical compression by a factor of $\frac{1}{4}$.

2. a) $y = f(x)$ becomes $y = f\left(\frac{1}{4}x\right)$.

b) $y = f(x)$ becomes $y = f(-2x)$.

c) $y = f(x)$ becomes $-\frac{1}{5}y = f(x)$ or $y = -5f(x)$.

d) $y = f(x)$ becomes $\frac{3}{2}y = f(x)$ or $y = \frac{2}{3}f(x)$.

3. a) $a = \frac{1}{2}$, $b = 3$ **b)** $a = -3$, $b = \frac{1}{2}$

c) $a = 4$, $b = -2$ **d)** $a = -\frac{1}{2}$, $b = -\frac{1}{4}$

4. No, the order does not matter for the transformations because they are independent; that is, they do not affect each other.

5. b) i) $y = \frac{1}{2}\sqrt{x}$ **ii)** $y = \frac{1}{2}\sqrt{x}$

c) Explanations may vary.

d) Answers may vary. $y = \sqrt{x} + 1$

6. a) The graph of $y = |2x|$ is the image of $y = |x|$ after a horizontal compression by a factor of $\frac{1}{2}$.

b) The graph of $y = |\frac{1}{2}x|$ is the image of $y = |x|$ after a horizontal expansion by a factor of 2.

c) The graph of $2y = |x|$ is the image of $y = |x|$ after a vertical compression by a factor of $\frac{1}{2}$.

d) The graph of $\frac{1}{2}y = |x|$ is the image of $y = |x|$ after a vertical expansion by a factor of 2.

7. a) The graph of $y = \dfrac{1}{2x}$ is the image of $y = \dfrac{1}{x}$ after a horizontal compression by a factor of $\frac{1}{2}$.

b) The graph of $y = \dfrac{1}{\frac{1}{2}x}$ is the image of $y = \dfrac{1}{x}$ after a horizontal expansion by a factor of 2.

c) The graph of $2y = \dfrac{1}{x}$ is the image of $y = \dfrac{1}{x}$ after a vertical compression by a factor of $\frac{1}{2}$.

d) The graph of $\frac{1}{2}y = \dfrac{1}{x}$ is the image of $y = \dfrac{1}{x}$ after a vertical expansion by a factor of 2.

8. a) The graph of $y = \sqrt{\frac{2}{3}x}$ is the image of $y = \sqrt{x}$ after a horizontal expansion by a factor of $\frac{3}{2}$.

b) The graph of $y = 6\left(-\frac{3}{2}x\right) - 3$ is the image of $y = 6x + 3$ after a horizontal compression by a factor of $\frac{2}{3}$ and then a reflection in the y-axis.

c) The graph of $y = \left(\frac{1}{2}x\right)^2 - 4\left(\frac{1}{2}x\right)$ is the image of

$y = x^2 - 4x$ after a horizontal expansion by a factor of 2.

9. b) $L \geq 0, T \geq 0$

10. a) $y = \sqrt[3]{x}$ **b)** $y = \sqrt[3]{10x}$ **c)** $y = \sqrt[3]{-2x}$

13. Explanations may vary.

15. a) The graph of $y = |-x + 1|$ is the image of $y = |x + 1|$ after a reflection in the y-axis. The graph of $y = |-2x + 1|$ is the image of $y = |2x + 1|$ after a reflection in the y-axis. The graph of $y = |-\frac{1}{2}x + 1|$ is the image of $y = |\frac{1}{2}x + 1|$ after a reflection in the y-axis.

b) The graph of $y = (-x)^2 + 1$ is the image of $y = x^2 + 1$ after a reflection in the y-axis. The graph of $y = (-2x)^2 + 1$ is the image of $y = (2x)^2 + 1$ after a reflection in the y-axis. The graph of $y = \left(-\frac{1}{2}x\right)^2 + 1$ is the image of $y = \left(\frac{1}{2}x\right)^2 + 1$ after a reflection in the y-axis.

c) The graph of $y = (-x)^3$ is the image of $y = x^3$ after a reflection in the y-axis. The graph of $y = (-2x)^3$ is the image of $y = (2x)^3$ after a reflection in the y-axis. The graph of $y = \left(-\frac{1}{2}x\right)^3$ is the image of $y = \left(\frac{1}{2}x\right)^3$ after a reflection in the y-axis.

16. b) i) $y = \sqrt{9 - \left(\frac{1}{2}x - 3\right)^2}$ **ii)** $y = \sqrt{9 - (3x - 3)^2}$
iii) $y = 2\sqrt{9 - (x - 3)^2}$ **iv)** $y = \frac{1}{3}\sqrt{9 - (x - 3)^2}$

17. a) The graph of $y = \sin 3x$ is a horizontal compression by a factor of $\frac{1}{3}$ of $y = \sin x$.

b) Transformations involving a compression (or expansion) describe by how much the graph is compressed (or expanded) and in which direction this occurs. It is independent of the shape of the graph.

20. Descriptions may vary.

Investigate, page 45

2. a) $y = (2x - 4)^2$
c) The graph of $y = x^2$ was translated 4 units right, then compressed horizontally by a factor of $\frac{1}{2}$.

3. a) $y = (2x - 8)^2$
c) The graph of $y = x^2$ was compressed horizontally by a factor of $\frac{1}{2}$, then translated 4 units right.

4. a) When translations are combined with expansions or compressions, the result depends on the order in which the transformations are applied.
b) Explanations may vary.

6. a) $y = 2x^2 + 6$
c) The graph of $y = x^2$ was translated 3 units up, then expanded vertically by a factor of 2.

7. a) $y = 2x^2 + 3$
c) The graph of $y = x^2$ was expanded vertically by a factor of 2, then translated 3 units up.

8. a) When translations are combined with expansions or compressions, the result depends on the order in which the transformations are applied.
b) Explanations may vary.

9. No; explanations may vary.

10. a) $y = 8(x - 4)^2 + 3$ **b)** $y = 8(x - 4)^2 + 3$
c) $y = 8(x - 2)^2 + 6$ **d)** $y = 8(x - 2)^2 + 6$

11. a) 24   **b)** 4; explanations may vary.

12. Yes; explanations may vary.

1.5 Exercises, page 51

1. a) There is a horizontal compression by a factor of $\frac{1}{3}$, then a translation 2 units right.
b) There is a translation 2 units right, then a horizontal compression by a factor of $\frac{1}{3}$.
c) There is a reflection in the y-axis, then a translation 4 units left.
d) There is a translation 4 units left, then a reflection in the y-axis.

2. a) There is a horizontal compression by a factor of $\frac{1}{2}$, then a vertical compression by a factor of $\frac{1}{2}$.
b) There is a horizontal compression by a factor of $\frac{1}{2}$, then a vertical compression by a factor of $\frac{1}{2}$, then a translation 6 units right.
c) There is a horizontal expansion by a factor of 2, a vertical compression by a factor of $\frac{1}{3}$, a reflection in the x-axis, and a translation 4 units up.
d) There is a horizontal expansion by a factor of 3, a vertical expansion by a factor of 2, translations of 1 unit left and 2 units down.

3. a) $y = f(x)$ becomes $2y = f(4x)$.
b) $y = f(x)$ becomes $y = f(4(x - 3))$.
c) $y = f(x)$ becomes $y = f\left(\frac{1}{2}(x + 3)\right)$.
d) $y = f(x)$ becomes $y = (4(x - 5))$.

5. Explanations may vary.

6. a) The graph of $y = \sqrt{4x + 8}$ is the image of $y = \sqrt{x}$ after a horizontal compression by a factor of $\frac{1}{4}$, then a translation of 2 units left.

7. b) The graph of $y = \left(\frac{1}{2}(x - 4)\right)^2$ is the image of $y = x^2$ after a horizontal expansion by a factor of 2, then a translation of 4 units right.

8. a) The graph of $y = \frac{1}{2(x - 3)}$ is the image of $y = \frac{1}{x}$ after a horizontal compression by a factor of $\frac{1}{2}$, then a translation 3 units right.
b) The graph of $y = \sqrt{2x - 5}$ is a translation of 5 units right, then a horizontal compression by a factor of $\frac{1}{2}$.
c) The graph of $y = |\frac{2}{3}x - 6|$ is the image of $y = |x|$ after a translation of 6 units right, then a horizontal expansion by a factor of $\frac{3}{2}$.

9. b) i) $f(x) = \sqrt{4 - (2x - 2)^2}$ **ii)** $f(x) = 3\sqrt{4 - (x - 2)^2}$
iii) $f(x) = 3\sqrt{4 - (2x - 2)^2}$

10. b) i) $f(x) = \sqrt{4 - \left(\frac{1}{2}x - 2\right)^2}$ **ii)** $f(x) = \sqrt{4 - (x + 1)^2}$

iii) $f(x) = \sqrt{4 - \left(\frac{1}{2}(x-1)\right)^2}$ iv) $f(x) = \sqrt{4 - \left(\frac{1}{2}x+1\right)^2}$

11. a) The graph of $y = \dfrac{1}{(2(x-3))^2} + 2$ is the image of
$y = \dfrac{1}{x^2+1}$ after a horizontal compression by a factor of
$\frac{1}{2}$, then translations of 3 units right and 2 units up.

12. a) There is a vertical expansion by a factor of 3, then a
translation of 4 units up.
b) $y = 3(f(x) + 4)$

Exploring with a Graphing Calculator: What Is a Line of Best Fit?, page 53

1. A(0.02, 35), D(0.04, 18), F(0.06, 12), H(0.08, 9), I(0.10, 7),
K(0.12, 6)

2. 129.49

4. Explanations may vary. The line of best fit is the one that
has the lowest sum of the squares of the distances between
the data and the line.

1.6 Exercises, page 57

2. a) $x = 2, y = 0$ **b)** $x = 0, x = 2, y = 0$
c) $x = 0, y = 0$

3. Explanations may vary.
a) $y = \dfrac{1}{x^2-4x}$ **b)** $y = |2x-4|$ **c)** $y = \dfrac{1}{\sqrt{2x-4}}$
d) $y = \dfrac{1}{2x-4}$ **e)** $y = |x^2-4x|$

5. Explanations may vary.

6. a) $x = 1, y = 0$ **b)** $x = \pm1, y = 0$ **c)** $x = 1, y = 0$

7. to 11. Explanations may vary.

12. None of parts a to d are functions, because they do not pass
the vertical line test.

13. Explanations may vary. These answers assume the graph of
the function has no asymptotes and has intercepts.
a) The parts of the graph that are to the left of the y-axis are
reflected in the line $y = b$, where b is the y-intercept of
the function.
b) The parts of the graph that are below the x-axis are
reflected in the x-axis.
c) The image graph has asymptotes at $x = 0$ and $y = b$,
where b is the y-intercept of the function.
d) The image graph has asymptotes at $x = a$, where a is the
x-intercept of the function and $y = 0$.

14. Explanations may vary.
a) $y = \dfrac{1}{|f(x)|}$ is the reciprocal of the absolute value of $f(x)$.
$y = \left|\dfrac{1}{f(x)}\right|$ is the absolute value of the reciprocal of $f(x)$.
b) Same
c) $y = \dfrac{1}{\sqrt{f(x)}}$ is the reciprocal of the square root of $f(x)$.
$y = \sqrt{\dfrac{1}{f(x)}}$ is the square root of the reciprocal of $f(x)$.
d) Same

Mathematical Modelling: When Will Women's Pay Equal Men's?, page 60

2. d) 2052 for the straight line, 2014 for the parabola, no

3. Answers may vary. 2041 for the power function

4. a) Descriptions may vary.
b) Year: 1980, 1985, 1990, 1995; Percent: 156, 155, 146.8,
140.1
c) 2031 for the straight line, 2009 for the parabola, 2041 for
the power function

5. Answers may vary.

6. b) 2033

7. No; answers may vary. The cubic regression suggests
equality occurred in 1967.

Problem Solving: Visualizing Foreign Exchange Rates, page 62

All explanations may vary.

1. The graph on the left shows the number of Indonesian
rupiahs for every Canadian dollar between April 1, 1997,
and March 31, 1998. The graph on the right shows the
number of Canadian dollars for every U.S. dollar between
April 1, 1977, and March 31, 1998.

2. a) The number of rupiahs for every Canadian dollar
increased from about 3000 to 9500.
b) The number of Canadian dollars for every U.S. dollar
increased from about $1.39 to $1.46. This is a 7.9%
increase, which is not enough to show a collapse in the
economy.

3. a) Descriptions may vary. The vertical axis could be the
percent change in the exchange rate, with April 1, 1997,
as the base amount.
b) Yes

5. Descriptions may vary.

1 Review, page 63

1. a) 8.499 88 m, 8.5 m, 8.500 12 m

2. a) $y = |x - k|, k = 0, \pm1, \pm2, \pm3, \pm4, \pm5$
b) $y = \sqrt{k^2 - (x-k)^2}, k = \pm0.5, \pm1, \pm1.5, \ldots, \pm10$

3. a) The graph is reflected in the y-axis and in the x-axis.
b) The graph is reflected in the line $y = -x$.

4. b) i) $0 \le h \le 500, 0 \le v \le 5\sqrt{5}$
ii) $0 \le h \le 350, 0 \le v \le 3\sqrt{14}$
iii) $0 \le h \le 200, 0 \le v \le 7\sqrt{2}$

6. The graph of $y = 4\cos(2(x-3)) + 1$ is the image of
$y = \cos x$ after a vertical expansion by a factor of 4, a
horizontal compression by a factor of $\frac{1}{2}$, and translations of
3 units right and 1 unit up.

Chapter 2
2.1 Exercises, page 69

Estimates may vary.

1. a) $1500   b) 12.5 years
   c) Descriptions may vary.

2. a) 2.2%   b) Descriptions may vary.

3. b) 5.3 million   c) In year 2016

4. Estimates may vary.
   a) 13 or 14 panes   b) Descriptions may vary.

5. Descriptions may vary.

6. b) About 11.5 h

7. b) $1500   c) 12 years   d) Explanations may vary.

8. a), b) Explanations may vary.

9. b) 34 million. Assumptions may vary.
   c) Late in 1998   d) 45%

10. a) $h = 3.5(1.84)^n$ c) Middle of 1999

11. a) $P = 100(0.975)^d$
   c) Descriptions may vary.   d) 77.6%
   e) About 27.4 m

12. b) 68.4%   c) Rest of the world at 195%

Exploring with a Graphing Calculator: Exponential Curves of Best Fit, page 72

1. b) $y = 34.74(1.1266)^x$ c) 12.66%
   d) Answers may vary. 684 billion

2. a) $t = 8.72(0.93)^T$ b) 2.0 s
   c) The time is halved.

3. a) Alberta: $y = 2.278(1.014)^x$, where x is number of years
   since 1981
   B.C.: $y = 2.758(1.022)^x$
   b) Alberta: 3.457 million   c) Alberta: 2050
   B.C.: 5.298 million   B.C.: 2016

Investigate, page 74

2. a) Explanations may vary. A power with base 10 and log x
   is equal to x.
   b) Exponent

2.2 Exercises, page 77

1. Explanations may vary.
   a) 1.372 91   b) 2.255 27   c) 3.707 57
   d) −0.346 79   e) −1.142 67   f) −2.537 60

2. Explanations may vary.

3. a) 0, 1, 2, 3, …, n b) 0, 1, 2, 3, …, n
   c) 0, 1, 2, 3, …, n

4. a) 4   b) 2   c) 3   d) 2
   e) −1   f) 1   g) 0   h) 3

5. Explanations may vary.

6. a) $2^3 = 8$ b) $2^5 = 32$ c) $6^2 = 36$
   d) $5^4 = 625$ e) $16^{\frac{1}{2}}$ f) $2^{-2} = \frac{1}{4}$

7. a) $\log_7 49 = 2$ b) $\log_3 243 = 5$ c) $\log_2 1 = 0$
   d) $\log_5 0.2 = -1$ e) $\log_2 \frac{1}{8} = -3$ f) $\log_{25} 5 = 0.5$

8. a) $\log 1, \log 0.1, \log 0.01, \log 0.001, \ldots,$
   $\log 10^{-n}; \log_3 1, \log_3\left(\frac{1}{3}\right), \log_3\left(\frac{1}{9}\right), \log_3\left(\frac{1}{27}\right), \ldots,$
   $\log_3 3^{-n}; \log_a 1, \log_a a^{-1}, \log_a a^{-2}, \log_a a^{-3}, \ldots, \log_a a^{-n}$
   b) All logarithms are 0, −1, −2, −3, …, −n

9. a) i) 0.176   ii) 0.125   iii) 0.097   iv) 0.079
   v) 0.067   vi) 0.058   vii) 0.051   viii) 0.046

2.3 Exercises, page 83

1. Explanations may vary.
   a) 0.699, 1.699, 2.699, 3.699, 4.699
   b) 0.699, 1.398, 2.097, 2.796, 3.495

2. Explanations may vary.
   a) All are approximately 1.556.
   b) All are approximately 0.301.

3. Answers may vary.

4. Explanations may vary.
   a), b) 0.301, 0.602, 0.903, 1.204, 1.505

5. Answers may vary.

6. a) 2   b) 1   c) 3   d) 3   e) 3   f) 4

7. a) 3.459 43   b) 2.578 90   c) 0.898 24

8. a) i) 0.630 93, 1.261 86, 1.892 79, 2.523 72
      ii) 0.630 93, 0.315 46, 0.210 31, 0.157 73
   b) Explanations may vary.
      i) Each root is a multiple of the first root.
      ii) Each root is a factor of the first root.

9. a) 2.365 21   b) −0.415 04   c) −0.564 58
   d) −0.464 97   e) −1.547 95   f) 0.769 12

10. a) $3^{1.771\ 24}$ b) $2^{2.321\ 93}$ c) $2^{4.857\ 98}$
   d) $8^{2.088\ 93}$ e) $0.5^{-1.584\ 96}$ f) $6^{-0.445\ 66}$

11. a) 1.464 97   b) 3.584 96   c) 2.236 54
   d) 1.584 96   e) 4.321 93   f) 2.321 93

12. a) $\log x, 2\log x, 3\log x, 4\log x, 5\log x$
   b) $\log x, 1 + \log x, 2 + \log x, 3 + \log x, 4 + \log x$

13. About 10.6 years

14. a) i) 2.726 83   ii) 2.726 83   iii) 2.726 83
   b) The quotients are equal.
   c) Explanations may vary.   d) 2.726 83

15. Proofs may vary.

16. Proofs may vary.

17. a) 4   b) $\frac{5}{3}$ c) 1.5
   d) 2.5   e) 1   f) −1.5

18. a) $3 + \log x + \log y$ b) $2\log x + \log y$
   c) $\log x - 2\log y$

19. a) $\log y = 1 + \log x$ b) $\log y = 2 + \log x$
   c) $\log y = 3 + \log x$ d) $\log y = -\log x$
   e) $\log y = 2\log x$ f) $\log y = -2\log x$

20. Explanations may vary.
   a) 79.485   b) 80   c) 30569

21. Answers may vary.
   a) If the bacteria double every 19 minutes, there will be
   about 76×10^{21} bacteria at the end of the day.
   b) Assumptions may vary.

1. b) About 9 years

c) Predictions may vary. Another 9 years

2. a) $1259.71

b) Predictions may vary. About 9 years

3. The doubling time applies at all times during the growth.

4. b) About 5 h

c) Predictions may vary. Another 5 h

5. a) About 75.69%

b) Predictions may vary. Another 5 h

c) Answers may vary.

6. The time to decay to one-half its original amount applies at all times during the decay.

2.4 Exercises, page 92

1. a) 7.0 years   **b)** 11.9 years   **c)** 18.9 years

2. a) i) The year 2012   **ii)** Late in 2021   **iii)** Late in 2038
b) Answers may vary.

3. a) $P = 828\,500(1.022)^y$ **b)** About 32 years
c) During 2003   **d)** $P = 828\,500(2)^{\frac{y}{32}}$

4. a) $P = 100(0.98)^d$ **b)** About 34 days
c) $P = 100(0.5)^{\frac{d}{34}}$

5. a) 10.0 years   **b)** 9.8 years   **c)** 9.7 years   **d)** 9.6 years

6. a) 10.0 years   **b)** 10.0 years   **c)** 10.0 years   **d)** 10.0 years

7. Explanations may vary.

8. a) Red: $P = 100(0.65)^d$, green: $P = 100(0.95)^d$, blue: $P = 100(0.975)^d$
b) Red: 1.6 m, green: 13.5 m, blue: 27.4 m
c) Red: $P = 100(2)^{-\frac{d}{1.6}}$, green: $P = 100(2)^{-\frac{d}{13.5}}$, blue: $P = 100(2)^{-\frac{d}{27.4}}$
d) Red: 10.7 m, green: 89.8 m, blue: 182 m

9. a) 8.8 g   **b)** 15.4 g   **c)** 4.1 g   **d)** 2.0 g

10. a) 4.3 mg   **b)** 79.3 h

11. Substances may vary.
b) 3.125%   **c)** 0.000 095%

12. Explanations may vary.
a) $2^{0.034216}$ **b)** $1.0124^{56.24483}$

13. a) Early in year 2017
b) i) Middle of year 2036   **ii)** Late in year 2063
iii) Early in year 2129

2.5 Exercises, page 98

1. a) 20 000   **b)** 200 000   **c)** 2 000 000

2. a) i) $5P_0$ **ii)** $5^{\frac{5}{4}}P_0$ **iii)** $5^{\frac{3}{2}}P_0$
iv) $5^{\frac{7}{4}}P_0$ **v)** $25P_0$
b) About 2.2 times

3. About 1.9 times

4. About 2.6 times

5. a) 10 times   **b)** 100 times   **c)** 1000 times
d) 10 000 times   **e)** 100 000 times   **f)** 1 000 000 times

6. About 5 times

7. a) i) About 501 times
ii) About 3162 times
iii) About 6 times
b) Answers may vary. It is along the San Andreas fault.

8. Explanations may vary.

9. Examples may vary.
a) Morocco 1960 and Columbia 1983
b) Guatemala 1976 and San Francisco 1989
c) B.C. coast 1700 and Mexico 1985
d) Peru 1970 and Italy 1976

10. a) 6 or 7 times as frequent
b) 36 to 49 times as frequent
c) 216 to 343 times as frequent

11. a) i) About 31 times
ii) About 961 times
b) About 165 870 times

12. a) 3.1×10^{15} **b)** 9.61×10^{16} J   **c)** 5.35×10^{17} J

13. a) 10^4 times   **b)** 10^7 times   **c)** 10^{10} times

14. 10^4 times

15. a) $L = L_0\, 10^{\frac{d}{10}}$ **b)** About 316 times
c) 10^4 times

16. a) About 8 times   **b)** About 7943

17. a) i) 30 min   **ii)** About 1.9 min

b) $T = 8 \times 0.5^{-\frac{(L-90)}{5}}$, where T is the acceptable exposure time in hours, L is the sound in dB

18. a) 100 times   **b)** 10^5 times
c) 10^4 times   **d)** 10^6 times

19. a) 10 times   **b)** Answers may vary.

20. a) About 4 times   **b)** About 4 times

21. a) About 4 times   **b)** 7.8
c) Explanations may vary.

Mathematical Modelling: Modelling the Growth of Computer Technology, page 102

1. b) $C = 2^n$, $0 \le n \le 10$

2. b) i) $C = 1000(2)^{\frac{n}{1.5}}$, $0 \le n \le 22$
ii) Late in 1999

3. No; it is much less. Explanations may vary.

4. a) $I = 60000(2)^{\frac{n}{1.5}}$
b) Yes; explanations may vary.

5. a) i) About 0.0025 of the cost   **ii)** About 0.05 of the cost
b) i) About 20 times the cost   **ii)** About 400 times the cost

6. Assumptions may vary.

7. Reasons may vary.

2.6 Exercises, page 110

1. a) iv   **b)** iii   **c)** i   **d)** ii

5. b) Predictions may vary.

6. Predictions may vary.

7. Predictions may vary.

8. Descriptions may vary.

9. b) 55°C   **c)** About 8.5 min
  d) The graph of $T = 79 \times 0.85^t + 20$ is the image of $y = 0.85^x$ after a vertical expansion by a factor of 79 and a vertical translation 20 units up.

10. a) $10^{0.301}$
  b) About 0.301; explanations may vary.
  c) 3.32

11. a) i) $5^{1.431}$ **ii)** $5^{0.431}$
  b) $10^x = 5^{1.431x}$, $2^x = 5^{0.431x}$

12. $P = 29.6 \times 2^{\frac{n}{56}}$ is the image of $y = 2^x$ after a vertical expansion by a factor of 29.6 and a horizontal expansion by a factor of 56.

13. $P = 100 \times 0.5^{\frac{n}{5}}$ is the image of $y = 0.5^x$ after a vertical expansion by a factor of 100 and a horizontal expansion by a factor of 5.

15. b) Domain: all real numbers, range: $y > 0$
  c) $y = 0$ **d)** y-intercept: $\frac{1}{64}$, x-intercept: none

16. a) Domain: all real numbers, range: $y > 0$, asymptote: $y = 0$, y-intercept: 2^{-12}, x-intercept: none
  b) Domain: all real numbers, range: $y > 0$, asymptote: $y = 0$, y-intercept: 3^{-5}, x-intercept: none
  c) Domain: all real numbers, range: $y > 0$, asymptote: $y = 0$, y-intercept: 64, x-intercept: none
  d) Domain: all real numbers, range: $y > 0$, asymptote: $y = 0$, y-intercept: 100, x-intercept: none

17. a) $P = 2.76(1.022)^{n+10}$ **b)** $P = 2.76(1.022)^{n-1981}$
  c) Part a: the population in year 1991; part b: the population in year 0.

19. About 7:22 P.M.

20. a) About 42 min   **b)** No; explanations may vary.

21. Answers may vary.

2.7 Exercises, page 115

1. a) Yes, 2   **b)** Not geometric   **c)** Yes, $-\frac{1}{2}$
  d) Yes, $\frac{1}{10}$ **e)** Not geometric   **f)** Yes, $-\frac{1}{3}$

2. a) 3; 81, 243, 729   **b)** −3; 405, −1215, 3645
  c) 2; 48, 96, 192   **d)** $\frac{1}{3}; \frac{2}{27}, \frac{2}{81}, \frac{2}{243}$
  e) $\frac{1}{4}; \frac{9}{64}, \frac{9}{256}, \frac{9}{1024}$ **f)** −4; 128, −512, 2048

3. a) 5:40 P.M.   **b)** 5:40 P.M.   **c)** 6:40 P.M.

4. a) i) 32   **ii)** 1024   **iii)** 2^{20} **iv)** 2^{40}
  b) Explanations may vary. People do not live for 1000 years.

5. Explanations may vary. The n in the equation for the nth term of a geometric sequence represents the term number. The first term, when $n = 1$, is a, which agrees with the formula $t_n = ar^{n-1}$. In the exponential function $y = Ab^x$, x has no restrictions.

6. Explanations may vary.
  a) No, there is no common ratio.
  b) Yes, the common ratio is 2.

7. a) $t_n = 2^n$, 1024   **b)** $t_n = 5(2)^{n-1}$, 20 480

c) $t_n = -3(-5)^{n-1}$, 234 375   **d)** $t_n = 12\left(\frac{1}{2}\right)^{n-1}$, $\frac{3}{512}$
e) $t_n = 6\left(-\frac{1}{3}\right)^{n-1}$, $\frac{2}{2187}$ **f)** $t_n = 3(6)^{n-1}$, 139 968

8. a) 0.64 mm, 0.97 mm, 1.47 mm, 2.24 mm, 3.41 mm
  b) $a = 0.6384$, $r = 1.52$
  c) 1.52; it represents the growth at each stage.

9. a) 100, 0.95   **b)** $t_n = 100(0.95)^{n-1}$
  c) Explanations may vary. The n in the equation for the nth term of the geometric sequence represents the term number. The first term, when $n = 1$, is 100, which agrees with the formula $t_n = 100(0.95)^{n-1}$. In the exponential function, n represents the layers of glass.

10. a) 1.26 m, 0.63   **b)** $t_n = 1.26(0.63)^{n-1}$
  c) 20 cm   **d)** 5

11. $\sqrt{2}$; explanations may vary.

12. a) i) AB = 1, BC = x, CD = x^2, DE = x^3, …
  ii) AB = 1, CD = x^2, EF = x^4, …
  iii) BC = x, DE = x^3, FG = x^5, …
  b) Answers may vary. AC = $\sqrt{1 - x^2}$, CE = $x^2\sqrt{1 - x^2}$, EG = $x^4\sqrt{1 - x^2}$, …

13. a) 75%, 56.25%, 42.1875%, 31.640 625%
  b) 16 steps

14. a) 9, −6, 4, $-\frac{8}{3}$, … or $\frac{9}{5}, \frac{6}{5}, \frac{4}{5}, \frac{8}{15}$, …
  b) 1, 3, 9, 27, …

15. Explanations may vary.

Exploring with a Graphing Calculator: Graphing Sequences, page 119

1. a) 17.83 m   **b)** 13

2. a) $t = 0.45(0.83)^n$ **c)** 0.048 m   **d)** 33

3. a) Explanations may vary.
  b) 820.4 m

2.8 Exercises, page 124

1. a) 63   **b)** 1092   **c)** 682   **d)** 77.5

2. Explanations may vary.

3. a) 1562   **b)** 484   **c)** 93
  d) 46.5   **e)** 605   **f)** 55

4. a) 78 732   **b)** 118 096

5. a) −78 732   **b)** 59 048

6. a) 5115   **b)** −1705   **c)** About 1.50
  d) About 0.75   **e)** About 9.99   **f)** About 3.33

7. 397 mg

8. a) 63   **b)** Explanations may vary.

9. a) $1 048 575
  b) Answers and explanations may vary.

10. a) 7th level   **b)** 128   **c)** 254
  d) 2^{n-1} **e)** 9th level

11. Answers may vary.

12. a) $r^6 - 1$

b) Explanations may vary. The middle terms in each expansion are opposites, and are eliminated.

c) i) $\frac{r^3 - 1}{r - 1}$ **ii)** $\frac{r^4 - 1}{r - 1}$ **iii)** $\frac{r^5 - 1}{r - 1}$

d) $\frac{r^n - 1}{r - 1}$ **e)** Explanations may vary.

13. b) Explanations may vary. $\frac{1}{2^{n-1}}$ gets closer and closer to zero.

14. b) Explanations may vary. $\frac{1}{3^{n-1}}$ gets closer and closer to zero.

2.9 Exercises, page 130

1. No; explanations may vary. An infinite geometric series with $|r| > 1$ does not have a sum.

2. a) 16   **b)** 81   **c)** $\frac{80}{7}$

d) $\frac{250}{9}$ **e)** No sum   **f)** $-\frac{64}{7}$

3. a) $\frac{32}{3}$ **b)** 6.4   **c)** 20   **d)** $\frac{20}{3}$

e) 7.5   **f)** 3.75   **g)** 120   **h)** 10

4. a) 8   **b)** 0.031 25

5. a) 500 000

b), c) Explanations and assumptions may vary.

6. a) 7.94 m   **b)** 8.81 m

c), d) Explanations may vary.

7. Proofs may vary.

8. a) i) 72 mm, 96 mm, 128 mm, $\frac{512}{3}$ mm

ii) Not possible

b) i) 249.4 mm^2, 332.6 mm^2, 369.5 mm^2, 385.9 mm^2

ii) 399.1 mm^2

c) The perimeter is infinite and area is finite.

9. i) $\frac{1}{1-x}$ units   **ii)** $\frac{1}{1-x^2}$ units   **iii)** $\frac{x}{1-x^2}$ units

b) Answers may vary. For AC, CE, EG, ... $\frac{\sqrt{1-x^2}}{1-x^2}$ units

Mathematics File: Sigma Notation, page 132

1. a) $5 + 6 + 7 + 8 + 9 + 10$ **b)** $2 + 4 + 8 + 16 + 32$

c) $-2 + 4 - 8 + 16 - 32$

d) $5 + 10 + 20 + 40 + 80 + 160$

2. a) $\sum_{k=1}^{6} \frac{1}{2^{k-1}}$ **b)** $\sum_{k=1}^{7} 2(-3)^{k-1}$

3. a) 189   **b)** -63 **c)** 728   **d)** -364

4. a) $\frac{1}{9}$ **b)** $\frac{1}{3}$ **c)** $\frac{2}{3}$ **d)** 1

5. a) 0   **b)** -1

6. Examples may vary. $\prod_{i=1}^{6} i = 1 \times 2 \times 3 \times 4 \times 5 \times 6$

2.10 Exercises, page 138

1. a) $y = \log x$ **b)** $y = \log_3 x$ **c)** $f^{-1}(x) = \log_7 x$

d) $f^{-1}(x) = \log_{0.4} x$ **e)** $g^{-1}(x) = \log_{\frac{3}{2}} x$ **f)** $h^{-1}(x) = \log_{15} x$

2. a) $y = 10^x$ **b)** $y = 2^x$ **c)** $f^{-1}(x) = 6^x$

5. Predictions may vary.

7. b) i) 0.69 m/s   **ii)** 1.25 m/s   **iii)** 1.67 m/s

c) About 5500

8. b) i) 57.4%   **ii)** 52.0%   **iii)** 27.1%   **iv)** 11.9%

c) 77.4 min

9. Explanations may vary.

10. a) $n = \frac{\log A - 3}{\log 1.08}$

b) Explanations may vary.

i) 2.9 years   **ii)** -13.6

d) Domain: $0 < A \le 1250$, range: $n \le 2.9$

11. a) $P = 1\ 832\ 000(1.025)^n$ **b)** $n = \frac{\log P - \log 1\ 832\ 000}{\log 1.025}$

c) i) 3.55 years   **ii)** 20 years

d) Answers may vary.

13. b) Domain: $x > 2$, range: all real numbers

c) $x = 2$

d) Vertical intercept: none, horizontal intercept: $\frac{7}{3}$

14. a) Domain: $x > 3$, range: all real numbers, asymptote: $x = 3$, vertical intercept: none, horizontal intercept: $\frac{13}{4}$

b) Domain: $x > \frac{5}{2}$, range: all real numbers, asymptote: $x = \frac{5}{2}$, vertical intercept: none, horizontal intercept: 3

c) Domain: $x > 2$, range: all real numbers, asymptote: $x = 2$, vertical intercept: none, horizontal intercept: $\frac{7}{3}$

d) Domain: $x > -\frac{2}{3}$, range: all real numbers, asymptote: $x = -\frac{2}{3}$, vertical intercept: 2.41, horizontal intercept: $-\frac{1}{3}$

15. a) $x = \frac{\log y - \log 5}{\log 2}$ **b)** $x = \log y - \log 1.3$

c) $x = \frac{\log y - \log 8.2}{\log 1.03}$ **d)** $x = \frac{\log 6.4 - \log y}{\log 2}$

e) $x = \frac{\log y - \log 3.5}{\log 2.7}$ **f)** $x = \frac{\log y - \log 2.75}{\log 2 - \log 3}$

18. a) $(x_1 x_2, y_1 + y_2)$, $(\frac{x_1}{x_2}, y_1 - y_2)$, $(x_1^2, 2y_1)$, $(\frac{1}{x_1}, -y_1)$

19. $(0.548, 0.548)$

Mathematics File: Expressing Exponential Functions Using the Same Base, page 141

1. a) 2.7183

b) i) 7.3891   **ii)** 20.0855   **iii)** 1

iv) 0.3679   **v)** 0.1353

2. Explanations may vary.

a) 0   **b)** 0.6931   **c)** 1   **d)** 2.3026   **e)** -0.6931

3. Explanations may vary.

a) 0   **b)** 0.6931   **c)** 1   **d)** 2.3026   **e)** -0.6931

4. b) Explanations may vary. Each graph is the inverse of the other.

5. a) 0.022, $P = 2.76e^{0.022n}$ **b)** About 5.3 million; yes

c) Yes

6. Answers may vary.

7. a) $P = 2.5e^{0.033n}$ **b)** 42 years

c) Answers may vary. A plague, not enough food to feed the population

8. a) $P = 100e^{-0.14n}$ **b)** Graphs are the same.

c) The graph of $P = 100e^{-0.14n}$ is the image of $y = e^x$ after a vertical expansion by a factor of 100, a horizontal expansion by a factor of approximately 7.14 and a reflection in the y-axis.

9. a) 5.86 kPa   **b)** 17 600 m   **c)** $h = \frac{\ln 130 - \ln P}{0.000\ 155}$

Problem Solving: Evaluating Exponential Functions, page 144

Answers may vary.

1. a) $y = x + 1$ **b)** $-0.144 < x < 0.138$

2. a) $y = \frac{1}{2}x^2 + x + 1$ **b)** $-0.404 < x < 0.378$

3. Descriptions may vary. Evaluate $y = \frac{1}{2}x^2 + x + 1$ for an approximate value of $y = e^x$, when x is close to 0.

4. a) 2   **b)** 6   **c)** 120   **d)** 720   **e)** 5040

5. $1 + x + \frac{1}{2}x^2 + \frac{1}{6}x^3 + \frac{1}{24}x^4 + \frac{1}{120}x^5 + \frac{1}{720}x^6 + \frac{1}{5040}x^7$

6. a) Answers may vary.

7. a) 1   **b)** 2.718   **c)** 0.368   **d)** 7.389   **e)** 0.135

8. a) $e^x = \sum\limits_{k=0}^{\infty} \frac{1}{k!} x^k$

b) Answers may vary. To simplify expressions involving factorials

9. $1 + (2.3x) + \frac{1}{2!}(2.3x)^2 + \frac{1}{3!}(2.3x)^3 + \ldots$

Investigate, page 146

1. 0.2   **2.** 0.2   **3.** 0.2

4. 0.2   **5.** 0.2   **6.** 0.217

2.11 Exercises, page 149

1. a) 1   **b)** 4   **c)** 7

d) -1 **e)** 2   **f)** 0

g) 1   **h)** 0   **i)** $\frac{4}{3}$

j) 0   **k)** $-\frac{1}{2}$ **l)** $\frac{1}{2}$

2. a), b), c) 2

3. a), b), c) 3

4. Explanations may vary.

5. a) $-\frac{1}{2}$ **b)** $\frac{1}{2}$ **c)** 1

d) -1 **e)** 1   **f)** $\frac{1}{3}$

6. a) 2   **b)** 3   **c)** 3

d) 2   **e)** 0   **f)** $-\frac{1}{2}$

7. a) 1   **b)** 1.585   **c)** 4.508

8. a) 0   **b)** $\frac{\log 2}{\log 4 - \log 3}$ **c)** $\frac{\log 2}{\log 3 - \log 4}$

d) $\frac{\log 24}{\log 4}$ **e)** 2   **f)** $\frac{3 \log 3}{2 \log 3 - \log 2}$

9. a) $\frac{\log a - \log c}{\log c - \log b}$ **b)** $\frac{\log a - 2 \log c}{\log c - \log b}$ **c)** $\frac{\log a - 3 \log c}{\log c - \log b}$

2.12 Exercises, page 152

1. a) 100   **b)** 100 000   **c)** $\frac{1}{1000}$

d) 8   **e)** 27   **f)** 64

2. a) 2   **b)** 3   **c)** 4   **d)** 5

3. a) 3   **b)** 4   **c)** 5   **d)** 6

4. a) 3   **b)** 6   **c)** 101

5. a) 8   **b)** 9   **c)** 2

d) 6   **e)** 2   **f)** 5

6. a) 100   **b)** 18   **c)** 3

d) 4   **e)** 3   **f)** 2

7. b) $\log_a \frac{1}{x^2}$ is defined for all real numbers except $x = 0$. $\log_a x$ is defined for $x > 0$.

8. a) $x > 2$ **b)** $x > 0$ **c)** $x > 5$

10. $\frac{1}{\log_a x} + \frac{1}{\log_b x} = \frac{1}{\log_{ab} x}$

Linking Ideas: Mathematics and Science, Carbon Dating, page 154

1. 52.85%   **2.** 26,000 years

3. 78.61%   **4.** 693 years

5. a) $n = \frac{5760(\log P - 2)}{\log(0.5)}$

b) i) 1896 years   **ii)** 3096 years   **iii)** 3972 years

iv) 11 859 years   **v)** 15 600 years   **vi)** 31 717 years

2 Review, page 155

1. a) $A = A_0 (0.1)^P$ **b)** 6.3 times

2. $747.26

3. 22.0 million

4. 79%

5. 9.9 years

6. 2621

7. Answers may vary. For 1999:

a) 1.58×10^{10} **b)** 6.60×10^{10}

8. 1.08%

9. a) $d \frac{\log 3}{\log 2}$ **b)** $2d$

10. a) i) $P = \frac{A}{(1 + i)^n}$ **ii)** $n = \frac{\log A - \log P}{\log (1 + i)}$ **iii)** $i = \left(\frac{A}{P}\right)^{\frac{1}{n}} - 1$

b) i) $A = \frac{y}{b^x}$ **ii)** $x = \frac{\log y - \log A}{\log b}$ **iii)** $b = \left(\frac{y}{A}\right)^{\frac{1}{x}}$

Chapter 3

Mathematical Modelling: Predicting Tide Levels, page 156

• 5:09 A.M. or 5:40 P.M.

• 6 h

• Estimates may vary.

a) 2 m   **b)** 2 m   **c)** 2 m   **d)** 2.5 m

3.1 Exercises, page 161

1. a) Estimates may vary.

i) 5:30 P.M.   **ii)** 9:00 P.M.   **iii)** 5:30 P.M.

b) i) April 21 and August 21
ii) March 21 and September 21
iii) February 21 and October 21
iv) December 21

2. Answers may vary.

3. Estimates may vary. 11 years

4. a) 150 m, 75m   **b)** 6 h

5. Answers may vary.

6. 5 s

7. Explanations may vary.

8. Estimates may vary.
  a) Periodic with period 8   **b)** Periodic with period 6
  c) Periodic with period 6
  d), e) Not periodic, because the graph does not repeat in a regular way
  f) Periodic with period 3

Investigate, page 163

2. 57.295 78°

3. a) π **b)** 3.141 59, yes

3.2 Exercises, page 167

1. a) $\frac{\pi}{6}$ radians   **b)** $\frac{\pi}{3}$ radians   **c)** $\frac{\pi}{2}$ radians
  d) $\frac{2\pi}{3}$ radians   **e)** $\frac{5\pi}{4}$ radians   **f)** π radians
  g) $\frac{7\pi}{6}$ radians   **h)** $\frac{4\pi}{3}$ radians   **i)** $\frac{3\pi}{2}$ radians
  j) $\frac{5\pi}{3}$ radians   **k)** $\frac{11\pi}{6}$ radians   **l)** 2π radians

2. a) $\frac{\pi}{4}$ radians   **b)** $\frac{3\pi}{4}$ radians
  c) $\frac{5\pi}{4}$ radians   **d)** $\frac{7\pi}{4}$ radians

3. Explanations may vary. For exercise 1a, 30° was multiplied by $\frac{\pi}{180°}$ to get $\frac{\pi}{6}$ radians.

4. a) 30°   **b)** 60°   **c)** 90°   **d)** 120°
  e) 150°   **f)** 180°   **g)** 45°   **h)** 90°
  i) 135°   **j)** 180°   **k)** 225°   **l)** 270°

5. a) 2π radians   **b)** π radians   **c)** $\frac{\pi}{2}$ radians

6. a) 1.75 radians   **b)** 3.93 radians   **c)** 1.00 radian
  d) 2.18 radians   **e)** 1.31 radians   **f)** 0.33 radians
  g) 1.13 radians   **h)** 0.43 radians   **i)** 2.62 radians
  j) 0.52 radians   **k)** 1.00 radian   **l)** 1.57 radians

7. a) 114.6°   **b)** 286.5°   **c)** 183.3°   **d)** 103.1°
  e) 40.1°   **f)** 80.2°   **g)** 383.9°   **h)** 360.0°

8. a) 90°   **b)** 330°   **c)** 120°   **d)** 210°
  e) 300°   **f)** 270°   **g)** 315°   **h)** 360°

9. Explanations may vary. For exercise 8a, $\frac{\pi}{2}$ was multiplied by $\frac{180°}{\pi}$ to get 90°.

10. a) 10.0 cm   **b)** 15.0 cm   **c)** 9.0 cm
  d) 30.5 cm   **e)** 21.0 cm   **f)** 3.0 cm

11. a) 28.3 cm   **b)** 15.7 cm   **c)** 22.0 cm   **d)** 34.6 cm
  e) 50.3 cm   **f)** 37.7 cm   **g)** 64.9 cm   **h)** 41.9 cm

13. a) i) $\sqrt{2}r$ **ii)** $\frac{\pi}{2}r$
  b) $\frac{\pi}{2\sqrt{2}}$ or 1.11 times as long

14. a) 0.5 radians, 28.6°   **b)** 1.17 radians, 66.8°
  c) 2.08 radians, 119.4°   **d)** 2.73 radians, 156.6°

15. a) $A = \frac{r^2\theta}{2}$ **b)** $A = \frac{\pi r^2\theta}{360°}$

16. a) 86.83 radians/s   **b)** $\frac{500x}{9d}$ radians/s

Linking Ideas: Mathematics and Science, Refraction, page 169

1. $n_{1\rightarrow 2} = 1.15$ **2.** $n_{1\rightarrow 2} = 0.77$

3. 0.50 m

3.3 Exercises, page 173

1. a) 180°, π radians   **b)** 450°, $\frac{5\pi}{2}$ radians
  c) −90°, $-\frac{\pi}{2}$ radians   **d)** −270°, $-\frac{3\pi}{2}$ radians

3. Answers may vary.
  a) −310°, 410°   **b)** 240°, 600°
  c) 195°, 555°   **d)** −120°, 600°
  e) $-\frac{3\pi}{2}$ radians, $\frac{5\pi}{2}$ radians   **f)** $\frac{7\pi}{4}$ radians, $\frac{15\pi}{4}$ radians
  g) $-\frac{4\pi}{3}$ radians, $\frac{8\pi}{3}$ radians   **h)** $\frac{\pi}{2}$ radians, $\frac{5\pi}{2}$ radians

4. Answers may vary.
  a) −300°, 420°   **b)** 150°, −570°
  c) $-\frac{3\pi}{2}$ radians, $\frac{13\pi}{4}$ radians   **d)** $\frac{3\pi}{2}$ radians, $-\frac{5\pi}{2}$ radians

5. Answers may vary.
  a) $-\pi$ radians, 3π radians   **b)** $-\frac{3\pi}{2}$ radians, $\frac{5\pi}{2}$ radians
  c) $\frac{5\pi}{3}$ radians, $-\frac{7\pi}{3}$ radians   **d)** 0 radians, 2π radians

6. Explanations will vary. For exercise 5a, 2π was subtracted from π to get $-\pi$. 2π was added to π to get 3π.

7. a) 1   **b)** 1st quadrant

8. a) i) 1   **ii)** 2nd quadrant
  b) i) 1   **ii)** 1st quadrant
  c) i) 2   **ii)** 2nd quadrant
  d) i) 2   **ii)** 1st quadrant

13. a) $-45° + n(360°)$ **b)** $150° + n(360°)$
  c) $240° + n(360°)$ **d)** $-30° + n(360°)$
  e) $(\pi + 2\pi n)$ radians   **f)** $\left(-\frac{\pi}{4} + 2\pi n\right)$ radians
  g) $\left(\frac{5\pi}{2} + 2\pi n\right)$ radians   **h)** $(-1 + 2\pi n)$ radians

14. b) Domain: $\theta < 0$ or $\theta \geq 2\pi$; range: $0 \leq \alpha < 2\pi$; period: 2π.

3.4 Exercises, page 179

1. a) 0.8, 0.6   **b)** 0.8, −0.6   **c)** −0.6, −0.8   **d)** −0.6, 0.8

2. a) 0.7660   **b)** −0.6428   **c)** −0.1736   **d)** −0.9397
  e) 0.2588   **f)** −0.5736   **g)** 0.9744   **h)** −0.9945

3. a) 0.3894   **b)** 0.8253   **c)** 0.9490   **d)** 0.9738
  e) −0.9900   **f)** −0.5332   **g)** −0.7791   **h)** −0.7808

5. a) 0.5736   **b)** Answers may vary. 305°, −55°, −305°

6. a) 0.8192   **b)** Answers may vary. 55°, −235°, −305°

7. a) −0.7660   **b)** Answers may vary. −220°, 140°, −140°

8. a) 0.9490
  b) Answers may vary. 1.8916, −4.3916, 7.5332

9. a) Positive   **b)** Negative   **c)** Negative
  d) Negative   **e)** Positive

10. For exercise 9a, the terminal arm of angle 145° lies in the second quadrant. Since the y-coordinate of a point in the second quadrant is positive, sin 145° is positive.

11. a) i) (0.866, 0.5), (−0.5, 0.866), (−0.866, −0.5),
(0.5, −0.866)
ii) 30°, 120°, 210°, 300°
iii) $\cos 30° = 0.866$, $\sin 30° = 0.5$; $\cos 120° = -0.5$,
$\sin 120° = 0.866$; $\cos 210° = -0.866$, $\sin 210° = -0.5$;
$\cos 300° = 0.5$, $\sin 300° = -0.866$
b) i) (0.342, 0.94), (−0.94, 0.342), (−0.342, −0.94),
(0.94, −0.342)
ii) 70°, 160°, 250°, 340°
iii) $\cos 70° = 0.342$, $\sin 70° = 0.94$; $\cos 160° = -0.94$,
$\sin 160° = 0.342$; $\cos 250° = -0.342$, $\sin 250° = -0.94$;
$\cos 340° = 0.94$, $\sin 340° = -0.342$

12. Explanations may vary.

13. a) −0.866, 0.5 **b)** 0.866, −0.5
c) 0.5, 0.866 **d)** −0.5, −0.866

14. a) −0.342, 0.94 **b)** 0.342, −0.94
c) 0.94, 0.342 **d)** −0.94, −0.342

15. Explanations may vary.

16. b) Answers may vary. $\left(\frac{2}{\sqrt{5}}, \frac{1}{\sqrt{5}}\right)$ or (2, 1)

17. b) Answers may vary. $\left(-\frac{2}{\sqrt{5}}, \frac{1}{\sqrt{5}}\right)$ or (−2, 1)

18. b) Answers may vary. $\left(-\frac{4}{5}, \frac{3}{5}\right)$ or (−4, 3)

19. a) Answers may vary. For the TI-83, 999 999 999 999° or
999 999 999 999 radians
b) No

20. a) $\sin \theta = -\frac{7}{\sqrt{58}}$, $\cos \theta = -\frac{3}{\sqrt{58}}$

b) $\sin \theta = -\frac{7}{\sqrt{58}}$, $\cos \theta = -\frac{3}{\sqrt{58}}$

Exploring with a Graphing Calculator: Coordinates of Points on the Unit Circle, page 182

All explanations and predictions may vary.

5. a) −0.951, 0.309 **b)** −0.951, −0.309
c) 0.951, −0.309 **d)** 0.309, 0.951
e) −0.309, 0.951 **f)** 0.309, −0.951

9. a) −0.985, 0.174 **b)** −0.985, −0.174
c) 0.984, −0.174 **d)** 0.174, 0.985
e) −0.174, 0.985 **f)** 0.174, −0.985

3.5 Exercises, page 188

1. Geometric relationships provide the exact values for sines
and cosines of special angles. A calculator gives an
approximation to the sines and cosines of these angles.

2. b) i) $\frac{\pi}{4}$ **ii)** $\frac{\pi}{6}$ **iii)** $\frac{\pi}{4}$ **iv)** $\frac{\pi}{3}$
v) $\frac{\pi}{6}$ **vi)** $\frac{\pi}{3}$ **vii)** $\frac{\pi}{4}$ **viii)** $\frac{\pi}{3}$

3. a) $\frac{1}{\sqrt{2}}$ **b)** $\frac{1}{\sqrt{2}}$ **c)** $\frac{1}{\sqrt{2}}$ **d)** $-\frac{1}{\sqrt{2}}$
e) $\frac{1}{2}$ **f)** $-\frac{1}{2}$ **g)** $-\frac{1}{2}$ **h)** $-\frac{1}{2}$
i) $\frac{\sqrt{3}}{2}$ **j)** $\frac{\sqrt{3}}{2}$ **k)** $\frac{\sqrt{3}}{2}$ **l)** $-\frac{\sqrt{3}}{2}$

4. a) $\frac{1}{\sqrt{2}}$ **b)** $-\frac{1}{\sqrt{2}}$ **c)** $-\frac{1}{\sqrt{2}}$ **d)** $\frac{1}{\sqrt{2}}$

e) $\frac{\sqrt{3}}{2}$ **f)** $-\frac{\sqrt{3}}{2}$ **g)** $-\frac{\sqrt{3}}{2}$ **h** $\frac{\sqrt{3}}{2}$
i) $\frac{1}{2}$ **j)** $-\frac{1}{2}$ **k)** $-\frac{1}{2}$ **l)** $\frac{1}{2}$

6. a) $\frac{\sqrt{3}}{2}$ **b)** $\frac{1}{2}$ **c)** $\frac{1}{\sqrt{2}}$ **d)** $\frac{1}{\sqrt{2}}$
e) $\frac{\sqrt{3}}{2}$ **f)** $-\frac{1}{2}$ **g)** $\frac{1}{\sqrt{2}}$ **h)** $-\frac{1}{\sqrt{2}}$

7. b) i) Yes **ii)** Yes

8. a) i) (0.866, 0.5), (−0.866, 0.5), (0, −1)
ii) 30°, 150°, 270°
iii) $\cos 30° = 0.866$, $\sin 30° = 0.5$;
$\cos 150° = -0.866$, $\sin 150° = 0.5$;
$\cos 270° = 0$, $\sin 270° = -1$
b) i) (0.707, 0.707), (−0.707, 0.707), (−0.707, −0.707),
(0.707, −0.707)
ii) 45°, 135°, 225°, 315°
iii) $\cos 45° = 0.707$, $\sin 45° = 0.707$;
$\cos 135° = -0.707$, $\sin 135° = 0.707$;
$\sin 225° = -0.707$, $\cos 225° = -0.707$;
$\cos 315° = 0.707$, $\sin 315° = -0.707$
c) i) Starting at the marked vertex and moving
counterclockwise: (0.5, 0.866), (−0.5, 0.866), (−1, 0),
(−0.5, −0.866), (0.5, −0.866), (1, 0)
ii) 60°, 120°, 180°, 240°, 300°, 360°
iii) $\cos 60° = 0.5$, $\sin 60° = 0.866$;
$\cos 120° = -0.5$, $\sin 120° = 0.866$;
$\cos 180° = -1$, $\sin 180° = 0$;
$\cos 240° = -0.5$, $\sin 240° = -0.866$;
$\cos 300° = -0.5$, $\sin 300° = -0.866$;
$\cos 360° = 1$, $\sin 360° = 0$
d) i) Starting at the marked vertex and moving
counterclockwise: (0.924, 0.383), (0.383, 0.924),
(−0.383, 0.924), (−0.924, 0.383), (−0.924, −0.383),
(−0.383, −0.924), (0.383, −0.924), (0.924, −0.383)
ii) 22.5°, 67.5°, 112.5°, 157.5°, 202.5°, 247.5°, 292.5°, 337.5°
iii) $\cos 22.5° = 0.924$, $\sin 22.5° = 0.383$;
$\cos 67.5° = 0.383$, $\sin 67.5° = 0.924$;
$\cos 112.5° = -0.383$, $\sin 112.5° = 0.924$;
$\cos 157.5° = -0.924$, $\sin 157.5° = 0.383$;
$\cos 202.5° = -0.924$, $\sin 202.5° = -0.383$;
$\cos 247.5° = -0.383$, $\sin 247.5° = -0.924$;
$\cos 292.5° = 0.383$, $\sin 292.5° = -0.924$;
$\cos 337.5° = 0.924$, $\sin 337.5° = -0.383$

9. Explanations may vary.

10. All values will be listed starting at the marked vertex and
moving counterclockwise.

a) i) triangle: $\frac{\pi}{6}, \frac{5\pi}{6}, \frac{3\pi}{2}$
ii) $\cos \frac{\pi}{6} = \frac{\sqrt{3}}{2}$, $\sin \frac{\pi}{6} = \frac{1}{2}$; $\cos \frac{5\pi}{6} = -\frac{\sqrt{3}}{2}$,
$\sin \frac{5\pi}{6} = \frac{1}{2}$; $\cos \frac{3\pi}{2} = 0$, $\sin \frac{3\pi}{2} = -1$
i) square: $\frac{\pi}{4}, \frac{3\pi}{4}, \frac{5\pi}{4}, \frac{7\pi}{4}$
ii) $\cos \frac{\pi}{4} = \frac{1}{\sqrt{2}}$, $\sin \frac{\pi}{4} = \frac{1}{\sqrt{2}}$; $\cos \frac{3\pi}{4} = -\frac{1}{\sqrt{2}}$,
$\sin \frac{3\pi}{4} = \frac{1}{\sqrt{2}}$; $\cos \frac{5\pi}{4} = -\frac{1}{\sqrt{2}}$, $\sin \frac{5\pi}{4} = -\frac{1}{\sqrt{2}}$;
$\cos \frac{7\pi}{4} = \frac{1}{\sqrt{2}}$, $\sin \frac{7\pi}{4} = -\frac{1}{\sqrt{2}}$
i) hexagon: $\frac{\pi}{3}, \frac{2\pi}{3}, \pi, \frac{4\pi}{3}, \frac{5\pi}{3}, 2\pi$

ii) $\cos \frac{\pi}{3} = \frac{1}{2}$, $\sin \frac{\pi}{3} = \frac{\sqrt{3}}{2}$ $\cos \frac{2\pi}{3} = -\frac{1}{2}$,

$\sin \frac{2\pi}{3} = \frac{\sqrt{3}}{2}$; $\cos \pi = -1$, $\sin \pi = 0$;

$\cos \frac{4\pi}{3} = -\frac{1}{2}$, $\sin \frac{4\pi}{3} = -\frac{\sqrt{3}}{2}$; $\cos \frac{5\pi}{3} = \frac{1}{2}$,

$\sin \frac{5\pi}{3} = -\frac{\sqrt{3}}{2}$; $\cos 2\pi = 1$, $\sin 2\pi = 0$

b) i) $\frac{\pi}{8}, \frac{3\pi}{8}, \frac{5\pi}{8}, \frac{7\pi}{8}, \frac{9\pi}{8}, \frac{11\pi}{8}, \frac{13\pi}{8}, \frac{15\pi}{8}$

12. a) $\sin(\pi + \theta) = -y$, $\cos(\pi + \theta) = -x$
b) $\sin(2\pi - \theta) = -y$, $\cos(2\pi - \theta) = x$

Exploring with a Graphing Calculator: Graphing Sines and Cosines, page 190

4. 45°, 225°; explanations may vary.

Exploring with a Graphing Calculator: Graphing $y = \sin \theta$ and $y = \cos \theta$, page 191

1. Answers may vary.

2. a) Yes, 0.5   **b)** 30°, −210°, −330°
c) −30°, −150°, 210°, 330°
d) Yes, 0.87; 120°, 60°, −240°, −300°; −120°, −60°, 240°, 300°; Yes, 0.71; 135°, 45°, −225°, −315°; −135°, −45°, 225°, 315°

3. a) −360°, −180°, 0°, 180°, 360°
b) Maximums: (−270°, 1), (90°, 1);
Minimums: (−90°, −1), (270°, −1)

4. a) 0.5   **b)** 390°, 30°, −210°, −330°
c) −390°, −150°, −30°, 210°, 330°
d) 0.87; 120°, 420°, 60°, −240°, −300°; −120°, 420°, −60°, 240°, 300°; 0.71; 405°, 135°, 45°, −225°, −315°; −135°, −45°, 225°, 315°, 405°

5. For 150°, 0.87; same cosine: 210°, −150°, −210°; opposite cosine: 30°, 330°, −30°, −330°
For 120°, −0.5; same cosine: 240°, −120°, −240°; opposite cosine: 60°, 300°, −60°, −300°
For 135°, 0.71; same cosine: 225°, −135°, −225°; opposite cosine: 45°, 315°, −45°, −315°;
θ-intercepts: −270°, −90°, 90°, 270°;
Maximums: (−360°, 1), (0°, 1), (360°, 1);
Minimums: (−180°, −1), (180°, −1)
For window settings −470° < θ < 470°, the answers are the same as above with a few additional angles. Opposite value of cosine 150°: −390°, 390°; opposite value of cosine 120°: −420°, 420°; opposite value of cosine 135°: −405°, 405°

6. Answers may vary.

7. a) 0.5   **b)** $\frac{\pi}{6}, -\frac{7\pi}{6}, -\frac{11\pi}{6}$ **c)** $-\frac{\pi}{6}, -\frac{5\pi}{6}, \frac{7\pi}{6}, \frac{11\pi}{6}$
d) 0.87; $\frac{2\pi}{3}, \frac{\pi}{3}, -\frac{4\pi}{3}, -\frac{5\pi}{3}, \frac{2\pi}{3}, -\frac{\pi}{3}, \frac{4\pi}{3}, \frac{5\pi}{3}$; 0.71; $\frac{3\pi}{4}, \frac{\pi}{4}, -\frac{5\pi}{4}, -\frac{7\pi}{4}; -\frac{3\pi}{4}, -\frac{\pi}{4}, \frac{5\pi}{4}, \frac{7\pi}{4}$

8. a) $-2\pi, -\pi, 0, \pi, 2\pi$
b) Maximums: $(-\frac{3\pi}{2}, 1), (\frac{\pi}{2}, 1)$; Minimums: $(-\frac{\pi}{2}, -1), (\frac{3\pi}{2}, -1)$

9. a) $470° = \frac{47\pi}{18}$ **b)** 0.5

c) $\frac{13\pi}{6}, \frac{\pi}{6}, -\frac{7\pi}{6}, -\frac{11\pi}{6}$
d) $-\frac{13\pi}{6}, -\frac{\pi}{6}, -\frac{5\pi}{6}, \frac{7\pi}{6}, \frac{11\pi}{6}$
e) 0.87; $\frac{7\pi}{3}, \frac{2\pi}{3}, \frac{\pi}{3}, -\frac{4\pi}{3}, -\frac{5\pi}{3}, -\frac{7\pi}{3}, \frac{2\pi}{3}, -\frac{\pi}{3}, \frac{4\pi}{3}, \frac{5\pi}{3}$

10. For $\frac{5\pi}{6}$: −0.87; same cosine $\frac{7\pi}{6}, -\frac{5\pi}{6}, -\frac{7\pi}{6}$; opposite cosine: $\frac{\pi}{6}, \frac{11\pi}{6}, -\frac{\pi}{6}, -\frac{11\pi}{6}$
For $\frac{2\pi}{3}$: −0.5; same cosine: $\frac{4\pi}{3}, -\frac{2\pi}{3}, -\frac{4\pi}{3}$; opposite cosine: $\frac{\pi}{3}, \frac{5\pi}{3}, -\frac{\pi}{3}, -\frac{5\pi}{3}$
For $\frac{3\pi}{4}$: −0.71; same cosine: $\frac{5\pi}{4}, -\frac{3\pi}{4}, -\frac{5\pi}{4}$; opposite cosine: $\frac{\pi}{4}, \frac{7\pi}{4}, -\frac{\pi}{4}, -\frac{7\pi}{4}$;
θ-intercepts: $-\frac{3\pi}{2}, -\frac{\pi}{2}, \frac{\pi}{2}, \frac{3\pi}{2}$
Maximums: (−2π, 1), (0, 1), (2π, 1)
Minimums: (−π, −1), (π, −1)
For window settings $-\frac{47\pi}{18} < \theta < \frac{47\pi}{18}$, the answers are the same as above with a few additional angles. Opposite value of cosine $\frac{5\pi}{6}$: $-\frac{13\pi}{6}, \frac{13\pi}{6}$; opposite value of cosine $\frac{2\pi}{3}$: $-\frac{7\pi}{3}, \frac{7\pi}{3}$

3.6 Exercises, page 196

4. Answers may vary. 140°, 400°, −220°, −320°
5. Answers may vary. 320°, 400°, −40°, −320°
6. Explanations may vary. For exercise 5, 360° was added and subtracted from 40° to get 400° and −320° respectively. Since the graph is symmetrical about the y-axis, the opposite angles −40° and 320° were also determined.

7. a) 0.5   **b)** 150°, 390°, −210°, −330°
8. a) 0.866   **b)** 330°, 390°, −30°, −330°, −390°

10. a) $y = 1$ when $\theta = \frac{\pi}{2}$ and $-\frac{3\pi}{2}$
b) $y = -1$ when $\theta = \frac{3\pi}{2}$ and $-\frac{\pi}{2}$
c) $-2\pi \le \theta \le 2\pi$ **d)** $-1 \le y \le 1$
e) 0   **f)** 0, ±π, ±2π

11. a) $y = 1$ when $\theta = 0$ and $\theta = ±2\pi$
b) $y = -1$ when $\theta = ±\pi$ **c)** $-2\pi \le \theta \le 2\pi$
d) $-1 \le y \le 1$ **e)** 1   **f)** $±\frac{\pi}{2}, ±\frac{3\pi}{2}$

12. a), b), c) They differ by $\frac{\pi}{2}$ units.
d) They are the same.

13. a) Descriptions may vary.
b) (45°, 0.707), (−135°, −0.707), (−315°, 0.707), (225°, −0.707)
c) $\cos \theta > \sin \theta$ for $0° \le \theta < 45°$, $\sin \theta > \cos \theta$ for $45° < \theta \le 90°$.

14. The graph will have about 115 cycles between −360° and 360°. Explanations may vary.

15. a) The graph would have the same shape. The horizontal axis would have degrees instead of radians as units.
b) Yes. All properties of the graphs are the same, with the appropriate units for θ.

16. a) $\cos 40° > \sin 40°$; explanations may vary.
b) True for all the points with the x-coordinate in the

intervals $-135° + n360° < \theta < 45° + n360°$; explanations may vary.

Mathematical Modelling: Predicting Tide Levels, page 200

1. b) (8.24, 1.97)   c) (20.6, 1.97)   f) 12.36 h
  g) 0.82 m

2. a) 2.08 m   b) 2.46 m   c) 2.04 m   d) 1.48 m

3. 2.72 m

4. $y = 2.3 \sin 2\pi \left(\frac{x - 80}{365} \right) + 19.2$; 9:15 P.M.; 6:25 P.M.

3.7 Exercises, page 209

1. a) 2   b) −0.6   c) −0.6   d) 0.7

2. a) $\frac{1}{2}, \frac{\sqrt{3}}{2}, \frac{1}{\sqrt{3}}$ b) $-\frac{\sqrt{3}}{2}, -\frac{1}{2}, \sqrt{3}$

  c) $-\frac{1}{2}, \frac{\sqrt{3}}{2}, -\frac{1}{\sqrt{3}}$ d) 0, 1, 0

  e) $-\frac{\sqrt{3}}{2}, \frac{1}{2}, -\sqrt{3}$ f) $\frac{1}{\sqrt{2}}, \frac{1}{\sqrt{2}}, 1$

  g) $\frac{1}{\sqrt{2}}, \frac{1}{\sqrt{2}}, 1$ h) $\frac{1}{\sqrt{2}}, -\frac{1}{\sqrt{2}}, -1$

  i) 1, 0, undefined

3. a) 4th   b) 3rd   c) 3rd   d) 2nd

4. a) 0.4663   b) Answers may vary. 205°, −155°, −335°

5. a) −2.7463   b) Answers may vary. 8.20, 5.06, −4.36

6. No.

7. a) $\sin \theta = y$, $\cos \theta = x$ b) $\tan \theta = \frac{y}{x}$

8. b) (−7, 3)

9. b) (2, −3)

10. b) (−5, −2)

11. b) $\tan \theta = \frac{2}{3}$
  c) Tangent of the angle θ is equal to the slope of the line.

12. Explanations may vary.

Exploring with a Graphing Calculator: Graphing $y = \tan \theta$, page 211

1. a) Yes; explanations may vary.
  b) Same value: 240°, −120°, −300°;
   opposite value: −240°, −60°, 120°, 300°

2. Same value: 420°, 240°, −120°, −300°;
  opposite value: −420°, −240°, −60°, 120°, 300°

3. a) Yes; explanations may vary.
  b) Same value: $\frac{4\pi}{3}, -\frac{2\pi}{3}, -\frac{5\pi}{3}$,
   opposite value: $-\frac{4\pi}{3}, -\frac{\pi}{3}, \frac{2\pi}{3}, \frac{5\pi}{3}$

4. Same value: $\frac{7\pi}{3}, \frac{4\pi}{3}, -\frac{2\pi}{3}, -\frac{5\pi}{3}$,
  opposite value: $-\frac{7\pi}{3}, -\frac{4\pi}{3}, -\frac{\pi}{3}, \frac{2\pi}{3}, \frac{5\pi}{3}$

3.8 Exercises, page 213

2. −320°, −140°, 220°, 400°

3. Explanations may vary.

4. $\tan 40° = \frac{\sin 40°}{\cos 40°}$

6. a) $\theta \neq \frac{n\pi}{2}$, where n is an odd integer
  b) All real numbers   c) 0
  d) $\theta = n\pi$, where n is an integer

7. a) $h = d \tan \theta$

8. Descriptions and explanations may vary.

Problem Solving: Trigonometric-Like Functions, page 215

1. Answers may vary. For the square with vertical and horizontal diagonals:
$$x = \frac{\cos \theta}{|\cos \theta| + |\sin \theta|}, \, y = \frac{\sin \theta}{|\cos \theta| + |\sin \theta|}$$

3. and 4. Answers may vary.

3.9 Exercises, page 218

1. Descriptions may vary.

2. a) 1.0642   b) −4.1336   c) 0.0875
  d) 28.6537   e) −0.6494   f) 1.0002
  g) −1.0024   h) −11.4301

3. a) −5.6102   b) −12.7670   c) −0.6684   d) 1.1198
  e) 1.6709   f) 0.0794   g) 1.0006   h) −1.6071

4. −320°, −220°, 140°, 400°

5. Explanations may vary. 360° was added and subtracted from 40° to get 400° and 320°, respectively. Since the graph is symmetrical about the line $x = 90°$, 50° was added to 90° to get 140°. Finally, 360° was subtracted from 140° to get −220°.

6. They are reciprocals.

8. Answers may vary. −320°, −40°, 320°, 400°

9. −320°, −140°, 220°, 400°

10. Explanations may vary. For exercise 8, 360° was added and subtracted from 40° to get 400° and −320° respectively. Since the graph is symmetrical about the y-axis, the opposite angles −40° and 320° were also determined.

11. They are reciprocals.

12. They are reciprocals.

3 Review, page 220

1. a) $\frac{2\pi}{365}$ radians   b) 2 564 917 km   c) 106 872 km/h

2. a) Explanations may vary. The columns for sin i and sin r have incorrect values. Therefore, the column for the index of refraction has incorrect values.
  c) About 1.5

3. a) $-\frac{\pi}{2}$ radians, −90°   b) $-\frac{5\pi}{2}$ radians, −450°
  c) $\frac{7\pi}{2}$ radians, 630°

4. The answer assumes the bicycle moves forward turning

right, as in the diagram. Other assumptions would create slight variations in the answers.
- **a)** Yes
- **b)** $0 < \theta < 90°$
- **c)** **i)** r_1 and r_2 approach infinity.
  - **ii)** r_1 approaches 1 and r_2 approaches 0.
- **d)** Explanations may vary.
- **e)** $r_1 = \csc\theta, r_2 = \cot\theta$
- **g)** The properties of the graphs correspond to the answers of parts a, b, and c.

5. Descriptions may vary.
- **a)** Q gets farther and farther away. When the terminal arm of θ reaches R, OQ has an infinite distance.
- **b)** OQ has an infinite distance when the terminal arm is at R. As the terminal arm moves from R to P_4, Q appears in the fourth quadrant.

6. Descriptions may vary.
- **a)** The graph would have the same appearance. The horizontal axis would be in degrees and the asymptotes would be $x = 90° \pm n(180°)$.
- **b)** Yes; explanations may vary.

3 Cumulative Review, page 222

1. b) Explanations may vary.   **d)** All are functions.

2. a) There is a horizontal compression by a factor of $\frac{1}{2}$.
- **b)** There is a vertical compression by a factor of $\frac{1}{4}$ and a reflection in the x-axis.
- **c)** There is a horizontal expansion by a factor of 3.
- **d)** There is a vertical expansion by a factor of 4.
- **e)** There are vertical and horizontal compressions by a factor of $\frac{1}{3}$.
- **f)** There is a reflection in the y-axis and a vertical expansion by a factor of 2.

3. b) Domain: $0 \le E \le 100$, range: $0 \le v \le \frac{10}{\sqrt{3}}$

4. a) 1.44, 1.74, 2.04, 2.34, 2.64, 2.94, 3.25, 3.55
- **b)** The logarithms are an arithmetic sequence with a common difference of 0.3.

5. a) i) $3, \frac{1}{3}$ **ii)** $2, \frac{1}{2}$
- **b)** Proofs may vary. $(\log_b a)(\log_a b) = 1$

6. Explanations may vary. x

7. a) 1, 1, 1, 1
- **b)** $1 + \log 6, 2 + \log 6, 3 + \log 6, 4 + \log 6$

8. a) $n = 11.9$ **b)** $n = 35.7$ **c)** $n = 13.5$
- **d)** $n = 11.6$ **e)** $n = 11.7$ **f)** $n = 11.7$

9. Explanations may vary.

10. About 316 times

11. a) $c = 4026.6 \times 1.12^n$ **b)** $c = 4.0913 \times 10^{-95} \times 1.12^n$
- **c)** The mass of all the satellites in orbit in 1991 and 1991 years ago

12. a) Both equal 48.   **b)** $t_{22} = 192$

13. a) $\frac{3}{2}$ **b)** -2 **c)** 6

14. a) 1.732   **b)** $-300°, -120°, 240°, 420°$

15. Descriptions may vary.

Chapter 4

Mathematical Modelling: Predicting Hours of Daylight, page 224

- The sun is closer to Red Deer during the summer because of the tilt of Earth.
  Estimates may vary.
  - **a)** 11 h   **b)** 16.5 h   **c)** 9 h

Investigate, page 226

4. Explanations may vary.
- **a)** Domain: all real numbers; range: $-1 \le x \le 1$
- **b)** Period: 6.28   **c)** 1.57, 4.71, 7.85, …

5. Explanations may vary.
- **a)** Domain: all real numbers; range: $-1 \le x \le 1$
- **b)** Period: 6.28   **c)** 0, 3.14, 6.28, …

4.1 Exercises, page 232

Estimates may vary.
1. a) 0.4   **b)** 0.5   **c)** -0.9 **d)** -0.7
2. a) 0.4   **b)** 0.45   **c)** -0.9 **d)** -0.7
3. a) 0.389   **b)** 0.454   **c)** -0.904 **d)** -0.706

4. a) Answers may vary. The same
- **b)** Explanations may vary.

5. a) 0.479, 0.878   **b)** 0.842, 0.540   **c)** 0.909, -0.416
- **d)** 0.335, -0.942 **e)** $-0.916, -0.401$ **f)** $-0.544, -0.839$
- **g)** $-0.479, 0.878$ **h)** 0.530, -0.848

6. b) The graph of $y = \sin x$ is the image of $y = \cos x$ after a translation of $\frac{\pi}{2}$ units right.
- **c)** 2π
- **d)** $(-5.498, 0.707), (-2.356, -0.707), (0.785, 0.707), (3.927, -0.707)$

7. a) A line with a very small positive slope through $(0, 0)$. Explanations may vary.

8. Answers may vary. For image 1:
- **a)** $1, 1 + \frac{\pi}{2}, 1 + \pi, 1 + \frac{3\pi}{2}$
- **b)** $(0.54, 0.841), (-0.841, 0.54), (-0.54, -0.841), (0.841, -0.54)$

9. Explanations may vary.

10. Answers may vary. $-5.347, -4.078, 2.206, 7.219$

11. Answers may vary. $-5.347, -0.936, 5.347, 7.219$

12. Explanations may vary. For exercise 11, 6.283 was added and subtracted from 0.936 to get 7.219 and -5.347, respectively. Since the graph is symmetrical about the y-axis, the opposite angles -0.936 and 5.347 were also determined.

13. a) 0.877   **b)** $-5.214, -4.211, 2.073$

14. a) 0.481   **b)** $-5.214, -1.069, 5.214$
- **c)** Explanations may vary.

15. a) $\sin 0.936 > \cos 0.936$; explanations may vary.
- **b)** True for all the points with the x-coordinate in the intervals $\frac{\pi}{4} + 2n\pi < x < \frac{5\pi}{4} + 2n\pi$; explanations may vary.

16. b) Answers and explanations may vary.

17. a) Explanations may vary. The square was rotated 3 radians, this is very close to π radians, which is required for the vertices to lie on the coordinate axes.

b) No, the numbers on the axes are multiples of π, which is an irrational number.

18. Proofs may vary.

Exploring with a Graphing Calculator: Graphing $y = \sin x$ and $y = \cos x$, page 236

1. b) $-6.283, -3.142, 0, 3.142, 6.283$
c) $(-4.712, 1), (1.571, 1), (7.854, 1)$
d) $(-7.854, -1), (-1.571, -1), (4.712, -1)$
e) 6.283　　　　　　**f)** No; explanations may vary.

2. b) $-6.283, -3.142, 0, 3.142, 6.283$
c) $(-4.712, 1), (1.571, 1), (7.854, 1)$
d) $(-7.854, -1), (-1.571, -1), (4.712, -1)$
e) 6.283　　　　　　**f)** Yes; explanations may vary.

3. Explanations may vary.

4. They are the same functions. Explanations may vary.

Investigate, page 237

1. b) The graph of $y = 2\sin x + 1$ is the image of $y = \sin x$ after a vertical expansion by a factor of 2 and a vertical translation 1 unit up.
c) Yes; explanations may vary.

2. Explanations may vary.
a) a is the amplitude and d is the vertical displacement.
b) The amplitude increases when $|a|$ increases and decreases when $|a|$ decreases. When a changes sign, there is also a reflection in the x-axis. The graph is translated up, when d increases and down when d decreases.

3. The graph of $y = 2\cos x + 1$ is the image of $y = \cos x$ after a vertical expansion by a factor of 2 and a vertical translation 1 unit up. In $y = a\cos x + d$, a is the amplitude and d is the vertical displacement. Explanations may vary.

5. a) The graph of $y = \sin\left(x - \frac{\pi}{3}\right)$ is the image of $y = \sin x$ after a horizontal translation $\frac{\pi}{3}$ units right.
b) Yes; explanations may vary.

6. a) c provides information about the horizontal translation.
b) Explanations may vary. The graph is translated to the right when c increases and to the left when c decreases.

7. The graph of $y = \cos\left(x - \frac{\pi}{3}\right)$ is the image of $y = \cos x$ after a horizontal translation $\frac{\pi}{3}$ units right. In $y = \cos(x - c)$, c provides information about the horizontal translation.
The graph is translated to the right when c increases and to the left when c decreases.

4.2 Exercises, page 245

1. a) $a = 1, c = -2, d = 0$　　**b)** $a = 3, c = 0, d = -2$
c) $a = \frac{1}{2}, c = 0, d = -3$　　**d)** $a = 2, c = -\frac{\pi}{3}, d = -2$

2. Explanations may vary.

3. Explanations may vary.
a) The graph is reflected in the x-axis and translated 2 units up.
b) The graph is translated $\frac{3\pi}{4}$ units right and 3 units up.
c) The graph is vertically expanded by a factor of 2, translated 3 units left and 1 unit up.
d) The graph is vertically compressed by a factor of $\frac{1}{2}$, translated 2 units left and 4 units down.

4. $(0.6, -2.175), (0.6, -1.175), (0.6, -0.175), (0.6, 0.825),$ $(0.6, 1.825), (0.6, 2.825), (0.6, 3.825)$

5. $(0.6, -2.476), (0.6, -1.651), (0.6, -0.825), (0.6, 0),$ $(0.6, 0.825), (0.6, 1.651), (0.6, 2.476)$

6. a) 0.565
b) $(0.6, -2.435), (0.6, -1.435), (0.6, -0.435), (0.6, 0.565),$ $(0.6, 1.565), (0.6, 2.565), (0.6, 3.565)$

7. a) 0.565
b) $(0.6, -1.694), (0.6, -1.129), (0.6, -0.565), (0.6, 0),$ $(0.6, 0.565), (0.6, 1.129), (0.6, 1.694)$

8. $(-1.2, 0.697), (-0.2, 0.697), (0, 0.697), (1.8, 0.697),$ $(2.8, 0.697)$

9. a) 0.717
b) $(-1.2, 0.717), (-0.2, 0.717), (0.8, 0.717), (1.8, 0.717),$ $(2.8, 0.717)$

11. Predictions may vary.　　　　**12.** Predictions may vary.

13. For parts a and b, the cosine graphs are the image of the sine graphs after a horizontal translation $\frac{\pi}{2}$ units left. For part c the results would be the same.

14. Answers may vary. $\frac{\pi}{3}, -\frac{5\pi}{3}; y = \sin\left(x - \frac{\pi}{3}\right),$ $y = \sin\left(x + \frac{5\pi}{3}\right)$

15. Answers may vary. $\frac{5\pi}{6}, -\frac{7\pi}{6}; y = \cos\left(x - \frac{5\pi}{6}\right),$ $y = \cos\left(x + \frac{7\pi}{6}\right)$

16. Explanations may vary.

18. a) $y = \cos x + 0.5$　　　**b)** $y = 3\sin x - 1$
c) $y = -\cos x - 1$　　　**d)** $y = \frac{2}{3}\cos x + 3$

19. a) Domain: all real numbers, range: $-5 \leq y \leq 5$, phase shift: 0, amplitude: 5, maximum: 5, minimum: -5
b) Domain: all real numbers, range: $-3 \leq y \leq 3$, phase shift: 0, amplitude: 3, maximum: 3, minimum: -3
c) Domain: all real numbers, range: $3.75 \leq y \leq 4.25$, phase shift: 0, amplitude: 0.25, maximum: 4.25, minimum: 3.75
d) Domain: all real numbers, range: $-5 \leq y \leq -1$, phase shift: 0, amplitude: 2, maximum: -1, minimum: -5
e) Domain: all real numbers, range: $-6 \leq y \leq 2$, phase shift: 0, amplitude: 4, maximum: 2, minimum: -6
f) Domain: all real numbers, range: $1.5 \leq y \leq 4.5$, phase shift: 0, amplitude: 1.5, maximum: 4.5, minimum: 1.5
g) Domain: all real numbers, range: $-1.5 \leq y \leq -0.5$, phase shift: 0, amplitude: 0.5, maximum: -0.5, minimum: -1.5
h) Domain: all real numbers, range: $0 \leq y \leq 4$, phase shift: 0, amplitude: 2, maximum: 4, minimum: 0
i) Domain: all real numbers, range: $-2.5 \leq y \leq 4.5$, phase shift: 0, amplitude: 3.5, maximum: 4.5, minimum: -2.5
j) Domain: all real numbers, range: $-1 \leq y \leq 1$, phase shift:

$\frac{\pi}{4}$, amplitude: 1, maximum: 1, minimum: −1

  k) Domain: all real numbers, range: −1 ≤ y ≤ 1, phase shift: $\frac{4\pi}{3}$, amplitude: 1, maximum: 1, minimum: −1

  l) Domain: all real numbers, range: −2 ≤ y ≤ 2, phase shift: −$\frac{5\pi}{6}$, amplitude: 2, maximum: 2, minimum: −2

  m) Domain: all real numbers, range: 0 ≤ y ≤ 6, phase shift: $\frac{\pi}{6}$, amplitude: 3, maximum: 6, minimum: 0

  n) Domain: all real numbers, range: −4 ≤ y ≤ 0, phase shift: −$\frac{5\pi}{3}$, amplitude: 2, maximum: 0, minimum: −4

  o) Domain: all real numbers, range: −3 ≤ y ≤ 7, phase shift: $\frac{7\pi}{6}$, amplitude: 5, maximum: 7, minimum: −3

20. Vertical displacement = $\frac{M+m}{2}$

21. Explanations may vary.

22. Explanations may vary.

23. a) $\cos\left(x - \frac{\pi}{2}\right) = \sin x$ **b)** $c = \frac{\pi}{2} + 2n\pi$

24. a) $\sin\left(x + \frac{\pi}{2}\right) = \cos x$ **b)** $c = -\frac{\pi}{2} + 2n\pi$

25. a) 4   **b)** 0.2   **c)** 9

     d) 3   **e)** 2   **f)** 10

26. When the cosine function is used, a phase shift is not needed. That is, $y = 2\sin\left(x + \frac{\pi}{2}\right)$ may be written as $y = 2\cos x$.

27. $y = \sin x - 1$, $y = \sin x + 1$; 2 functions; explanations may vary.

28. a) $a + d$, when $x = \frac{\pi}{2} + 2n\pi + c$

   b) $d - a$, when $x = 2n\pi + c$

29. a) $y = \sin\left(x + \frac{\pi}{2}\right) + 2$

   b) No, the sine function is periodic.

Investigate, page 249

1. b) Explanations may vary. The graph of $y = \sin 2x$ is the image of $y = \sin x$ after a horizontal compression by a factor of $\frac{1}{2}$.

   c) The period of $y = \sin 2x$ is π, $\frac{1}{2}$ the period of $y = \sin x$.

2. b) Explanations may vary. The graph of $y = \sin \frac{1}{2}x$ is the image of $y = \sin x$ after a horizontal expansion by a factor of 2.

   c) The period of $y = \sin \frac{1}{2}x$ is 4π, 4 times the period of $y = \sin x$.

3. Yes.

4. Explanations may vary.

   a) b provides information about the period of the function.

   b) The period decreases when $|b|$ increases, and increases when $|b|$ decreases. When b changes sign, there is also a reflection in the y-axis.

5. Explanations may vary. The graph of $y = \cos 2x$ is the image of $y = \cos x$ after a horizontal compression by a factor of $\frac{1}{2}$. The period of $y = \cos 2x$ is π; that is, it's $\frac{1}{2}$ the period of $y = \cos x$. The graph of $y = \cos \frac{1}{2}x$ is the image of $y = \cos x$ after a horizontal expansion by a factor of 2.

The period of $y = \cos \frac{1}{2}x$ is 4π; that is, it's 2 times the period of $y = \cos x$.

4.3 Exercises, page 254

1. a) π, $y = \sin 2x$ **b)** $\frac{2\pi}{3}$, $y = \sin 3x$

   c) 10π, $y = \sin \frac{1}{5}x$ **d)** 4π, $y = \sin \frac{1}{2}x$

2. a) $a = 1$, $b = 2$, $c = 0$, $d = 0$

   b) $a = 1$, $b = \frac{1}{5}$, $c = 0$, $d = 0$

   c) $a = 4$, $b = 3$, $c = 0$, $d = 0$

   d) $a = 3$, $b = \frac{1}{2}$, $c = \frac{\pi}{2}$, $d = 0$

3. Explanations may vary.

   a) There is a horizontal compression by a factor of $\frac{1}{2}$.

   b) There is a horizontal expansion by a factor of 2.

   c) There is a vertical expansion by a factor of 2, a horizontal compression by a factor of $\frac{1}{2}$, and a translation of 2 units right.

   d) There is a vertical expansion by a factor of 3, a horizontal expansion by a factor of 2, and translations of 4 units left and 3 units up.

4. Answers may vary. (0.6, 0.362), (2.4, 0.362)

5. a) 0.717

   b) Answers may vary. (0.4, 0.717), (1.6, 0.717)

6. b) Descriptions may vary. The period decreases when $|b|$ increases, and increases when $|b|$ decreases. When b changes sign, there is also a reflection in the y-axis.

7. a) i) 2   **ii)** 2π   **iii)** $\frac{\pi}{4}$

     iv) 5, $\frac{\pi}{4} + 2n\pi$ **v)** 1, $\frac{5\pi}{4} + 2n\pi$

     vi) Domain: all real numbers, range: 1 ≤ y ≤ 5

     vii) 3 units up

   b) i) 3   **ii)** π   **iii)** $\frac{\pi}{2}$

     iv) 6, $\frac{3\pi}{4} + n\pi$ **v)** 0, $\frac{\pi}{4} + n\pi$

     vi) Domain: all real numbers, range: 0 ≤ y ≤ 6

     vii) 3 units up

   c) i) 5   **ii)** 2π   **iii)** $\frac{\pi}{6}$

     iv) 20, $\frac{\pi}{6} + 2n\pi$ **v)** 10, $\frac{7\pi}{6} + 2n\pi$

     vi) Domain: all real numbers, range: 10 ≤ y ≤ 20

     vii) 15 units up

8. a) $y = 2\cos\left(x - \frac{\pi}{4}\right) + 3$ **b)** $y = 3\sin 2\left(x - \frac{\pi}{2}\right) + 3$

   c) $y = 5\cos\left(x - \frac{\pi}{6}\right) + 15$

9. a) 2, π   **b)** 3, 4π   **c)** 4, π

   d) 4, 4π   **e)** 5, π   **f)** 3, $\frac{2\pi}{3}$

10. Explanations may vary.

11. a) 5, $\frac{2\pi}{3}$, π   **b)** 2, $\frac{2\pi}{3}$, $\frac{\pi}{2}$

   c) 2.5, $\frac{\pi}{3}$, $\frac{2\pi}{3}$ **d)** 0.5, $\frac{2\pi}{5}$, −$\frac{5\pi}{4}$

12. a) $y = \cos 2x$ **b)** $y = \cos \frac{1}{10}x$

   c) $y = \cos \frac{1}{4}x$ **d)** $y = \cos 3x$

13. Explanations may vary.

   a) The amplitude increases if $|a|$ increases, and decreases if $|a|$ decreases. When a changes sign, there is also a reflection in the x-axis.

b) The period decreases if $|b|$ increases, and increases if $|b|$ decreases. When b changes sign, there is also a reflection in the y-axis.
c) The graph is translated to the right when c increases and the left when c decreases.
d) The graph is translated up when d increases and down when d decreases.

14. a) $\frac{\pi}{3}, \pi$ **b)** 4, 2

15. a) $\frac{\pi}{4}, \pi$ **b)** 3, 3

16. a) $2, \pi, -\frac{\pi}{6}$ **b)** $5, \pi, \frac{\pi}{4}$ **c)** $3, \pi, \frac{\pi}{4}$ **d)** $5, \pi, -\frac{\pi}{6}$

17. a) $\frac{\pi}{2}, \pi$ **b)** $\frac{\pi}{3}, \frac{2\pi}{3}$ **c)** $\frac{\pi}{3}, \frac{2\pi}{3}$ **d)** $-\frac{\pi}{4}, \frac{\pi}{2}$

18. Explanations may vary.

20. a) $y = \sin(1 + 0.1n)x, n = 0, 1, 2, ..., 9$
b) $y = \cos(1 + 0.1n)x, n = 0, 1, 2, ..., 9$

21. Answers may vary.

22. Explanations may vary.

23. a) Maximum value: $a + d$ when $x = \dfrac{\frac{\pi}{2} + 2n\pi}{b} + c$

b) Minimum value: $d - a$ when $x = \dfrac{\frac{3\pi}{2} + 2n\pi}{b} + c$

24. a) Maximum value: $a + d$ when $x = \dfrac{\frac{\pi}{2} + 2n\pi}{b} + c$

b) Minimum value: $d - a$ when $x = \dfrac{\frac{3\pi}{2} + 2n\pi}{b} + c$

25. a and c; explanations may vary.

26. a and b; explanations may vary.

Linking Ideas: Mathematics and Technology, Dynamic Graphs, page 259

1. b) $0, \frac{\pi}{4}, \frac{\pi}{2}, \frac{3\pi}{4}, \pi, \frac{5\pi}{4}, \frac{3\pi}{2}, \frac{7\pi}{4}$; 0, 0.785, 1.571, 2.356, 3.142, 3.927, 4.712, 5.498
c) $\frac{\pi}{2} + 2n\pi$ **d)** $\frac{3\pi}{2} + 2n\pi$

2. The first graph on the left on page 259.
c) $y = \sin 5x, y = \sin 4x, y = \sin 3x, y = \sin 2x, y = \sin x,$ $y = \sin\frac{1}{2}x, y = \sin\frac{1}{4}x, y = \sin\frac{1}{10}x$

Investigate, page 260

1. a) 1   **b)** 2π **c)** $y = \sin 2\pi x, y = \cos 2\pi x$
2. a) 2   **b)** 3   **c)** 4   **d)** 5
3. a) $y = \cos\frac{2\pi}{2}x$ **b)** $y = \cos\frac{2\pi}{3}x$
c) $y = \cos\frac{2\pi}{4}x$ **d)** $y = \cos\frac{2\pi}{5}x$
4. p; explanations may vary. By observation, when the numerator is 2π, the denominator is the period.

4.4 Exercises, page 265

1. a) 3, 5, 1, 4   **b)** 2, 4, 5, −6
2. a) $10, y = \sin\frac{2\pi}{10}x$ **b)** $2, y = \sin\frac{2\pi}{2}x$

c) $0.025, y = \sin\frac{2\pi}{0.025}x$ **d)** $0.25, y = \sin\frac{2\pi}{0.25}x$

3. a) i) 6   **ii)** 4   **iii)** 0
    iv) 14, when $x = 4n$ **v)** 2, when $x = 2 + 4n$
    vi) Domain: all real numbers, range: $2 \leq y \leq 14$
    vii) 8
b) i) 10   **ii)** 40   **iii)** −5
    iv) 30, $x = -5 + 40n$ **v)** 10, when $x = 15 + 40n$
    vi) Domain: all real numbers, range: $10 \leq y \leq 30$
    vii) 20

4. a) $y = 6\cos\frac{2\pi}{4}t + 8$ **b)** $y = 10\cos 2\pi\frac{(t + 5)}{40} + 20$

5. a) $y = 3\cos\frac{2\pi}{7}x + 2$ **b)** $y = -2\cos\frac{2\pi}{5}x + 3$
c) $y = 0.2\cos\frac{2\pi}{0.1}x + 0.1$ **d)** $y = 0.005\cos\frac{2\pi}{0.001}x$

6. The period is the reciprocal of k, or $\frac{1}{k}$.

7. a) $y = 5\cos 2\pi(x - 9) - 4$ **b)** $y = 12\cos 2\pi\frac{(x + 3)}{0.5} + 1.5$
c) $y = 2.4\cos 2\pi\frac{(x - 19)}{27} + 15.1$

8. Answers may vary. 0.4, 1.1, 2.1, 2.4

9. Answers may vary. 0.6, 4.6, 5.4, 8.6

10. Explanations may vary. For exercise 8, the period of the function is 1. Therefore, 1 was added and subtracted from 1.4 to get 2.4 and 0.4, respectively. Since the graph is symmetrical about the line $x = 1.25$, 0.15 was subtracted from 1.4 to get 1.1. Finally, 1 was added to 1.1 to get 2.1.

11. Explanations may vary. The two functions have different periods.

12. a) −1   **b)** 1.75, 2.75, 3.75
13. a) 0.924   **b)** 1.25, 4.75, 5.25

14. a) Domain: all real numbers, range: $2 \leq y \leq 6$
b) Domain: all real numbers, range: $-6 \leq y \leq 0$
c) Domain: all real numbers, range: $-6 \leq y \leq -1.2$
d) Domain: all real numbers, range: $6.5 \leq y \leq 13.5$

15. a) Maximum: −1 at $t = 1 + 3n$, minimum: −5 at $t = 2.5 + 3n$
b) Maximum: 0 at $t = -0.75 + 5n$, minimum: −8 at $t = 0.5 + 5n$
c) Maximum: 8 at $t = 1.75 + 3n$, minimum: 4 at $t = 0.25 + 3n$
d) Maximum: 7 at $t = 3 + 6n$, minimum: −3 at $t = 6n$

16. a) $y = 6\sin 2\pi\frac{(x - 9)}{5} + 17$
b) $y = 4.3\sin 2\pi\frac{(x - 4.7)}{3.9} + 12.9$

17. Answers may vary. $y = 2\sin 2\pi\frac{(x - 1)}{2} + 1$, amplitude: 2, vertical displacement: 1, phase shift: 1, period: 2

18. a) $y = \pm\cos\pi x, y = \pm\cos\frac{\pi}{2}x$
b) $y = \pm\sin\frac{\pi x}{2} + 1, y = \pm\frac{2}{3}\sin\frac{\pi x}{2} + 1,$
$y = \pm\frac{1}{3}\sin\frac{\pi x}{2} + 1, y = \pm\sin\frac{\pi x}{2} - 1,$
$y = \pm\frac{2}{3}\sin\frac{\pi x}{2} - 1, y = \pm\frac{1}{3}\sin\frac{\pi x}{2} - 1$

20. $V = 2000\sin\frac{2\pi}{10}t + 3000$

21. b) $y = 40\sin\frac{\pi}{5}t$

22. a) 40 cm, 0 cm, 0.05 s   **b)** 72 000

23. a) $y = 10\sin\frac{2\pi}{0.15}t + 12$ **b)** 0.05, 0.10

c) $y = 10\sin 2\pi\frac{(t - 0.05)}{0.15} + 12$, $y = 10\sin 2\pi\frac{(t - 0.10)}{0.15} + 12$

d) Answers may vary. Pairs of pistons are moving simultaneously.

e) About 260 cm/s

25. a) $y = 0.5\sin\frac{2\pi}{100}x$, $y = 0.5\sin\frac{2\pi}{50}x$, $y = 0.5\sin\frac{3\pi}{50}x$; $y = 0.5\sin\frac{2\pi}{25}x$

26. b and c; explanations may vary.

Mathematical Modelling: Predicting Hours of Daylight, page 270

1. $a = 4.56$, $c = 172$, $p = 365$, $d = 12.24$

2. $h = 4.56\cos 2\pi\frac{(n - 172)}{365} + 12.24$

4. a) 10.65 h   **b)** 16.42 h   **c)** 8.69 h

5. a) 12.18 h, 16.21 h, 7.68 h   **b)** Answers may vary.

6. a) On day 245, there are 13.65 hours of daylight.

b) 10.65 h, 16.42 h, 8.69 h

7. a) i) $h = 2.31\cos 2\pi\frac{(n - 172)}{365} + 18.71$

ii) $h = 2.25\cos 2\pi\frac{(n + 10)}{365} + 6.47$

8. The graphs remain the same.

4.5 Exercises, page 276

1. a) i) 12.4 h   **ii)** 1.8 m   **iii)** 4.00 m

b) 4.9 m, 7:06 A.M. and 7:30 P.M.

c) 4.0 m, 1.7 m   **d)** 3:02 A.M.

2. a) i) Amplitude   **ii)** Period

iii) Phase shift   **iv)** Vertical displacement

b) 1.8 m, 10:48 A.M.   **c)** 2.3 m

d) Answers may vary. 1:16 A.M.

3. a) 12.4 h

b) $h = 6\cos 2\pi\frac{(t - 8)}{12.4} + 14$ or $h = 6\sin 2\pi\frac{(t - 4.9)}{12.4} + 14$

c) 17.2 m   **d)** 12:32 P.M.

4. a) $h = 4.6\cos 2\pi\frac{(t - 4.5)}{12.4} + 5$ **b)** 7.1 m

c) The depth of the water is 0.7 m greater at 2:45 P.M.

5. a) $h = 1.4\cos 2\pi\frac{(t - 7.75)}{12.4} + 5$ **b)** 3.9 m

c) The depth of the water is 2.5 m less at 2:45 P.M.

6. b) $h = 0.4\cos 2\pi\frac{(t - 0.6)}{1.2} + 0.5$ or

$y = 0.4\sin 2\pi\frac{(t - 0.3)}{1.2} + 0.5$

c) i) 0.5 m   **ii)** 0.1 m

d) 0.43 s

7. b) $h = 25\cos 2\pi\frac{(t - 25)}{50} + 26$

c) i) 18 m   **ii)** 46 m   **iii)** 18 m   **iv)** 18 m

d) 23 s

8. a) $\frac{1}{60}$ s, 60 cycles per second

b) 160 V   **c)** $y = 160\sin 120\pi t$

9. a) 0.0038 s, 263 cycles per second

b) $V = 0.05\sin 526\pi t$

10. b) $h = 16.5\cos\frac{2\pi}{5}t + 29$

c) i) 45.5 cm   **ii)** 15.7 cm   **iii)** 15.7 cm

d) Estimates may vary. 0.67 s

11. a) $d = 2.5\cos 2\pi\frac{(t - 172)}{365} + 149.7$

b) i) 148.8 million kilometres

ii) 151.3 million kilometres

iii) 150.5 million kilometres

12. Yes; explanations may vary. Ensure all angles are in degrees.

13. Yes; explanations may vary. Using the sine function, a different phase shift is needed.

Problem Solving: Visualizing Graphs of Composite Functions, page 280

All predictions may vary.

1. Starting at the left: $y = |\sin x|$, $y = \sin|x|$

3. Starting at the left: $y = \sin x^2$, $y = (\sin x)^2$

6. Starting at the left: $y = \sin x^3$, $y = (\sin x)^3$

9. Starting at the left: $y = 2^{\sin x}$, $y = \sin 2^x$

11. a) Explanations may vary.   **b)** Answers may vary.

Exploring with a Graphing Calculator: The Sinusoid of Best Fit, page 283

1. b) Amplitude: 4.562, phase shift: 80.159, period: 369.6

2. Answers may vary. For exercise 6,
$y = 0.4\sin(5.236x - 1.571) + 0.5$

3. a) $y = \sin 1.571x$ **b)** Explanations may vary.

Investigate, page 284

3. Descriptions may vary.

5. a) About 1.6 and 4.7

b) Domain: $x \neq 1.6$, $x \neq 4.7$, range: all real numbers

c) π **d)** $n\pi$ **e)** $x = \frac{\pi}{2} + n\pi$

4.6 Exercises, page 287

1. a) The graph of $y = \tan x + 1$ is the image of $y = \tan x$ after a vertical translation 1 unit up.

c) Translate the graph of $y = \tan x$ vertically d units to obtain the image, $y = \tan x + d$.

2. a) Domain: $x \neq \frac{\pi}{2} + n\pi$, range: all real numbers, period: π, asymptotes: $x = \frac{\pi}{2} + n\pi$

b) Domain: $x \neq \frac{\pi}{2} + n\pi$, range: all real numbers, period: π, asymptotes: $x = \frac{\pi}{2} + n\pi$

3. a) The graph of $y = 3\tan x$ is the image of $y = \tan x$ after a vertical expansion by a factor of 3.

c) When $|a| > 1$, vertically expand the graph of $y = \tan x$ by a factor of a to obtain the image $y = |a|\tan x$. When $|a| < 1$, vertically compress the graph of $y = \tan x$ by a factor of a to obtain the graph of $y = |a|\tan x$. When $a < 0$, a reflection in the x-axis is also needed to obtain

$y = a \tan x$.

4. a) Domain: $x \neq \frac{\pi}{2} + n\pi$, range: all real numbers, period: π, asymptotes: $x = \frac{\pi}{2} + n\pi$

b) Domain: $x \neq \frac{\pi}{2} + n\pi$, range: all real numbers, period: π, asymptotes: $x = \frac{\pi}{2} + n\pi$

5. a) The graph of $y = \tan\left(x - \frac{\pi}{4}\right)$ is the image of $y = \tan x$ after a phase shift of $\frac{\pi}{4}$.

c) Translate the graph of $y = \tan x$ horizontally c units to obtain the image $= \tan(x - c)$.

6. a) Domain: $x \neq \frac{\pi}{4} + n\pi$, range: all real numbers, period: π, asymptotes: $x = \frac{\pi}{4} + n\pi$

b) Domain: $x \neq n\pi$, range: all real numbers, period: π, asymptotes: $x = n\pi$

7. a) The graph of $y = \tan 2x$ is the image of $y = \tan x$ after a horizontal compression by a factor of $\frac{1}{2}$.

c) When $|b| > 1$, horizontally compress the graph of $y = \tan x$ by a factor of $\frac{1}{|b|}$ to obtain the image $y = \tan|b|x$. When $|b| < 1$, horizontally expand the graph of $y = \tan x$ by a factor of $\frac{1}{|b|}$ to obtain the image $y = \tan|b|x$. When $b < 0$, a reflection in the y-axis is also needed to obtain $y = \tan bx$.

8. a) Domain: $x \neq \frac{\pi}{8} + \frac{n\pi}{4}$, range: all real numbers, period: $\frac{\pi}{4}$, asymptotes: $x = \frac{\pi}{8} + \frac{n\pi}{4}$

b) Domain: $x \neq \pi + 2n\pi$, range: all real numbers, period: 2π, asymptotes: $x = \pi + 2n\pi$

9. a) The graph of $y = \tan \frac{\pi}{2}x$ is the image of $y = \tan x$ after a horizontal compression by a factor of $\frac{2}{\pi}$.

c) Explanations may vary.

10. a) Domain: $x \neq 2 + 4n$, range: all real numbers, period: 4, asymptotes: $x = 2 + 4n$

b) Domain: $x \neq 0.25 + 0.5n$, range: all real numbers, period: 0.5, asymptotes: $x = 0.25 + 0.5n$

11. a) -0.747 **b)** 0.264 **c)** -14.101 **d)** 3.381
e) 0 **f)** -1 **g)** 1 **h)** Undefined

12. $-5.347, -2.206, 4.078, 7.219$

13. $-4.6, -2.6, -0.6, 3.4$

14. Explanations may vary. For exercise 12, the period of the function is π. Therefore, π, or 3.1416 (to ensure 3 decimal place accuracy) was added twice to 0.936, to get 4.078 and 7.219. To get -2.206 and -5.347, 3.1416 was subtracted twice from 0.936.

15. a) 1.823 **b)** $-2.073, 4.211, 7.352$

16. a) 3.078 **b)** $-1.2, 2.8, 4.8$

17. a) $d = 3 \tan \pi t$ for $0 \leq t < 0.5$

18. c) $\tan x > \sin x$ when x is close to 0. Explanations may vary.

19. a) $y = \pm \tan x$, $y = \pm \tan 0.5x$, $y = \pm \tan 2x$
b) $y = \pm \tan 0.25\pi x$, $y = \pm \tan 0.25\pi x + 1$, $y = \pm \tan 0.25\pi x - 1$

4.7 Exercises, page 291

2. Period: 2π, domain: $x \neq n\pi$, range: $y \leq -1$ or $y \geq 1$, asymptotes: $x = n\pi$

4. Period: 2π, domain: $x \neq \frac{\pi}{2} + n\pi$, range: $y \leq -1$ or $y \geq 1$, asymptotes: $x = \frac{\pi}{2} + n\pi$, y-intercept: 1

6. Period: π, domain: $x \neq n\pi$, range: all real numbers, x-intercepts: $\frac{\pi}{2} \pm n\pi$, asymptotes: $x = n\pi$

7. a) Domain: $x \neq \frac{n\pi}{2}$, range: $y \leq -1$ or $y \geq 1$, period: π, asymptotes: $x = \frac{n\pi}{2}$

b) Domain: $x \neq \frac{n\pi}{4}$, range: $y \leq -3$ or $y \geq 3$, period: $\frac{\pi}{2}$, asymptotes: $x = \frac{n\pi}{4}$

c) Domain: $x \neq \frac{\pi}{6} + \frac{n\pi}{4}$, range: $y \leq -3$ or $y \geq 3$, period: $\frac{\pi}{2}$, asymptotes: $x = \frac{\pi}{6} + \frac{n\pi}{4}$

d) Domain: $x \neq \frac{n\pi}{3} + \frac{\pi}{6}$, range: $y \leq -1$ or $y \geq 1$, period: $\frac{2\pi}{3}$, asymptotes: $x = \frac{n\pi}{3} + \frac{\pi}{6}$

e) Domain: $x \neq \pi + 2n\pi$, range: $y \leq -2$ or $y \geq 2$, period: 4π, asymptotes: $x = \pi + 2n\pi$

f) Domain: $x \neq \frac{n\pi}{2} - \frac{\pi}{12}$, range: $y \leq -4$ or $y \geq 4$, period: π, asymptotes: $x = \frac{n\pi}{2} - \frac{\pi}{12}$

g) Domain: $x \neq \frac{n\pi}{2}$, range: all real numbers, period: $\frac{\pi}{2}$, asymptotes: $x = \frac{n\pi}{2}$

h) Domain: $x \neq 2n\pi$, range: all real numbers, period: 2π, asymptotes: $x = 2n\pi$

i) Domain: $x \neq \frac{n\pi}{3}$, range: all real numbers, period: $\frac{\pi}{3}$, asymptotes: $x = \frac{n\pi}{3}$

8. a) Domain: $x \neq \frac{n}{2}$, range: $y \leq -1$ or $y \geq 1$, period: 1, asymptotes: $x = \frac{n}{2}$

b) Domain: $x \neq n$, range: $y \leq -2$ or $y \geq 2$, period: 2, asymptotes: $x = n$

c) Domain: $x \neq \frac{3n}{2}$, range: $y \leq -3$ or $y \geq 3$, period: 3, asymptotes: $x = \frac{3n}{2}$

d) Domain: $x \neq \frac{n}{4} + \frac{1}{8}$, range: $y \leq -1$ or $y \geq 1$, period: $\frac{1}{2}$, asymptotes: $x = \frac{n}{4} + \frac{1}{8}$

e) Domain: $x \neq \frac{n}{3} + \frac{1}{6}$, range: $y \leq -3$ or $y \geq 3$, period: $\frac{2}{3}$, asymptotes: $x = \frac{n}{3} + \frac{1}{6}$

f) Domain: $x \neq \frac{5n}{2} + \frac{5}{4}$, range: $y \leq -3$ or $y \geq 3$, period: 5, asymptotes: $x = \frac{5n}{2} + \frac{5}{4}$

g) Domain: $x \neq n$, range: all real numbers, period: 1, asymptotes: $x = n$

h) Domain: $x \neq \frac{n}{2}$, range: all real numbers, period: $\frac{1}{2}$, asymptotes: $x = \frac{n}{2}$

i) Domain: $x \neq 10n$, range: all real numbers, period: 10, asymptotes: $x = 10n$

9. Explanations may vary.

10. Answers may vary.
a) $y = \csc \frac{\pi}{3}x - 1$ **b)** $y = \sec \frac{\pi}{3}x - 1$

11. Explanations may vary.

12. a) $d = |100\cot\pi\frac{(t-6)}{12}|$

4 Review, page 293

1. Answers may vary.
   a) $y = 1.5\sin x$; vertical displacement: 0, amplitude: 1.5, maximum: 1.5, minimum: -1.5, domain: all real numbers, range: $-1.5 \le y \le 1.5$, y-intercept: 0
   b) $y = 0.5\cos x$, vertical displacement: 0, amplitude: 0.5, maximum: 0.5, minimum: -0.5, domain: all real numbers, range: $-0.5 \le y \le 0.5$, y-intercept: 0.5
   c) $y = \cos(x - \frac{\pi}{2}) + 0.5$, vertical displacement: 0.5, amplitude: 1, maximum: 1.5, minimum: -0.5, domain: all real numbers, range: $-0.5 \le y \le 1.5$, y-intercept: 0.5
   d) $y = \cos x - 1$, vertical displacement: -1, amplitude: 1, maximum: 0, minimum: -2, domain: all real numbers, range: $-2 \le y \le 0$, y-intercept: 0

2. **a)** $c = \pi, y = \cos(x - \pi)$ **b)** $c = \frac{\pi}{4}, y = \cos(x - \frac{\pi}{4})$

3. Explanations may vary. The graph of the sine function is the image of the cosine function after a horizontal translation of $\frac{\pi}{2}$ units right. That is, $y = \sin x$ may be written as $y = \cos(x - \frac{\pi}{2})$.

4. Explanations may vary. The graph of the cosine function is the image of the sine function after a horizontal translation of $\frac{\pi}{2}$ units left. That is, $y = \cos x$ may be written as $y = \sin(x + \frac{\pi}{2})$.

6. **b)** $y = 16.5\cos\frac{2\pi}{5}(x - 2.5) + 29$

4 Cumulative Review, page 294

1. **b)** 6 h, 5 h, 3 h   **c)** It is halved.

2. **a)** The graph of $y = 2x^2 + 5$ is the image of $y = 2x^2$ after a vertical translation of 5 units up.
   b) The graph of $y = 2(x - 5)^2$ is the image of $y = 2x^2$ after a horizontal translation of 5 units right.
   c) The graph of $y = 2(x + 5)^2$ is the image of $y = 2x^2$ after a horizontal translation of 5 units left.

3. **a)** 4000   **b)** 45 255

4. **a)** 10.6 h   **b)** 20.7 h

5. **a)** $\log_2\left(\frac{xy}{z}\right)$ **b)** $\log_4\left(\frac{x^2}{y}\right)$ **c)** $\log_7(x^3y^5)$
   d) $\log_5(\sqrt{x}y^3)$ **e)** $\log[(2x - 3)(y + 5)]$
   f) $\log_8\frac{(x+y)^3}{x-y}$

6. **a)** $\frac{189}{4}$ **b)** $\frac{9}{8}$

7. 4 m

9. **a)** Negative   **b)** Negative   **c)** Positive
   d) Negative   **e)** Negative   **f)** Positive
   g) Positive   **h)** Positive

10. **a)** $y = \sin(x - \frac{\pi}{6})$ **b)** $y = \sin(x + \frac{\pi}{4})$
   c) $y = \sin(x = \frac{\pi}{3})$ **d)** $y = \sin(x - \frac{\pi}{2})$

11. **a)** $y = \cos(x - \frac{2\pi}{3})$ **b)** $y = \cos(x - \frac{\pi}{4})$
   c) $y = \cos(x - \frac{\pi}{6})$ **d)** $y = -\cos x$

12. **a)** $d = \tan(66.5 - l)$

Chapter 5

5.1 Exercises, page 302

1. **a)** 0.9273, 2.2143   **b)** 3.6084, 5.8164   **c)** 0.7227, 5.5605
   d) 4.4597, 1.8235   **e)** 0.9828, 4.1244   **f)** 1.8925, 5.0341

3. **a)** 0.5824, 2.5592   **b)** 0.6094, -0.6094 **c)** -2.7828, 0.3588
   d) -2.0657, -1.0759 **e)** 2.2916, -2.2916 **f)** -1.3986, 1.7430

5. **a)** 0.6435, 2.4981   **b)** 1.2310, -1.2310 **c)** 1.1071, -2.0344
   d) No solution   **e)** 1.5708   **f)** -0.7854, 2.3562
   g) No solution   **h)** No solution

6. **a)** $0, \frac{\pi}{2}, \pi, \frac{3\pi}{2}$ **b)** $0, \frac{\pi}{3}, \frac{2\pi}{3}, \pi, \frac{4\pi}{3}, \frac{5\pi}{3}$
   c) $0, \frac{\pi}{4}, \frac{\pi}{2}, \frac{3\pi}{4}, \pi, \frac{5\pi}{4}, \frac{3\pi}{2}, \frac{7\pi}{4}$ **d)** 0
   e) 0   **f)** 0   **g)** $\frac{\pi}{6}, \frac{5\pi}{6}, \frac{7\pi}{6}, \frac{11\pi}{6}$
   h) 0.1545, 1.2017, 2.2489, 3.2961, 4.3433, 5.3905
   i) 0.2618, 1.3090, 1.8326, 2.8798, 3.4034, 4.4506, 4.9742, 6.0214
   j) 0.4240, 1.1468, 3.5656, 4.2884
   k) 0.2827, 0.7645, 2.3771, 2.8589, 4.4715, 4.9533
   l) 0.2120, 0.5734, 1.7828, 2.1442, 3.3536, 3.7150, 4.9244, 5.2858

7. **a)** 3.6652, 5.7596; 3.6652 + $2n\pi$, 5.7596 + $2n\pi$
   b) 2.0944, 4.1888; 2.0944 + $2n\pi$, 4.1888 + $2n\pi$
   c) 1.0472, 5.2360; 1.0472 + $2n\pi$, 5.2360 + $2n\pi$
   d) 0.7227, 5.5605; 0.7227 + $2n\pi$, 5.5605 + $2n\pi$
   e) 3.9897, 5.4351; 3.9897 + $2n\pi$, 5.4351 + $2n\pi$
   f) 0.8481, 2.2935; 0.8481 + $2n\pi$, 2.2935 + $2n\pi$
   g) 2.6180, 5.7596; 2.6180 + $2n\pi$, 5.7596 + $2n\pi$
   h) 2.3562, 5.4978; 2.3562 + $n\pi$
   i) $\frac{\pi}{2}, \frac{3\pi}{2}$; $\frac{\pi}{2} + n\pi$
   j) All values of x

8. **a)** 2.6416, 5.7832; 2.6416 + $n\pi$
   b) 1.5, 3.5, 5.5; 1.5 + $2n$ **c)** 0.5, 1.5, 2.5; 0.5 + n
   d) All values of x **e)** $\frac{\pi}{4}, \frac{5\pi}{4}; \frac{\pi}{4} + n\pi$
   f) $\frac{\pi}{8}, \frac{5\pi}{8}, \frac{9\pi}{8}, \frac{13\pi}{8}; \frac{\pi}{8} + \frac{n\pi}{2}$
   g) $\frac{\pi}{6}, \frac{5\pi}{6}, \frac{3\pi}{2}; \frac{\pi}{6} + 2n\pi, \frac{5\pi}{6} + 2n\pi, \frac{3\pi}{2} + 2n\pi$
   h) $\frac{\pi}{6}, \frac{\pi}{2}, \frac{5\pi}{6}, \frac{3\pi}{2}; \frac{\pi}{6} + 2n\pi, \frac{\pi}{2} + 2n\pi, \frac{5\pi}{6} + 2n\pi, \frac{3\pi}{2} + 2n\pi$

9. Explanations may vary.

10. **a)** 0.3076, 2.8340; 0.3076 + $2n\pi$, 2.8340 + $2n\pi$; 0.4271, 2.7145; 0.4271 + $2n\pi$, 2.7145 + $2n\pi$; 0.6662, 2.4754; 0.6662 + $2n\pi$, 2.4754 + $2n\pi$; $\frac{\pi}{2}, \frac{3\pi}{2}; \frac{\pi}{2} + n\pi$
   b) 0.4492, 2.6924, 4.0167, 5.4081; 0.4492 + $2n\pi$, 2.6924 + $2n\pi$, 4.0167 + $2n\pi$, 5.4081 + $2n\pi$; $\frac{\pi}{6}, \frac{5\pi}{6}, \frac{3\pi}{2}$; $\frac{\pi}{6} + 2n\pi, \frac{5\pi}{6} + 2n\pi, \frac{3\pi}{2} + 2n\pi$; 0.6662, 2.4754; 0.6662 + $2n\pi$, 2.4754 + $2n\pi$; $\frac{\pi}{2}; \frac{\pi}{2} + 2n\pi$

11. **b)** 0, 1.895, -1.895

12. **a)** 1.030   **b)** 0, $-4.275 - n\pi$, 4.275 + $n\pi$, $n \ge 0$

13. **a)** 4; explanations may vary.
   b) -4.6036, -1.8452, 0.6100, 1.0980, 4.8164; 5 roots

14. **a)** 1; explanations may vary.
   b) -6.2829, -3.1520; 2 roots

15. 3; explanations may vary.

16. b) $\left(-\frac{4\pi}{3}, -0.5\right), \left(\frac{4\pi}{3}, -0.5\right), (0, 1)$

17. Explanations may vary.
   a) One root; $x = 0$ **b)** One root; $x = 0.7391$
   c) Infinite number of roots; $x = 0$, $x = -4.493 - n\pi$,
     $x = 4.493 + n\pi$, $n \geq 0$

18. No

19. b) $x = n\pi$ **c)** Explanations may vary.

20. b) $x = 0$, $x = \frac{\pi}{2} + n\pi$ **c)** Explanations may vary.

Exploring with a Graphing Calculator: Investigating $y = x - 2\sin x$ and $y = x - 2\cos x$, page 305

Explanations and predictions may vary.

5. $\left(-\frac{3\pi}{4}, -\frac{3\pi}{4} + \sqrt{2}\right), \left(\frac{\pi}{4}, \frac{\pi}{4} - \sqrt{2}\right)$
In general, $x = \frac{\pi}{4} + n\pi$, $y = \frac{\pi}{4} + n\pi - 2\sin\left(\frac{\pi}{4} + n\pi\right)$

Mathematical Modelling: When Will They be in the Dark?, page 306

1. Cosine function: $a = 2.033$, $c = 355$, $p = 365$, $d = 18.983$
   Sine function: $a = 2.033$, $c = 263.5$, $p = 365$, $d = 18.983$

2. $s = 2.033\cos 2\pi\left(\frac{n-355}{365}\right) + 18.983$;
   $s = 2.033\sin 2\pi\left(\frac{n-263.5}{365}\right) + 18.983$

4. a) $s = 1.983\cos 2\pi\left(\frac{n-355}{365}\right) + 6.767$;
     $s = 1.983\sin 2\pi\left(\frac{n-263.5}{365}\right) + 6.767$

5. a) $y = 7.25$, $y = 16.50$

6. a) $(67, 7.25), (278, 7.25)$
   b) October 5 to March 8; 154 days   **c)** No dates

7. to 11. Answers may vary.

5.2 Exercises, page 313

1. Explanations may vary.
   a) 2   **b)** 4   **c)** 6
   d) 4   **e)** 8   **f)** 16

2. b) $\frac{\pi}{4}, \frac{3\pi}{4}, \frac{5\pi}{4}, \frac{7\pi}{4}$ **c)** $\frac{\pi}{4} + \frac{n\pi}{2}$

3. b) $0, \frac{\pi}{2}, \pi, \frac{3\pi}{2}$ **c)** $\frac{n\pi}{2}$

4. b) $0, \frac{\pi}{2}, \pi, \frac{3\pi}{2}$ **c)** $\frac{n\pi}{2}$

5. a) $\frac{2\pi}{3}, \frac{4\pi}{3}$ **b)** $\frac{3\pi}{2}$ **c)** $\frac{\pi}{6}, \frac{7\pi}{6}$
   d) $\frac{\pi}{6}, \frac{11\pi}{6}$ **e)** $\frac{\pi}{2}, \frac{3\pi}{2}$ **f)** $\frac{\pi}{3}, \frac{2\pi}{3}$

6. a) $\frac{\pi}{3}, \frac{5\pi}{3}; \frac{\pi}{6}, \frac{5\pi}{6}, \frac{7\pi}{6}, \frac{11\pi}{6}; \frac{11\pi}{9}, \frac{\pi}{9}, \frac{5\pi}{9}, \frac{7\pi}{9}, \frac{13\pi}{9}, \frac{17\pi}{9}; \frac{\pi}{12}, \frac{5\pi}{12},$
    $\frac{7\pi}{12}, \frac{11\pi}{12}, \frac{13\pi}{12}, \frac{17\pi}{12}, \frac{19\pi}{12}, \frac{23\pi}{12}$

   b) $\frac{\pi}{3}, \frac{2\pi}{3}; \frac{\pi}{6}, \frac{7\pi}{6}, \frac{4\pi}{3}; \frac{\pi}{9}, \frac{2\pi}{9}, \frac{7\pi}{9}, \frac{8\pi}{9}, \frac{13\pi}{9}, \frac{14\pi}{9}; \frac{\pi}{12}, \frac{\pi}{6},$
    $\frac{7\pi}{12}, \frac{2\pi}{3}, \frac{13\pi}{12}, \frac{7\pi}{6}, \frac{19\pi}{12}, \frac{5\pi}{3}$

   c) $\frac{\pi}{3}, \frac{4\pi}{3}; \frac{\pi}{6}, \frac{2\pi}{3}, \frac{7\pi}{6}, \frac{5\pi}{3}; \frac{\pi}{9}, \frac{4\pi}{9}, \frac{7\pi}{9}, \frac{10\pi}{9}, \frac{13\pi}{9}, \frac{16\pi}{9}; \frac{\pi}{12}, \frac{\pi}{3},$
    $\frac{7\pi}{12}, \frac{5\pi}{6}, \frac{13\pi}{12}, \frac{4\pi}{3}, \frac{19\pi}{12}, \frac{11\pi}{6}$

7. a) $\cos x(\cos x + 2)$ **b)** $(\sin x + 2)(\sin x + 3)$
   c) $(2\sin x - 3)(\sin x + 2)$ **d)** $(3\cos x + 1)(\cos x - 1)$
   e) $(2\cos x - 1)(\cos x - 3)$ **f)** $(2\sin x + 1)(3\sin x - 1)$

8. a) $\frac{\pi}{2}, \frac{3\pi}{2}$ **b)** No solution
   c) No solution   **d)** 0, 1.9106, 4.3726
   e) $\frac{\pi}{3}, \frac{5\pi}{3}$ **f)** $\frac{7\pi}{6}, \frac{11\pi}{6}, 0.3398, 2.8018$

9. a) $0, \frac{\pi}{2}, \pi$ **b)** $\frac{\pi}{2}, \pi, \frac{3\pi}{2}$ **c)** $\frac{\pi}{6}, \frac{5\pi}{6}, \frac{3\pi}{2}$
   d) $\frac{\pi}{2}, \frac{7\pi}{6}, \frac{11\pi}{6}$ **e)** π **f)** $\frac{\pi}{2}$
   g) $\pi, \frac{2\pi}{3}, \frac{4\pi}{3}$ **h)** $0, \frac{\pi}{3}, \frac{5\pi}{3}$ **i)** $\frac{3\pi}{2}$
   j) No solution   **k)** $\frac{\pi}{3}, \frac{5\pi}{3}$ **l)** $\frac{\pi}{6}, \frac{5\pi}{6}, \frac{7\pi}{6}, \frac{11\pi}{6}$

11. a) $\frac{\pi}{4}, \frac{5\pi}{4}$ **b)** $\frac{\pi}{8}, \frac{5\pi}{8}, \frac{9\pi}{8}, \frac{13\pi}{8}$
   c) $\frac{\pi}{16}, \frac{5\pi}{16}, \frac{9\pi}{16}, \frac{13\pi}{16}, \frac{17\pi}{16}, \frac{21\pi}{16}, \frac{25\pi}{16}, \frac{29\pi}{16}$
   d) $\frac{\pi}{3}, \frac{4\pi}{3}$
   e) $\frac{\pi}{6}, \frac{5\pi}{12}, \frac{2\pi}{3}, \frac{11\pi}{12}, \frac{7\pi}{4}, \frac{17\pi}{12}, \frac{5\pi}{3}, \frac{23\pi}{12}$
   f) $\frac{\pi}{12}, \frac{7\pi}{12}, \frac{13\pi}{12}, \frac{19\pi}{12}$

12. a) $0, \frac{\pi}{4}, \pi, \frac{5\pi}{4}$ **b)** $0, \frac{3\pi}{4}, \pi, \frac{7\pi}{4}$
   c) $\frac{\pi}{4}, \frac{3\pi}{4}, \frac{5\pi}{4}, \frac{7\pi}{4}$ **d)** $\frac{\pi}{4}, \frac{4\pi}{3}, \frac{5\pi}{4}, \frac{4\pi}{3}$

13. Answers may vary.
   a) $\cos x = 1$ **b)** $\sin\left(x + \frac{\pi}{3}\right) = 1$
   c) $\sin x = 1$ **d)** $\tan\left(\frac{x}{2}\right) = \frac{1}{\sqrt{3}}$
   e) $\sin\left(x + \frac{\pi}{4}\right) = 1$ **f)** $\sin\left(x - \frac{\pi}{4}\right) = 1$

14. Answers may vary.
   a) $\left(\sin\left(x + \frac{\pi}{3}\right) - 1\right)(\sin x - 1) = 0$
   b) $\left(\sin\left(x - \frac{\pi}{6}\right) - 1\right)\left(\sin\left(x + \frac{\pi}{4}\right) - 1\right) = 0$
   c) $\left(\sin\left(x + \frac{\pi}{4}\right) - 1\right)\left(\sin\left(x - \frac{\pi}{4}\right) - 1\right) = 0$

15. $\frac{\pi}{5}, \frac{3\pi}{5}, \frac{7\pi}{5}, \frac{9\pi}{5}$

16. a) $x = \frac{7\pi}{6} + 2n\pi$, $x = \frac{11\pi}{6} + 2n\pi$
   b) $x = 0.2450 + n\pi$, $x = \frac{\pi}{4} + n\pi$

5.3 Exercises, page 319

1. to 3. Descriptions may vary.

5. c) Explanations may vary.

6. c) $x \neq n\pi$ **7. c)** $x \neq \frac{\pi}{2} + n\pi$

8. c) All values of x

9. a) Not an identity   **b)** Identity
   c) Not an identity   **d)** Not an identity
   e) Identity   **f)** Identity

10. Explanations may vary.

11. Answers may vary.
   b) For $\sin x = (\cos x)(\tan x)$: $x \neq \frac{\pi}{2} + n\pi$
   c) Explanations may vary.

15. a) $\cos x = -\cos(\pi - x)$, $\cos\left(\frac{\pi}{2} - x\right) = -\cos\left(\frac{\pi}{2} + x\right)$

   b) $\cos(\pi - x) = \cos(\pi + x)$, $\cos\left(\frac{\pi}{2} + x\right) = \cos\left(\frac{3\pi}{2} - x\right)$

   c) $\cos x = \cos(2\pi - x)$, $\cos\left(\frac{\pi}{2} - x\right) = \cos\left(\frac{3\pi}{2} + x\right)$

16. b) π, 1   **c)** $\sin^2 x - \cos^2 x = -\cos 2x$

17. a) 0, π **b)** $\frac{\pi}{2}$, $\frac{3\pi}{2}$ **c)** 0, π **d)** $\frac{\pi}{2}$, $\frac{3\pi}{2}$

18. $\csc x = \frac{1}{(\cos x)(\tan x)}$, $\sec x = \frac{\tan x}{\sin x}$, $\cot x = \frac{\cos x}{\sin x}$

19. a) Yes   **b)** Yes

5.4 Exercises, page 326

All proofs may vary.

6. Answers may vary. For exercise 1a: $x \neq \frac{\pi}{2} + n\pi$;
   For exercise 2b: $x \neq \frac{\pi}{2} + n\pi$; For exercise 3c: $x \neq \frac{\pi}{2} + n\pi$;
   For exercise 4d: $x \neq \frac{n\pi}{2}$, $x \neq (4n-1)\frac{\pi}{4}$

9. a) $\frac{1 + \cos x}{1 + \sec x} = \cos x$ **b)** $\frac{1 + \tan x}{1 + \cot x} = \tan x$

10. b) $\frac{1 + \sin x}{1 - \sin x} + \frac{1 + \csc x}{1 - \csc x} = 0$; $\frac{1 + \tan x}{1 - \tan x} + \frac{1 + \cot x}{1 - \cot x} = 0$

11. b) $\frac{\sin x}{1 + \cos x} + \frac{\sin x}{1 - \cos x} = 2\csc x$;

    $\frac{\tan x}{1 + \cos x} + \frac{\tan x}{1 - \cos x} = 2\sec x\csc x$

13. a) Identity   **b)** Identity   **c)** Not an identity
     d) Identity   **e)** Not an identity   **f)** Identity

14. b) $\frac{\sin x + \tan x}{\csc x + \cot x} = \sin x \tan x$
     c) Answers may vary. $\frac{\sin x + \sec x}{\csc x + \cos x} = \sin x \sec x$
     d) 10; explanations may vary.

15. b) $\frac{1}{1 + \cos x} + \frac{1}{1 - \cos x} = 2\csc^2 x$

16. b) $\frac{\cot x}{\csc x + 1} = \frac{\csc x - 1}{\cot x}$
     c) Answers may vary. $\frac{\cos x}{1 - \sin x} = \frac{1 + \sin x}{\cos x}$

17. a) $y = \sin x^2$ **b)** $y = \sin^2 x$
     c) $y = \cos x^2$ **d)** $y = \cos^2 x$

18. a) π, $\frac{1}{2}$ **b)** $y = \sin^2 x$
     c) $\sin^2 x = 1 - \cos^2 x$; $\sin^2 x = \cos^2 x \tan^2 x$

19. $\frac{1}{1 + \tan x} + \frac{1}{1 - \tan x} = \frac{2}{1 - \tan^2 x}$;
    $\frac{1}{1 + \csc x} + \frac{1}{1 - \csc x} = -2\tan^2 x$;
    $\frac{1}{1 + \sec x} + \frac{1}{1 - \sec x} = -2\cot^2 x$;
    $\frac{1}{1 + \cot x} + \frac{1}{1 - \cot x} = \frac{2}{1 - \cot^2 x}$

20. Answers may vary.
   $(\sin x)(\cot x)(\sec x) = (\cos x)(\tan x)(\csc x)$

22. Answers may vary.
   a) $\sin x = \cos\left(x - \frac{\pi}{2.01}\right)$

   c) $\sin x = \cos\left(x - \frac{\pi}{2.000\,000\,001}\right)$

23. Explanations may vary.
   b) No

Investigate, page 329

All explanations may vary.
1. a) $3(x + y) = 3x + 3y$
2. Not equal
3. Not equal
4. When the addition sign is replaced with a subtraction sign, there are corresponding expressions. Sum may also be replaced with difference. $3(x - y) = 3x - 3y$ is the only equation true for all values of x and y.

5.5 Exercises, page 333

4. a) $-\frac{1}{2}$ **b)** $-\frac{1}{2}$ **c)** $-\frac{1}{2}$

5. a) $\frac{\sqrt{2}}{2}$ **b)** $\frac{\sqrt{2}}{2}$

7. a) Graph 4   **b)** Graph 1   **c)** Graph 3   **d)** Graph 2

8. Explanations may vary.

9. Graph 3: $y = \cos(x + 1)$, $y = \cos x \cos 1 - \sin x \sin 1$;
   Graph 2: $y = \sin(x + 1)$, $y = \sin x \cos 1 + \cos x \sin 1$

10. to 12. Proofs may vary.

13. a) $\cos(\pi - x) = -\cos x$ **b)** $\sin(\pi - x) = \sin x$
     c) $\sin(\pi + x) = -\sin x$ **d)** $\cos\left(x - \frac{\pi}{2}\right) = \sin x$

14. a), b) $\frac{\sqrt{6} - \sqrt{2}}{4}$

15. a), b) $\frac{\sqrt{6} + \sqrt{2}}{4}$

16. Counterexamples may vary. $x = \pi$, $y = \frac{\pi}{2}$ or $x = \frac{\pi}{2}$, $y = \frac{\pi}{4}$

17. a) Proofs may vary.
   b) $\sin\left(x + \frac{4\pi}{6}\right) = -\frac{1}{2}\sin x + \frac{\sqrt{3}}{2}\cos x$;
    $\sin\left(x + \frac{5\pi}{6}\right) = -\frac{\sqrt{3}}{2}\sin x + \frac{1}{2}\cos x$;
    $\sin\left(x + \frac{6\pi}{6}\right) = -1\sin x + 0\cos x$

18. a) $\sin\left(x + \frac{0\pi}{4}\right) = 1\sin x + 0\cos x$;
    $\sin\left(x + \frac{\pi}{4}\right) = \frac{\sqrt{2}}{2}\sin x + \frac{\sqrt{2}}{2}\cos x$;
    $\sin\left(x + \frac{2\pi}{4}\right) = 0\sin x + 1\cos x$;
    $\sin\left(x + \frac{3\pi}{4}\right) = -\frac{\sqrt{2}}{2}\sin x + \frac{\sqrt{2}}{2}\cos x$

   b) $\sin\left(x + \frac{0\pi}{2}\right) = 1\sin x + 0\cos x$;
    $\sin\left(x + \frac{\pi}{2}\right) = 0\sin x + 1\cos x$;
    $\sin\left(x + \frac{2\pi}{2}\right) = -1\sin x + 0\cos x$;
    $\sin\left(x + \frac{3\pi}{2}\right) = 0\sin x - 1\cos x$

   c) $\sin(x + 0\pi) = 1\sin x + 0\cos x$;
    $\sin(x + \pi) = -1\sin x + 0\cos x$;
    $\sin(x + 2\pi) = 1\sin x + 0\cos x$;
    $\sin(x + 3\pi) = -1\sin x + 0\cos x$

19. a) $(x\cos 1 - y\sin 1, y\cos 1 + x\sin 1)$
   b) Yes; explanations may vary.

20. Starting at P and moving counterclockwise:

a) $Q\left(\dfrac{-x-\sqrt{3}y}{2}, \dfrac{\sqrt{3}x-y}{2}\right), R\left(\dfrac{-x+\sqrt{3}y}{2}, \dfrac{-\sqrt{3}x-y}{2}\right)$

b) $Q(-y, x), R(-x, -y), S(y, -x)$

c) $Q\left(x\cos\dfrac{2\pi}{5} - y\sin\dfrac{2\pi}{5}, \ y\cos\dfrac{2\pi}{5} + x\sin\dfrac{2\pi}{5}\right),$
$R\left(x\cos\dfrac{4\pi}{5} - y\sin\dfrac{4\pi}{5}, \ y\cos\dfrac{4\pi}{5} + x\sin\dfrac{4\pi}{5}\right),$
$S\left(x\cos\dfrac{6\pi}{5} - y\sin\dfrac{6\pi}{5}, \ y\cos\dfrac{6\pi}{5} + x\sin\dfrac{6\pi}{5}\right),$
$T\left(x\cos\dfrac{8\pi}{5} - y\sin\dfrac{8\pi}{5}, \ y\cos\dfrac{8\pi}{5} + x\sin\dfrac{8\pi}{5}\right)$

d) $Q\left(\dfrac{x-\sqrt{3}y}{2}, \dfrac{\sqrt{3}x+y}{2}\right), R\left(\dfrac{-x-\sqrt{3}y}{2}, \dfrac{\sqrt{3}x-y}{2}\right),$
$S(-x, -y), T\left(\dfrac{-x+\sqrt{3}y}{2}, \dfrac{-\sqrt{3}x-y}{2}\right),$
$U\left(\dfrac{x+\sqrt{3}y}{2}, \dfrac{-\sqrt{3}x+y}{2}\right)$

21. Explanations may vary.

22. b) $\sin\left(\dfrac{\pi}{6} + x\right) + \sin\left(\dfrac{\pi}{6} - x\right) = \cos x$

c) $\sin\left(\dfrac{\pi}{3} + x\right) + \sin\left(\dfrac{\pi}{3} - x\right) = \sqrt{3}\cos x$

d) $\sin(x + y) + \sin(x - y) = 2\sin x\cos y$

23. Answers may vary.

24. a) True for $x = 0$ or $y = 0$ or $x = -y$
b) True for $x = 2n\pi \pm \dfrac{\pi}{3}$ and $y = -x$
c) True for $x = 0$ or $y = 0$ or $x = -y$

25. Explanations may vary.

26. a) $k = 2n\pi$
b) No values of k; explanations may vary.

27. Proofs may vary.

28. $\cos(x + y) = \cos x\cos y - \sin x\sin y$;
$\sin(x + y) = \sin x\cos y + \cos x\sin y$

Exploring with a Graphing Calculator: Products Involving sin *x* and cos *x*, page 337

1. b) $\pi, \dfrac{1}{2}$ **c)** $f(x) = \dfrac{1}{2}\sin 2x$

2. a) $\sin 2x = 2\sin x\cos x$

3. b) $\pi, \dfrac{1}{2}, \dfrac{1}{2}$ **c)** $g(x) = \dfrac{1}{2}\cos 2x + \dfrac{1}{2}$

4. a) $\cos 2x = 2\cos^2 x - 1$

5. b) $\pi, \dfrac{1}{2}, \dfrac{1}{2}$ **c)** $h(x) = \dfrac{1}{2} - \dfrac{1}{2}\cos 2x$

6. a) $\cos 2x = 1 - 2\sin^2 x$

Investigate, page 338

1. $\sin 2x = 2\sin x\cos x$

3. a) Yes   **b)** No   **c)** Yes

4. $\cos 2x = \cos^2 x - \sin^2 x$

6. a) Yes   **b)** No   **c)** Yes

7. $\cos 2x = 1 - 2\sin^2 x$; $\cos 2x = 2\cos^2 x - 1$

5.6 Exercises, page 342

3. b) Explanations may vary. The graphs are different.

4. b) Explanations may vary. The graphs are different.

5. a) $\sin 2, \sin 4, \sin 6, \sin 8$ **b)** $\cos 2, \cos 4, \cos 6, \cos 8$
c) $\dfrac{1}{2}\sin 6, \dfrac{1}{2}\sin 12, \dfrac{1}{2}\sin 24, \dfrac{1}{2}\sin 48$

6. a) $\sin\dfrac{\pi}{3}$ **b)** $\cos\dfrac{\pi}{5}$ **c)** $\cos 1$
d) $\cos 6$ **e)** $\cos 2$ **f)** $-\cos 8$

7. a) Graph 4   **b)** Graph 1   **c)** Graph 3   **d)** Graph 2

8. Explanations may vary.

9. Graph 3: $y = \cos 2x, y = 2\cos^2 x - 1$; graph 2: $y = \sin 2x,$
$y = 2\sin x\cos x$

11. a) $\cos^2 x = \dfrac{1 + \cos 2x}{2}$

12. Explanations may vary.

13. Proofs may vary.

14. a) $\dfrac{\sqrt{2 - \sqrt{3}}}{2}$ **b)** $\dfrac{\sqrt{2 - \sqrt{2}}}{2}$

15. a) $\dfrac{\sqrt{2 + \sqrt{3}}}{2}$ **b)** $\dfrac{\sqrt{2 + \sqrt{2}}}{2}$

16. a) $\dfrac{\sqrt{3}}{2}, \dfrac{1}{2}$ **b)** $\dfrac{\sqrt{3}}{2}, -\dfrac{1}{2}$

17. a) $1, 0$ **b)** $0, -1$

18. Explanations may vary.

19. b) $\dfrac{2\tan x}{1 + \tan^2 x} = \sin 2x$

20. b) $\dfrac{2\cot x}{1 + \cot^2 x} = \sin 2x$

21. a) $\dfrac{\sin x}{1 + \sin^2 x} = \dfrac{\csc x}{1 + \csc^2 x}, \dfrac{\cos x}{1 + \cos^2 x} = \dfrac{\sec x}{1 + \sec^2 x}$

22. a) $(1 - 2y^2, 2xy)$
b) Yes; explanations may vary.

23. a) $\dfrac{1 - \cos 2x}{\sin 2x} = \tan x$ **b)** $\dfrac{\sin 2x}{1 + \cos 2x} = \tan x$
c) $\dfrac{\sin 2x}{1 - \cos 2x} = \cot x$

24. b) Explanations may vary. Using $\cos 2x = \cos^2 x - \sin^2 x$ minimizes the algebra needed.

25. $-\dfrac{3}{4}$

26. a) $x = \dfrac{\pi}{4} + n\pi$ **b)** $x = \dfrac{\pi}{12} + \dfrac{n\pi}{2}, x = \dfrac{5\pi}{12} + \dfrac{n\pi}{2}$

27. a) $x = n\pi$ **b)** $x = \dfrac{\pi}{2} + n\pi$

28. a), b) $0, \dfrac{\pi}{3}, \pi, \dfrac{5\pi}{3}$

29. a), b) $0, \dfrac{2\pi}{3}, \dfrac{4\pi}{3}$

30. a) The expansion of a binomial squared
b) Explanations may vary.

31. a) $\sin 3x = 3\sin x - 4\sin^3 x$; $\cos 3x = 4\cos^3 x - 3\cos x$
b) Explanations may vary.

Problem Solving: Evaluating Trigonometric Functions, page 346

1. a) $y = x$ **b)** $-0.3925 < x < 0.3925$
c) Explanations may vary.

2. a) a is negative.   **b)** $y = -0.5x^2 + 1$
   c) $-0.7028 < x < 0.7028$

3. Descriptions may vary.

4. $\sin x = x - \frac{1}{6}x^3 + \frac{1}{120}x^5 - \frac{1}{5040}x^7 + \frac{1}{362\,880}x^9$
$- \frac{1}{39\,916\,800}x^{11}$;
$\cos x = 1 - \frac{1}{2}x^2 + \frac{1}{24}x^4 - \frac{1}{720}x^6 + \frac{1}{40\,320}x^8 - \frac{1}{3\,628\,800}x^{10}$

5. a) 0.3624   **b)** 0.9320   **c)** −0.9427   **d)** 0.3349

6. a) Answers may vary.

7. a) Answers may vary.

8. a) Explanations may vary.

9. 13   **10.** 13   **11.** $k = \sqrt{-1}$

5 Review, page 349

1. 3.57, 5.86; $3.57 + 2n\pi$, $5.86 + 2n\pi$

2. a) $\frac{\pi}{2}, \frac{3\pi}{2}$ **b)** $\frac{3\pi}{2}$ **c)** $\frac{2\pi}{3}, \frac{4\pi}{3}$

3. Answers may vary.
  a) $\cos x = -1$ **b)** $\sin x = 1$
  c) $\sin\left(x - \frac{2\pi}{3}\right) = 1$ **d)** $\cos\left(x + \frac{\pi}{3}\right) = -1$
  e) $\cos x = 1$

4. Proofs may vary.

5. Explanations may vary.

6. a) $\sin\frac{7\pi}{12}$ **b)** $\cos 0.9$
  c) $\cos\left(-\frac{\pi}{3}\right)$, or $\cos\frac{\pi}{3}$ **d)** $\sin 0.8$

7. a) $\frac{20 - 3\sqrt{119}}{60}$ **b)** $\frac{15 - 4\sqrt{119}}{60}$
  c) $\frac{24}{25}$ **d)** $-\frac{47}{72}$

8. Answers may vary.
  a) $\frac{\sqrt{2 - \sqrt{3}}}{2}$ **b)** $-\frac{\sqrt{2 + \sqrt{3}}}{2}$
  c) $-\frac{\sqrt{2 - \sqrt{3}}}{2}$ **d)** $-\frac{\sqrt{2 - \sqrt{3}}}{2}$

9. Explanations may vary.

10. a) 0.2527, 2.8889, 0.5236, 2.6180
  b) 2.3005, 3.9827
  c) 1.5708, 3.6652, 4.7124, 5.7596
  d) 1.1503, 1.9913, 4.2919, 5.1329

12. Answers may vary.
  a) $\sin 3x = 3\sin x - 4\sin^3 x$; $\cos 3x = 4\cos^3 x - 3\cos x$
  b) Explanations may vary. The functions are the same on
    each side of the equation.

5 Cumulative Review, page 350

1. b) Domain: all real numbers, range: $y \geq 4$
  c) No x-intercept, y-intercept: 8

2. a) $y = f(x)$ becomes $y = f(x) + 4$.
  b) $y = f(x)$ becomes $y = f(x + 7)$.
  c) $y = f(x)$ becomes $y = f(x - 5) - 3$.
  d) $y = f(x)$ becomes $y = f(x + 6) - 12$.

3. a) The graph of $y = \frac{1}{2x} + 4$ is the image of $y = \frac{1}{2x}$ after a
    vertical translation of 4 units up.
  b) The graph of $y = |x - 3|$ is the image of $y = |x|$ after a
    horizontal translation of 3 units right.

5. Answers may vary.

6. a) $a = 3$, $b = 4$ **b)** $a = 4$, $b = -\frac{1}{2}$
  c) $a = \frac{1}{5}$, $b = 3$ **d)** $a = -\frac{1}{6}$, $b = \frac{1}{3}$

7. b) $x = -2$, $y = 0$

8. a) $P = 300 \times 2^{\frac{d}{5}}$ **b)** 25 days

9. a) −1   **b)** 3   **c)** 0.5

10. a) 8   **b)** −2   **c)** 32

11. a) 76.4 days   **b)** 116.4 days

13. a) 297.94°   **b)** 74.48°   **c)** 171.89°

14. a) 8.4 cm   **b)** 25.4 cm

16. Answers may vary.
  a) 330°, 690°   **b)** −250°, 470°
  c) $\frac{7\pi}{5}$ radians, $\frac{17\pi}{5}$ radians   **d)** $-\frac{\pi}{3}$ radians, $\frac{11\pi}{3}$ radians

17. Explanations may vary. For part a: 360° was added to −30° to
  get 330°. 360° was added to 330° to get 690°.

19. $y = 2\sin\frac{2\pi}{3}(x - 4) - 5$

20. −4.946, −1.805, 4.479, 7.620

Chapter 6

Investigate, page 354

1. a) EMY, EMP, EJY, EJP, ECY, ECP, HMY, HMP, HJY, HJP,
    HCY, HCP
  b) 12
  c) Multiply the number of choices by each other.

2. a) 3   **b)** 24   **c)** 24   **d)** 1728

3. 41 472   **4.** mn

6.1 Exercises, page 356

1. a) S1, P1, H1; S1, P1, H2; S1, P1, H3; S1, P2, H1; S1, P2,
    H2; S1, P2, H3; S2, P1, H1; S2, P1, H2; S2, P1, H3; S2,
    P2, H1; S2, P2, H2; S2, P2, H3; S3, P1, H1; S3, P1, H2;
    S3, P1, H3; S3, P2, H1; S3, P2, H2; S3, P2, H3; S4, P1,
    H1; S4, P1, H2; S4, P1, H3; S4, P2, H1; S4, P2, H2; S4,
    P2, H3; 24
  b) 24; yes

2. 15   **3.** 80

4. Answers may vary. For exercise 2, 1 shirt and 1 pair of pants
  comprise an outfit. For exercise 3, 1 blouse, 1 skirt, and 1
  sweater comprise an outfit.

5. 45

6. a) 28   **b)** 32

7. a) 60   **b)** 120

8. a) 2   **b)** 32   **c)** $\frac{1}{32}$

9. a) 4   **b)** 1024   **c)** $\frac{1}{1024}$

10. Explanations may vary.

11. a) 24   **b)** 960   **c)** 144   **d)** 40 320

12. a) 7, including minivans
  b) Compact: 11, luxury: 17, mid-sized: 2, minivans: 6, small
    cars: 10, sport utility: 2, sport cars: 9

c) 403 920

13. a) 2 176 782 336
   b) Explanations may vary. Only the number of licence plates that are needed would be produced.

14. 2 238 976 116

15. a) 1.126×10^{15}
   b) Assuming a world population of 6 billion: 187 650 keys per person

16. There are 10^{14} different sonnets.

17. 18

18. a) 208 860   **b)** 198 360

19. a) Examples may vary. To promote and support the growth and improvement of the health department of Canada.
   b) 25 088

20. a) 5.52×10^{26}
   b) Assuming a world population of 6 billion: 2.93×10^9 years
   c) 5.52×10^{19} km

Investigate, page 360

1. a) ABC, ACB, BAC, BCA, CAB, CBA
   b) ABCD, ABDC, ACBD, ACDB, ADBC, ADCB, BACD, BADC, BCAD, BCDA, BDAC, BDCA, CABD, CADB, CBAD, CBDA, CDAB, CDBA, DABC, DACB, DBAC, DBCA, DCAB, DCBA
   c) 120
   d) If the number of letters is n, the number of permutations is $n(n-1)(n-2)(n-3) \dots (3)(2)(1)$.

2. a) 120   **b)** 120   **c)** 120

3. a) AB, AC, AD, AE, BA, BC, BD, BE, CA, CB, CD, CE, DA, DB, DC, DE, EA, EB, EC, ED
   b) 60

4. 20

5. b) Answers may vary. $_3P_3 = 6$, $_4P_4 = 24$
   c) Explanations may vary. Number of arrangements of n different objects taken r at a time.

6. $_nP_r = \frac{n!}{(n-r)!}$, $n \geq r$, n and r are whole numbers.

6.2 Exercises, page 364

1. a) 720   **b)** 5040   **c)** 362 880
   d) 6   **e)** 30   **f)** 120

2. a) 1   **b)** 2, 2   **c)** 3, 6, 6
   d) 4, 12, 24, 24   **e)** 5, 20, 60, 120, 120

3. Descriptions may vary.

4. a) 6   **b)** 24   **c)** 120

5. a) 6; 12; 20   **b)** 6; 24; 60

6. a) 24   **b)** 120   **c)** 720
   d) 5040   **e)** 40 320   **f)** 362 880

7. In the same order as exercise 6:
   a) 12, 20, 30, 42, 56, 72
   b) 24, 60, 120, 210 336, 504
   c) 24, 120, 360, 840, 1680, 3024

8. 120

9. 5040

10. 120

11. Explanations may vary.
   c) $_6P_9$ is not defined because you cannot take 9 objects from 6 objects.
   d) $_{-6}P_3$ is not defined because you cannot take 3 objects from −6 objects.
   e) $_6P_{2.5}$ is not defined because you cannot break an object into 2 to take 2.5 objects.

12. 24

13. 5040

14. 116 280

15. a) 680   **b)** 39   **c)** 39 270

16. 3024

17. Answers may vary.

18. a) 4   **b)** 5   **c)** 6   **d)** 10
   e) 4   **f)** 5   **g)** 6   **h)** 10

19. a) 2   **b)** 3   **c)** 4   **d)** 5 or 6

20. Explanations may vary.

21. a) $(n+2)(n+1)$ **b)** $\frac{1}{n(n-1)(n-2)}$
   c) $n(n+1)$ **d)** $(n+4)(n+3)$

Investigate, page 366

1. a) 24
   b) No, there will be fewer permutations. Explanations may vary.

2. a) AABC, AACB, ABAC, ABCA, ACAB, ACBA, BAAC, BACA, BCAA, CAAB, CABA, CBAA; 12
   b) The number of permutations is halved.

3. a) AAAB, AABA, ABAA, BAAA; 4
   b) There are $\frac{1}{6}$ as many permutations.

4. a) AABB, ABAB, ABBA, BAAB, BABA, BBAA; 6
   b) There are $\frac{1}{4}$ as many permutations.

5. The number of permutations of n objects taken n at a time if there are a alike of one kind, b alike of another kind is: $\frac{n!}{a!b!}$

6. a) i) 20   **ii)** 5   **iii)** 10
   b) i) ABBBC, ABBCB, ABCBB, ACBBB, BABBC, BABCB, BACBB, BBABC, BBACB, BBBAC, BBBCA, BBCAB, BBCBA, CABBB, CBABB, CBBAB, CBBBA, BCABB, BCBAB, BCBBA
   ii) ABBBB, BABBB, BBABB, BBBAB, BBBBA
   iii) AABBB, ABABB, ABBAB, ABBBA, BAABB, BABAB, BABBA, BBAAB, BBABA, BBBAA

6.3 Exercises, page 368

1. a) 30   **b)** 3360   **c)** 6 652 800   **d)** 37 800

2. a) 12   **b)** 60   **c)** 180
   d) 60   **e)** 420   **f)** 105

3. 1260

4. 560

5. 1260

6. a) 1   **b)** 5   **c)** 10
d) 10   **e)** 5   **f)** 1

7. 32; explanations may vary.

8. 10

9. 12 600; explanations may vary.

10. a) 420
    b) There would be 60 different itineraries.
    c) Explanations may vary.

11. 20; explanations may vary.

12. a) 70   **b)** 56
    c) i) 184 756   **ii)** $\frac{(2x)!}{x!x!}$ **iii)** 125 970   **iv)** $\frac{(x+y)!}{x!y!}$

13. a) 200   **b)** 240

14. Explanations may vary.
    a) i) 6   **ii)** 30   **iii)** 90
    b) i) 5.55×10^{12} **ii)** $\frac{(3x)!}{x!x!x!}$ **iii)** 3.78×10^{12}
    iv) $\frac{(x+y+z)!}{x!y!z!}$

15. 3360; explanations may vary.

6.4 Exercises, page 374

1. a) 1   **b)** 1, 1   **c)** 1, 2, 1
    d) 1, 3, 3, 1   **e)** 1, 4, 6, 4, 1   **f)** 1, 5, 10, 10, 5, 1

2. a) Descriptions may vary.
    b) 1, 6, 15, 20, 15, 6, 1; 1, 7, 21, 35, 35, 21, 7, 1

3. 15

4. a) 210   **b)** 210
    c) The number of ways to choose 6 students from 10
      students is the same as the number of ways to reject 4
      students from 10 students.

5. a) 4   **b)** 10   **c)** 20   **d)** 35

6. Answers may vary. For part a: OYE, OYN, OEN, YEN

7. 3 268 760

8. 84

9. a) 5   **b)** 10   **c)** 10   **d)** 5   **e)** 1

10. a) 1   **b)** 10   **c)** 45   **d)** 55   **e)** 66
    f) 120   **g)** 165   **h)** 220   **i)** 210   **j)** 330

11. a) 1   **b)** n **c)** $\frac{n(n-1)}{2}$
    d) $\frac{n(n-1)(n-2)}{6}$ **e)** $\frac{n(n-1)(n-2)(n-3)}{24}$

12. a) 10
    b) Let a, b, c, d, and e represent the students. abc, abd, abe,
      acd, ace, ade, bcd, bce, bde, cde

13. 30

14. 700

15. a) 120   **b)** 720
    c) In part a, choosing 3 people from 10 people is a
      combination calculation. In part b, choosing 3 people,
      with specific designations, from 10 people is a
      permutation calculation.

16. a) 8 996 462 475   **b)** 165   **c)** 15 024 516

17. a) 4368   **b)** 376 992   **c)** 1287   **d)** 65 780

18. 360

19. a) 845 000   **b)** 1 299 480   **c)** 2 144 480

20. a) Lotto 649   **b)** Lotto 649 is about 4.5 times more likely.

21. a) 80 089 128   **b)** $\frac{1}{80\ 089\ 128} \doteq 1.25 \times 10^{-8}$

22. Explanations may vary.

23. a) 26; explanations may vary.
    b) $(26, 4.96 \times 10^{14})$; the y-coordinate represents the number
      of different hands when dealt 26 cards from 52.
    d) $0 \le x \le 52$; you may only be dealt 0 to 52 cards from a
      deck of 52.
    e) $y = {}_{52}C_x$

24. a) 5   **b)** 5   **c)** 7   **d)** 8
    e) 2 or 3   **f)** 2 or 4   **g)** 3   **h)** 3 or 7

25. a) 56   **b)** 28   **c)** 20

26. a) $\frac{n(n-1)(n-2)}{6}$ **b)** $\frac{n(n-1)}{2}$ **c)** $\frac{n(n-3)}{2}$

27. Explanations may vary. ${}_7C_6 = 7$, ${}_8C_6 = 28$, ${}_9C_6 = 84$

28. a) 10 752   **b)** 14 608   **c)** 9856

29. 47

30. 56

31. a) 11   **b)** 25   **c)** 15

32. a) 9   **b)** 24   **c)** 45

33. a) $\frac{1}{534\ 168\ 600}$
    b) 0.0237 (Assuming all contest cards are sold.)

Mathematical Modelling: Analyzing Club Keno, page 379

1. a) i) 10   **ii)** 2.164×10^{12}
    b) 2.164×10^{13} **c)** 3.535×10^{18} **d)** 6.12×10^{-6}

2. a) i) 1   **ii)** 8.627×10^{13}
    b) 8.627×10^{13} **c)** 2.44×10^{-5}

3. Picking 7 and matching 7

4. $\frac{{}_xC_y \times {}_{80-x}C_{20-y}}{{}_{80}C_{20}}$

5. a) 1.354×10^{-4} **b)** 0.0183   **c)** 0.213   **d)** 0.06

6. a) $0.60
    b) i) $0.50   **ii)** $0.56   **iii)** $0.58

8. a) 5   **b)** Answers may vary.

9. a) The expectations would be greater.

10. 3.39×10^{-16}

11. a) Answers may vary.
    b) i) 1.84×10^{-5} **ii)** 9.69×10^{-4} **iii)** 0.0177

6.5 Exercises, page 384

1. to 4. Proofs may vary.

5. a) 49
    b) By observation, subtract 1 from the row number.

6. a) 1176
    b) The third number in the nth row is $\frac{(n-1)(n-2)}{2}$.
      Explanations may vary.

7. Explanations may vary.

8. a) Sums: 3, 6, 10, or 15; explanations may vary.

b) $_{n+1}C_2 = \dfrac{n(n+1)}{2}$

9. Explanations may vary.

b) Sums: 4, 10, or 20

c) $_{n+2}C_3 = \dfrac{n(n+1)(n+2)}{6}$

10. a) Situation 1: **i)** 1, 2, 1   **ii)** 4   **iii)** $\dfrac{1}{4}, \dfrac{1}{2}, \dfrac{1}{4}$

    Situation 2: **i)** 1, 3, 3, 1   **ii)** 8   **iii)** $\dfrac{1}{8}, \dfrac{3}{8}, \dfrac{3}{8}, \dfrac{1}{8}$

    Situation 3: **i)** 1, 4, 6, 4, 1 **ii)** 16

        iii) $\dfrac{1}{16}, \dfrac{1}{4}, \dfrac{3}{8}, \dfrac{1}{4}, \dfrac{1}{16}$

    Situation 4: **i)** 1, 5, 10, 10, 5, 1   **ii)** 32

        iii) $\dfrac{1}{32}, \dfrac{5}{32}, \dfrac{5}{16}, \dfrac{5}{16}, \dfrac{5}{32}, \dfrac{1}{32}$

b) Answers may vary. The balls do not travel sideways nor upwards.

11. Explanations may vary. At each junction the ball has 2 choices.

12. Explanations may vary.

a) 70   **b)** 210

13. a) i) 1, 1   **ii)** 1, 2, 1   **iii)** 1, 3, 3, 1   **iv)** 1, 4, 6, 4, 1

b) Explanations may vary.

15. a) 66   **b)** 54

16. 170

17. Number of diagonals $= \dfrac{n(n-3)}{2}$; explanations may vary.

18. a) The sums are powers of 2.

b) The number of ways you can choose items from a set of 5. That is, there are 32 subsets of a set with 5 items. (See exercise 19)

c), d) Explanations may vary.

19. a) i) ϕ **ii)** $\{a\}, \{b\}, \{c\}, \{d\}$

    iii) $\{a, b\}, \{a, c\}, \{a, d\}, \{b, c\}, \{b, d\}, \{c, d\}$

    iv) $\{a, b, c\}, \{a, b, d\}, \{a, c, d\}, \{b, c, d\}$

    v) $\{a, b, c, d\}$

b) i) 1   **ii)** 4   **iii)** 6   **iv)** 4

    v) 1; these numbers are found in the 5th row of Pascal's triangle.

c) 16; explanations may vary

20. 2^n; explanations may vary.

21. n even: $_{n-1}C_{\frac{n}{2}}$; n odd: $_{n-1}C_{\frac{n-1}{2}}$

Problem Solving: The Pigeonhole Principle, page 389

1. a) 13   **b)** 25   **c)** 37

2. 367

4. a) 3   **b)** 5   **c)** 14

5. 3

6. to 11. Proofs may vary.

Exploring with a Graphing Calculator: Graphing the Functions $y = {_n}C_x$ and $y = {_x}C_r$, page 390

1. All predictions and explanations may vary.

c) Domain of $y = {_n}C_x : 0 \le x \le n$, x is a whole number

3. c) Domain of $y = {_x}C_r : x \ge r$, x is a whole number

5. a) Polynomial functions

b) $y_1 = x$, $y_2 = \dfrac{x(x-1)}{2}$, $y_3 = \dfrac{x(x-1)(x-2)}{6}$,

    $y_4 = \dfrac{x(x-1)(x-2)(x-3)}{24}$

6. No; explanations may vary.

Investigate, page 391

1. Descriptions may vary.

2. a) $x^6 + 6x^5 + 15x^4 + 20x^3 + 15x^2 + 6x + 1$

b) $x^7 + 7x^6 + 21x^5 + 35x^4 + 35x^3 + 21x^2 + 7x + 1$

c) $x^8 + 8x^7 + 28x^6 + 56x^5 + 70x^4 + 56x^3 + 28x^2 + 8x + 1$

3. a) $a + b$

b) $a^2 + 2ab + b^2$

c) $a^3 + 3a^2b + 3ab^2 + b^3$

d) $a^4 + 4a^3b + 6a^2b^2 + 4ab^3 + b^4$

e) $a^5 + 5a^4b + 10a^3b^2 + 10a^2b^3 + 5ab^4 + b^5$

f) $a^6 + 6a^5b + 15a^4b^2 + 20a^3b^3 + 15a^2b^4 + 6ab^5 + b^6$

4. a) $x^3 + 6x^2 + 12x + 8$

b) $x^4 + 8x^3 + 24x^2 + 32x + 16$

c) $x^5 + 10x^4 + 40x^3 + 80x^2 + 80x + 32$

d) $x^3 - 3x^2 + 3x - 1$

e) $x^4 - 4x^3 + 6x^2 - 4x + 1$

f) $x^5 - 5x^4 + 10x^3 - 10x^2 + 5x - 1$

g) $x^3 - 9x^2 + 27x - 27$

h) $x^4 - 12x^3 + 54x^2 - 108x + 81$

i) $x^5 - 15x^4 + 90x^3 - 270x^2 + 405x - 243$

5. Explanations may vary.

6.6 Exercises, page 396

1. Explanations may vary.

2. a) $x^5 + 5x^4 + 10x^3 + 10x^2 + 5x + 1$

b) $x^6 + 6x^5 + 15x^4 + 20x^3 + 15x^2 + 6x + 1$

c) $x^7 - 7x^6 + 21x^5 - 35x^4 + 35x^3 - 21x^2 + 7x - 1$

d) $x^8 - 8x^7 + 28x^6 - 56x^5 + 70x^4 - 56x^3 + 21x^2 - 8x + 1$

e) $x^5 + 5x^4y + 10x^3y^2 + 10x^2y^3 + 5xy^4 + y^5$

f) $x^6 + 6x^5y + 15x^4y^2 + 20x^3y^3 + 15x^2y^4 + 6xy^5 + y^6$

g) $x^7 - 7x^6y + 21x^5y^2 - 35x^4y^3 + 35x^3y^4 - 21x^2y^5 + 7xy^6 - y^7$

h) $x^8 - 8x^7y + 28x^6y^2 - 56x^5y^3 + 70x^4y^4 - 56x^3y^5 + 28x^2y^6 - 8xy^7 + y^8$

3. a) 10, 11   **b)** $(x + y)^{10}$ **c)** n is even.

4. a) $x^3 + 6x^2 + 12x + 8$

b) $x^4 - 12x^3 + 54x^2 - 108x + 81$

c) $1 + 6x + 15x^2 + 20x^3 + 15x^4 + 6x^5 + x^6$

d) $32 - 80x + 80x^2 - 40x^3 + 10x^4 - x^5$

e) $a^4 - 8a^3b + 24a^2b^2 - 32ab^3 + 16b^4$

f) $8a^3 + 36a^2b + 54ab^2 + 27b^3$

g) $243a^5 - 405a^4 + 270a^3 - 90a^2 + 15a - 1$

h) $729a^6 - 2916a^5b + 4860a^4b^2 - 4320a^3b^3 + 2160a^2b^4 - 576ab^5 + 64b^6$

5. a) $c^{10} + 10c^9d + 45c^8d^2 + 120c^7d^3$

b) $x^{12} + 24x^{11} + 264x^{10} + 1760x^9$

c) $256 - 1024x + 1792x^2 - 1792x^3$

d) $1 - 18x + 144x^2 - 672x^3$

6. a) $-15\,360x^3$ **b)** $-35\,750x^{10}$ **c)** $-112\,640a^9$
d) $292\,864x^3$ **e)** There is no middle term.

7. a) $2x^6 + 30x^4y^2 + 30x^2y^4 + 2y^6$
b) $12x^5y + 40x^3y^3 + 12xy^5$
c) $x^{10} - 5x^8 + 10x^6 - 10x^4 + 5x^2 - 1$
d) $1 - \dfrac{7}{x} + \dfrac{21}{x^2} - \dfrac{35}{x^3} + \dfrac{35}{x^4} - \dfrac{21}{x^5} + \dfrac{7}{x^6} - \dfrac{1}{x^7}$

8. $(a+b)^n = \displaystyle\sum_{k=0}^{n} {}_nC_k a^{n-k}b^k$, where ${}_nC_k = \dfrac{n!}{k!(n-k)!}$

9. a) $(1-x)^{-1} = 1 + x + x^2 + x^3 + \dots, x \neq 1$;
$(1-x)^{0.5} = 1 - \dfrac{1}{2}x - \dfrac{1}{8}x^2 - \dfrac{1}{16}x^3 + \dots, x < 1$

6 Review, page 397

1. 336

2. 657 720

3. a) 7 200 000 **b)** 10 376 000

4. 840

5. 120

6. a) 18 564 **b)** 3150 **c)** 5880

7. a) 1, 3, 6, 10, 15
b) In the 3rd diagonal; explanations may vary.
c) $L = {}_nC_2$

6 Cumulative Review, page 398

1. a) The graph is translated 5 units right.
b) The graph is tranlated 7 units left.
c) The graph is translated 4 units up.
d) The graph is translated 2 units down.
e) The graph is translated 2 units left and 1 unit down.
f) The graph is translated 3 units right and 6 units up.

2. b) $f(2x) = 3(4x^2 - 1); f(\tfrac{1}{3}x) = 3(\tfrac{1}{9}x^2 - 1)$

3. a) $\log x + 2\log y$ **b)** $\log x + \tfrac{1}{2}\log y$
c) $1 + 3\log x + 2\log y$ **d)** $\tfrac{1}{3}\log x + \tfrac{2}{3}\log y$
e) $\log x - \tfrac{1}{2}\log y$ **f)** $2\log x - \tfrac{1}{3}\log y$

4. 7.94 times

5. 29 524

6. t_{10}

7. a) $25° + n(360°)$ **b)** $-48° + n(360°)$
c) $-\dfrac{\pi}{4} + 2n\pi$ radians **d)** $\dfrac{7\pi}{2} + 2n\pi$

8. a) $-\dfrac{\sqrt{3}}{2}$ **b)** -1 **c)** $\dfrac{\sqrt{3}}{2}$ **d)** $\dfrac{\sqrt{3}}{2}$
9. a) $-\dfrac{1}{2}$ **b)** 0 **c)** $-\dfrac{1}{2}$ **d)** $\dfrac{1}{2}$

10. a) Amplitude: 3, period: 7, phase shift: -4, vertical displacement: -3
b) Amplitude: 4, period: 6, phase shift: 1, vertical displacement: 2

11. a) Domain: $x \neq \dfrac{n\pi}{3}$, range: $y \geq 1$ or $y \leq -1$, period: $\dfrac{2\pi}{3}$, asymptotes: $x = \dfrac{n\pi}{3}$
b) Domain: $x \neq \dfrac{n}{3}$, range: $y \geq 1$ or $y \leq -1$, period: $\dfrac{2}{3}$, asymptotes: $x = \dfrac{n}{3}$

12. a) 11.8 h **b)** $4\cos\dfrac{2\pi}{11.8}(t - 19.4) + 14$
c) 13 m
d) 3:40 A.M.; 11:32 A.M.; 3:28 P.M.; 11:20 P.M.

14. 12

15. a) 648 **b)** 4 536 **c)** 27 216

16. a) Explanations may vary. **b)** 24; Explanations may vary.

17. Yes; explanations may vary. When 3 or more points are collinear it is more difficult to visualize, yet the number of ways 2 points can be chosen from n points is still ${}_nC_2$.

Chapter 7

7.1 Exercises, page 405

1. $\dfrac{3}{8}$ **2.** $\dfrac{9}{19}$

3. a) $\dfrac{2}{3}$ **b)** $\dfrac{1}{2}$

4. a) Answers may vary. **b)** $\dfrac{1}{4}$

5. a) $\dfrac{1}{16}$
b) Just as likely; explanations may vary.

6. a) $\dfrac{1}{4}$ **b)** $\dfrac{3}{20}$

7. Explanations may vary.
a) Consider a roll of a 1, 2, or 3 the first outcome, and a roll of a 4, 5, or 6 the second outcome.
b) Consider a roll of a 1 or 2 the first outcome, a roll of a 3 or 4 the second outcome, and a roll of a 5 or 6 the third outcome.
c) Disregard the roll of a 5 or 6. The rolls of 1, 2, 3, and 4 are the four outcomes.
d) Disregard the roll of a 6. The rolls of 1 to 5 are the five outcomes.

8. a) $\dfrac{1}{64}$ **b)** Answers may vary.

9. a) $\dfrac{1}{2}$ **b)** $\dfrac{3}{13}$ **c)** $\dfrac{1}{4}$ **d)** $\dfrac{1}{13}$

10. Answers may vary.

11. a) i) $\dfrac{6}{25}$ **ii)** $\dfrac{12}{25}$ **iii)** 0 **iv)** $\dfrac{2}{25}$
b) Answers may vary. It's a fair deck of cards. The cards are not replaced once they have been dealt.

12. a) $\dfrac{17}{57}$ **b)** $\dfrac{30}{57}$ **c)** $\dfrac{49}{57}$

13. c) Obtaining at least one 6 on four rolls of a die is more likely.

14. Answers may vary. The probability is closer to the theoretical probability than in exercise 4.

Linking Ideas: Probability and Exponential Functions, Simulating Exponential Decay, page 408

All explanations may vary.
3. a) $y = 96.35(0.58)^x$
b) When the equation is solved for $x = 0$, $y = 96.35$. Therefore, the curve does not pass through (0, 100).

4. Answers may vary.

6. b) $y = 100(0.5)^x$

7. a) 1

b) The number of trials needed to halve the number of coins is 1.

8. a) $y = 110.8(0.78)^x$ **b)** $y = 100\left(\frac{5}{6}\right)^x$

9. a) 3.80

b) It takes about 4 trials to halve the number of dice.

10. Answers may vary.

Investigate, page 410

3. a) (1, 3), (2, 4), (3, 5), (4, 6); 4

b) (1, 6), (2, 5), (3, 4), (4, 3), (5, 2), (6, 1); 6

c) (1, 5), (2, 4), (3, 3), (4, 2), (5, 1); 5

d) (1, 6), (2, 5), (3, 4), (4, 3), (5, 2), (6, 1), (1, 3), (2, 4), (3, 5), (4, 6); 10

e) (1, 6), (2, 5), (3, 4), (4, 3), (5, 2), (6, 1), (1, 5), (2, 4), (3, 3), (4, 2), (5, 1); 11

f) (1, 5), (2, 4), (3, 3), (5, 1), (1, 3), (2, 4), (3, 5), (4, 6); 8

g) None   **h)** None   **i)** (2, 4); 1

4. a) $\frac{1}{9}$ **b)** $\frac{1}{6}$ **c)** $\frac{5}{36}$ **d)** $\frac{5}{18}$ **e)** $\frac{11}{36}$

f) $\frac{2}{9}$ **g)** 0   **h)** 0   **i)** $\frac{1}{36}$

5. S and V can occur at the same time. T cannot occur with S or with V.

6. Answers and explanations may vary.

7.2 Exercises, page 414

1. 11, 21, 31, 41, 51, 61, 12, 22, 32, 42, 52, 62, 13, 23, 33, 43, 53, 63, 14, 24, 34, 44, 54, 64

2. H0, H1, H2, H3, H4, H5, H6, H7, H8, H9, T0, T1, T2, T3, T4, T5, T6, T7, T8, T9

3. 0.6. The probability of no precipitation is 0.6.

4. 0.9

5. a) Any number from 3 to 12 eggs are broken.

b) Either 0, 1, or 2 eggs are broken.

c) Either 0 or 1 egg is broken.

6. a) $\frac{1}{6}$ **b)** $\frac{1}{12}$ **c)** $\frac{1}{4}$ **d)** 0

7. a) $\frac{1}{2}$ **b)** $\frac{1}{6}$ **c)** $\frac{1}{6}$ **d)** $\frac{5}{18}$

8. a) i) H, T

ii) HH, HT, TH, TT

iii) HHH, HHT, HTH, HTT, THH, THT, TTH, TTT

iv) HHHH, HHHT, HHTH, HHTT, HTHH, HTHT, HTTH, HTTT, THHH, THHT, THTH, THTT, TTHH, TTHT, TTTH, TTTT

9. a) Same as exercise 8 a iii

b) Fewer than 2 heads   **c)** $\frac{1}{2}, \frac{1}{2}$

10. a) Same as exercise 8 a iii

b) More than 2 heads   **c)** $\frac{7}{8}, \frac{1}{8}$

11. a) Descriptions may vary. The sample space is the list of the 52 cards in a deck.

b) i) $\frac{1}{13}$ **ii)** $\frac{1}{2}$ **iii)** $\frac{1}{4}$

c) i) $\frac{3}{4}$ **ii)** $\frac{7}{13}$ **iii)** $\frac{1}{52}$ **iv)** $\frac{3}{4}$

12. $\frac{4}{9}$

13. a) ABC, ABO, ABP, ACO, ACP, AOP, BCO, BCP, BOP, COP

b) BCO, BCP, BOP, COP

c) ABC, ABO, ABP, ACO, ACP, AOP

d) $\frac{2}{5}, \frac{3}{5}$

14. a) S, A, K, T, C, H, E, W, N

b) i) $\frac{2}{3}$ **ii)** $\frac{1}{6}$ **iii)** $\frac{1}{4}$

c) i) $\frac{1}{3}$ **ii)** $\frac{1}{6}$ **iii)** $\frac{11}{12}$

15. a) Africa, Asia, Europe, North and Central America, Oceania, South America

b) $\frac{1}{6}$

c) French Polynesia, Fiji, Australia, Papua New Guinea, Guam, New Caledonia, Kiribati, Samoa, Solomon Islands, Tonga, Vanuatu, New Zealand

d) $\frac{1}{12}$

16. $\frac{1}{15}, \frac{2}{15}, \frac{4}{15}, \frac{8}{15}$

17. a) (0, 1), (9, 0), (8, 9), (7, 8), (6, 7), (5, 6), (4, 5), (3, 4), (2, 3), (1, 2)

b) 10

c) Anita will eat with her grandmother 90% of the time and with her mother 10% of the time.

18. A+, A−, B+, B−, AB+, AB−, O+, O−

Mathematics File: Non-Transitive Dice, page 417

1. b) Die B; $\frac{2}{3}$

2. b) Die C; $\frac{2}{3}$

3. b) Die D; $\frac{2}{3}$

4. b) Die A; $\frac{2}{3}$

5. Answers may vary.

6. Explanations may vary.

Problem Solving: A Three-Game Series, page 418

1. Answers may vary. "m-t-m," the teacher is a better player than her mother; therefore, it would be better to have 2 potential matches with her mother than her teacher.

2. b) "t-m-t"; no

3. Answers may vary.

c) "t-m-t"; no

4. a) Win-win or lose-win-win

b) Answers may vary. For P (m) = 0.6, P (t) = 0.3, "m-t-m": 0.252, "t-m-t": 0.306

c) "t-m-t"; no

5. Explanations may vary.

6. a) The probability of winning the "m-t-m" series is $mt + (1 - m)tm$. The probability of winning the "t-m-t" series is $tm + (1 - t)mt$.

b) Explanations may vary.

7. Answers may vary. The answers obtained in exercises 2 and 3 are close to the theoretical probabilities on substitution of the values chosen. The answers obtained in exercise 4 are the same as the theoretical probabilities on substitution of the values chosen.

8. No; explanations may vary.

9. Explanations may vary.

7.3 Exercises, page 423

1. a) $\frac{7}{12}$ **b)** $\frac{1}{3}$ **c)** 1   **d)** $\frac{11}{18}$ **e)** $\frac{19}{36}$

2. 53%   **3.** 20%   **4.** 10%   **5.** 16.8%

6. a) The probability the event A or its complement occurs is 1.
   b) The probability the event A and its complement occurs is 0.

7. a) $\frac{1}{2}$ **b)** $\frac{4}{13}$ **c)** $\frac{11}{26}$ **d)** $\frac{4}{13}$

8. a) $\frac{3}{20}$ **b)** $\frac{2}{5}$ **c)** $\frac{1}{4}$

9. a) 59   **b)** 8   **c)** $\frac{33}{59}$ **d)** $\frac{30}{59}$

10. a) 0.95   **b)** 0.4   **c)** 0.30

11. a) 3 and 5   **b)** 5 and 6   **c)** 5 and 7
   d) 2 to 7 inclusive   **e)** 3 to 8 inclusive
   f) 2, 3, 5, 6, 7, and 8   **g)** 5
   h) 2 to 8 inclusive

12. P(A or B or C) = P(A) + P(B) + P(C) − P(A and B) − P(A and C) − P(B and C) + P(A and B and C)

7.4 Exercises, page 429

1. Explanations may vary.
   a) Dependent; no card replacement
   b) Independent; card replacement
   c) Independent; coin and die are distinct
   d) Independent; first roll does not influence second roll
   e) Independent; each toss does not affect the next toss
   f) Independent; the result of the first game does not affect the second game

2. No   **3.** Yes   **4.** No   **5.** $\frac{1}{2}$

6. a) $\frac{5}{91}$ **b)** $\frac{48}{91}$

7. a) i) $\frac{1}{3}$ **ii)** $\frac{4}{15}$ **iii)** $\frac{2}{9}$ **iv)** $\frac{8}{45}$
   b) 1; explanations may vary. Part a includes all the outcomes; therefore, the sum of the probabilities equals 1.

8. $\frac{4}{17}$

9. a) $\frac{11}{850}$ **b)** $\frac{1}{5525}$ **c)** $\frac{2}{17}$ **d)** $\frac{11}{1105}$

10. $\frac{7}{100}$

11. a) $\frac{6}{25}$ **b)** $\frac{13}{25}$ **c)** $\frac{6}{25}$

12. 0.063   **13.** $\frac{3}{8}$ **14.** $\frac{1}{14}$

15. a) i) $\frac{480}{2652} = \frac{40}{221}$ **ii)** $\frac{1560}{2652} = \frac{10}{17}$

b) 1; explanations may vary. Part a includes all the outcomes; therefore, the sum of the probabilities equals 1.

16. a) $\frac{156}{2652} = \frac{1}{17}$ **b)** $\frac{1482}{2652} = \frac{247}{442}$ **c)** $\frac{1014}{2652} = \frac{13}{34}$

17. a) $\frac{650}{2652} = \frac{25}{102} \cdot \frac{25}{102} \cdot \frac{26}{51}$ **b)** $\frac{132}{2652} = \frac{11}{221} \cdot \frac{10}{17} \cdot \frac{80}{221}$
   c) $\frac{12}{2652} = \frac{1}{221} \cdot \frac{188}{221} \cdot \frac{32}{221}$

18. b) 0.335, 0.665

19. a) $y = \left(\frac{5}{6}\right)^n$, $y = 1 - \left(\frac{5}{6}\right)^n$
   b) For $y = \left(\frac{5}{6}\right)^n$, domain: all whole numbers; asymptote: $y = 0$. For $y = 1 - \left(\frac{5}{6}\right)^n$, domain: all whole numbers; asymptote: $y = 1$
   c) For $y = \left(\frac{5}{6}\right)^n$, $0 < y \leq 1$. For $y = 1 - \left(\frac{5}{6}\right)^n$, $0 \leq y < 1$.

20. a) 0.162   **b)** 26

21. a) $f(n) = \left(\frac{12}{13}\right)^n$ **b)** Descriptions may vary.

22. $\frac{2}{7}$

23. a) $\frac{1}{36}$ **b)** $\frac{5}{18}$ **c)** $\frac{17}{36}$

24. a) 0.1897   **b)** 0.2577

25. Answers may vary.

26. a) $\frac{1}{22}$ **b)** $\frac{1}{220}$ **c)** $\frac{1}{55}$

27. a) $\frac{3}{22}$ **b)** $\frac{3}{55}$ **c)** $\frac{3}{22}$ **d)** $\frac{3}{44}$ **e)** $\frac{3}{11}$

28. a) 0.2   **b)** 0.72   **c)** 0.5   **d)** 0.625

29. a) 0.491   **b)** At least 1 six in 4 rolls of 1 die is more likely.

30. a) $\frac{8}{45}$ **b)** $\frac{1}{3}$ **c)** $\frac{22}{45}$

Mathematical Modelling: Interpreting Medical Test Results, page 434

1. a) 5000   **b)** 995 000

2. a) 4900   **b)** 100

3. a) 975 100   **b)** 19 900

4. a) 24 800   **b)** 4900   **c)** 0.198

6. 0.0001

7. Answers may vary. Positive results are 99.5% accurate and negative results are 98% accurate.

7.5 Exercises, page 439

1. a) $\frac{1}{13}$ **b)** $\frac{1}{4}$ **c)** $\frac{1}{2}$
   d) $\frac{3}{26}$ **e)** $\frac{1}{26}$ **f)** $\frac{1}{52}$

2. a) 0.0045   **b)** 0.0045

3. a) 0.0577   **b)** 0.0577
   c) Explanations may vary. Both parts a and b determine the probability for the same final outcome. That is, when choosing 2 cards, the probability that (at least) 1 card is a spade and (at least) 1 card is a face card is 0.0577.

4. Explanations may vary.
   a) The probability that a chip is defective is less. The

probability that a defective chip came from factory 1 is less.

b) The probability that a chip is defective is greater. The probability that a defective chip came from factory 1 is greater.

c) The probability that a chip is defective is greater. The probability that a defective chip came from factory 1 is less.

5. a) $\frac{5}{22}$ **b)** $\frac{3}{5}$

6. 0.757

7. a) $\frac{3}{5}$ **b)** $\frac{1}{3}$

8. a) 0.52 **b)** 0.54

9. $\frac{19}{36}$

10. a) Explanations may vary. There are 2 steps. The first step is choosing a box, within each box, the number of nickels and dimes is not equal.

b) Answers may vary. Place 3 nickels and no dimes in one box. Place 2 nickels and 2 dimes in another box. The remaining box contains the 3 remaining dimes.

c) Answers may vary.
  i) Place 1 nickel in each of 2 boxes. The remaining box contains 3 nickels and 5 dimes.
  ii) Place 1 dime in each of 2 boxes. The remaining box contains 3 dimes and 5 nickels.

11. 20%

12. a) 94.28% **b)** 0.04% **c)** 86.71%

7.6 Exercises, page 444

1. a) 0.010 **b)** 0.141 **c)** 0.424

2. P(0) = 0.071, P(1) = 0.354, P(2) = 0.424, P(3) = 0.141, P(4) = 0.010

3. a) 56 **b)** 15 **c)** $\frac{15}{56}$

4. a) 1365 **b)** 0.114

5. a) 0.160 **b)** 0.070 **c)** 0.140

6. 0.250

7. 0.429

8. a) 9.23×10^{-6} **b)** 4.95×10^{-4} **c)** 3.85×10^{-7}
  d) 1.85×10^{-5} **e)** 1.54×10^{-6}

9. 1.54×10^{-6}

10. 2.40×10^{-4}

11. a) **i)** 0.304 **ii)** 0.439 **iii)** 0.213
  iv) 0.041 **v)** 0.003

b) 1; explanations may vary. Part a includes all the outcomes; therefore, the sum of the probabilities equals 1.

12. Explanations may vary.
  a) The individual probabilities will increase, decrease, decrease, decrease, and decrease. The sum of the probabilities will remain 1.
  b) The individual probabilities will decrease, decrease, increase, increase, and increase. The sum of the probabilities will remain 1.

13. b) Explanations may vary. Calculate the number of ways to choose 5 spades from 13, multiply by the number of ways to choose 8 cards from the remaining 39 and divide by the number of ways to choose 13 cards from 52.

14. a) $y = \frac{{}_{13}C_n \times {}_{39}C_{13-n}}{{}_{52}C_{13}}$
  b) $0 \leq n \leq 13$, n is a whole number
  c) Descriptions may vary. y is between 0 and 0.286 inclusive and has exactly 14 possible values corresponding to the 14 values for n.

15. a) 0.286 **b)** 1.17×10^{-3} **c)** 1.58×10^{-12}

16. a) 0.4 **b)** 0.3 **c)** 0.3

17. a) 0.2963 **b)** 0.0081 **c)** 0.0446

18. a) **i)** 7.15×10^{-8}
  ii) 1.84×10^{-5}, 9.69×10^{-4}, 0.0177, 0.132, 0.413, 0.436
  b) Very close to 1; explanations may vary. Part a includes all the outcomes; therefore, the sum of the probabilities is 1. The difference is due to rounding.

19. 90% **20.** 0.0357

21. a) $\frac{1}{3}$ **b)** $\frac{2}{15}$ **c)** $\frac{8}{15}$

22. a) 0.129 **b)** 0.0135 **c)** 0.986

23. a) $\frac{125}{3125} = \frac{1}{25}$ **b)** $\frac{800}{3125} = \frac{32}{125}$ **c)** $\frac{120}{3125} = \frac{24}{625}$

24. 0.216 **25.** Yes; explanations may vary.

Exploring with a Graphing Calculator: What Are Your Chances of Guessing Correct Answers on Tests?, page 448

1. a) 4 and 5; 2 and 8; explanations may vary.
  b) Yes; explanations may vary.

2. b) Answers may vary. 0, 0.01, 0.05, 0.1, 0.24, 0.22, 0.2, 0.11, 0.06, 0.01, 0

3. Answers may vary. 0.60

4. to 6. Descriptions may vary.

7. a) Descriptions may vary.
  c) Answers may vary. 0.06, 0.19, 0.28, 0.25, 0.14, 0.06, 0.01, 0, 0, 0

8. Answers may vary. 0.07

7.7 Exercises, page 457

1. Answers may vary.
  a) Male, female **b)** Right-handed, left-handed
  c) Pass, fail **d)** Above 70%, 70% and below
  e) Drugs that require prescription, over-the-counter drugs
  f) Fixed return, variable return

2. 2.14×10^{-5}, 6.43×10^{-4}, 8.04×10^{-3}, 5.36×10^{-2}, 2.01×10^{-1}, 4.02×10^{-1}, 3.35×10^{-1}; 1. The sum of the probabilities is 1. Any difference is due to rounding.

3. a) 0.125 **b)** 0.375 **c)** 0.375 **d)** 0.125

4. a) 0.1664 **b)** 0.4084 **c)** 0.3341 **d)** 0.0911

5. a) $\frac{1}{32}$ **b)** $\frac{5}{32}$ **c)** $\frac{5}{16}$
  d) $\frac{5}{16}$ **e)** $\frac{5}{32}$ **f)** $\frac{1}{32}$

6. a) 0.7776   **b)** 0.2592   **c)** 0.3456

d) 0.2304   **e)** 0.0768   **f)** 0.0102

7. 0.268   **8.** 0.261   **9.** 0.164   **10.** 0.174   **11.** 0.255

12. a) 415   **b)** 288   **c)** 0.6940   **d)** 0.9973   **e)** 0.9669

13. Answers may vary.

14. 0.318

15. a) i) 0.5   **ii)** 0.176   **iii)** 0.056

b) Explanations may vary. As the number of tosses increases, the number of possible outcomes increase. Therefore, the probability for any one outcome decreases.

17. a) 0.125   **b)** 64

18. Explanations and examples may vary.

19. 2.96×10^{-5} **20.** 0.274   **21.** 0.153

22. Explanations may vary.

a) Few questions   **b)** Yes

7 Review, page 460

1. Answers may vary. 0.2

2. Answers may vary. 0.03

3. $\frac{1}{36}$

4. a) 0.2524   **b)** 0.4563   **c)** 0.0777   **d)** 0.0874

5. a) $\frac{1}{28}$ **b)** $\frac{1}{21}$ **c)** $\frac{14}{105}$

6. 3%

7. 0.0292

8. 0.2708

7 Cumulative Review, page 461

1. a) The graph of $y = -x^3 - 2$ is the image of $y = x^3 - 2$ after a reflection in the y-axis.

b) The graph of $y = 2 - x^3$ is the image of $y = x^3 - 2$ after a reflection in the x-axis.

c) The graph of $x = y^3 - 2$ is the image of $y = x^3 - 2$ after a reflection in the line $y = x$.

2. a) $y = f(x)$ becomes $y = f(-x)$.

b) $y = f(x)$ becomes $x = f(y)$.

c) $y = f(x)$ becomes $y = -f(x)$.

d) $y = f(x)$ becomes $y = -f(-x)$.

3. a) 1   **b)** 2   **c)** 2

d) 2   **e)** 3   **f)** 3

4. a) 1000 times   **b)** About 3.16×10^{12} times

c) About 316 times

5. a) $\frac{\sqrt{3}}{2}, -\frac{1}{2}, -\sqrt{3}$ **b)** $-\frac{1}{2}, \frac{\sqrt{3}}{2}, -\frac{1}{\sqrt{3}}$

c) $\frac{1}{\sqrt{2}}, -\frac{1}{\sqrt{2}}, -1$

6. a) 0.3894, 0.9211   **b)** 0.9093, −0.4161

c) 0.7568, −0.6536   **d)** 0.7055, 0.7087

7. a) 1.3734, 4.5150   **c)** $1.3734 + n\pi$

8. a) $-\frac{63}{65}$ **b)** $\frac{56}{65}$ **c)** $-\frac{7}{25}$ **d)** $-\frac{120}{169}$

9. a) 1000   **b)** 200

Chapter 8

8.1 Exercises, page 471

1. Explanations may vary.

a) Distribution 2; all the outcomes have equal probabilities.

b) Distribution 1; there are two possible outcomes, either a boy or a girl. Distribution 4; there are two possible outcomes, either under 17 or 17 and over.

c) Distribution 3; the probabilities are not equal and there are an unlimited number of outcomes.

2. b) No; all probabilities are not equal.

c) No; there are more than 2 outcomes.

3. a) HHH, HHT, HTH, HTT, THH, THT, TTH, TTT

4. Answers may vary.

5. a) Predictions may vary.

6. Predictions and explanations may vary.

7. Answers may vary. Construct a spinner with 4 equal sections, 3 of which are red, and 1 of which is black. Spin the spinner 10 times.

8. b) Explanations may vary.

9. a) Answers may vary.

b) No; there are more than 2 outcomes.

10. Yes; the sum of all the probabilities is 1.

Investigate, page 473

1. and 2. Answers and explanations may vary.

3. Carton A: mean = 3, median = 3, mode = 3; carton B: mean = 3, median = 3, mode = 3; answers and explanations may vary.

4. Answers and explanations may vary.

8.2 Exercises, page 477

1. a) 30, 7.07   **b)** 24, 4.24   **c)** 20, 3.29   **d)** 26.4, 12.29

2. Greatest standard deviation: b; least standard deviation: c

3. a) 0.93   **b)** 1.41   **c)** 0.63

4. People at a shopping mall; there will be greater variation in the results.

5. a) 10.964, 1.307   **b)** 11.350, 1.445

6. People at a hockey game; there will be greater variation in the results.

7. Greatest standard deviation: a; there are more times far from the mean. Least standard deviation: c; there are more times close to the mean.

8. a) Brand X: 5.6 years, Brand Y: 5.6 years

b) Brand X: 1.33 years, Brand Y: 0.78 years

c) Brand Y; there is less deviation from the mean.

9. Answers may vary.

Investigate, page 479

1. and 2. Answers may vary.

3. The mean value is getting closer to 6.

4. It is equal to *np*.

8.3 Exercises, page 483

1. a) i) 5.64 ii) 5.86 iii) 5.958
 c) i) 1.597 ii) 1.569 iii) 1.672

2. Answers may vary.

3. a) 60, 4.899 **b)** 224, 8.198 **c)** 100, 9.747

4. a) 1.667, 1.179 **b)** 8.333, 1.179

5. a) Predictions may vary. The mean will increase and the standard deviation will decrease.
 b) Answers may vary. For part a:
  i) 3.333, 1.667 ii) 10, 2.887 iii) 50, 6.455
 c) Descriptions may vary.

8.4 Exercises, page 489

1. Explanations may vary.
 a) ii   **b)** iii   **c)** i

2. b) i) 100% ii) 0% iii) 68% iv) 95%
 v) 34% **vi)** 13.5% **vii)** 84% **viii)** 2.5%
 c) 3 g

3. a) 2 **b)** 2.5% **c)** 12 cm
 d) Explanations may vary; do not have the data

4. a) 2280, 446.5 **b)** 35
 c) 37

5. a) 2.5% **b)** 0.15% **c)** 75 **d)** 97.5%

8.5 Exercises, page 492

1. a) 0.9896 **b)** 0.1093 **c)** 0.9699
 d) 0.8414 **e)** 0.9786 **f)** 0.4908

2. a) 0.1859 **b)** 0.0313 **c)** 0.0334
 d) 0.9452 **e)** 0.0735 **f)** 0.0082

4. A graphing calculator was used to determine the area.
 a) 0.6827, 0.0797, 0.0080 **b)** 1, 0.3829, 0.0399
 c) 0.0606, 0.2417, 0.3829, 0.2417, 0.0606
 d) 0.9452, 0.0548, 0.8904, 0.9452, 0.0548

6. A graphing calculator was used to determine the area.
 a) 0.3829 **b)** 0.2417 **c)** 0.2417 **d)** 0.0606 **e)** 0.0606

7. A graphing calculator was used to determine the area.
 a) 0.6247 **b)** 0.0703 **c)** 0.0070

9. c) Answers may vary.

10. a) 0.705 402, same result **b)** 2.87105×10^{-7}

8.6 Exercises, page 500

Answers to exercises 1 to 11 were obtained using the table on pages 494, 495 for Areas under a Standard Normal Curve. Answers obtained using the calculator function might vary slightly due to rounding.

1. a) i) 50% ii) 84.13% iii) 15.87% iv) 68.27%
 b) About 186 000 **c)** 10.56%

2. b) 1.98%

3. a) 41.92% **b)** 76.98% **c)** 0.47%

4. a) 0.47% **b)** 99.53%

5. a) 11 **b)** 11

6. No; explanations may vary.

7. Answers may vary.

8. a) 25.10% **b)** 85.02% **c)** 95.99%
 d) 4.01% **e)** 75.49% **f)** 0.36%

9. a) 70.813, 17.207 **b)** 0.2433 **c)** 22.79%

10. a) i) 7.03%, 3.52%, 0.70% ii) 2.60%, 1.30%, 0.26%
 b) i) 0   ii) 0

11. a) 0.52 **b)** 1.04 **c)** 1.04
 d) 0.39 **e)** 0.52 **f)** 1.28

12. a) 1 **b)** 0.5 **c)** 0.84

13. Explanations may vary.

Mathematical Modelling: Converting Scores, page 504

Answers to exercises 1 to 12 were obtained using the calculator function for areas under a Standard Normal Curve. Answers obtained using the table on pages 494, 495 might vary slightly due to rounding.

1. 38.75%

2. a) 56, 14   **b)** 66.59%

3. 22

4. Answers and explanations may vary.

5. a) 88.49%
 b) i) 73.98%   ii) 68
 c) i) 54   ii) 83

6. 88.49%

7. a) 7.3   **b)** 67
 c) i) 56   ii) 77

8. Instructions may vary.

9. Answers and explanations may vary.

10. a) 628
 b) i) 349   ii) 914   **c)** 69.15%

11. a) 78   **b)** 550
 c) i) 439   ii) 662

12. Descriptions may vary.

8.7 Exercises, page 510

Answers to exercises 1 to 9 were obtained using the calculator function for areas under a Standard Normal Curve. Answers obtained using the table on pages 494, 495 might vary slightly due to rounding.

1. a) 240, 12 **b)** 200, 10 **c)** 12, 3.29 **d)** 10, 3.15

2. 25.12%

3. 7.74%

4. a) 65.72% **b)** 96.43% **c)** 100%
 d) Explanations may vary.

5. a) 39.67% **b)** 74.99% **c)** 99.95%
 d) Explanations may vary.

6. a) Predictions may vary. The probabilities would be less.
c) Not affected

7. 96.35%

8. a) 16.04%　　**b)** 3.40%　　**c)** 0%　　**d)** 3.63%

9. a) Parts a and d can be calculated to be 0.160 and 0.035, respectively, using the binomial distribution and the TI-83.
b) Parts b and c can not be calculated using the binomial distribution and the TI-83. Their calculations cause an overflow error.

Problem Solving: Using the Functions Toolkit, page 512

1. a) i) σ　　　**ii)** σ
b) μ
c) $y = \frac{1}{\sigma}f(\frac{x-\mu}{\sigma})$

2. a) $0 \le x \le 30$　　**b)** 5, 2.04; 25, 2.04　　**d)** $y = f(x - 20)$

3. a) $y = \frac{1}{\sigma\sqrt{2\pi}}e^{-\frac{1}{2}(\frac{x-\mu}{\sigma})^2}$

b) $y = \frac{1}{2.04\sqrt{2\pi}}e^{-\frac{1}{2}(\frac{x-5}{2.04})^2}$, $y = \frac{1}{2.04\sqrt{2\pi}}e^{-\frac{1}{2}(\frac{x-25}{2.04})^2}$

8 Review, page 513

1. Uniform

2. Answers may vary. Construct a spinner with 10 equal sections, 3 of which are red, and 7 of which are black. Spin the spinner 8 times.

3. a) Crispy Chips: mean = 80, standard deviation = 6.90; Special Spuds: mean = 80, standard deviation = 5.45
b) Special Spuds

4. Answers may vary.

5. a) i) Barkley: 27.60; Ewing: 26.59
ii) Barkley: 23.04, 3.67; Ewing: 23.43, 2.38
c) Patrick Ewing

6. 16%

8 Cumulative Review, page 514

1. a) The graph of $y = 4x^2$ is the image of $y = x^2$ after a vertical expansion by a factor of 4.
b) The graph of $y = \frac{1}{4}x^2$ is the image of $y = x^2$ after a vertical compression by a factor of $\frac{1}{4}$.
c) The graph of $y = (-x)^2$ is the same as the graph of $y = x^2$.
d) The graph of $2y = x^2$ is the image of $y = x^2$ after a vertical compression by a factor of $\frac{1}{2}$.

2. a) $A = 2000(1.05)^n$　　**c)** 8.3 years

3. 97.15%

4. $t_6 = 15\,552$, $t_n = 2(6)^{n-1}$

5. t_8

6. a) $y = 3^x$　　**b)** $y = 0.5^x$　　**c)** $f^{-1}(x) = 7^x$

7. a) 27.9 cm　　**b)** 52.3 cm

8. a) $-60° + n(360°)$　　**b)** $115° + n(360°)$
c) $-\frac{3\pi}{5} + 2n\pi$　　**d)** $\frac{2\pi}{3} + 2n\pi$

10. a) $\cos\frac{\pi}{3}$　　　　**b)** $\sin 1.6$
c) $\cos 0.7$　　　　**d)** $\cos\frac{\pi}{2}$

11. a) 3.4814, 4.7127, 5.9433　　**b)** 1.9106, 4.3726
c) 4.7124　　**d)** 1.8235, 4.4597

12. a) $3.4814 + 2n\pi$, $4.7127 + 2n\pi$, $5.9433 + 2n\pi$
b) $1.9106 + 2n\pi$, $4.3726 + 2n\pi$
c) $4.7124 + 2n\pi$
d) $1.8235 + 2n\pi$, $4.4597 + 2n\pi$

13. a) $35x^4$　　**b)** $-x^5$　　**c)** $5670x^4$
d) $60x^2$　　**e)** $109\,824x^7$

14. a) 74 613　　**b)** 26 400

15. a) 20 358 520　　**b)** 1716　　**c)** 18 418 010

16. 275

17. a) 189.123, 33.115　　**b)** 2.28%
c) 84.03%
d) 5.26%, 89.47%; explanations may vary.

18. a) 0.15　　**b)** 0.11　　**c)** 0.07

19. 18 s

20. b) They are the same to 4 decimal places.

Chapter 9

Investigate, page 518

All descriptions may vary.
1. a) Circle
b) The circle gets larger or smaller.
c) The circle gets smaller.

2. a) An ellipse has an oval shape.
b) The ellipse gets larger.
c) The ellipse gets larger.

3. a) A hyperbola has the shape of a boomerang with infinite sides.
b) The edge of the shadow corresponding to the "arms" of the hyperbola get closer.
c) The "arms" of the hyperbola get closer.

4. a) 45°
b) Parabola
c) It makes an angle of 45° with the wall.

5. a) The ellipse gets larger or smaller.
b) The ellipse gets smaller.
c) The hyperbola gets larger or smaller. As the flashlight gets closer to the wall, the hyperbola gets smaller.

6. The curve near the vertex of a parabola is more rounded.

9.1 Exercises, page 522

All descriptions may vary.
1. A hyperbola with the light slightly diffused

2. a) The circle gets larger.
b) The circle gets smaller.
c) The circle becomes a point.

3. a) Same as exercise 2, after replacing circle with ellipse.
b) Parts a and b: the parabola doesn't change.
Part c: the parabola becomes a line that is a generator.

c) Part a: the distance between the two branches gets larger.
Part b: the distance between the two branches gets smaller.
Part c: the hyperbola becomes two intersecting lines.

4. Yes; explanations may vary.

5. Explanations may vary.

6. a) At an angle greater than 45° with the floor
b) At an angle equal to 45° with the floor
c) At an angle less than 45° with the floor

7. a) Hyperbola
b) Ellipse, hyperbola, or parabola
c) Hyperbola   **d)** Circle

8. Descriptions may vary. A point of intersection, one line of intersection, and two intersecting lines of intersection

9. Explanations may vary.

10. Descriptions and explanations may vary.
a) Circle or ellipse   **b)** No

11. a) i) $\theta = 0°$ **ii)** $0 \le \theta < 60°$
iii) $\theta = 60°$ **iv)** $60° < \theta \le 90°$
b) $90° < \theta < 120°$

12. a) $\theta = 0, 180°$
b) $0 < \theta < 45°, 135° < \theta < 225°, 315° < \theta < 360°$
c) $\theta = 45°, 135°, 225°, 315°$
d) $45° < \theta < 135°, 225° < \theta < 315°$

13. a) $\beta = 0, 180°$
b) $0 < \beta < 45°, 135° < \beta < 225°, 315° < \beta < 360°$
c) $\beta = 45°, 135°, 225°, 315°$
d) $45° < \beta < 135°, 225° < \beta < 315°$

14. Descriptions may vary.   **15.** Explanations may vary.

Investigate, page 526

1. a) A circle with radius 4; explanations may vary.
b) Yes

2. a) k provides information about the radius. Explanations may vary.
b) When k changes and $k > 0$, the radius of the circle changes. When $k < 0$, the relation cannot be graphed. When $k = 0$, the graph is the point $(0, 0)$.

4. a) Hyperbola; descriptions may vary.
b) Yes

5. a) $|k|$ provides information about the distance between the two branches. The sign of k provides information about the direction of the opening of the branches.
b) When $|k|$ increases, the distance between the branches increases. When $|k|$ decreases, the distance between the branches decreases. When k changes sign, there is also a reflection in the line $y = x$. When $k = 0$, the graph is the lines $y = \pm x$.

6. Descriptions may vary.

9.2 Exercises, page 532

1. a) Yes   **b)** No   **c)** Yes   **d)** No
2. a) $(\pm 5, 0)$; 10; $y = \pm x$ **b)** $(\pm 8, 0)$; 16; $y = \pm x$

c) $(0, \pm 9)$; 18; $y = \pm x$ **d)** $(\pm\sqrt{2}, 0)$; $2\sqrt{2}$; $y = \pm x$
e) $(0, \pm\sqrt{5})$; $2\sqrt{5}$; $y = \pm x$ **f)** $(0, \pm 2\sqrt{5})$; $4\sqrt{5}$; $y = \pm x$

3. a) $x^2 - y^2 = 49$ **b)** $x^2 - y^2 = -16$
c) $x^2 - y^2 = -36$ **d)** $x^2 - y^2 = 100$

5. a) $x^2 + y^2 = \left(\frac{n}{2}\right)^2$, where $n = 1, 2, 3, ..., 12$
b) $x^2 - y^2 = \pm r^2$, where $r = 1, 2, 3, ..., 6$

7. Predictions may vary.
a) i) Circles with centre $(0, 0)$ and radii 4, 2, 1, $\frac{1}{2}$, and $\frac{1}{4}$
ii) Rectangular hyperbolas with centre $(0, 0)$ and vertices: $(\pm 4, 0)$, $(\pm 2, 0)$, $(\pm 1, 0)$, $\left(\pm\frac{1}{2}, 0\right)$, and $\left(\pm\frac{1}{4}, 0\right)$
c) The hyperbolas become the lines $y = \pm x$. The circles become the point $(0, 0)$.
d) The circles do not exist and the hyperbolas have their vertices on the y-axis.
e) Explanations may vary.

8. Explanations may vary. The circles and all answers about the circles would remain the same. The vertices of the original hyperbolas would be on the y-axis. The vertices of the hyperbolas in part d would be on the x-axis.

9. b) i) A circle with centre $(0, 0)$ and radius 3
ii) The point $(0, 0)$ **iii)** Cannot be graphed

10. b) Descriptions may vary. When $k = 9$, the vertices are $(3, 0)$ and $(-3, 0)$. As k decreases but remains positive, the vertices remain on the x-axis, but approach $(0, 0)$. As soon as k becomes negative, the vertices change to the y-axis, starting at $(0, \sqrt{k})$ and $(0, -\sqrt{k})$, and moving away from $(0, 0)$ to $(0, 3)$ and $(0, -3)$ when $k = 9$.

11. The maximum angle formed by the generators at the vertex is 90°.

12. No; explanations may vary.

13. a) $A = B, A > 0, B > 0, C < 0$ or $A = B, A < 0, B < 0, C > 0$
b) i) $A = -B$, C has the same sign as B
ii) $A = -B$, C has the same sign as A

Investigate, page 535

1. The graph of $\left(\frac{x}{3}\right)^2 + \left(\frac{y}{2}\right)^2 = 1$ is the image of $x^2 + y^2 = 1$ after a horizontal expansion by a factor of 3 and a vertical expansion by a factor of 2.

2. a provides information about the horizontal expansion. b provides information about the vertical expansion. Explanations may vary.

3. $4x^2 + 9y^2 - 36 = 0$; $b^2x^2 + a^2y^2 - a^2b^2 = 0$

4. Descriptions may vary.

5. a) The graph of $\left(\frac{x}{3}\right)^2 - y^2 = 1$ is the image of $x^2 - y^2 = 1$ after a horizontal expansion by a factor of 3. The graph of $x^2 - \left(\frac{y}{2}\right)^2 = 1$ is the image of $x^2 - y^2 = 1$ after a vertical expansion by a factor of 2.

b) The graph of $\left(\frac{x}{9}\right)^2 - \left(\frac{y}{4}\right)^2 = 1$ is the image of $x^2 - y^2 = 1$ after a horizontal expansion by a factor of 3 and a vertical expansion by a factor of 2.

c) Answers may vary. The graph of $\left(\frac{x}{25}\right)^2 - \left(\frac{y}{16}\right)^2 = 1$ is the image of $x^2 - y^2 = 1$ after a horizontal expansion by a factor of 5 and a vertical expansion by a factor of 4.

6. a provides information about the horizontal expansion. b provides information about the vertical expansion. $(a, 0)$ and $(-a, 0)$ are the vertices. Explanations may vary.

7. $x^2 - 9y^2 - 9 = 0$; $4x^2 - y^2 - 4 = 0$; $4x^2 - 9y^2 - 36 = 0$; $b^2x^2 - a^2y^2 - a^2b^2 = 0$

8. Descriptions may vary.

9. Same as 5 after interchanging the appropriate equations.

10. a provides information about the horizontal expansion. b provides information about the vertical expansion. $(0, b)$ and $(0, -b)$ are the vertices. Explanations may vary.

11. $x^2 - 9y^2 + 9 = 0$; $4x^2 - y^2 + 4 = 0$; $4x^2 - 9y^2 + 36 = 0$; $b^2x^2 - a^2y^2 + a^2b^2 = 0$

12. Descriptions may vary.

13. a) The graph of $y = (2x)^2$ is the image of $y = x^2$ after a horizontal compression by a factor of $\frac{1}{2}$.
b) Answers may vary. The graph of $y = (5x)^2$ is the image of $y = x^2$ after a horizontal compression by a factor of $\frac{1}{5}$.
c) k provides information about the horizontal compression. Explanations may vary.

14. a) The graph of $x = (2y)^2$ is the image of $x = y^2$ after a vertical compression by a factor of $\frac{1}{2}$.
b) Answers may vary. The graph of $x = (5y)^2$ is the image of $x = y^2$ after a vertical compression by a factor of $\frac{1}{5}$.
c) k provides information about the vertical compression. Explanations may vary.

15. $4x^2 - y = 0$; $k^2x^2 - y = 0$; $x - 4y^2 = 0$, $x - k^2y^2 = 0$

16. Descriptions may vary.

9.3 Exercises, page 542

1. a) $x^2 + y^2 - 9 = 0$ **b)** $x^2 - y^2 + 16 = 0$
c) $x^2 - 9y^2 + 9 = 0$ **d)** $4x^2 + y^2 - 4 = 0$
e) $4x^2 - 25y^2 - 100 = 0$ **f)** $4x^2 + 16y^2 - 64 = 0$

3. Explanations may vary.
a) The major axis gets longer and the minor axis remains the same.

5. Explanations may vary.
a) The minor axis gets shorter and the major axis remains the same.

6. c) Ellipse   **d)** It becomes a hyperbola.

7. c) Hyperbola
d) When the second denominator is replaced, the graph becomes an ellipse. When the first denominator is replaced, the relation cannot be graphed.

8. Predictions may vary.
a) Ellipses   **b)** Hyperbolas

10. a) $\frac{x^2}{a^2} + \frac{y^2}{16} = 1$, where $a = 2, 4, 6, 8$
b) $\frac{x^2}{16} + \frac{y^2}{9} = 1$, $\frac{x^2}{16} - \frac{y^2}{9} = 1$, $\frac{x^2}{16} - \frac{y^2}{9} = -1$, $y = \pm\frac{3}{4}x$

11. a) $\frac{x^2}{a^2} + \frac{y^2}{b^2} = 1$, where $a = 1, 2, 3, \ldots, 12$, $b = 0.5, 1, 1.5, \ldots, 6$
b) $\frac{x^2}{a^2} - \frac{y^2}{b^2} = \pm 1$, where $a = 1, 2, 3, \ldots, 12$, $b = 0.5, 1, 1.5, \ldots, 6$

13. i) None   **ii)** $\frac{x^2}{9} + \frac{y^2}{36} = 1$
iii) $\frac{x^2}{5} - \frac{y^2}{10} = 1$, $\frac{x^2}{20} - \frac{y^2}{25} = -1$, $\frac{x^2}{4} - \frac{y^2}{4} = 1$
iv) $y^2 = -4x$, $x^2 = \frac{1}{3}y$

14. Descriptions may vary.
a) Ellipse   **b)** Ellipse   **c)** Hyperbola
d) Hyperbola   **e)** Ellipse   **f)** Hyperbola
g) Parabola   **h)** Parabola   **i)** Ellipse

15. Descriptions may vary.
a) Ellipse   **b)** Ellipse   **c)** Hyperbola
d) Hyperbola   **e)** Ellipse   **f)** Hyperbola
g) Ellipse   **h)** Hyperbola

16. b) Explanations may vary.

17. a) $\frac{x^2}{a^2} + \frac{y^2}{(7-a)^2} = 1$, where $a = 0.25, 0.5, 0.75, \ldots, 6.75$
b) $\frac{x^2}{a^2} + \frac{y^2}{(48-a)^2} = 1$, where $a = 1, 2, 3, \ldots, 47$

18. a) $4\left(\frac{a^2b^2}{a^2 + b^2}\right)$
b) $\frac{4a^2b^2}{b^2 - a^2}$, when $b > a$, or not possible

9.4 Exercises, page 548

1. a) $\frac{(x-3)^2}{4} - (y-2)^2 = 1$ **b)** $\frac{(x-3)^2}{9} + \frac{(y+2)^2}{4} = 1$

2. Descriptions may vary.

3. Descriptions may vary.
a) Ellipse with centre $(4, -1)$, vertices $(1, -1)$ and $(7, -1)$, and coordinates of the endpoints of the minor axis $(4, -3)$, $(4, 1)$
b) Ellipse with centre $(4, -1)$, vertices $(4, -4)$ and $(4, 2)$, and coordinates of the endpoints of the major axis $(2, -1)$ and $(6, -1)$
c) Hyperbola with centre $(4, -1)$, vertices $(1, -1)$ and $(7, -1)$, and asymptotes with slopes $\pm\frac{2}{3}$
d) Hyperbola with centre $(4, -1)$, vertices $(2, -1)$ and $(6, -1)$, and asymptotes with slopes $\pm\frac{3}{2}$
e) Hyperbola with centre $(4, -1)$, vertices $(4, -3)$ and $(4, 1)$, and asymptotes with slopes $\pm\frac{2}{3}$
f) Hyperbola with centre $(4, -1)$, vertices $(4, -4)$ and $(4, 2)$, and asymptotes with slopes $\pm\frac{3}{2}$

6. a) $(x-3)^2 + y^2 = r^2$, where $r = 1, 2, \ldots, 6$; $(x+3)^2 + y^2 = r^2$, where $r = 1, 2, \ldots, 6$
b) $(x-h)^2 + y^2 = h^2$, where $h = \pm 1, \pm 2, \pm 3$; $x^2 + (y-k) = k^2$, where $k = \pm 1, \pm 2, \pm 3$
c) $\frac{(x-h)^2}{36} + \frac{(y-k)^2}{9} = 1$, where, $(h, k) = (0, 0), (\pm 12, 0)$, $(\pm 6, 3)$ $(\pm 6, -3)$, $(0, \pm 6)$, $(\pm 12, 6)$, $(\pm 12, -6)$
d) $(x-a)^2 - y^2 = 9$, where $a = 0, \pm 1, \pm 2, \pm 3$

9. a) $\frac{(x-h)^2}{4} + \frac{y^2}{64} = 1$, where $h = -20, -19, \ldots, 0, \ldots, 19, 20$
b) $(x-h)^2 - y^2 = 1$ for $h = -10, -9, \ldots, 0, \ldots, 9, 10$

Investigate, page 550

All explanations may vary.

1. b) Any value for the coefficient of x^2, greater than 0, does not change the type of conic section. That is, the graphs in part a, i and ii are ellipses. When $A = 0$, the graph is 2 horizontal lines.

2. a) i) $4x^2 + y^2 - 16 = 0$ **ii)** $9x^2 + y^2 - 16 = 0$
   b) i) $\frac{1}{4}x^2 + y^2 - 16 = 0$ **ii)** $\frac{1}{9}x^2 + y^2 - 16 = 0$

3. a) Circle when $F < 0$
   b) Ellipse when $A > 0$ and $F < 0$, hyperbola when $A < 0$ and $F \neq 0$
   c) None

5. c) Any value for the coefficient of x or y does not change the type of conic section. That is, all of the graphs in parts a and b are circles.

6. c) Any value for the coefficient of x or y does not change the type of conic section. That is, all of the graphs in parts a and b are ellipses.

7. Answers may vary.

8. c) Any value, greater than 0, for the coefficient of x^2 does not change the type of conic section. That is, the graphs in parts a and b, i and ii are hyperbolas. When $A = 0$, the graph doesn't exist in part a and the graph is 2 horizontal lines in part b.

9. a) i) $4x^2 + y^2 - 16 = 0$, $4x^2 + y^2 + 16 = 0$
   ii) $9x^2 + y^2 - 16 = 0$, $9x^2 + y^2 + 16 = 0$
   b) i) $\frac{1}{4}x^2 + y^2 - 16 = 0$, $\frac{1}{4}x^2 + y^2 + 16 = 0$
   ii) $\frac{1}{9}x^2 + y^2 - 16 = 0$, $\frac{1}{9}x^2 + y^2 + 16 = 0$

10. a) Hyperbola
   b) Ellipse when $A < 0$ and $F > 0$; hyperbola when $A > 0$
   c) None

11. Answers may vary.

12. c) Any value for the coefficient of x or y does not change the type of conic section. That is, all the graphs in parts a and b are hyperbolas.

13. c) Same as exercise 12c

14. Answers may vary.

15. c) Any value for the coefficient of x or y does not change the type of conic section. That is, all the graphs in parts a and c are parabolas.

16. b) The major axis of the graph is no longer horizontal or vertical.

9.5 Exercises, page 555

All explanations may vary.
1. c) Ellipse
   d) The graph becomes a hyperbola.

2. b) The graph does not exist.

3. b) Hyperbola; the graph becomes an ellipse.

4. b) The graph becomes a hyperbola with a vertical transverse axis.

5. b) Any value, greater than 0, for the coefficient of y^2 does

not change the type of conic. That is, the graphs of part a, i and ii are ellipses. When $C = 0$, the graph is 2 vertical lines.

6. a) i) $x^2 + 4y^2 - 16 = 0$ **ii)** $x^2 + 9y^2 - 16 = 0$
   b) i) $x^2 + \frac{1}{4}y^2 - 16 = 0$ **ii)** $x^2 + \frac{1}{9}y^2 - 16 = 0$

7. c) Any value, less than 0, for the coefficient of y^2 does not change the type of conic. That is, the graphs in part a, i and ii are hyperbolas. When $C = 0$, the graph is 2 vertical lines.

8. a) i) $x^2 - 4y^2 - 16 = 0$, $x^2 - 4y^2 + 16 = 0$
   ii) $x^2 - 9y^2 - 16 = 0$, $x^2 - 9y^2 + 16 = 0$
   b) i) $x^2 - \frac{1}{4}y^2 - 16 = 0$, $x^2 - \frac{1}{4}y^2 + 16 = 0$
   ii) $x^2 - \frac{1}{9}y^2 - 16 = 0$, $x^2 - \frac{1}{9}y^2 + 16 = 0$

9. b) Ellipse with centre $(0, 0)$, coordinates of the vertices (also x-intercepts) $(\pm 3, 0)$, and coordinates of the endpoints of the minor axis (also y-intercepts) $(0, \pm 1)$
   c) i) $36x^2 + 9y^2 - 36 = 0$
   ii) $(x + 2)^2 + 9(y - 1)^2 - 9 = 0$

10. Predictions may vary.
   a) i) Ellipses with centre $(0, 0)$, vertices $(0, \pm 5,)$, and endpoints of the minor axes $(\pm 5, 0)$, $(\pm \sqrt{2.5}, 0)$, $(\pm \sqrt{0.25}, 0)$, $(\pm \sqrt{.025}, 0)$
   ii) Ellipses with centre $(0, 0)$, vertices $(\pm 5, 0)$, $(\pm 5\sqrt{10}, 0)$, $(\pm 50, 0)$, $(\pm 50\sqrt{10}, 0)$, and endpoints of the minor axes $(0, \pm 5)$
   c) The ellipse becomes two horizontal lines.
   d) The ellipse becomes a hyperbola.
   e) Explanations may vary.

11. Predictions may vary.
   a) i) Ellipses with centre $(0, 0)$, vertices $(\pm 5, 0)$, and endpoints of the minor axes $(0, \pm 5)$, $(0, \pm \sqrt{2.5})$, $(0, \pm \sqrt{0.25})$, $(0, \pm \sqrt{.025})$
   ii) Ellipses with centre $(0, 0)$, vertices $(0, \pm 5)$, $(0, \pm 5\sqrt{10})$, $(0, \pm 50)$, $(0, \pm 50\sqrt{10})$, and endpoints of the minor axes $(\pm 5, 0)$
   c) The ellipse becomes two vertical lines.
   d), e) Same as in exercise 10 parts d and e

12. Predictions may vary.
   a) i) Hyperbolas with centre $(0, 0)$, vertices $(\pm 5, 0)$, $(\pm \sqrt{2.5}, 0)$, $(\pm \sqrt{0.25}, 0)$, $(\pm \sqrt{.025}, 0)$, and asymptotes $y = \pm x$, $y = \pm \sqrt{10}x$, $y = \pm 10x$, $y = \pm 10\sqrt{10}x$
   ii) Hyperbolas with centre $(0, 0)$, vertices $(0, \pm 5)$, $(0, \pm 5\sqrt{10})$, $(0, \pm 50)$, $(0, \pm 50\sqrt{10})$, and asymptotes $y = \pm x$, $y = \pm \frac{1}{\sqrt{10}}x$, $y = \pm \frac{1}{10}x$, $y = \pm \frac{1}{10\sqrt{10}}x$
   c), d) No graph
   e) Explanations may vary.

13. Predictions may vary.
   a) i) Hyperbolas with centre $(0, 0)$, vertices $(0, \pm 5)$, $(0, \pm \sqrt{2.5})$, $(0, \pm \sqrt{0.25})$, $(0, \pm \sqrt{.025})$, and asymptotes $y = \pm x$, $y = \pm \frac{1}{\sqrt{10}}x$, $y = \pm \frac{1}{10}x$, $y = \pm \frac{1}{10\sqrt{10}}x$
   ii) Hyperbolas with centre $(0, 0)$, vertices $(\pm 5, 0)$, $(\pm 5, \sqrt{10})$, $(\pm 50, 0)$, $(\pm 50, 10\sqrt{10})$, and asymptotes $y = \pm x$, $y = \pm \sqrt{10}x$, $y = \pm 10x$, $y = \pm 10\sqrt{10}x$
   c) The hyperbola becomes two horizontal lines.
   d) The hyperbola becomes an ellipse.
   e) Explanations may vary.

17. i) $3x^2 + 3y^2 - 5 = 0$ **ii)** $x^2 + 4y^2 - 8 = 0$
iii) $4x^2 - 4y^2 - 9 = 0$, $2x^2 - 3y^2 + 6 = 0$
iv) $x^2 - 6y - 3 = 0$, $x = 6y^2$

18. a) $A = C$, $D = E = 0$, F has the opposite sign of A and C
b) $A = -C$, $D = E = 0$, F has the same sign as C
c) $A = -C$, $D = E = 0$, F has the same sign as A
d) $A = 0$, $C \neq 0$, $E = 0$, $F \neq 0$
e) $A > C$, $A, C > 0$, $D = E = 0$, $F < 0$ or $A < C$, $A, C < 0$, $D = E = 0$, $F > 0$

19. a) $Ax^3 + By^3 + Cx^2y + Dxy^2 + Exy + Fx^2 + Gy^2 + Hx + Iy + J = 0$

20. a) i) A horizontal line through $\left(0, \frac{-C}{B}\right)$
ii) A vertical line through $\left(-\frac{C}{A}, 0\right)$
b) i) A horizontal line through $\left(0, \frac{-F}{E}\right)$
ii) Two horizontal lines
iii) A vertical line through $\left(\frac{-E}{D}, 0\right)$
iv) Two vertical lines
c) Explanations may vary.

Mathematical Modelling: Designing a Dish Antenna, page 558

5. b) 2

6. a) $x = 2y$ **b)** $(2p, p)$
c) Answers may vary.

7. a) H(30, 7) **b)** $a = \frac{7}{900}$; $y = \frac{7}{900}x^2$
c) $p = \frac{225}{7}$ **d)** $(0, \frac{225}{7})$

8. a) H$(\frac{d}{2}, h)$ **b)** $a = \frac{4h}{d^2}$; $y = \frac{4h}{d^2}x^2$
c) $p = \frac{d^2}{16h}$ **d)** $(0, \frac{d^2}{16h})$

9.6 Exercises, page 563

1. a) $3x^2 + y^2 - 6x + 4y - 4 = 0$; $A = 3$, $C = 1$, $D = -6$, $E = 4$, $F = -4$
b) $x^2 - 2y^2 + 10x + 4y + 13 = 0$; $A = 1$, $C = -2$, $D = 10$, $E = 4$, $F = 13$
c) $3x^2 - 24x - y + 50 = 0$; $A = 3$, $C = 0$, $D = -24$, $E = -1$, $F = 50$
d) $4x^2 + 9y^2 + 16x - 54y + 61 = 0$; $A = 4$, $C = 9$, $D = 16$, $E = -54$, $F = 61$
e) $3x^2 - 6y^2 + 6x - 24y - 3 = 0$; $A = 3$, $C = -6$, $D = 6$, $E = -24$, $F = -3$
f) $2y^2 + x - 12y + 22 = 0$; $A = 0$, $C = 2$, $D = 1$, $E = -12$, $F = 22$

2. a) Ellipse **b)** Hyperbola **c)** Parabola **d)** Circle

3. a) $\frac{(x+3)^2}{2} + \frac{(y-1)^2}{4} = 1$; ellipse
b) $\frac{(x-1)^2}{9} + \frac{(y+2)^2}{4} = 1$; ellipse
c) $\frac{(x-2)^2}{9} - (y-1)^2 = 1$; hyperbola
d) $\frac{(x-1)^2}{4} - y^2 = 1$; hyperbola
e) $x + \frac{13}{4} = \frac{1}{4}(y+4)^2$; parabola

f) $y + 1 = -\frac{1}{3}(x+1)^2$; parabola
4. a) $x^2 - 6x - 6y - 9 = 0$
6. No; explanations may vary.

Problem Solving: Why Is an Ellipse an Ellipse?, page 564

1. to 3. Explanations may vary.
4. b) $\sqrt{(x-c)^2 + y^2} + \sqrt{(x+c)^2 + y^2} = 2a$
c) $\frac{x^2}{a^2} + \frac{y^2}{a^2 - c^2} = 1$
5. to 7. Explanations may vary.

9 Review, page 568

1. a) $x^2 - \frac{y^2}{a^2} = 1$, where $a = 0.25, 0.5, 0.75, \dots, 4$
b) $\frac{x^2}{a^2} - y^2 = 1$ and $x^2 - \frac{y^2}{a^2} = -1$, where $a = 1, 2, 3, 4, 5, 6$

2. a) Ellipse with centre $(0, 0)$, coordinates of the vertices (also y-intercepts) $(0, \pm\sqrt{6})$, and coordinates of the endpoints of the minor axis (also x-intercepts) $(\pm\sqrt{2}, 0)$
c) i) $3x^2 + 16y^2 - 24 = 0$
ii) $3x^2 - 18x + y^2 - 4y + 25 = 0$

3. a) Circle **b)** Ellipse **c)** Hyperbola
d) Parabola **e)** Ellipse **f)** Hyperbola

4. Explanations may vary.

5. Explanations may vary.

6. a) $\frac{(x-4)^2}{16} + \frac{(y+3)^2}{16} = 1$; circle
b) $\frac{(x-1)^2}{4} + (y+2)^2 = 1$; ellipse
c) $\frac{(x+3)^2}{4} + \frac{(y-2)^2}{3} = 1$; ellipse
d) $\frac{(x-6)^2}{4} - \frac{y^2}{6} = 1$; hyperbola
e) $x + 2 = \frac{1}{8}(y-4)^2$; parabola
f) $6(x+2)^2 = y + 5$; parabola

7. a) Point $(1, -4)$ **b)** No graph
c) Two straight lines
d) Ellipse with slanted major and minor axes

9 Cumulative Review, page 569

1. a) $y = f(x)$ becomes $y = 5f(x)$.
b) $y = f(x)$ becomes $y = f(-3x)$.
c) $y = f(x)$ becomes $y = f(\frac{x}{3})$.
d) $y = f(x)$ becomes $y = -\frac{1}{2}f(x)$.

2. a) $[H^+] = 10^{-7}$, $[OH^-] = 10^{-7}$
b) Since $[H^+]$ for pure water is 10^{-7}, solving $pH = -\log 10^{-7}$, provides the answer. That is, the pH value for pure water is 7.
c) $10^{-3.1}$

3. a) 13.66 m
b) Estimates may vary. 13.6 m
c) Explanations may vary.

4. a) $\tan 2x = \frac{2\tan x}{1 - \tan^2 x}$

b) $\sin 4x = 2\sin 2x - 4\sin 2x \sin^2 x$;
$\cos 4x = 1 - 8\cos^2 x + 8\cos^4 x$

5. a) $y = 3.5\cos\frac{2\pi}{2.5}(x + 6) - 2$

b) $y = 0.5\cos\frac{\pi}{3}(x - 2) + 3$

6. a) Domain: $x \neq \frac{\pi}{4} + \frac{n\pi}{2}$, range: all real numbers, period: $\frac{\pi}{2}$,
asymptotes: $x = \frac{\pi}{4} + \frac{n\pi}{2}$

b) Domain: $x \neq \frac{5}{8} + 2.5n$, range: $y \geq 0.5$, or $y \leq -0.5$,
period: 2.5, asymptotes: $x = \frac{5}{8} + 2.5n$

9. Answers may vary.

a) $\frac{\sqrt{2} + \sqrt{6}}{4}$

b) $\frac{\sqrt{2} - \sqrt{6}}{4}$

c) $\frac{\sqrt{2} + \sqrt{6}}{4}$

d) $\frac{\sqrt{2} + \sqrt{6}}{4}$

10. a) $0, \pi$

b) $\frac{\pi}{2}, \frac{3\pi}{2}$

11. a) 45 360

b) 1260

12. 280

13. $2^{50} - 50 \times 2^{49}x + 1225 \times 2^{48}x^2 - 19\,600 \times 2^{47}x^3 +$
$230\,300 \times 2^{46}x^4$

14. a) $-160x^3$

b) 1

c) $495x^4y^8$

15. Answers may vary.

16. a) 0.246

b) 0.205

c) 0.172

17. a) \$35 636.09, \$48 729.47

b) 46.41% (using tables)

c) 29.82%

d) The data is not normally distributed.

18. b) 0.3989

c) (2.5066, 0), (−2.5066, 0)

19. a) $x^2 + (y - \frac{k}{2})^2 = \frac{k^2}{4}$ for $k = 0, \pm 1, \pm 2, \ldots, \pm 27$

b) $(x - h)^2 + \frac{y^2}{36} = 1$ for $h = 0, \pm 2, \pm 4, \ldots, \pm 10$

20. Answers may vary.

21. a) Ellipse with centre (0, 0), x-intercepts: ± 3,
y-intercepts: ± 4

b) Hyperbola with centre (0, 0), x-intercepts: ± 5,
asymptotes: $y = \pm\frac{2}{5}x$

c) Hyperbola with centre (0, 0), y-intercepts: ± 4,
asymptotes: $y = \pm\frac{3}{5}x$

d) Ellipse with centre (0, 0), x-intercepts: $\pm 2\sqrt{3}$,
y-intercepts: ± 2

e) Ellipse with centre (0, 0), x-intercepts: $\pm\sqrt{5}$,
y-intercepts: $\pm 2\sqrt{5}$

f) Ellipse with centre (0, 0), x-intercepts: $\pm 2\sqrt{3}$,
y-intercepts: ± 3

g) Hyperbola with centre (0, 0), x-intercepts: $\pm\sqrt{6}$,
asymptotes: $y = \pm\sqrt{3}x$

h) Hyperbola with centre (0, 0), x-intercepts: $\pm 2\sqrt{2}$,
asymptotes: $y = \pm\frac{\sqrt{5}}{2}x$

i) Hyperbola with centre (0, 0), x-intercepts: $\pm\sqrt{6}$,
asymptotes: $y = \pm\frac{2\sqrt{3}}{3}x$

GLOSSARY

absolute maximum point: a point on a graph whose *y*-coordinate is greater than those of all other points on the graph

absolute minimum point: a point on a graph whose *y*-coordinate is less than those of all other points on the graph

absolute value: the distance between any real number and 0 on a number line; for example, $|-3| = 3$, $|3| = 3$

absolute value equation: an equation containing the variable inside the absolute value sign

absolute value inequality: an inequality containing the variable inside the absolute value sign

accumulated amount: the value of the principal plus interest

acute angle: an angle measuring less than 90°

acute triangle: a triangle with three acute angles

algebraic expression: a mathematical expression containing a variable: for example, $6x - 4$ is an algebraic expression

altitude: the perpendicular distance from the base of a figure to the opposite side or vertex; also the height of an aircraft above the ground

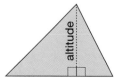

amplitude of a function: the distance from the central axis to the minimum or maximum value of the periodic function

angle: the figure formed by two rays from the same endpoint

angle bisector: the line that divides an angle into two equal angles

angular velocity: the angle per unit time through which an object rotates about the centre of a circle

approximation: a number close to the exact value of an expression; the symbol $\doteq$ means "is approximately equal to"

area: the number of square units needed to cover a region

arithmetic sequence: a sequence of numbers in which each term after the first term is calculated by adding the same number to the preceding term; for example, in the sequence 1, 4, 7, 10, …, each number is calculated by adding 3 to the previous number

arithmetic series: the indicated sum of the terms of an arithmetic sequence

asymptote: a line that a curve approaches, but never reaches

average: a single number that represents a set of numbers; see *mean*, *median*, and *mode*

axis of symmetry: see *line symmetry*

bar graph: a graph that displays data by using horizontal or vertical bars whose heights are proportional to the frequencies of the numbers or intervals they represent

bar notation: the use of a horizontal bar over decimal digits to indicate that they repeat; for example, $1.\overline{3}$ means 1.333 333 …

base: the side of a polygon or the face of a solid from which the height is measured; the factor repeated in a power

binomial: a polynomial with two terms; for example, $3x - 8$

bisector: a line that divides a line segment in two equal parts
The broken line is a bisector of AB.

binomial distribution: a probability distribution in which the probabilities of all the outcomes are equal

binomial experiment: an experiment that has two outcomes

branches: the two distinct parts of a hyperbola

broken-line graph: a graph that displays data by using points joined by line segments

carbon dating: the method used to estimate the age of ancient specimens by measuring the radioactivity of the carbon-14 it contains and comparing it with that of living matter

central axis: the line halfway between the minimum and maximum values of a periodic function

chord: a line segment whose endpoints lie on a circle

circle: the curve that results when a plane intersects a cone, parallel to the base of the cone, or, a set of points in a plane that are a given distance from a fixed point (the centre)

circumference: the distance around a circle, and sometimes the circle itself

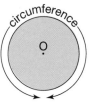

coefficient: the numerical factor of a term; for example, in the terms $3x$ and $3x^2$, the coefficient is 3

collinear points: points that lie on the same line

combination: a selection from a group of objects without regard to order

combinatorics: a branch of mathematics involving permutations and combinations

common denominator: a number that is a multiple of each of the given denominators; for example, 12 is a common denominator for the fractions $\frac{1}{3}$, $\frac{5}{4}$, $\frac{7}{12}$

common difference: the number obtained by subtracting any term from the next term in an arithmetic sequence

common factor: a number that is a factor of each of the given numbers; for example, 3 is a common factor of 15, 9, and 21

common ratio: the ratio formed by dividing any term after the first one in a geometric sequence by the preceding term

complementary angles: two angles whose sum is 90°

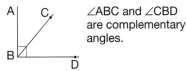

∠ABC and ∠CBD are complementary angles.

complement of an event A: the entire set of outcomes that are not favourable to A

complete the square: add or subtract constants to rewrite a quadratic expression $ax^2 + bx + c$ as $a(x - p)^2 + q$

composite function: the resulting function when two functions are applied in succession

composite number: a number with three or more factors; for example, 8 is a composite number because its factors are 1, 2, 4, and 8

compound interest: see *interest*; if the interest due is added to the principal and thereafter earns interest, the interest earned is compound interest

compounding period: the time interval for which interest is calculated

conditional probability: the probability that an event will occur given that another event has occurred

congruent: figures that have the same size and shape, but not necessarily the same orientation

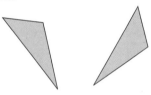

cone: the surface generated when a line is rotated about a fixed point P on the line; see page 519

conic sections: the curves that result when a plane intersects a cone

consecutive integers: integers that come one after the other without any integers missing; for example, 34, 35, 36 are consecutive integers, so are −2, −1, 0, and 1

constant: a particular number

constant function: $f(x) = c$, where c is a constant

constant of proportionality: the constant k in a direct variation of the form $y = kx$; the slope of the graph of this equation

constant slope property: the slopes of all segments of a line are equal

constant term: a number

continuity correction: an adjustment of 0.5 added to or subtracted from the endpoints when the normal distribution is used to approximate the binomial distribution in probability calculations

continuous function: a function whose graph can be drawn without lifting the pencil from the paper

coordinate axes: the x- and y-axes on a grid that represents a plane

coordinate plane: a two-dimensional surface on which a coordinate system has been set up

coordinates: the numbers in an ordered pair that locate a point in the plane

corresponding angles in similar triangles: two angles, one in each triangle, that are equal

cosecant of θ: the reciprocal of the sine of θ

cosine of θ: the first coordinate of point P, where P is on the unit circle, centre (0, 0), and θ is the measure of the angle in standard position

cotangent of θ: the reciprocal of the tangent of θ

coterminal angles: two or more angles in standard position that have the same terminal arm

counterexample: an example that shows a conjecture to be false

cube: a solid with six congruent, square faces

cube root: a number which, when raised to the power 3, results in a given number; for example, 3 is the cube root of 27, and −3 is the cube root of −27

cubic units: units that measure volume

curve of best fit: a curve that passes as close as possible to a set of plotted points

cyclic quadrilateral: a quadrilateral whose vertices lie on a circle

cylinder: a solid with two parallel, congruent, circular bases

data: facts or information

database: facts or information supplied by computer software

decibel: the unit used to measure the loudness of a sound

denominator: the term below the line in a fraction

density: the mass of a unit volume of a substance

dependent events: the occurrence of one event is affected by the occurrence of another event

diagonal: a line segment that joins two vertices of a figure, but is not a side

diameter: the distance across a circle, measured through the centre; a line segment through the centre of the circle with its endpoints on the circle

difference of squares: a polynomial that can be expressed in the form $x^2 - y^2$; the product of two monomials that are the sum and difference of the same two quantities, $(x + y)(x - y)$

digit: any of the symbols used to write numerals; for example, in the base-ten system the digits are 0, 1, 2, 3, 4, 5, 6, 7, 8, and 9

discontinuous function: a function whose graph cannot be drawn without lifting the pencil from the paper

discriminant: the discriminant of the quadratic equation $ax^2 + bx + c = 0$ is $b^2 - 4ac$

dispersion: the extent to which data cluster around the mean, often measured by the standard deviation

distribution: see *frequency distribution* and *probability distribution*

domain of a function: the set of x-values (or valid input numbers) represented by the graph or the equation of a function

doubling time: the term used in exponential growth problems; the time it takes the quantity of an item to double

ellipse: the closed curve that results when a plane intersects a cone, or the curve obtained by expanding or compressing a circle in perpendicular directions

entire radical: an expression of the form $\sqrt{x}$; for example, $\sqrt{20}$ is an entire radical

equation: a mathematical statement that two expressions are equal

equidistant: the same distance apart

equilateral triangle: a triangle with three equal sides

evaluate: to substitute a value for each variable in an expression and simplify the result

even number: an integer that has 2 as a factor; for example, 2, 4, −6

even function: a function with the property $f(-x) = f(x)$

event: any set of outcomes of an experiment

expanding: multiplying a polynomial by a polynomial

experiment: an operation, carried out under controlled conditions, that is used to test or establish a hypothesis

exponent: a number, shown in a smaller size and raised, that tells how many times the number before it is used as a factor; for example, 2 is the exponent in 6^2

expression: a meaningful combination of mathematical symbols, such as a polynomial

extraneous root: a root of an equation, obtained algebraically from the original equation, that is not a root of the original equation

experimental probability: probabilities determined using sampling or a simulation

exponential function: $y = Ab^x$, where A and b are constants, and $b > 0$

extrapolate: to estimate a value beyond known values

extremes: the highest and lowest values in a set of numbers

factor: to factor means to write as a product; to factor a given integer means to write it as a product of integers, the integers in the product are the factors of the given integer; to factor a polynomial with integer coefficients means to write it as a product of polynomials with integer coefficients

fifth root: a number which, when raised to the power 5, results in a given number; for example, 3 is the fifth root of 243 since $3^5 = 243$, and -3 is the fifth root of -243 since $(-3)^5 = -243$

foci of an ellipse: the two points F_1 and F_2 on the major axis of an ellipse such that $PF_1 + PF_2$ is constant for all points P on the ellipse

focus: the point in a parabolic dish antenna where the receiver is located; parallel signals are reflected off the surface and concentrated at this point

formula: a rule that is expressed as an equation

fourth root: a number which, when raised to the power 4, results in a given number; for example, -2 and 2 are the fourth roots of 16

45-45-90 triangle: a triangle with angles 45°, 45°, and 90°

fraction: an indicated quotient of two quantities

frequency: the number of times a particular number occurs in a set of data

frequency distribution: a function often described using a table of values or a histogram, that provides the frequency for every outcome of an experiment

function: a rule that gives a single output number for every valid input number

future value: see *accumulated amount*

general form of an equation: the form of the equation that has all the terms on the left side

generators of a cone: the lines that are rotated to form the surface of the cone; see page 519

geometric sequence: a sequence of numbers in which each term after the first term is calculated by multiplying the preceding term by the same number; for example, in the sequence 1, 3, 9, 27, 81, …, each number is calculated by multiplying the preceding term by 3

geometric series: the indicated sum of the terms of a geometric sequence

greatest common factor: the greatest factor that 2 or more monomials have in common; $4x^2$ is the greatest common factor of $8x^3 + 16x^2y - 64x^4$

half-life: the time it takes a quantity of an item to be halved

hemisphere: half a sphere

hexagon: a six-sided polygon

histogram: a vertical bar graph where the height (or area) of each bar is proportional to the frequency or probability of the number or interval it represents

horizontal intercept: the horizontal coordinate of the point where the graph of a line or a function intersects the horizontal axis

hyperbola: the curve that results when a plane intersects both nappes of a cone

hypotenuse: the side that is opposite the right angle in a right triangle

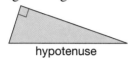
hypotenuse

identity: an equation that is satisfied for all values of the variable for which both sides of the equation are defined

image: the graph that results from applying a transformation to a graph

independent events: two or more events for which the occurrence or nonoccurrence of one does not affect the occurrence of the others

inductive reasoning: reasoning based on the results of an experiment

inequality: a statement that one quantity is greater than (or less than) another quantity

initial arm: the horizontal ray of an angle in standard position

integers: the set of numbers... −3, −2, −1, 0, +1, +2, +3, ...

interest: money that is paid for the use of money, usually according to a predetermined percent

interpolate: to estimate a value between two known values

intersecting lines: lines that meet or cross; lines that have one point in common

interval: a regular distance or space between values

inverse of a function: a relation whose rule is obtained from that of a function by interchanging x and y

irrational number: a number that cannot be written in the form $\frac{m}{n}$, where m and n are integers ($n \neq 0$)

isosceles triangle: a triangle with at least two equal sides

lattice point: on a coordinate grid, a point at the intersection of two grid lines

leading coefficient: the coefficient of the highest power of a polynomial expression

like radicals: radicals that have the same radical part; for example, $\sqrt{5}$, $3\sqrt{5}$, and $-7\sqrt{5}$

like terms: terms that have the same variables; for example, $4x$ and $-3x$ are like terms

line of best fit: a line that passes as close as possible to a set of plotted points

line segment: the part of a line between two points on the line, including the two points

line symmetry: a figure that maps onto itself when it is reflected in a line is said to have line symmetry; for example, line l is the line of symmetry for figure ABCD

linear equation: an equation that represents a straight line

linear function: a function whose defining equation can be written in the form $y = mx + b$, where m and b are constants

logarithm: an exponent

$\log_a x$: the exponent that a has when x is written as a power of a ($a > 0$, $a \neq 1$)

major arc: a part of a circle that is greater than half a circle

major axis: the longest chord in an ellipse

mass: the amount of matter in an object

mean: the sum of a set of numbers divided by the number of numbers in the set

median: the middle number when data are arranged in numerical order

measures of central tendency: statistics that describe the "middle" of the data: mean, median, and mode

midpoint: the point that divides a line segment into two equal parts

minor arc: a part of a circle that is less than half a circle

minor axis: the chord through the centre perpendicular to the major axis

mixed radical: an expression of the form $a\sqrt{x}$; for example, $2\sqrt{5}$ is a mixed radical

mode: the number that occurs most often in a set of numbers

monomial: a polynomial with one term; for example, each of 14 and $5x^2$ is a monomial

multiple: the product of a given number and a natural number; for example, some multiples of 8 are 8, 16, 24, …

multiplicative inverse: a number and its reciprocal; the product of multiplicative inverses is 1; for example, $3 \times \frac{1}{3} = 1$

mutually exclusive events: events that cannot occur at the same time

nappes of a cone: the 2 symmetrical parts of a cone on either side of the vertex; see page 519

natural numbers: the set of numbers 1, 2, 3, 4, 5, …

negative number: a number less than 0

negative reciprocals: two numbers that have a product of -1; for example, $\frac{3}{4}$ is the negative reciprocal of $-\frac{4}{3}$, and vice versa

normal distribution: a probability function with mean μ and a standard deviation of σ; the graph is symmetrical about the mean; abides by the 68-95-99 rule

normally distributed: referring to populations that have a normal distribution

numeracy: the ability to read, understand, and use numbers

numerator: the term above the line in a fraction

obtuse angle: an angle greater than 90° and less than 180°

obtuse triangle: a triangle with one angle greater than 90°

octagon: an eight-sided polygon

odd number: an integer that does not have 2 as a factor; for example, 1, 3, −7

odd function: a function with the property $f(-x) = -f(x)$

operation: a mathematical process or action such as addition, subtraction, multiplication, or division

opposite angles: the equal angles that are formed by two intersecting lines

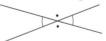

opposite number: a number whose sum with a given number is 0; for example, 3 and −3 are opposites

opposites: two numbers whose sum is zero; each number is the opposite of the other

order of operations: the rules that are followed when simplifying or evaluating an expression

ordered pair: a pair of numbers, written as (*x*, *y*), that represents a point on a coordinate grid

outcome: a possible result of an experiment or a possible answer to a survey question

parabola: the curve that results when a plane intersects a cone parallel to a generator, or, the name given to the shape of the graph of a quadratic function

parallel lines: lines in the same plane that do not intersect

parallelogram: a quadrilateral with both pairs of opposite sides parallel

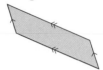

pentagon: a five-sided polygon

per capita: for each person

percent: the number of parts per 100; the numerator of a fraction with denominator 100

perfect square: a number that is the square of a whole number; a polynomial that is the square of another polynomial

perimeter: the distance around a closed figure

period: the length of the shortest part of the graph that repeats, measured along the horizontal axis

periodic function: a function that repeats in a regular way, or, a function that has a number *p* such that $f(x + p) = f(x)$ for all values of *x* in the domain

permutation: an arrangement of a set of objects

perpendicular: intersecting at right angles

pH scale: measures the acidity or alkalinity of a solution

pi (π): the ratio of the circumference of a circle to its diameter; $\pi \doteq 3.1416$

point of intersection: a point that lies on two or more figures

point symmetry: a figure that maps onto itself after a rotation of 180° about a point is said to have point symmetry

polygon: a closed figure that consists of line segments; for example, triangles and quadrilaterals are polygons

polynomial: a mathematical expression with one or more terms, in which the exponents are whole numbers and the coefficients are real numbers

polynomial function: a function where $f(x)$ is a polynomial expression

population density: the average number of people for each square unit of land

positive number: a number greater than 0

power: an expression of the form a^n, where *a* is called the base and *n* is called the exponent; it represents a product of equal factors; for example, $4 \times 4 \times 4$ can be expressed as 4^3

power function: a function that has the form $f(x) = x^a$

prime number: a whole number with exactly two factors, itself and 1; for example, 3, 5, 7, 11, 29, 31, and 43

primary trigonometric functions: sine, cosine, and tangent of θ

prism: a solid that has two congruent and parallel faces (the *bases*), and other faces that are parallelograms

probability: the likelihood of an event, calculated by dividing the number of favourable outcomes of the event by the total number of outcomes

probability distribution: a function that provides the probability for every outcome of an experiment

proportion: a statement that two ratios are equal

Pythagorean Theorem: for any right triangle, the area of the square on the hypotenuse is equal to the sum of the areas of the squares on the other two sides

quadrant: one of the four regions into which coordinate axes divide a plane

quadratic equation: an equation in which the variable is squared; for example, $x^2 + 5x + 6 = 0$ is a quadratic equation

quadratic function: a function with defining equation $f(x) = ax^2 + bx + c$, where a, b, c are constants and a cannot be zero

quadrilateral: a four-sided polygon

radian: a unit for measuring angles; 1 radian is about 57.295 78°

radical: the root of a number; for example, $\sqrt{400}$, $\sqrt[3]{8}$, and so on

radical equation: an equation where the variable occurs under a radical sign

radical sign: the symbol $\sqrt{}$ that denotes the positive square root of a number

radius (plural, **radii**): the distance from the centre of a circle to any point on the circumference, or a line segment joining the centre of a circle to any point on the circumference

random sample: a sampling in which all members of the population have an equal chance of being selected

range: the difference between the highest and lowest values (the *extremes*) in a set of data

range of a function: the set of y-values (or output numbers) represented by the graph or the equation of a function

rate: a certain quantity or amount of one thing considered in relation to a unit of another thing

ratio: a comparison of two or more quantities with the same unit

rational equation: an equation where the variable occurs in the denominator of a rational expression

rational expression: an algebraic expression that can be written as the quotient of two polynomials; for example, $\frac{3x^2 + 5}{2x + 7}$

rational function: a function where $f(x)$ is a rational expression

rational number: a number that can be written in the form $\frac{m}{n}$, where m and n are integers $(n \neq 0)$

rationalize the denominator: write the denominator as a rational number, to replace the irrational number; for example, $\frac{6}{\sqrt{2}}$ is written $\frac{6\sqrt{2}}{2}$, or $3\sqrt{2}$

real numbers: the set of rational numbers and the set of irrational numbers; that is, all numbers that can be expressed as decimals

reciprocal of a function $f(x)$: the function $\frac{1}{f(x)}$, where $f(x)$ is not zero

reciprocals: two numbers whose product is 1; for example, $\frac{3}{4}$ and $\frac{4}{3}$ are reciprocals, 2 and $\frac{1}{2}$ are reciprocals

rectangle: a quadrilateral that has four right angles

rectangular hyperbola: a hyperbola that has perpendicular asymptotes

reference angle: the acute angle between the terminal arm and the *x*-axis

regular decagon: a polygon that has ten equal sides and ten equal angles

regular hexagon: a polygon that has six equal sides and six equal angles

regular octagon: a polygon that has eight equal sides and eight equal angles

regular polygon: a polygon that has all sides equal and all angles equal

relation: a rule that produces one or more output numbers for every valid input number

relative frequency: the ratio of the number of times a particular outcome occurred to the number of times the experiment was conducted

relative maximum point: a point on a graph whose *y*-coordinate is greater than those of the neighbouring points on the graph

relative minimum point: a point on a graph whose *y*-coordinate is less than those of the neighbouring points on the graph

rhombus: a parallelogram with four equal sides

Richter scale: a scale that compares the intensity of earthquakes

right angle: a 90° angle

right triangle: a triangle that has one right angle

rise: the vertical distance between 2 points

root of an equation: a value of the variable that satisfies the equation

rounding: approximating a number to the next highest (or lowest) number; for example, rounding 3.46 to the tenth is 3.5, and rounding 4.34 is 4.3

run: the horizontal distance between 2 points

sample/sampling: a representative portion of a population

sample space: a list of all the possible outcomes of an experiment

scalene triangle: a triangle with no two sides equal

scatterplot: a graph of data that is a series of points

scientific notation: a number expressed as the product of a number greater than −10 and less than −1 or greater than 1 and less than 10, and a power of 10; for example, 4700 is written as 4.7×10^3

secant of *θ*: the reciprocal of sine of *θ*

sector: the figure formed by an arc of a circle, the radii at the ends of the arc, and all enclosed points

sector angle: see *sector;* the angle at the centre of the circle between two radii

segment of a circle: the figure formed by an arc of a circle, the chord joining the endpoints of the arc, and all enclosed points

seismograph: an instrument for recording the direction, intensity, and duration of earthquakes

seismometer: see *seismograph*

semicircle: half a circle

significant digits: the meaningful digits of a number representing a measurement

sigma: the Greek letter corresponding to S, σ in lowercase or Σ capitalized

sigma notation: a method using Σ, the capital Greek letter, sigma, to represent a sum

similar figures: figures with the same shape, but not necessarily the same size

simple harmonic motion: a periodic up-and-down motion commonly created by a spring

sine of θ: the second coordinate of P, where P is on the unit circle, centre (0, 0) and θ is the measure of the angle in standard position

sinusoidal functions: functions of curves that look like waves

sinusoidal regression: determining the equation of the sinusoid of best fit for given data

sinusoids: curves that look like waves

68-95-99 rule: about 68% of the population are within 1 standard deviation of the mean; about 95% of the population are within 2 standard deviations of the mean; about 99.7% of the population are within 3 standard deviations of the mean

slope: the measure of steepness of a line; calculated as slope $= \frac{\text{rise}}{\text{run}}$

slope y-intercept form: the equation of a line in the form $y = mx + b$

sphere: the set of points in space that are a given distance from a fixed point (the centre)

spreadsheet: a computer-generated arrangement of data in rows and columns, where a change in one value results in appropriate calculated changes in the other values

square: a rectangle with four equal sides

square of a number: the product of a number multiplied by itself; for example, 25 is the square of 5

square root: a number which, when multiplied by itself, results in a given number; for example, 5 and −5 are the square roots of 25

standard deviation: a measure of the extent to which data cluster around the mean

standard form of an equation: the form of the equation that is useful for graphing

standard normal distribution: a probability function with mean 0 and a standard deviation of 1; the graph is symmetrical about the mean; obeys the 68-95-99 rule

standard position: an angle is in standard position when the initial arm starts at (0, 0) and lies on the positive x-axis; the terminal arm is rotated about (0, 0)

statistic: a number that describes a distribution, such as the mean or the standard deviation

statistics: the branch of mathematics that deals with the collection, organization, and interpretation of data

straight angle: an angle measuring 180°

straightedge: a strip of wood, metal, or plastic with a straight edge, but no markings

supplementary angles: two angles whose sum is 180°

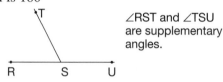

∠RST and ∠TSU are supplementary angles.

survey: an investigation of a topic to find out people's views

symmetrical: possessing symmetry; see *line symmetry;* see *point symmetry*

tangent of θ: the y-coordinate of Q, where P is on the unit circle and AQ is perpendicular to OA

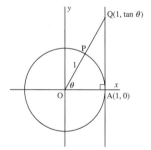

terminal arm: the final ray when defining an angle

theorem: a conclusion reached by deductive reasoning

theoretical probability: probability determined using combinations

three-dimensional: having length, width, and depth or height

30-60-90 triangle: a triangle with angles 30°, 60°, and 90°

trajectory: the curved path of an object moving through space

translation: a transformation that moves all points in the plane in a given direction through a given distance

transitivity: a logical property which, applied to dice, states that if A is better than B and B is better than C, then A is better than C

transverse axis: the line segment joining the vertices of a hyperbola

trapezoid: a quadrilateral that has only one pair of parallel sides

tree diagram: a branching diagram used to show all possible outcomes of an experiment

triangle: a three-sided polygon

trigonometric equation: an equation involving one or more trigonometric functions of a variable

trinomial: a polynomial with three terms; for example, $3x^2 + 6x + 9$

truncating: approximating a number by cutting off the end digits; for example, 3.141 8 is truncated to 3.141, or to 3.14, or to 3.1, and so on

two-dimensional: having length and width, but no thickness, height, or depth

unit circle: a circle with radius 1 unit

unit rate: the quantity associated with a single unit of another quantity; for example, 6 m in 1 s is a unit rate

unlike radicals: radicals that have different radical parts; for example, $\sqrt{5}$ and $\sqrt{11}$

unlike terms: terms that have different variables, or the same variable but different exponents; for example, $3x$, $-4y$, $3x^2$

variable: a letter or symbol representing a quantity that can vary

Venn diagram: a diagram where the elements of sets are represented by points within closed loops

vertex (plural, **vertices**): the corner of a figure or a solid

vertex of a cone: the point where all the generators intersect; see page 519

vertex of a parabola: the point where the axis of symmetry of a parabola intersects the parabola

vertical intercept: the vertical coordinate of the point where the graph of a line or a function intersects the vertical axis

vertical line test: if no two points on a graph can be joined by a vertical line, then the graph represents a function

vertices of a hyperbola: the endpoints of the transverse axis

vertices of an ellipse: the endpoints of the major axis

volume: the amount of space occupied by an object

whole numbers: the set of numbers 0, 1, 2, 3,...

x-axis: the horizontal number line on a coordinate grid

x-intercept: the x-coordinate where the graph of a line or a function intersects the x-axis

y-axis: the vertical number line on a coordinate grid

y-intercept: the y-coordinate where the graph of a line or a function intersects the y-axis

zero exponent: any number, a, that has the exponent 0 $(a \neq 0)$, is equal to 1; for example, $(-6)^0 = 1$

zero of a function: a value of the variable for which the function has value zero

INDEX

PHOTO CREDITS AND ACKNOWLEDGMENTS

The publisher wishes to thank the following sources for photographs, illustrations, articles, and other materials used in this book. Care has been taken to determine and locate ownership of copyright material used in the test. We will gladly receive information enabling us to rectify any errors or omissions in credits.

PHOTOS

p. 2-3 Dave Starrett; **p. 23** CORBIS; **p. 24** Zigy Kalunzy/Tony Stone Images; **p. 62** (top right) Paul Chesley/Tony Stone Images; **p. 63** CP Picture Archives (Toronto Sun/Craig Robertson); **p. 64-65** Dave Starrett, (bkgd) Corel Stock Photo Library; **p. 68** CORBIS/Bettmann; **p. 85** CORBIS/Hulton Deutsch Collection; **p. 88** Ken Fisher/Tony Stone Images; **p. 91** CP Picture Archives (CP PHOTO); **p. 95** (bottom) First Light; **p. 120** Chuck O'Rear/West Light/First Light; **p. 130** Darwin Wiggett/First Light; **p. 154** CP Picture Archives (AP PHOTO/Barrie M. Schwortz); **p. 156-157** (bkgd) Mark Kaarremaa/Imageplay Photography; **p. 156** (inset) Darwin R. Wiggett/First Light; **p. 157** (top inset) Darwin R. Wiggett/First Light; **p. 157** (bottom inset) Mark Kaarremaa/Imageplay Photography; **p. 164** Dave Starrett; **p. 222** Terry Vine/Tony Stone Images; **p. 224-225** (bottom) Tony Maxwell/Mach II Stock Exchange, (top) D. Boone/Westlight/First Light; **p. 240** Richard Simpson/Tony Stone Images; **p. 241** (top) Daniel J. Cox/ Tony Stone Images, (bottom) Art Wolfe/Tony Stone Images; **p. 250** (bottom) NOAO/TSADO/Tom Stack&Associates/First Light; **p. 269** (top) Don Bonsey/Tony Stone Images; **p. 272** (top right and top left) John Sylvester/First Light; **p. 279** (top) Dave Starrett; **p. 296-297** Julie Habel/West Light/First Light; **p. 297** (inset) T. Stewart/First Light; **p. 321** Ian Crysler; **p. 358** (top right) Ian Crysler, (key card supplied courtesy of the Excelsior Hotel/Hong Kong) (bottom right) Corel Stock Photo Library; **p. 359, 361** Dave Starrett; **p. 362** Euan Myles/Tony Stone Images; **p. 369** (girl and boy) Dave Starrett; **p. 371** Ian Crysler; **p. 382** Library of Congress/CORBIS; **p. 400-401** Digital Vision Ltd.; **p. 405, 418** Ian Crysler; **p. 424** CP Picture Archives; **p. 462** (top right) Peter Cade/Tony Stone Images; **p. 462-463** Rob Gage-FPG/Masterfile; **p. 463** (top right) David Young-Wolff/Tony Stone Images; **p. 502** Ian Crysler; **p. 516** (bottom right) Bob Thomason/Tony Stone Images; **p. 516-517** (bkgd) Corel Stock Photo Library, (top) Frank Rossotto/First Light; **p. 517** (middle right) David Prichard/First Light, (bottom) Bob Thomason/Tony Stone Images; **p. 519** Dave Starrett; **p. 520** (top right) Pat O'Hara/Tony Stone Images, (bottom right) William James Warren/First Light; **p. 522** (top) Dave Starrett; **p. 523** (bottom) Edmonton Space & Science Centre; **p. 566** Dave Starrett

ILLUSTRATIONS

Steve Attoe: 352-353
Mike Herman: 6-7, 90, 169, 214, 221, 274, 369, 373, 403, 523 (top)
Dave McKay: 277, 279 (bottom), 289, 295, 385-386, 408, 410, 422, 427, 429, 440, 521
Jack McMaster: 157 (inset, middle right), 354, 375, 388
Jun Park: 268, 516, 522 (bottom), 565
Pronk&Associates: 62 (top right), 520 (from original reference courtesy of David J. Tholen), 524

TEXT

p. 2, 60 From "Trend sees 2020 vision of women's pay" by John Kettle from The Globe and Mail, July 2, 1998. Reprinted with permission of John Kettle.
p. 71 Three graphs from "Builders of the electronic mall" from The Globe and Mail, July 11, 1998. Reprinted with permission from the Globe and Mail.
p. 359 From "Our mission: the best mission statement" from The Toronto Star, August 1, 1998. Copyright David Martin. Reprinted with permission - The Toronto Star Syndicate.